T0230093

Lecture Notes in Computer Science 739

Edited by G. Goos and J. Hartmanis

Advisory Board: W. Brauer   D. Gries   J. Stoer

Hideki Imai   Ronald L. Rivest
Tsutomu Matsumoto (Eds.)

Advances in Cryptology – ASIACRYPT '91

International Conference on the
Theory and Application of Cryptology
Fujiyoshida, Japan, November 11-14, 1991
Proceedings

Springer-Verlag
Berlin Heidelberg New York
London Paris Tokyo
Hong Kong Barcelona
Budapest

Series Editors

Gerhard Goos
Universität Karlsruhe
Postfach 69 80
Vincenz-Priessnitz-Straße 1
D-76131 Karlsruhe, Germany

Juris Hartmanis
Cornell University
Department of Computer Science
4130 Upson Hall
Ithaca, NY 14853, USA

Volume Editors

Hideki Imai
Tsutomu Matsumoto
Division of Electrical and Computer Engineering, Yokohama National University
156 Tokiwadai, Hodogaya, Yokohama 240, Japan

Ronald L. Rivest
Massachusetts Institute of Technology, Laboratory for Computer Science
Cambridge, Massachusetts 02139, USA

CR Subject Classification (1991): E.3-4, D.4.6, G.2.1, C.2.0, K.6.5

ISBN 3-540-57332-1 Springer-Verlag Berlin Heidelberg New York
ISBN 0-387-57332-1 Springer-Verlag New York Berlin Heidelberg

© Springer-Verlag Berlin Heidelberg 1993
Printed in Germany

Typesetting: Camera-ready by author
Printing and binding: Druckhaus Beltz, Hemsbach/Bergstr.
45/3140-543210 - Printed on acid-free paper

Contents

Session 1: Invited Lecture 1
Chair: Ronald L. Rivest

The Transition from Mechanisms to Electronic Computers, 1940 to 1950   1
   Donald W. Davies

Session 2: Differential Cryptanalysis and DES-Like Cryptosystems
Chair: Ronald ·L. Rivest

Cryptanalysis of LOKI   22
   Lars Ramkilde Knudsen

Improving Resistance to Differential Cryptanalysis and the Redesign of LOKI   36
   Lawrence Brown, Matthew Kwan, Josef Pieprzyk, and Jennifer Seberry

A Method to Estimate the Number of Ciphertext Pairs for Differential Cryptanalysis   51
   Hiroshi Miyano

Construction of DES-Like S-Boxes Based on Boolean Functions Satisfying the SAC   59
   Kwangjo Kim

The Data Base of Selected Permutations   73
   Jun-Hui Yang, Zong-Duo Dai, and Ken-Cheng Zeng

Session 3: Hashing and Signature Schemes
Chair: Andrew Odlyzko

A Framework for the Design of One-Way Hash Functions Including Cryptanalysis of
Damgård's One-Way Function Based on a Cellular Automaton   82
   Joan Daemen, René Govaerts, and Joos Vandewalle

How to Construct a Family of Strong One-Way Permutations   97
   Babak Sadeghiyan, Yuliang Zheng, and Josef Pieprzyk

On Claw Free Families   111
   Wakaha Ogata and Kaoru Kurosawa

Sibling Intractable Function Families and Their Applications   124
   Yuliang Zheng, Thomas Hardjono, and Josef Pieprzyk

A Digital Multisignature Scheme Based on the Fiat-Shamir Scheme   139
   Kazuo Ohta and Tatsuaki Okamoto

Session 4: Secret Sharing, Threshold, and Authentication Codes
Chair: Chin-Chen Chang

A Generalized Secret Sharing Scheme with Cheater Detection   149
     Hung-Yu Lin and Lein Harn

Generalized Threshold Cryptosystems   159
     Chi-Sung Laih and Lein Harn

Feistel Type Authentication Codes   170
     Reihaneh Safavi-Naini

Session 5: Invited Lecture 2
Chair: Sang-Jae Moon

Research Activities on Cryptology in Korea   179
     Man Y. Rhee

Session 6: Block Ciphers — Foundations and Analysis
Chair: James L. Massey

On Necessary and Sufficient Conditions for the Construction of
Super Pseudorandom Permutations   194
     Babak Sadeghiyan and Josef Pieprzyk

A Construction of a Cipher from a Single Pseudorandom Permutation   210
     Shimon Even and Yishay Mansour

Optimal Perfect Randomizers   225
     Josef Pieprzyk and Babak Sadeghiyan

A General Purpose Technique for Locating Key Scheduling Weaknesses in
DES-like Cryptosystems   237
     Matthew Kwan and Josef Pieprzyk

Results of Switching-Closure-Test on FEAL   247
     Hikaru Morita, Kazuo Ohta, and Shoji Miyaguchi

Session 7: Invited Lecture 3
Chair: Ken-Cheng Zeng

IC-Cards and Telecommunication Services   253
     Jun-ichi Mizusawa

Session 8: Cryptanalysis and New Ciphers
Chair: Ingemar Ingemarsson

Cryptanalysis of Several Conference Key Distribution Schemes   265
    Atsushi Shimbo and Shin-ichi Kawamura

Revealing Information with Partial Period Correlations   277
    Andrew Klapper and Mark Goresky

Extended Majority Voting and Private-Key Algebraic-Code Encryptions   288
    Joost Meijers and Johan van Tilburg

A Secure Analog Speech Scrambler Using the Discrete Cosine Transform   299
    B. Goldburg, E. Dawson, and S. Sridharan

Session 9: Proof Systems and Interactive Protocols 1
Chair: Yvo G. Desmedt

An Oblivious Transfer Protocol and Its Application for the Exchange of Secrets   312
    Lein Harn and Hung-Yu Lin

4 Move Perfect ZKIP of Knowledge with No Assumption   321
    Takeshi Saito, Kaoru Kurosawa, and Kouichi Sakurai

On the Complexity of Constant Round ZKIP of Possession of Knowledge   331
    Toshiya Itoh and Kouichi Sakurai

On the Power of Two-Local Random Reductions   346
    Lance Fortnow and Mario Szegedy

A Note on One-Prover, Instance-Hiding Zero-Knowledge Proof Systems   352
    Joan Feigenbaum and Rafail Ostrovsky

Session 10: Proof Systems and Interactive Protocols 2
Chair: Eiji Okamoto

An Efficient Zero-Knowledge Scheme for the Discrete Logarithm Based on Smooth Numbers   360
    Yvo Desmedt and Mike Burmester

An Extension of Zero-Knowledge Proofs and Its Applications   368
    Tatsuaki Okamoto

Any Language in IP Has a Divertible ZKIP   382
    Toshiya Itoh, Kouichi Sakurai, and Hiroki Shizuya

A Multi-Purpose Proof System   397
    Chaosheng Shu, Tsutomu Matsumoto, and Hideki Imai

Formal Verification of Probabilistic Properties in Cryptographic Protocols   412
    Marie-Jeanne Toussaint

Session 11: Invited Lecture 4
Chair: Hideki Imai

Cryptography and Machine Learning   427
    Ronald L. Rivest

Session 12: Public-Key Ciphers — Foundations and Analysis
Chair: Tsutomu Matsumoto

Speeding Up Prime Number Generation   440
    Jørgen Brandt, Ivan Damgård, and Peter Landrock

Two Efficient Server-Aided Secret Computation Protocols Based on the Addition Sequence   450
    Chi-Sung Laih, Sung-Ming Yen, and Lein Harn

On Ordinary Elliptic Curve Cryptosystems   460
    Atsuko Miyaji

Cryptanalysis of Another Knapsack Cryptosystem   470
    Antoine Joux and Jacques Stern

Rump Session: Impromptu Talks
Chairs: Thomas A. Berson and Kenji Koyama

Collisions for Schnorr's Hash Function FFT-Hash Presented at Crypto'91   477
    Joan Daemen, Antoon Bosselaers, René Govaerts, and Joos Vandewalle

On NIST's Proposed Digital Signature Standard   481
    Ronald L. Rivest

A Known-Plaintext Attack of FEAL-4 Based on the System of Linear Equations on Difference   485
    Toshinobu Kaneko

Simultaneous Attacks in Differential Cryptanalysis (Getting More Pairs Per Encryption)   489
    Matthew Kwan

Privacy, Cryptographic Pseudonyms, and The State of Health   493
    Stig Fr. Mjølsnes

Limitations of the Even-Mansour Construction   495
    Joan Daemen

Author Index   499

Preface

ASIACRYPT '91 was the first international conference on the theory and application of cryptology to be held in the Asian area. It was held at Fujiyoshida, Yamanashi, Japan, overlooking beautiful Mt. Fuji, from November 11 to November 14, 1991.

The conference was modelled after the very successful CRYPTO and EUROCRYPT series of conferences sponsored by the International Association for Cryptologic Research (IACR). The IACR and the Institute of Electronics, Information and Communication Engineers were sponsors for ASIACRYPT '91.

The program committee published a call for papers and received 100 extended abstracts for consideration. Three of them were not reviewed since they arrived too late. Each of the other abstracts was sent to all the program committee members and carefully evaluated by at least 10 referees. The committee accepted 39 papers for presentation. In addition, the program committee invited four papers for special presentation as "invited talks." Unfortunately, three of the accepted papers were withdrawn by the authors before the conference.

The conference attracted 188 participants from 17 countries around the world. The technical presentations were well attended and enthusiastically received. Following the CRYPTO tradition, an evening "rump session" was held. This session, chaired by Thomas Berson and Kenji Koyama, included short presentations of recent results. The non-technical portion of the conference included a sightseeing trip to the base of Mt. Fuji, a Japanese barbecue lunch (*robatayaki*), and a banquet with drummers and a magic show.

After the conference the authors produced the full papers, in some cases with slight improvements and corrections, for inclusion here. For ease of reference by those who attended the conference, the papers are placed in the same order and under the same headings as they appeared at the conference. Because of the interest expressed in the rump session presentations, we have included short papers contributed by the rump session speakers at the end of this proceedings. Of the 12 rump session presentations, the 6 abstracts included here have gone through a thorough, if expedited, refereeing process.

It is our pleasure to thank all those who contributed to make these proceedings possible: the authors, program committee, organizing committee, IACR officers and directors, and all the attendees.

Yokohama, Japan   Hideki Imai
Cambridge, U.S.A.   Ronald L. Rivest
Yokohama, Japan   Tsutomu Matsumoto
August 1993

ASIACRYPT'91

Program Committee:

Hideki Imai (Co-Chair, Yokohama National University, Japan)
Ronald L. Rivest (Co-Chair, Massachusetts Institute of Technology, U.S.A.)
Tsutomu Matsumoto (Vice Chair, Yokohama National University, Japan)
Thomas A. Berson (Anagram Laboratories, U.S.A.)
Chin-Chen Chang (National Chung Cheng University, R.O.C.)
Yvo G. Desmedt (University of Wisconsin – Milwaukee, U.S.A.)
Shimon Even (Technion, Israel)
Shafi Goldwasser (Massachusetts Institute of Technology, U.S.A.)
Ingemar Ingemarsson (Linköping University, Sweden)
Kenji Koyama (NTT Corporation, Japan)
James L. Massey (ETH Zürich, Switzerland)
Sang-Jae Moon (Kyung Pook National University, Korea)
Eiji Okamoto (NEC Corporation, Japan)
Keng-Cheng Zeng (Academia Sinica, P.R.O.C.)

Organizing Committee:

Shigeo Tsujii (Chair, Tokyo Institute of Technology)
Yoshihiro Iwadare (Vice Chair, Nagoya University)
Masao Kasahara (Vice Chair, Kyoto Institute of Technology)
Kenji Koyama (Local Arrangement Chair, NTT)
Ryota Akiyama (Fujitsu)
Hideki Imai (Yokohama National University)
Toshiya Itoh (Tokyo Institute of Technology)
Shin-ichi Kawamura (Toshiba)
Naohisa Komatsu (Waseda University)
Sadami Kurihara (NTT)
Kaoru Kurosawa (Tokyo Institute of Technology)
Tsutomu Matsumoto (Yokohama National University)
Hideo Nakano (Osaka University)
Koji Nakao (KDD)
Kazuo Ohta (NTT)
Tatsuaki Okamoto (NTT)
Ryoui Onda (SECOM)
Kazuo Takaragi (Hitachi)
Kazue Tanaka (NEC)
Atsuhiro Yamagishi (Mitsubishi)

THE TRANSITION FROM MECHANISMS TO ELECTRONIC COMPUTERS, 1940 TO 1950

Donald W. Davies, Independent Consultant
15 Hawkewood Road, Sunbury-on-Thames, Middlesex UK, TW16 6HL

Abstract

The peak of mechanical cryptography was reached in World War II, then electronics rapidly replaced these machines. A very remarkable technology then ended. Some of the best examples that I have found will be illustrated. The paper continues with some memories of building the first computer at NPL during 1947 to 1950.

The age of mechanisms

The difference engines and analytical engines designed by Charles Babbage would have been, if completed, one of the greatest achievements of the mechanical age. Computing devices remained mechanical (or electro-mechanical) for another 100 years. Today we are in the electronic age and it is interesting to look at the short period of transition from mechanisms to electronics, which began about 50 years ago. In this paper I shall consider only digital systems and my examples come from cryptography and my own memories of the first electonic computers.

Electronics can be pretty, but what you see is only distantly related to its function. At the peak of the mechanical age, the function of mechanisms was very clear; they could be seen working at human speeds. This led their designers and constructors to emphasize their function with shapes of striking beauty and with surface finishes that were often much more elaborate than strictly required. Not only steam engines and pumps had this quality - it can be seen in Babbage's designs and in his test assemblies.

It has often been assumed that Babbage did not complete his machines because the technology of the time was inadequate. The recently completed difference engine No.2 at the London Science Museum shows that Babbage's machines do work, when they are built with the materials and precision available to Babbage. The design needed several corrections and some counterbalancing springs. The very complex running carry mechanism works perfectly and spectacularly, and wheels that are not being stepped are firmly held. The only concerns of its designers and operators at the Science Museum are with lubrication and with wear. The part of the machine which impressed the printing plates has not yet been built and urgently needs sponsorship.

Calculating mechanisms often have repeated units such as counter wheels and registers but they cannot be organised simply by linking together large numbers of very simple devices in the way that gates and storage cell are used. The best that can be done is illustrated by Babbage's notation for mechanisms and his suggestions for some general-purpose mechanical principles. Conrad Zuse once described to me a mechanical binary store array with which he had proposed to make a mechanical 'minicomputer'. But these were exceptions and usually a digital mechanism is designed as a whole rather than assembled from identical subunits. In this respect, the precursors of the gates and cells of electronics were electro-mechanical systems such as telephone exchanges which used relays and rotary switches as subunits, and appeared briefly in cryptography.

To illustrate this period of transition I will first describe two cryptographic mechanisms used by Germany in WWII, then some of my own experience with the first electronic computers.

On-line ciphers of World War II

The Enigma machine is very well known. This was operated off-line, producing a written ciphertext which was then manually transmitted. In the Defence Museum in Oslo there are printer attachments for enigma machines, remote displays and a large commutator called 'Enigma-Uhr' which could be wired to the plugboard to give hourly changes of key. Fortunately for the Allies, this last device came into use very late. By upsetting the involution property of the plugboard, the Enigma-Uhr would have given a major problem to the cryptanalysts.

There were two on-line machines in wide use by the German forces.

One was known as SZ40 or SZ42, where SZ stands for Schluessel Zusatz meaning cipher attachment. As its name implies, it operated 'in-line' in a teleprinter circuit and did not have its own keyboard or printer. The maker was Lorenz. It was installed with a teleprinter and radio equipment in a vehicle designed for the warfare of rapid movement or 'blitzkrieg' planned by Germany. This machine was used at the level of high command, making its messages very valuable to the allies.

The second on-line machine was the T52, essentially a standard teleprinter working together with a built-in cipher unit. This was made by Siemens and Halske. For most of the war it was used with transmission by cable, but at a late stage it enciphered radio messages and the Allied cryptographers began to take an interest in it.

Eric Huttenhain, who developed cryptanalytic machines for the cipher bureau of the OKW told me that his group made a comparison of the security level of these two machine, but he would not tell me the outcome. In the event, the SZ was put in the most sensitive place. I believe the T52 as eventually developed was stronger.

The T52 was very bulky and heavy, compared with the SZ. The SZ needed a teleprinter but this was a separate unit, easier to instal in a vehicle. Perhaps these considerations led to the decision that SZ should be in the mobile system.

The Lorenz Schluessel Zusatz

This machine fits well into our theme because the cipher unit was almost completely mechanical. It is a mechanism of great elegance. Though the principle is simple, designing it mechanically was difficult. I have not, at the time of writing, had enough time with a working model to understand the mechanism completely. Two exist, at the Oslo museum and the DeutschesMuseum in Munich.

From the many photographs I have obtained it is clear that the SZ went through many changes and improvements. This also appears in the official history, which mentions changes in the cipher called 'Fish'. But I have been unable to match the various clues and produce a coherent account of its varieties. We can guess that SZ40 and SZ42 were introduced in 1940 and 1942, but the significant changes were later.

4

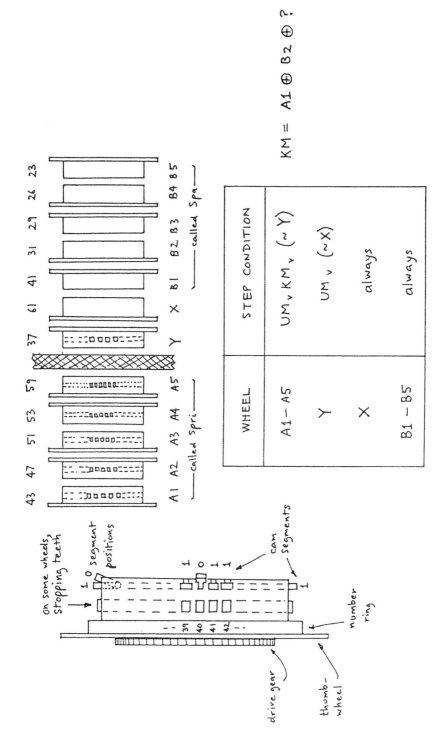

WHEEL	STEP CONDITION
A1 — A5	$UM \vee KM \vee (\sim Y)$
Y	$UM \vee (\sim X)$
X	always
B1 — B5	always

$$KM = A1 \oplus B2 \oplus ?.$$

Figure 1: Cipher Wheels

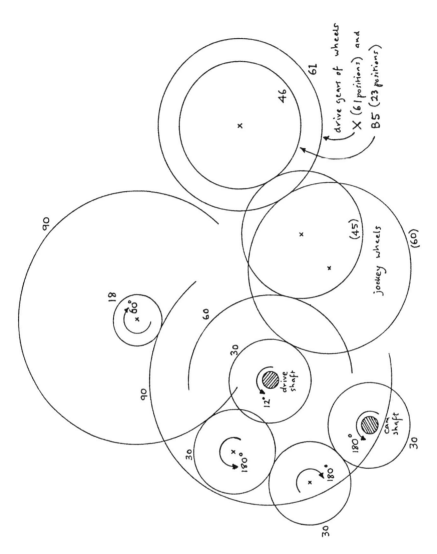

Figure 2: Gear Train

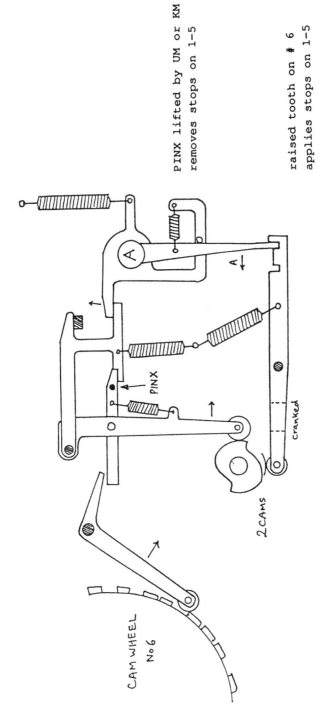

PINX lifted by UM or KM
removes stops on 1-5

raised tooth on # 6
applies stops on 1-5

Figure 3

The outer case of the machine contained both the mechanical cipher unit and electromechanical devices to convert from the start/stop 5 unit telegraph signal to 5 parallel signals and back from the parallel channels to the serial signal.

Figure 1 shows the 12 cipher wheels, which had relatively prime numbers of teeth, namely 23, 26, 29, 31, 41, 61, 37, 59, 53, 51, 47, 43 reading from the right. Functionally there were two sets of 5 wheels (coresponding to the 5 channels) at each end and special wheels of 61 and 37 teeth in the middle.

A very distinctive feature of the SZ is that the binary output of each wheel can be set by moving to one side or another individual hinged teeth. We can imagine that setting these patterns would be tedious and error prone, so it would not be done frequently.

The cipher priniple was simply to add the output of one A wheel and one B wheel, modulo 2 to each channel. If every wheel moved one step between characters, this is a conventional Vernam cipher and rather weak. Clearly some extra principle is needed. This is the intermittent stopping of all the A wheels.

To drive the wheels at their various rates there is an ingenious set of gears and jockey wheel of diferent sizes shown in Figure 2. Wheels which might need to stop are driven through a slipping mechanism inside the wheel. On the edge of these wheels are square teeth which can be engaged by a latch to stop them for one or more character times. Incidentally, Babbage deplored the use of slipping like this and insisted that all movements were positively driven. His machines could possibly jam but would never make errors by bouncing or overshooting. I wonder if the SZ had any such problems.

In the simplest form the SZ worked in this way: The X wheel always moved. Its output controlled the stopping of wheel Y. The output of the Y wheel controlled the stopping of all the A wheels together. The later developments controlled the stopping of the A wheels also by outputs from individual A and D wheels.

Figure 3 shows part of the mechanism which uses the output of the Y wheel to lift an interposer and either turn shaft A or not, powered by

8

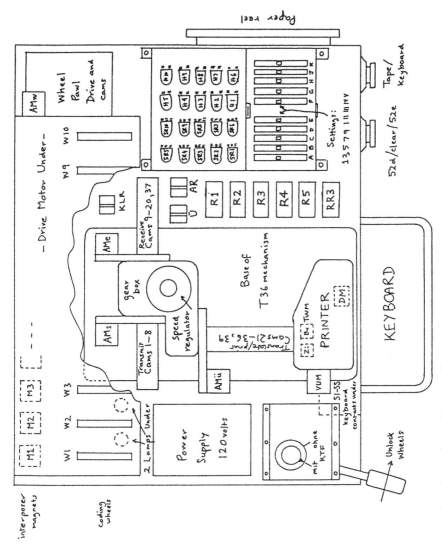

Figure 4: Plan View of the T52e

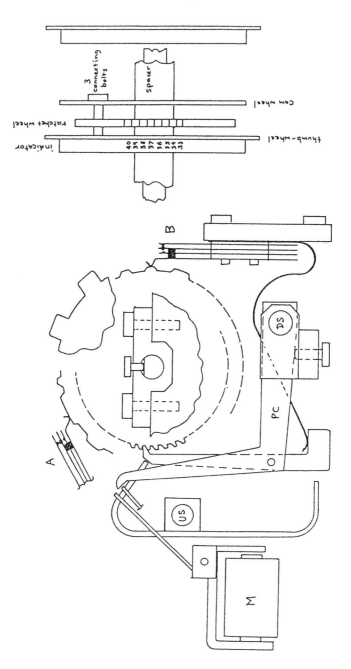

Figure 5: Interposer Mechanism, Cam-wheels and Contacts

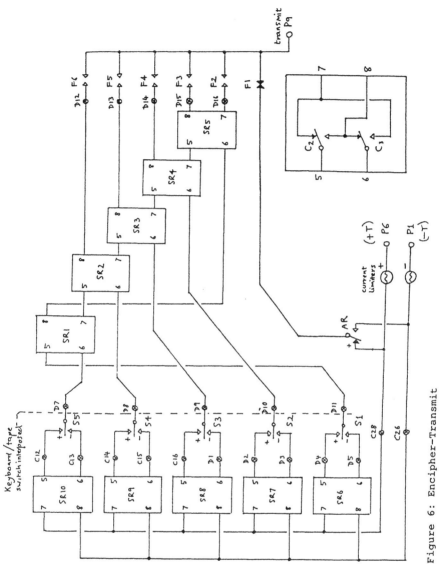

Figure 6: Encipher-Transmit

a cam. The output of A is locked by another cam. Shaft A drives the mechanisms on each of the A wheels which stop their movement.

The Siemens and Halske T52

The very large baseplate of the T52 holds the mechanism of a T36 teleprinter, with it's keyboard, tape reader and printer, together with all the cipher equipment. Figure 4 is a plan view. The largest addition is a set of 10 cipher wheels at the back. The other main additions in the figure are 20 relays and 10 rotary switches on which the basic key is entered.

There were five models labelled a to e and the one I shall describe is model e. The machines were sometimes modified in the field without changing their label, but I have been able to get a detailed picture of each model. Models a and b were logically identical, b having improved electrical filters, supposedly to reduce radio interference. Models c and d introduced the important feature of intermittent wheel motion. A model f was under development but was not made and no information about it has been found.

The 10 wheels have relatively prime numbers of teeth, respectively 47, 53, 59, 61, 64, 65, 67, 69, 71 and 73. The cam profiles were fixed for the whole time the T52 was used, as far as I know. A change of wheel is possible with simple tools but readjustment of contacts could have been a problem. In some models modified for use in Norway, wheels have been assembled in different orientations and there are very few original machines in existence.

Figure 5 shows how a wheel is driven by a ratchet and how magnet M stops the drive. Each has two sets of contact springs. One is used in the cipher transformation, the other drives the wheel stopping logic.

The 10 binary wheel outputs go first to the rotary switches on which the main key has been set, which permute them. Then they go to the relays which perform a linear (mod 2) tranformation on the 10 bits. 5 of the bits are added modulo 2 to the telegraph code, then the other 5 bits determine a permutation of the 5 elements of the code, as shown in figure 6. The machine has separate relay logic for encipherment and decipherment. When transmitting in cipher it simultaneously receives, deciphers and prints the character, giving a check on the operation of some (but not all) of the equipment.

12

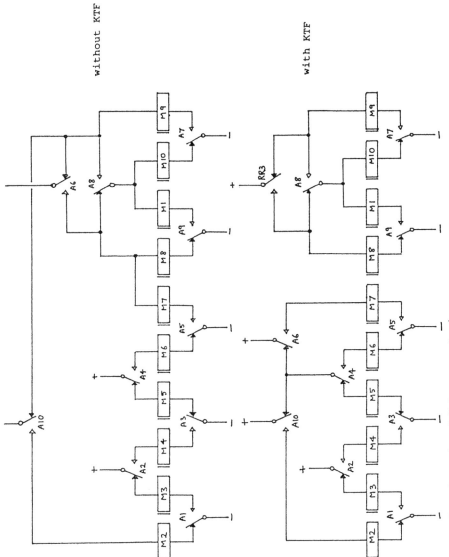

Figure 7: Schematic of Interposer Logic

Perhaps the most interesting feature is the wheel stopping logic, one version of which is shown in Figure 7. The wheels have a total of 8.9 x 10^{17} states. A random state transition table would give a cycle length of the order of 9.5 x 10^8 and it would be interesting to know what cycle lengths were actually obtained.

The main key was set by a plugboard in earlier models. Both the plugboard and the later rotary switches were in a locked box and were probably changed infrequently. The 'message key' was probably the initial wheel settings, but the method of transmitting them is unknown. Models a, b and c had an elaborate mechanism for returning all wheels to a chosen setting, perhaps for sending the message key.

Model c had an additional unit with 10 levers on which a 'message key' could be set. These performed yet another permutation on the outputs of the 10 wheels.

The last of their line?

The early history of the SZ is unknown but the T52 can be traced back to a patent in 1930. US patent 1,912,983 is closer to the eventual T52. It was developed as a commercial venture and I was told that one version was supplied to Hungary in 1932. Another person said the first deliveries of the T52 were in 1934. Bombing stopped production in Berlin in 1944 but a small facility in Kladov near Berlin tried to continue for a while until the Russian army arrived in May 1955.

On-line cipher machines for teleprinter messages had to encipher a 5-bit code in about 150 ms. Electro-mechanical technology of the 1930-1940 period implied a stream cipher driven by contacts from cams on wheels, or the use of uniselectors. My experience suggests that uniselectors would be difficult to maintain, the T52 would be reliable with skilled maintenance and the mainly mechanical SZ would work best in practice in military conditions.

I have no information about other on-line cipher machines of the WWII period. The two I have described represent, I believe, the most advanced level of on-line cryptography before electronics took over. In particular, the T52 suggests some interesting theoretical problems.

The first electronic computers

I joined a small team at the UK National Physical Laboratory (NPL) under the leadership of Alan Turing in 1947. By that time the logical design of ACE had reached version 7c, but nothing of significance had been built.

There were three pioneering computer projects in UK, the others were at Cambridge University and Manchester University. I shall speak only about my own experience.

For months our design team continued to refine the design, testing the order code by programming excercises and the logical design by stepwise tabulation of the states of triggers etc. This was extremely tedious and frustrating. Only a short time after I joined, Turing left the project but I did have one discussion with him about his 'computable numbers' paper. Rather it was an argument because I wanted to correct all the many errors in the formal part of the paper and Turing felt this was a waste of time.

There is evidence that when the project was officially approved, the possible future use of computers for cryptanalysis was in the mind of at least one member of the committee. Also the wartime experience in making Colossus may have led to the decision that the Post Office Research Station at Dollis Hill should build the machine. I should explain that the telephone network was part of the Post Office; Martlesham is the successor of Dollis Hill, where the Collossi had been built. Two of that team became our engineers. But the arrangement did not work well, hence our frustration. In the long run our design transferred to Dollis Hill became a computer which worked well in a defence establishment, but we cut ourselves loose from this scheme and built ACE ourselves.

Getting it started at NPL met further snags and it was only when finally the mathematicians and electronics engineers finally all moved into one large room and made a single team that construction really began. We lost years in this muddle.

One of our strokes of luck was that Harry Huskey, from the ENIAC team, joined us for a year. He started a project to build a 'test assembly'

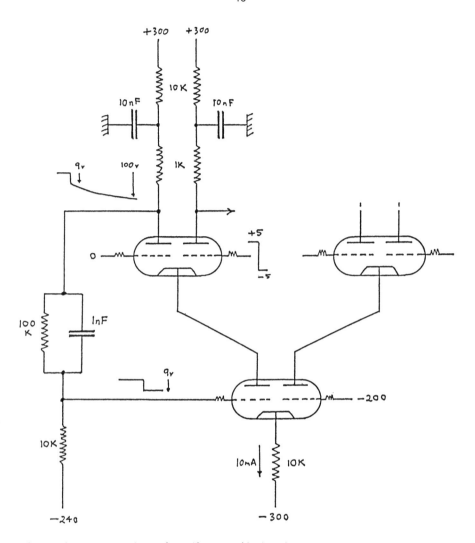

Figure 8: Current-Steering (long-tailed pair) due to Blumlein

which later, under the name 'Pilot ACE' was adopted as the main objective. It became the English Electric 'Deuce' and in the US used the design for the Bendix G15, which was an early commercial success. Huskey had one in his garage in Berkley when I visited him and this must be the first personal computer, though it was rather large and hot. Now this machine is on display in the Smithsonian Museum in Washington DC.

Engineering

ACE had only about 1000 thermionic tubes(valves in UK jargon) so I think it was the first RISC machine. Its memory (we called it the store) was about 8 mercury delay lines (long tanks) each holding 32 numbers of 32 bits, together with a small number of short mercury delay lines (short tanks). I later made a basic redesign of the long tanks, folding in half by two reflections and making the crystals fully adjustable.

The clock rate was 1 MHz, considered fast for valves but actually conservative with the circuit design we employed. The later full scale ACE ran easily at 1.5 MHz. Our circuit design came from the work of Blumlein at EMI, who developed the UK's prewar television equipment and wartime radar losing his life while testing radar in an aircraft.

The basic circuit is shown in Figure 8. It fed a constant current into the common cathode of two triodes and diverted to one or other anode. This is precisely the analogue of current steering or emitter coupled logic. We stacked these two high and with the necessary current defining resistors it meant voltages from -300v to +300v. Since all testing was done under power, we learnt to think before touching. I still work on live 240v circuits without fear, though with caution.

Then we needed couplings for the 10v signals from a top anode to a lower grid. Blumlein had a perfect answer which is shown in Figure 8, but it would take too long to analyse this circuit and its tolerancing here. Two engineers from EMI joined us with this technology. Without them we would have floundered. We also had to learn a good discipline for the timing of signals and that is another story.

ACE PILOT MODEL 1949 to 1956

List of Sources and Destinations

Number	Source	Destination
0	Input 32 bits	Control
1 to 10	Long Delay Lines 32 words each	
11	TS11 1 word	TS11
12	DS12 2 words	DS12
13	DS14 DIV 2	Add to DS14
14	DS14 Long Acc	DS14
15	TS15	TS15
16	TS16 Short Acc	TS16
17	NOT TS26	ADD to TS16
18	TS26 DIV 2	SUBTRACT from TS16
19	TS26 times 2	MULTIPLY
20	TS20	TS20
21	TS26 AND TS27	Set TCA
22	TS26 XOR TS27	spare
23	P17	TCB
24	P32	Jump if negative
25	P1	Jump if non zero
26,27	TS26, TS27	TS26, TS27
28	ZERO	OUTPUT 32 bits
29	ONES	Buzzer
30	TIL (card control)	Start Punch
31	spare	Start Reader

In 1954 DL9 handled magnetic drum transfers
which were controlled by two destinations

Figure 9

Programming

The order code had 2 store addresses for data and one short address for the next instruction. Each operation took data from a source and moved it to a destination. The source or destination value defined also the type of operation, for example some addresses belonged to the accumulator or multiplier or to tanks which precessed at each use (for sequential access). Figure 9 shows a list of the sources and destinations.

The really novel feature was that operations could transfer up to 32 words in sequence (the first vector operations) and when an operation ended the next instrucion could be loaded at once, if it was in the right place in a tank. These things could make the machine very fast, for its time, but to exploit them was a difficult problem for the programmer. It has been called 'optimum programming' but it is simply making the best of an awkward kind of memory with long latency.

Programming had two stages. First we wrote the program then we laid the instructions carefully in the store. This second phase was like solving a puzzle. For important subroutines, days might be spent trying to reduce the length of a loop to get it into one less circuit of a long tank (one less major cycle, or one millisecond.)

Input and output

Our input and output was on 80 column punched cards, but not usually in the standard way. We usually put one binary word on each row of the card, 12 to a card. Since this could in principle reduce a card to 'lace' there was no certainty that the card machine could handle it, but they always did. We found that chads could be pushed back into a card and were very firm. This was useful to correct minor errors in programs, which we punched ourselves in binary. I still remember the number up to 31, least significant bit first. The convention arose because the unit bit had to go first through an adder and appeared first on the monitor screen. We were quite shaken when we found the rest of the world put the msb first.

Our card machines read at 200 cards per minute and punched at 100, making them much faster than the paper tapes others were using, especially if each held 384 bits. With our small store we had to operate on big matrices, so we used cards as intermediate memory until our drum came along. The operators became skilled at loading blocks of cards in the right sequence and avoiding jams.

The drums were novel. Like the Manchester team we were able to synchronize the drum rotation accurately to the clock. We chose one revolution in 9 major cycles, precisely 9216 microseconds, making 6510.4 rpm. The drum surface must not be more than about 1 microsecond late or early. We also had moving heads driven by moving coil linear motors, after trying several other drive principles which would make an interestng story themselves. Mechanisms have not disappeared, of course, they tend to get simpler but faster.

What was it like?

I have been asked about our motivation. NPL had a Division specializing in numerical mathematics and its members, myself included, had struggled for years with heavy calculations that demanded man years (more accurately woman years) of a human computer's time. In this Mathematics Division were some acknowledged experts on numerical analysis. In some ways this was more highly developed for human computers than it is today because we could use judgement about how to approach a singularity or when to use an alternative iterative step. We were fully ready to exploit the machine when it worked and we ran one of the world's first computer services, early customers being aircraft designers with their new flutter problems.

At the start we all saw clearly the potential of the vast increase of computing speed and there was plenty of discussion of big numerical problems, such as weather forecasting. I do not remember our group talking about applications in commerce until much later.

My main impression was of isolation from the rest of the world. To describe properly what we were doing would have needed a long lecture about numerical work, programs, instructions, electronics and input and output. We could talk to the teams in Cambridge and Manchester (we met often in Cambridge) and to friends in the US but for the rest we tended to remain silent. Colleagues in analogue computing looked at our work and thought we were crazy.

In one sense we were crazy. Experience showed that our collections of

more than 1000 valves would never all work together. The completely
accurate working of everything seemed too much to hope for. The first
trials seemed to confirm the gloomy view. The simplest possible
program ran for a second before it failed. Next day it ran for 10
seconds after an improvement in timing, for example. There was often
enough an improvement to keep us motivated, but all thoughts of the
complex programs written in the pre-building stage were forgotten.

When the tolerancing in signal level and timing had been got right we
were at the mercy of valves. Heater failures were not such a problem
and stability of characteristics was made unimportant by Blumlein's
genius but their were other plagues. Our double triodes had grids
close to the cathode. They could become 'tap happy', causing momentary
faults when they had the slightest movement. The faults, I should
explain, never gave wrong results but would drop out of the program.
We found (or thought so) that small particles of cathode material were
lodging between cathode and grid. These valves should be replaced, so
we sometimes ran a test program and tapped all the valves one by one.
We felt this was not good engineering but it worked

Years later I visited IBM's first computer producion line, I think it
was the 650 magnetic drum machine. At the final stage of testing, blue
suited IBM engineers were tapping all the valves with a carefully
designed special hammer. I felt vindicated. We had been engineers
after all.

The progress from just working to becoming an important service to
industry was very gradual, with many backward slides. Always the best
test program was the latest really complex application. The mature
machine could always sail through the programs devised by engineers.
Jim Wikinson, our genius of linear algebra would say 'It's a poor
machine that won't run its test programs.'

I ran the first program that used a subroutine, just a ray tracing
program with a square root. Seeing happen what we had thought about
four years earlier was exciting. The reality of programming (more
important, of problem solution) was very different from the early
dream. Later I did the first simulation of traffic, both road traffic
and men receiving warnings and exiting a coal mine. But others took
over the programming art. Unfortunately the nice trick of optimum
programming was not very compatible with languages. Turing had been
the first to conceive a machine as having a language and being able to

Cryptanalysis of LOKI

Lars Ramkilde Knudsen
Aarhus Universitet* Datalogisk Afdeling
Ny Munkegade
DK-8000 Aarhus C.

Abstract

In [BrPiSe90] Brown, Pieprzyk and Seberry proposed a new encryption primitive, which encrypts and decrypts a 64-bit block of data using a 64-bit key. Furthermore they propose a way to build private versions of LOKI.

In this paper we show first that the keyspace of any LOKI-version is only 2^{60}, not 2^{64} as claimed. Therefore there are 15 equivalent keys for every key, that encrypts/decrypts texts the same way. An immediate consequence is, that the proposed Single Block Hash Mode is no good. It is very easy to find collisions.

Secondly we do differential cryptanalysis on LOKI and show that n-round LOKI, $n \leq 14$ is vulnerable to this kind of attack, at least in principle. We show that we cannot find a characteristic with a probability high enough to break LOKI with 16 rounds. However one **might** find a private LOKI-version, that is vulnerable to a differential attack for n = 16.

1 LOKI - a family of encryption primitives

In [BrPiSe90] Brown, Pieprzyk and Seberry proposed a new encryption primitive, which encrypts and decrypts a 64-bit block of data using a 64-bit key. LOKI is interface compatible with the DES (ISO DEA-1) and its structure is very much like the structure of DES. Therefore it is obvious to try to do a differential attack proposed by Biham-Shamir in [BiSha90] on LOKI.

The main difference between DES and LOKI is the S-boxes. All 4 LOKI S-boxes are equal, take a 12 bit input and produce a 8 bit output. The output is evaluated through exponentations (with a fixed exponent) in 16 different fields generated by 16 irreducible polynomials.

Another difference between DES and LOKI is that in the latter, the key is added (modulo 2) just before and after 16 rounds of F-iterations.

Furthermore the P-permutation in LOKI allows us to find good fixpoints for the F-function, which we cannot do in DES.

---

*The main part of this paper was made at the Technical University of Eindhoven, the Netherlands, during my stay there from 1.2.91 to 29.6.91.

interpret another language, but his creation, the logic design of ACE and its Pilot ACE paradoxically did not shine in this environment. Still, our small team made a contribution to Algol (and later to ADA) and went on to many achievements.

My interests moved to data communications, then to security and processors, once the centre of our thoughts, became something you can buy at the corner shop.

2 The keyschedule algorithm

The keysize in LOKI is 64 bits. The key is divided into two halves of 32 bits, KL and KR, respectively. Initially KL and KR are added (modulo 2) to the left and right halves of the plaintext, see Figure 1.

The 32 bits of KL are used as the keys in the odd rounds, that is round no. 1,3,5,.....,15, in the following way:

$$K_{2i+1} = ROL_{12}(K_{2i-1}), \quad i \in \{1,....,7\}$$
$$\text{and } K_1 = KL$$

where ROL_{12} is 12 rotations of the key to the left.

Finally $ROL_{12}(K_{15}) = KL$ is added to the right half of the plaintext.

The 32 bits of KR are used as the keys in the even rounds, in the following way:

$$K_{2i+2} = ROL_{12}(K_{2i}), \quad i \in \{1,....,7\}$$
$$\text{and } K_2 = KR$$

Finally $ROL_{12}(K_{16}) = KR$ is added to the left half of the plaintext.

The initial and final addition of the key has an unfortunate impact on the keyspace of LOKI, as we will show in the following.

From Figure 1, we see that the final addition of KR and KL can be pushed 'upwards in the tree'. If we try to push KL upwards, then every time we pass the F-function on the right side, we have to add KL to the corresponding key. If we push KL to the top of the tree every even subkey K_i is added to KL and the initial addition of KL disappears.

Of course the same trick goes for the final addition of KR, and we obtain a rearranged model of LOKI, Figure 2.

Instead of 16 subkeys, 8 consisting of 32 bits from the set $\{k_0,....,k_{31}\}$ and 8 consisting of 32 bits from the set $\{k_{32},....,k_{63}\}$, we have 16 **actual subkeys**, $AK_1, AK_2, ..., AK_{16}$, each consisting of 32 **keybit-xors** from the set

$$\{a \oplus b \mid a \in \{k_0,....,k_{31}\}, \ b \in \{k_{32},....,k_{63}\}\}$$

We have the following theorem:

Theorem 1 *The keyspace of LOKI is* 2^{60}.

That is, for any key K, all keys of the form K $\oplus$ hhhhhhhh hhhhhhhh$_x$, where h is one of the hexadecimal symbols 0,1,....,e,f, are equivalent.

Proof:

An arbitrary bit-xor in an actual subkey has the form $k_i \oplus k_j$, where it holds that i-j = 0 mod 4, due to the fact that 12 = 32 mod 4, and where the indices i and j refer to the bit numbering in the original key. It is obvious that when modifying the key with the repeated hex digit h, k_i and k_j will be xor'ed with the same bit

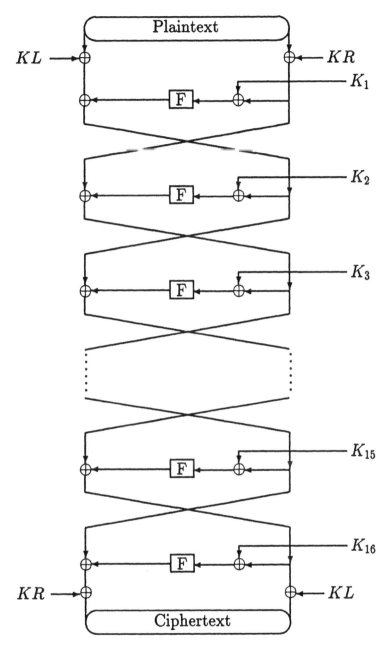

Figure 1: LOKI with 16 rounds

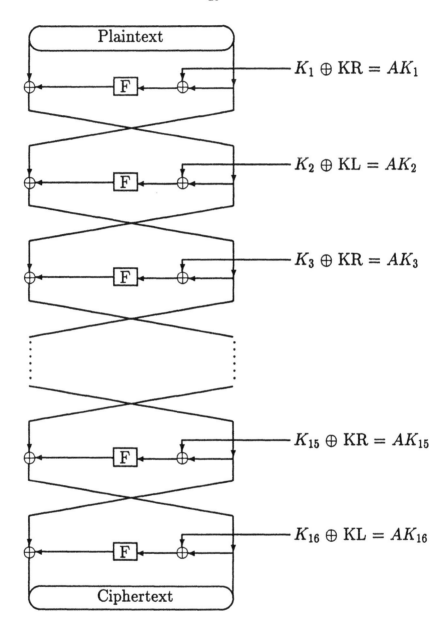

Figure 2: Rearranged LOKI with 16 rounds

from h, whence all $k_i \oplus k_j$ and therefore all actual subkeys are unchanged.   □

As an example suppose K = 13cd2452 d97e8b60$_x$ one gets the following 16 actual subkeys, regardless of the value of h:

AK_1 = cab3af32$_x$ AK_2 = cab3af32$_x$
AK_3 = 0b3baa5c$_x$ AK_4 = fb7b29c5$_x$
AK_5 = 8b6d4644$_x$ AK_6 = 73145ad9$_x$
AK_7 = e5acce41$_x$ AK_8 = 8425925f$_x$
AK_9 = fd2c98ad$_x$ AK_{10} = 98adfd2c$_x$
AK_{11} = f8425925$_x$ AK_{12} = 1e5acce4$_x$
AK_{13} = 145ad973$_x$ AK_{14} = 6d46448b$_x$
AK_{15} = 9c5fb7b2$_x$ AK_{16} = a5c0b3ba$_x$

2.1 Single Block Hash (SBH) Mode is no good

SBH described in [BrPiSe90] is no good, because of Theorem 1. SBH is described as follows:

$$H_0 = IV, \quad \text{initial hash value}$$
$$H_i = LOKI(m_i \oplus H_{i-1}, H_{i-1}) \oplus H_{i-1}$$
$$SBH = H_n$$

where LOKI(K,M) is the plaintext M encrypted using the key K. Assume we have a hashvalue H_n for M = $m_1, m_2, \ldots, m_n$, where every m_i is of length 64 bits. The hashvalue for

$$M^* = m_1 \oplus s_1, \ldots, m_n \oplus s_n,$$

where s_i = hhhhhhhh hhhhhhhh$_x$, h $\in$ {0,1,....,e,f}, is

$$H_0^* = IV = H_0,$$
$$H_1^* = LOKI(m_1 \oplus s_1 \oplus IV, IV) \oplus IV = LOKI(m_1 \oplus IV, IV) \oplus IV = H_1$$
$$H_2^* = LOKI(m_2 \oplus s_2 \oplus H_1^*, H_1^*) \oplus H_1^* = LOKI(m_2 \oplus H_1, H_1) \oplus H_1 = H_2$$

$$\cdots\cdots\cdots$$

$$H_i^* = LOKI(m_i \oplus s_i \oplus H_{i-1}^*, H_{i-1}^*) \oplus H_{i-1}^* = LOKI(m_i \oplus H_{i-1}, H_{i-1}) \oplus H_{i-1} = H_i$$

$$\cdots\cdots\cdots$$

$$H_n^* = H_n$$

It means, we can easily find 16^n messages, that will be hashed to the same hash value.

3 Differential cryptanalysis on (Standard) LOKI

In this section we will do differential cryptanalysis on LOKI. Differential crypt-analysis was introduced by Biham and Shamir [BiSha90] and is a method, which analyses the effect of particular differences in plaintext pairs on the differences of the resultant ciphertext pairs. That is, by choosing a certain difference in a plaintext pair, we can put probabilities on the different possible resultant ciphertext pairs. In the following we will use the notion of Biham-Shamir from [BiSha90]. For further details please consult these papers.

A table, which shows these probabilities is called a *pairs XOR distribution table*. For one S-box in LOKI it is a table with $2^{12} * 2^8 = 2^{20} = 1,048,576$ entries. The average value of the entries is 16 and of the 1,048,576 entries 99,9% are non-zero. That is, only 951 entries are zero. We write $X \rightarrow Y$, if an inputxor X can result in an inputxor Y.

The first step in a differential attack is to find good characteristics. In [BiSha90] Biham-Shamir state, that the best characteristics for a 16-round DES attack is obtained by concatenating a 2-round iterative characteristic with itself a certain number of times (see [BiSha90] page 27). The best 2-round iterative characteristic in DES has a probability of $\frac{1}{234}$ [BiSha90]. For LOKI we have:

Theorem 2 *To have two equal outputs of the F-function based on two different inputs, we have to have at least two neighbouring S-boxes with different inputs.*

Proof
Assume we have two equal outputs of the F-function based on two different inputs, which differ only in the input to one S-box. Due to the E-expansion the inputxor to that S-box must have the following form:

$$0000a_1a_2a_3a_40000 \quad (binary),$$

where $a_1a_2a_3a_4 \neq 0000$.

The two left-outermost and the two right-outermost bits determine the Sfn_r-function to be used in the evaluation. But these selectionbits are equal, which means the two inputs are evaluated through the same Sfn_r-function. If $00a_1a_2a_3a_400 = p \oplus q$ (i.e. $p \neq q$) then since the outputs are equal (x is an irreducible polynomial) we have:

$$p^{31} \, (mod \, x) = q^{31} \, (mod \, x) \Rightarrow p = q.$$

since $\gcd(31,255)=1$. A contradiction. □

There exists many iterative characteristics where two neighbouring S-boxes have different inputs. The best one has a probability of [BiSha91]

$$\frac{118}{2^{20}} \simeq 2^{-13,12}$$

No iterative characteristic where more than two S-boxes have different inputs is better than the abovementioned. However we can find 3-round iterative characteristics, which are better.

Definition 1 *A $\mathcal{F}$-fixpoint is an inputxor x, for which $\mathcal{F}(x) = x$, with some probability. $\mathcal{F}(x)$ is the random variable with distribution induced by putting $\mathcal{F}(x) = F(a) \oplus F(b)$, where a and b are uniformly chosen, such that $a \oplus b = x$.*

We obtain the best probabilities for a non-trivial input/outputxor combination, when the inputs differ only in the inputs to one S-box. We therefore first try to find $\mathcal{F}$-fixpoints, which differ only in the inputs to S-box 4.

Remark: In LOKI we have that xor-addition of pairs is linear in the E-ekspansion and the P-permutation. That is $E(X \oplus X^*) = E(X) \oplus E(X^*)$ and $P(X \oplus X^*) = P(X) \oplus P(X^*)$.

Let $B = b_{31}b_{30}.......b_1b_0$ be an inputxor to the F-function (before E). The 12 input-bitxors to S-box 4 are $b_3b_2b_1b_0b_{31}....b_{24}$. We have that $b_i = 0$ for i = 0,1,...,27 and that $b_{31}b_{30}b_{29}b_{28} \neq 0000$, because the inputs differ only in the input to S-box 4. Let $c_{31}c_{30}.......c_0$ be an outputxor from the F-function (before the P-permutation). $c_{31}c_{30}......c_{24}$ are outputbitxors from S-box 4. Therefore $c_j = 0$ for j = 0,1,.....,23, because the inputs to the S-boxes 3,2 and 1 are equal. After the P-permutation we have outputxor $C = c_{31}c_{23}c_{15}c_7........c_8c_0$. If B is a $\mathcal{F}$-fixpoint we must have C = B. It means that $b_{31} = c_{31} = 1$ and that $b_j = c_j = 0$ for j = 0,1,.....,30. Therefore we have a $\mathcal{F}$-fixpoint 80000000_x, if the combination $080_x \rightarrow 80_x$ is possible in LOKI. It is furthermore the only $\mathcal{F}$-fixpoint, where the inputs differ only to S-box 4.

By similar calculations we find that we have $\mathcal{F}$-fixpoints where the inputs differ to only S-box 3, 2 and 1, as follows:

00400000_x, if the combination $040_x \rightarrow 20_x$ is possible.

00002000_x, if the combination $020_x \rightarrow 08_x$ is possible.

00000010_x, if the combination $010_x \rightarrow 02_x$ is possible.

From *pairs XOR distribution table* we find that we have 3 fixpoints for LOKI where the inputs differ only to one S-box.

80000000_x with the probability $\frac{14}{4096}$

00400000_x with the probability $\frac{28}{4096}$

00000010_x with the probability $\frac{14}{4096}$

We have other $\mathcal{F}$-fixpoints for LOKI, where the inputs differ to more than one S-box. However none of these have a probability higher than the 3 abovementioned. Using these fixpoints we can build iterative characteristics. The following, which we will call **the fixpoint characteristic** is the best:

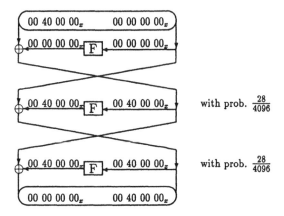

The total probability becomes $\frac{28^2}{4096^2} \simeq 2^{-14.4}$. That is if we have a plaintext pair, whose sum is $00400000\ 00000000_x$, then after 3 rounds of encryption the sum will be $00000000\ 00400000_x$ with probability $2^{-14.4}$.

We have checked, that for DES we have no fixpoints, where the inputs differ in less than 3 S-boxes. Although we may find fixpoints, where the inputs differ in 3 or more S-boxes, it cannot be used to build a better (iterative) characteristic, than the 2-round iterative characteristic given by Biham-Shamir [BiSha90].

3.1 Attacks on LOKI

For every attack we use the rearranged model of LOKI, that is, in every n-round LOKI we push the final addition of the key-halves 'upwards in the tree'. That gives n actual subkeys plus an initial addition of some keybit-xors, these being zero for n=16.

To find a complete 64-bit key (or 256 keybit-xors that determines 16 equivalent keys) we need to know 3 actual subkeys. In other words finding AK_n, AK_{n-1} and AK_{n-2} enables us to find all the keybit-xors we need. In [BiSha90] Biham-Shamir use an estimate, they call S/N-ratio, to find out how many pairs are needed for the attacks (on DES).

$$S/N = \frac{2^k * p}{\alpha * \beta},$$

where k = number of keybits we are looking for
       p = probability of the characteristic
       α = average count (of keys) per counted pair
       β = ratio of counted to all pairs.

The main difference of the S/N-ratio in DES and in LOKI is the calculation of α. Looking for keybits entering n S-boxes in DES yields an α of 4^n. In LOKI four of the keybit-xors entering S4 enter S3 as well. Other keybit-xors enter S1 and so forth. Looking for keybit-xors entering one S-box in LOKI gives an α of 16. Looking for keybit-xors entering two neighbouring S-boxes still gives an α of 16. The following tabel shows α for different search criteria. There are several

	α
One S-box	16
Two neighbouring S-boxes	16
Two not neighbouring S-boxes	16^2
Three S-boxes	16
Four S-boxes	1

types of attacks depending on the number of rounds that are not covered by the characteristics used in the attacks. That is, a xR-attack on a n-round cryptosystem is an attack where we use a (n-x)-round characteristic.

LOKI with 4 rounds can be broken in a way quite similar to the one given in [BiSha90] for breaking DES with 4 rounds using independent subkeys. We do 3R attacks using 4 characteristics, each one with a probability of 1 and find every keybit-xor using only 16 ciphertexts. We made 200 different tests, all of them showing that the right keybit-xors were the most suggested keyvalues.

LOKI with 6 rounds can be broken using 2 3-round characteristics with probabilities of 2^{-12} and 2^{-13}, respectively, to find AK_6. We made 20 tests looking for AK_6. In every test the right values of the keybitxors were the most suggested values using a total of 2^{15} pairs. For LOKI with 8 rounds the best attack we have found is a 3R attack, using the following one-round characteristic, the **start-characteristic**:

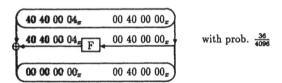

We concatenate it with the fixpoint characteristic plus one round with the trivial fixpoint $00\ 00\ 00\ 00_x$. The total probability is $2^{-14.4} \times \frac{36}{4096} \simeq 2^{-12.2}$. We need 2^{24} pairs for a successful attack. For n $\leq$ 9, 3R-attacks (see [BiSha90]) are possible, but for n > 9 the probabilities of the characteristics get too small. But for $9 < n < 14$ we can choose a 2R-attack, using the fixpoint characteristic and in some attacks the start-characteristic. Whether to use the start-characteristic or not is determined as follows:

If we need a r-round characteristic, we concatenate the fixpoint characteristic with itself $\frac{r}{3}$ times. If the last round of the obtained characteristic is a round with the fixpoint $00\ 40\ 00\ 00_x$, we remove this last round and use the start-characteristic as the first round, thus obtaining a r-round characteristic with a probability improved by a factor $\frac{36}{28}$. For n = 14 the best attack is a 1R attack using a 13-round characteristic built alone from the fixpoint characteristic.

We will show in more details how LOKI with 13 rounds can be broken using 2^{53} pairs in a 2R-attack. We concatenate the start-characteristic with 3 fixpoint characterstics plus one round with the trivial fixpoint $00\ 00\ 00\ 00_x$. The total probability of the characteristic becomes $\left(\frac{28}{4096}\right)^5 \times \frac{36}{4096} \simeq 2^{-50}$. We have:

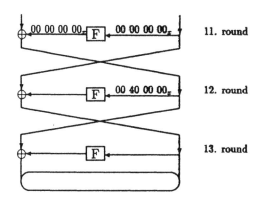

We know that 24 bits of the ciphertext pairs in the output of the 12. round should be equal. Furthermore we know the inputsum to the 11. round, which means we can check 24 bits of the right halves of the ciphertext pairs. If the bits are not as expected, we filter out the ciphertext pair.

From the ciphertexts we know the inputs to the 13. round and we know the expected outputsums of all the pairs in the 13. round. Using 2^{32} counters we get

$$S/N \simeq \frac{2^{32} * 2^{-50}}{1 * 2^{-24}} = 2^6$$

Using 2^{53} pairs, we get 2^{29} ($2^{53} * 2^{-24}$) pairs each suggesting one arbitrary 32 bit keyvalue. The right keyvalue is suggested about 8 times and with a high probability it is the one most suggested keyvalue. Now we can decrypt all pairs one round and do a 2R- or 1R-attack on the remaining 12-round cryptosystem using characteristics with higher probability than 2^{-50} and find AK_{12} and AK_{11} in a similar way.

Tabel 1 below contains estimates of how many pairs are needed for different kinds of attacks on LOKI. As indicated in the table we cannot break LOKI with 16 rounds doing a differential attack. The best attack we have found is to use a

Rounds	Char.Prob.	S/N	Attack	Number of keybit-xors	Pairs Needed
4	1	2^8	3R	All	8
6	$2^{-12}, 2^{-13}$	$2^4, 2^3$	3R	All	2^{15}
8	2^{-21}	7	3R	28	2^{24}
9	2^{-29}	8	3R	32	2^{32}
9	2^{-29}	2^{27}	2R	32	2^{31}
10	2^{-36}	2^{20}	2R	32	2^{38}
11	2^{-43}	2^{13}	2R	32	2^{45}
12	2^{-43}	2^{13}	2R	32	2^{45}
13	2^{-50}	2^6	2R	32	2^{53}
14	2^{-58}	2^{-2}	2R	32	Not possible
14	2^{-58}	2^6	1R	12	2^{60}
15	2^{-65}	2^{-1}	1R	12	Not possible

Table 1: Estimates of needed pairs for differential attacks on LOKI

15-round characteristic in a 1R attack. We concatenate the start-characteristic, 4 fixpoint characteristics, one round with the fixpoint $00\ 00\ 00\ 00_x$ and one round with the combination $00\ 40\ 00\ 00_x \rightarrow 40\ 40\ 00\ 04_x$. The total probability for the 15-round characteristic is $2^{-71.2}$ and the attack is not possible. A necessary condition for a successful 1R attack using a 15-round characteristic is that the probability of the characteristic is greater than 2^{-64}. We show that we cannot find such a characteristic. We need a few definitions:

Definition 2 *A round with the combination $00\ 00\ 00\ 00_x \rightarrow 00\ 00\ 00\ 00_x$ we call a zero-round.*
If the rounds $(i-1)$ and $(i+1)$ are zero-rounds, round i is of type A.
If the rounds $(i-1)$ and $(i+2)$ are zero-rounds, round i and round $(i+1)$ are of type B.

A round of type A must have the form $\phi_i \rightarrow 00\ 00\ 00\ 00_x$. The best probability of such a round is $2^{-13.12}$ [BiSha91]. The two rounds of type B must have the following form:

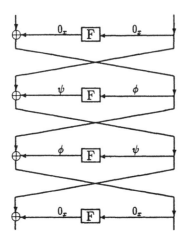

By consulting the *pairs XOR distribution table* we find that we obtain the best probability for the two rounds if $\phi = \psi = 00\ 40\ 00\ 00_x$, that is exactly the situation we obtain using the fixpoint characteristic. The best probability for two rounds of type B is therefore $2^{-14.4}$. Now we can prove the following theorem:

Theorem 3 *For LOKI we cannot find a 15-round characteristic with a probability greater than 2^{-64}.*

Proof
The best non-trivial input/outputxor combination in LOKI has a probability of 2^{-6}. It means that some of the 15 rounds have to be zero-rounds with probability 1. At least 5 rounds must be zero-rounds, because 5 is the lowest value of x, such that:

$$(2^{-6})^{15-x} > 2^{-64}$$

If two neighbouring rounds are zero-rounds then all rounds are zero-rounds and we get equal plaintexts resulting in equal ciphertexts, a trivial fact. It means that we can have at most 8 zero-rounds and must have at least 5 zero-rounds in the 15-round characteristic (15R).

8 zero-rounds: First round and then every other round are zero-rounds. The remaining 7 rounds are all of type A. We get:

$$P(15R) \leq (2^{-13,12})^7 = 2^{-91,84}$$

7 zero-rounds: We have at least 4 rounds of type A. The remaining 4 (nonzero) rounds have a probability of at most 2^{-6}. We get:

$$P(15R) \leq (2^{-13,12})^4 \times (2^{-6})^4 = 2^{-76,48}$$

6 zero-rounds: We have at least 1 round of type A and two situations to examine:

1. Only one round of type A, thereby 4×2 rounds of type B.

$$P(15R) \leq 2^{-13,12} \times (2^{-14,4})^4 = 2^{-70,72}$$

2. Two or more rounds of type A.

$$P(15R) \leq (2^{-13,12})^2 \times (2^{-6})^7 = 2^{-68,24}$$

5 zero-rounds: We have two situations to examine:

1. One round of type A.

$$P(15R) \leq 2^{-13,12} \times (2^{-6})^9 = 2^{-67,12}$$

2. No rounds of type A and thereby 2×2 rounds of type B.

$$P(15R) \leq (2^{-14,4})^2 \times (2^{-6})^6 = 2^{-64,8}$$

□

4 Private LOKI-ciphers

In the Appendices to [BrPiSe90], Brown, Pieprzyk and Seberry propose a way to build private versions of LOKI. They enumerate 30 different irreducible polynomials and 24 different exponents to be used in the building of 4 (not necessarily equal) S-boxes. The number of possible different LOKI-versions is huge,

$$(K_{30,16} * 16! * 24^{16})^4 \simeq 2^{576}$$

making it impossible to do differential cryptanalysis on all versions.

We examined 50 private versions of LOKI. The best characteristic we have found, is a fixpoint characteristic (with fixpoint 00 40 00 00$_x$) with a probability of $\frac{54^2}{4096^2} \simeq 2^{-12,5}$ for 3 rounds.

In that private LOKI-version it is possible to break a 15 round version using about 2^{53} pairs. Note that because the fixpoint only differs in the input to the third S-box S3, it doesn't matter how we build the S-boxes S4, S2 and S1 with respect to the mentioned attack.

It is not at all impossible that we can find several private versions of LOKI, for which a differential attack can be done with a complexity less than 2^{64}.

Acknowledgements

Thanks to Ivan Bjerre Damgård for the help in shortening the proof of Theorem 1 and for moral support.

References

[BrPiSe90] Lawrence Brown, Josef Pieprzyk, Jennifer Seberry. *LOKI - A Crypto-graphic Primitive for Authentication and Secrecy Applications.* Advances in Cryptology - AUSCRYPT '90. Springer Verlag, Lecture Notes 453, pp. 229-236, 1990.

[BiSha90] Eli Biham, Adi Shamir. *Differential Cryptanalysis of DES-like Cryptosystems.* Journal of Cryptology 1991.

[BiSha91] Eli Biham, Adi Shamir. *Differential Cryptanalysis of Snefru, Khafre, REDOC-II, LOKI and Lucifer.* Presented at CRYPTO '91.

Improving Resistance to Differential Cryptanalysis and the Redesign of LOKI

Lawrence BROWN   Matthew KWAN
Josef PIEPRZYK
Jennifer SEBERRY

Department of Computer Science,
University College, UNSW, Australian Defence Force Academy,
Canberra ACT 2600. Australia.

Abstract

Differential Cryptanalysis is currently the most powerful tool available for analysing block ciphers, and new block ciphers need to be
designed to resist it. It has been suggested that the use of S-boxes
based on bent functions, with a flat XOR profile, would be immune.
However our studies of differential cryptanalysis, particularly applied
to the LOKI cipher, have shown that this is not the case. In fact, this
results in a relatively easily broken scheme. We show that an XOR
profile with carefully placed zeroes is required. We also show that in
order to avoid some variant forms of differential cryptanalysis, permutation P needs to be chosen to prevent easy propagation of a constant
XOR value back into the same S-box. We redesign the LOKI cipher
to form LOKI91, to illustrate these results, as well as to correct the
key schedule to remove the formation of equivalent keys. We conclude
with an overview of the security of the new cipher.

1   Introduction

Cryptographic research is currently a very active field, with the need for
new encryption algorithms having spurred the design of several new block
ciphers [1]. The most powerful tool for analysing such block ciphers currently known is differential cryptanalysis. It has been used to find design
deficiencies in many of the new ciphers. Some new design criteria have been

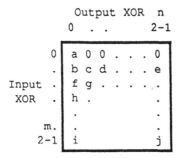

Figure 1: An XOR Profile

proposed which are claimed to provide immunity to differential cryptanalysis. These involve the use of S-boxes based on bent functions, selected so the resultant box has a flat XOR profile.

In this paper, after presenting a brief introduction to the key concepts in differential cryptanalysis, we will show that these new criteria do not provide the immunity claimed, but rather can result in the design of a scheme which may be relatively easily broken. What we believe is required, is an S-box with carefully placed zeroes, which significantly hinder the differential cryptanalysis process.

We continue by documenting our analysis of the LOKI cipher. We report on a previously discovered differential cryptanalysis attack, faster than exhaustive search up to 11 rounds, as well as on a new attack, using an alternate form of analysis, which is faster than exhaustive search up to 14 rounds. We also briefly note some design deficiencies in the key schedule which resulted in the generation of equivalent keys. This process highlighted some necessary additional design criteria needed to strengthen block ciphers against these attacks.

We conclude by describing the redesign of the LOKI cipher, using it to illustrate the application of these additional design principles, and make some comments on what we believe is the security of the new scheme.

2 Differential Cryptanalysis

2.1 Overview

Differential Cryptanalysis is a dynamic attack against a cipher, using a very large number of chosen plaintext pairs, which through a statistical analysis of the resulting ciphertext pairs, can be used to determine the key in use. In general, differential cryptanalysis is much faster than exhaustive search for a certain number of rounds in the cipher, however there is a breakeven point where it becomes slower than exhaustive search. The lower the number of rounds this is, the greater the security of the cipher. Differential Crypt-

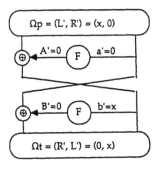

Figure 2: A 2-round Iterative Characteristic

analysis was first described by Biham and Shamir in [2], and in greater detail in [3]. These described the general technique, and its application to the analysis of the DES and the Generalised DES. Subsequent papers by them have detailed its application to FEAL and N-Hash [4], and to Snefru, Khafre, Redoc-II, LOKI, and Lucifer [5].

In Differential Cryptanalysis, a table showing the distribution of the XOR of input pairs against the XOR of output pairs is used to determine probabilities of a particular observed output pair being the result of some input pair. The general form of such a table is shown in Fig 1.

To attack a multi-round block cipher, the XOR profile is used to build n round characteristics, which have a given probability of occurring. These characteristics specify a particular input XOR, a possible output XOR, the necessary intermediate XOR's, and the probability of this occurring. In their original paper [3], Biham and Shamir describe 1,2,3 and 5 round characteristics which may be used to directly attack versions of DES up to 7 rounds. Knowing a characteristic, it is possible to infer information about the outputs for the next two rounds. To utilise this attack, a number of pairs of inputs, having the nominated input XOR, are tried, until an output XOR results which indicates that the pattern specified in the characteristic has occurred. Since an n round characteristic has a probability of occurrence, for most keys we can state on average, how many pairs of inputs need to be trialed before the characteristic is successfully matched. Once a suitable pair, known as a right pair, has been found, information on possible keys which could have been used, is deduced. Once this is done we have two plaintext-ciphertext pairs. We know from the ciphertext, the input to the last round. Knowing the input XOR and output XOR for this round, we can thus restrict the possible key bits used in this round, by considering those outputs with an XOR of zero, providing information on the outputs

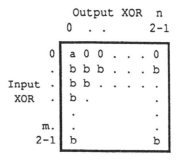

Figure 3: Flat XOR Profile

of some of the S-boxes. By then locating additional right pairs we can eventually either uniquely determine the key, or deduce sufficient bits of it that an exhaustive search of the rest may be done.

N round characteristics can be concatenated to form longer characteristics if the output of the first supplies the input to the second, with probabilities multiplied together. A particularly useful characteristic is one whose output is a swapped version of its input, and which hence may be iterated with itself. This may be used to analyse an arbitrary number of rounds of the cipher, with a steadily increasing work factor. A particularly useful form is one where a non-zero input XOR to the F function results in a zero output XOR. Such a characteristic is illustrated in Fig 2, and may be denoted as:

```
A. (x, 0) -> (0, x) always (ie Pr=1)
B. (0, x) -> (x, 0) with some probability p
```

It may be iterated as A B A B A B A B to 8 rounds for example, with characteristic probability p^4. This form of characteristic is then used in the analysis of arbitrary n round forms of a cipher, until the work factor exceeds exhaustive search. These techniques are described in detail in [3].

2.2 Why Flat XOR profiles Don't Work

Given the success of differential cryptanalysis in the analysis of block ciphers, it has become important to develop design criteria to improve the resistance of block ciphers to it, especially with several candidates having performed poorly. Dawson and Tavares [6] have proposed that the selection of S-boxes with equal probabilities for each output XOR given an input XOR (except input 0) would result in a cipher that was immune to differential cryptanalysis (see Fig. 3). However a careful study of Biham and Shamir's attack on the 8 round version of DES [3], confirmed by our own analyses of LOKI, have shown that this is not the case.

Indeed the selection of such S-boxes results in a cipher which is significantly easier to cryptanalyse than normal. This is done by constructing a 2 round iterative characteristic of the form in Fig. 2, with an input XOR that changes bits to one S-box only. We know we can do this, since the flat XOR profile implies that an output XOR of 0 for a specified input XOR will occur with $Pr(1/2^n)$. When iterated over 2k rounds, this will have a probability of $Pr(1/2^{(k-1)n})$, since you get the last round for free. Consider a 16 round DES style cryptosystem, but with S-boxes having a flat XOR profile of the form in Fig. 3 with $m = 6$, $n = 4$, and $k = 8$. This may be attacked by a 15-round characteristic, chosen to alter inputs to a single S-box only. This gives a probability to break with a given test pair of $Pr(1/2^{28})$, implying that about 2^{28} pairs need to be tried to break the cipher, far easier than by exhaustive search.

2.3 Significance of Permutation P

Although differential cryptanalysis may be done independent of permutation P, knowledge of a particular P may be used to construct some other useful n round characteristics for cryptanalysing a particular cipher. The most useful of these take the form of a 2 or 3 round characteristic which generate an output XOR identical to the input XOR, either directly in 2 rounds, or by oscillating between two alternate XOR values over 3 rounds. The 3 round characteristic is sensitive to the form of permutation P. This form of characteristic has been found in the original version of LOKI, as detailed below. It thus indicates that care is needed in the design of not just the S-boxes, but of all elements in function F, in order to reduce susceptibility to differential cryptanalysis.

3 Analysis of LOKI

3.1 Overview

LOKI is one of several recently proposed new block ciphers. It was originally detailed by Brown, Pieprzyk and Seberry in [7]. Its overall structure is shown in Fig. 4. We will refer to this version of the cipher as LOKI89 in the remainder of this paper. It is a 16 round Feistel style cipher, with a relatively straightforward key schedule, a non-linear function $F = P(S(E(R \oplus K)))$, and four identical 12-to-8 bit S-boxes. Permutation P in LOKI has a regular structure which distributes the 8 output bits from each S-box to boxes $[+3 +2 +1 0 +3 +2 +1 0]$ in the next round. Its S-box consists of 16 1-1 functions, based on exponentiation in a Galois field $GF(2^8)$ (see [8]), with the general structure shown in Fig. 4.

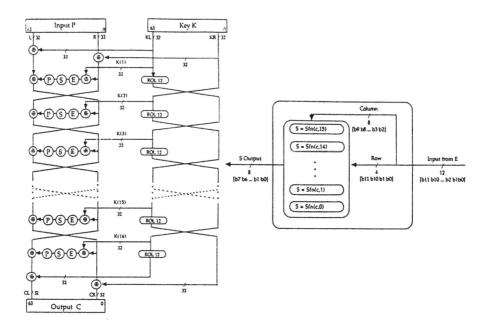

Figure 4: LOKI89 Overall Structure and S-box Detail

3.2   Security of LOKI89

Initial testing of the statistical and complexity properties of LOKI89 indicated that its properties were very similar to those exhibited by DES, FEAL8 and Lucifer [9], results that were very encouraging. Initial examination of the XOR profile of the LOKI89 S-box also suggested that it should be more resistant than DES to differential cryptanalysis.

LOKI89 was then analysed in detail using differential cryptanalysis. Biham [5] describes an attack, using a 2 round iterative characteristic with $Pr(118/2^{20}) \approx Pr(2^{-13.12})$, with non-zero inputs to 2 S-boxes resulting in the same output. There are four related variants (by rotation) of the form:

```
A. (00000510,00000000) -> (00000000,00000510)   always
B. (00000000,00000510) -> (00000510,00000000)   Pr(118/2^20)
```

This characteristic is iterated to 8 rounds with $Pr(2^{-52.48})$, allowing up to 10 rounds to be broken faster than by exhaustive key space search. This is a significantly better result than for the DES. The authors have verified this attack.

The authors have subsequently found an alternate 3 round iterative characteristic, attacking S-box 3, with an output XOR identical to the input XOR. Since permutation P in LOKI89 permutes some output bits back to the same S-box in the next round, this allows the characteristic to be iterated. It has the form:

```
A.  (00400000,00000000) -> (00000000,00400000)   always
B.  (00000000,00400000) -> (00400000,00400000)   Pr(28/4096)
C.  (00400000,00400000) -> (00400000,00000000)   Pr(28/4096)
```

Since $28/4096 \approx 2^{-7.2}$ if we concatenate these characteristics, we get a 13 round characteristic in the order A B C A B C A B C A B C A with a probability of $Pr(2^{-7.2*8}) \approx Pr(2^{-57.6})$. This may be used to attack a 14 round version of LOKI89, and requires $O(2^{59})$ pairs to succeed. This is of the same order as exhaustive search (which is of $O(2^{60})$, as detailed below), and is thus a more successful attack than that reported previously. It has been verified by Biham. This still leaves the full 16 round version of LOKI89 secure, but with a reduced margin against that originally believed.

Independently, the authors [10], Biham [5], and the members of the RIPE consortium have discovered a weakness in the LOKI89 key schedule. It results in the generation of 15 equivalent keys for any given key, effectively reducing the key-space to 2^{60} keys. A complementation property also exists which results in 256 (key, plain, cipher) triples being formed, related by $LOKI(P, K) \oplus pppppppppppppppp = LOKI(P \oplus pppppppppppppppp, K \oplus mmmmmmmmnnnnnnnn)$, where $p = m \oplus n$ for arbitrary hex values m, n. This may be used to reduce the complexity of a chosen plaintext attack by an additional factor of 16. These results were found by analysing the key schedule by regarding each S-box input as a linear function of the key and plaintext, and solving to form (key, plaintext) pairs which result in identical S-box input values. This lead to solving the following equations:

$$RD \oplus KRD \oplus n.ROT12(KLD) = 0 \qquad (1)$$

$$LD \oplus KLD \oplus n.ROT12(KRD) = 0 \qquad (2)$$

where $LD = L' \oplus L$, $RD = R' \oplus R$, $KLD = KL' \oplus KL$, and $KRD = KR' \oplus KR$ describe the difference between the related (key, plaintext) pairs. This method is detailed by Kwan in [10].

In the light of these results, the authors have devised some additional design guidelines to those originally used in the design of LOKI, and have applied them in the development of a new version, LOKI91.

4 Redesign of LOKI

4.1 Some Additional Design Guidelines

To improve the resistance of a cipher to differential cryptanalysis, and to remove problems with the key schedule, the following guidelines were used:

- analyse the key schedule to minimize the generation of equivalent key, or related (key, plaintext) pairs.

- minimise the probability that a non-zero input XOR results in a zero output XOR, or in an identical output XOR, particularly for inputs that differ in only 1 or 2 S-boxes.

- ensure the cipher has sufficient rounds so that exhaustive search is the optimal attack (ie have insufficient pairs to do differential cryptanalysis).

- ensure that there is no way to make all S-boxes give 0 outputs, to increase the ciphers security when used in hashing modes.

These criteria were used when selecting the changes made to the LOKI structure, detailed below.

4.2   Design of LOKI91

LOKI91 is the revised version, developed to address the results detailed above. Changes have been made to two aspects of the LOKI structure. Firstly the key schedule has been amended in several places to significantly reduce the number of weak keys. Secondly the function used in the S-boxes has been altered to improve its utility in hashing applications, and to improve its immunity to differential cryptanalysis. In more detail, the four changes made to the original design were:

1. change key schedule to swap halves after every second round

2. change key rotations to alternate between ROT13 and ROT12

3. remove initial and final XORs of key with plaintext and ciphertext

4. alter the S-box function used in the LOKI S-box (Fig. 4) to

$$Sfn(row, col) = (col + ((row * 17) \oplus ff_{16})\&ff_{16})^{31} \bmod g_{row} \quad (3)$$

where $+$ and $*$ refer to arithmetic addition and multiplication, $\oplus$ is addition modulo 2, and the exponentiation is performed in $GF(2^8)$. The generator polynomials used (g_{row}) are as for LOKI89 [7].

The overall structure of LOKI91 that results from these changes is shown in Fig. 5.

The key schedule changes remove all but a single bit complementation property, leaving an effective search space of 2^{63} keys under a chosen-plaintext attack. It reduces key equivalences to a single bit complementation property, similar to that of the DES where $LOKI(\overline{P}, \overline{K}) = \overline{LOKI(P, K)}$,

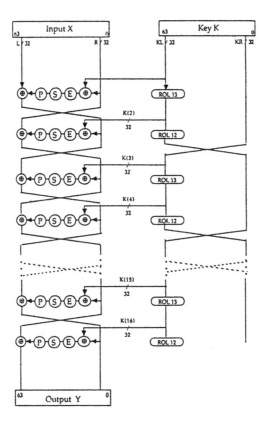

Figure 5: LOKI91 Overall Structure

by reducing the solutions to each of Eq. 1 and Eq. 2 to one, and removing the independence between them by altering the swapping to every two rounds. It also greatly reduces the number of weak and semi-weak keys to those shown in Table 1 (where an * denotes weak keys). The DES also has 16 weak and semi-weak keys.

The removal of the initial and final XORs became necessary with the change in the swap locations, since otherwise it would have resulted in cancellation of the keys bits at the input to E in some rounds. This change does affect the growth of ciphertext dependence on keys bits (see [11]), increasing it from 3 to 5 rounds. This still compares favorably with the DES which takes either 5 or 7 rounds, dependent on the type of dependency analysed. This change also greatly assisted in the reduction of the number of equivalent keys.

The new Sfn uses arithmetic addition and multiplication, as these are non-linear when used in $GF(2^8)$. The addition modulo two of the *row* with ff_{16} ensures that an all zero input gives a non-zero output, thus removing the major deficiency of LOKI89 when used as a hash function. The result of the addition and multiplication is masked with ff_{16} to restrict the value to lie with $GF(2^8)$, prior to the exponentiation in that field. The new function reduces the probabilities of occurance of n round iterative characteristics,

Encrypt Key	Decrypt Key	
0000000000000000	0000000000000000	*
00000000aaaaaaaa	aaaaaaaa00000000	
0000000055555555	5555555500000000	
00000000ffffffff	ffffffff00000000	
aaaaaaaa00000000	00000000aaaaaaaa	
aaaaaaaaaaaaaaaa	aaaaaaaaaaaaaaaa	*
aaaaaaaa55555555	55555555aaaaaaaa	
aaaaaaaaffffffff	ffffffffaaaaaaaa	
5555555500000000	0000000055555555	
55555555aaaaaaaa	aaaaaaaa55555555	
5555555555555555	5555555555555555	*
55555555ffffffff	ffffffff55555555	
ffffffff00000000	00000000ffffffff	
ffffffffaaaaaaaa	aaaaaaaaffffffff	
ffffffff55555555	55555555ffffffff	
ffffffffffffffff	ffffffffffffffff	*

Table 1: LOKI91 Weak and semi-weak key pairs

useful for differential cryptanalysis, to be as low as possible. With the new function, LOKI91 is theoretically breakable faster than an exhaustive key space search in:

- up to 10 rounds using a 2 round characteristic with the $f(x')-> 0'$ mapping occuring with $Pr(122/1048576)$

- up to 12 rounds using a 3 round characteristic with the $f(x')-> x'$ mapping occuring with $Pr(16/4096)$ (used twice)

At 16 rounds, cryptanalysis is generally impossible, as insufficient pairs are available to complete the analysis. It would require:

- 2^{80} pairs using the 3 round characterisitic, or

- 2^{92} pairs using the 2 round characteristic

compared to a total of 2^{63} possible plaintext pairs.

5 Conclusion

In this paper, we have shown that a flat XOR profile does not provide immunity to differential cryptanalysis, but in fact leads to a very insecure scheme. Instead a carefully chosen XOR profile, with suitably placed 0 entries is required to satisfy the new design guidelines we have identified. We also note an analysis of key schedules, which can be used to determine the number of equivalent keys. We conclude with the application of these results to the design of LOKI91.

6 Acknowledgements

Thank you to the members of the crypt group for their support and suggestions. This work has been supported by ARC grant A48830241, ATERB, and Telecom Australia research contract 7027.

Appendix A - Specification of LOKI91

Encryption

The overall structure of LOKI91 is shown in Fig. 5, and is specified as follows. The 64-bit input block X is partitioned into two 32-bit blocks L and R. Similarly, the 64-bit key is partitioned into two 32-bit blocks KL and KR.

$$
\begin{aligned}
L_0 &= L & KL_0 &= KL \\
R_0 &= R & KR_0 &= KR
\end{aligned}
\tag{4}
$$

The key-dependent computation consists (except for a final interchange of blocks) of 16 rounds (iterations) of a set of operations. Each iteration includes the calculation of the encryption function f. This is a concatenation of a modulo 2 addition and three functions E, S, and P. Function f takes as input the 32-bit right data half R_{i-1} and the 32-bit left key half KL_i produced by the key schedule KS (denoted K_i below), and which produces a 32-bit result which is added modulo 2 to the left data half L_{i-1}. The two data halves are then interchanged (except after the last round). Each round may thus be characterised as:

$$
\begin{aligned}
L_i &= R_{i-1} \\
R_i &= L_{i-1} \oplus f(R_{i-1}, KL_i) \\
f(R_{i-1}, K_i) &= P(S(E(R_{i-1} \oplus K_i)))
\end{aligned}
\tag{5}
$$

The component functions E, S, and P are described later.

The key schedule KS is responsible for deriving the sub-keys K_i, and is defined as follows: the 64-bit key K is partitioned into two 32-bit halves KL and KR. In each round i, the sub-key K_i is the current left half of the key KL_{i-1}. On odd numbered rounds (1, 3, 5, etc), this half is then rotated 12 bits to the left. On even numbered rounds (2, 4, 6, etc), this half is then rotated 13 bits to the left, and the key halves are interchanged.

3	2	1	0	31	30	29	28	27	26	25	24
27	26	25	24	23	22	21	20	19	18	17	16
19	18	17	16	15	14	13	12	11	10	9	8
11	10	9	8	7	6	5	4	3	2	1	0

Table 2: LOKI Expansion Function E

This may be defined for odd numbered rounds as:

$$K_i = KL_{i-1}$$
$$KL_i = ROL(KL_{i-1}, 13) \tag{6}$$
$$KR_i = KR_{i-1}$$

This may be defined for even numbered rounds as:

$$K_i = KL_{i-1}$$
$$KL_i = KR_{i-1} \tag{7}$$
$$KR_i = ROL(KL_{i-1}, 12)$$

Finally after the 16 rounds, the two output block halves L_{16} and R_{16} are then concatenated together to form the output block Y. This is defined as (note the swap of data halves to undo the final interchange in Eq.5):

$$Y = R_{16} \,|\, L_{16} \tag{8}$$

Decryption

The decryption computation is identical to that used for encryption, save that the partial keys used as input to the function f in each round are calculated in reverse order. The rotations are to the right, and an initial pre-rotation of 8 places is needed to form the key pattern.

Function f

The encryption function f is a concatenation of a modulo 2 addition and three functions E, S, and P, which takes as input the 32-bit right data half R_{i-1} and the 32-bit left key half KL_i, and produces a 32-bit result which is added modulo 2 to the left data half L_{i-1}.

$$f(R_{i-1}, K_i) = P(S(E(R_{i-1} \oplus K_i))) \tag{9}$$

The modulo 2 addition of the key and data halves ensures that the output of f will be a complex function of both of these values.

Row	gen_{row}	e_{row}
0	375	31
1	379	31
2	391	31
3	395	31
4	397	31
5	415	31
6	419	31
7	425	31
8	433	31
9	445	31
10	451	31
11	463	31
12	471	31
13	477	31
14	487	31
15	499	31

Table 3: LOKI S-box Irreducible Polynomials and Exponents

The expansion function E takes a 32-bit input and produces a 48-bit output block, composed of four 12-bit blocks which form the inputs to four S-boxes in function f. Function E selects consecutive blocks of twelve bits as inputs to S-boxes S(4), S(3), S(2), and S(1) respectively, as follows:

$[b_3\, b_2 \ldots b_0\, b_{31}\, b_{30} \ldots b_{24}]$

$[b_{27}\, b_{26} \ldots b_{16}]$

$[b_{19}\, b_{18} \ldots b_8]$

$[b_{11}\, b_{10} \ldots b_0]$

This is shown in full in Table 2 which specifies the source bit for outputs bits 47 to 0 respectively:

The substitution function S provides the confusion component in the LOKI cipher. It takes a 48-bit input and produces a 32-bit output. It is composed of four S-boxes, each of which takes a 12-bit input and produces an 8-bit output, which are concatenated together to form the 32-bit output of S. The 8-bit output from S(4) becomes the most significant byte (bits [31...24]), then the outputs from S(3) (bits[23...16]), S(2) (bits[15...8]), and S(1) (bits [7...0]). In LOKI91 the four S-boxes are identical. The form of each S-box is shown in Fig 4. The 12-bit input is partitioned into two segments: a 4-bit row value row formed from bits $[b_{11}\, b_{10}\, b_1\, b_0]$, and an 8-bit column value col formed from bits $[b_9\, b_8 \ldots b_3\, b_2]$. The row value row is used to select one of 16 S-functions $Sfn_{row}(col)$, which then take as input the

31	23	15	7	30	22	14	6
29	21	13	5	28	20	12	4
27	19	11	3	26	18	10	2
25	17	9	1	24	16	8	0

Table 4: LOKI Permutation P

column value col and produce an 8-bit output value. This is defined as:

$$Sfn_{row}(col) = (col + ((row * 17) \oplus ff_{16})r \& ff_{16})^{e_{row}} \mod g_{row} \qquad (10)$$

where gen_{row} is an irreducible polynomial in $GF(2^8)$, and e_{row} is the (constant 31) exponent used in forming $Sfn_{row}(col)$. The generators and exponents to be used in the 16 S-functions in LOKI91 are specified in Table 3. For ease of implementation in hardware, this function can also be written as:

$$Sfn_{row}(col) = (col + ((\overline{row})|(\overline{row} << 4)) \& ff_{16})^{31} \mod g_{row} \qquad (11)$$

The permutation function P provides diffusion of the outputs from the four S-boxes across the inputs of all S-boxes in the next round. It takes the 32-bit concatenated outputs from the S-boxes, and distributes them over all the inputs for the next round via a regular wire crossing which takes bits from the outputs of each S-box in turn, as defined in Table 4 which specifies the source bit for output bits 31 to 0 respectively.

Test Data

A single test triplet for the LOKI91 primitive is listed below.

```
#    Single LOKI91 Certification triplet
#    data is saved as (key, plaintext, ciphertext) hex triplets
#
3849674c2602319e 126898d55e911500 c86caec1e3b7b17e
```

References

[1] J. Seberry and J. Pieprzyk, *Cryptography: An Introduction to Computer Security.* Englewood Cliffs, NJ, Prentice Hall, 1989.

[2] E. Biham and A. Shamir, "Differential Cryptanalysis of DES-like Cryptosystems," *Journal of Cryptology*, 4, no. 1, 1991, to appear.

[3] E. Biham and A. Shamir, "Differential Cryptanalysis of DES-like Cryptosystems," Weizmann Institute of Science, Rehovot, Israel, Technical Report, 19 July 1990.

[4] E. Biham and A. Shamir, "Differential Cryptanalysis of Feal and N-Hash," in *Eurocrypt'91 Abstracts*, Brighton, UK, 8-11 April 1991.

[5] E. Biham and A. Shamir, "Differential Cryptanalysis Snefru, Kharfe, REDOC-II, LOKI and Lucifer," in *Abstracts Crypto'91*, Santa Barbara, Aug. 1991.

[6] M. H. Dawson and S. E. Tavares, "An Expanded Set of S-box Design Criteria Based On Information Theory and Its Relation to Differential-Like Attacks," in *Eurocrypt'91 Abstracts*, Brighton, UK, 8-11 April 1991.

[7] L. Brown, J. Pieprzyk and J. Seberry, "LOKI - A Cryptographic Primitive for Authentication and Secrecy Applications," in *Advances in Cryptology: Auscrypt '90* (Lecture Notes in Computer Science), vol. 453. Berlin: Springer Verlag, pp. 229–236, 1990.

[8] J. Pieprzyk, "Non-Linearity of Exponent Permutations," in *Advances in Cryptology - Eurocrypt'89* (Lecture Notes in Computer Science), vol. 434, J. J. Quisquater and J. Vanderwalle, Eds. Berlin: Springer Verlag, pp. 80–92, 1990.

[9] L. Brown, J. Pieprzyk, R. Safavi-Naini and J. Seberry, "A Generalised Testbed for Analysing Block and Stream Ciphers," in *Proceedings of the Seventh Internation IFIP TC11 Conference on Information Security*, W. Price and D. Lindsey, Eds. North-Holland, May 1991, to appear.

[10] M. Kwan and J. Pieprzyk, "A General Purpose Technique for Locating Key Scheduling Weaknesses in DES-style Cryptosystems," in *Advances in Cryptology - Asiacrypt'91* (Lecture Notes in Computer Science). Berlin: Springer Verlag, Nov 1991, to appear.

[11] L. Brown and J. Seberry, "Key Scheduling in DES Type Cryptosystems," in *Advances in Cryptology: Auscrypt '90* (Lecture Notes in Computer Science), vol. 453. Berlin: Springer Verlag, pp. 221–228, 1990.

A Method to Estimate the Number of Ciphertext Pairs for Differential Cryptanalysis

Hiroshi Miyano

C&C Information Technology Research Laboratories
NEC Corporation
4-1-1 Miyazaki, Miyamae-ward, Kawasaki 216, JAPAN
E-mail miyano@IBL.CL.nec.co.jp

ABSTRACT Differential cryptanalysis introduced by Biham and Shamir in 1990 is one of the most powerful attacks to DES-like cryptosystems. This attack presumes on some tendency of the target cryptosystem. So the efficiency of the attack depends upon the conspicuousness of this tendency. S/N ratio introduced in the paper is to evaluate this conspicuousness. In other words, the S/N ratio is a measure of the efficiency of the attack. Nevertheless, S/N ratio does NOT suggest how many pairs of ciphertexts are needed.

In this paper, we show how to estimate the number of necessary pairs of ciphertexts for the differential cryptanalysis. We also show that our estimation is adequate using the 8-round-DES as an example. Biham and Shamir also showed a counting scheme to save memories at the cost of efficiency. We show an algorithm to find the secret key saving memories at the less cost of efficiency.

1 Introduction

The Data Encryption Standard (DES)[1] was developed at IBM and adopted by the U.S. National Bureau of Standards (NBS) as the standard cryptosystem in 1977. It is a kind of iterated cryptosystems. It's key size is 56 bits and the block size is 64 bits. After DES, several iterated cryptosystems, or DES-like cryptosystems, are introduced, that are FEAL[2], LOKI[3], and so on.

Differential cryptanalysis introduced by Biham and Shamir[4] in 1990 is one of the most powerful attack to DES-like cryptosystems. This attack presumes on some tendency of the target cryptosystem. So the efficiency of the attack depends upon the conspicuousness of this tendency. S/N ratio introduced in [4] is to evaluate this conspicuousness. In other words, the S/N ratio is a measure of the efficiency of the attack. Nevertheless, S/N ratio does NOT suggest how many pairs of ciphertexts are needed for the cryptanalysis.

In section 2, we show how to estimate the number of necessary pairs of ciphertexts. We also show that our estimation is adequate using the 8-round-DES as an example.

By the way, it costs a huge number of memories for this cryptanalysis. In [4], a counting scheme is introduced to save memories, however, it pays some efficiency. In section 3, we show an algorithm of cryptanalysis of 8-round-DES which saves memories at the less cost of efficiency. Our method could be applied to the other DES-like cryptosystems.

2 Estimating the Number of Ciphertext Pairs

When two plaintexts whose XORed value is some particular value are encrypted by a DES-like crptosystem, the XORed value of the inputs of the final round tends to contain some particular pattern in some particular position. Differential cryptanalysis is based on this property. Concerned with the 6-round-DES for example, the bits of the XORed value of the inputs of 6th round to be XORed with the outputs of S_5, S_6, S_7, S_8 of F function of 6th round are all 0 with probability at least 25% when the XORed value of the plaintexts is 4008000004000000_x. (The subscript 'x' means that the value is expressed in the hexadecimal.)

A cryptanalyst analyzes the target cryptosystem to find such a tendency at first. Assume that this kind of tendency appears more conspicuously when the XORed value of two plaintexts is Ω_p. Call a couple of ciphertexts whose plaintexts' XOR is Ω_p a pair. If the target pattern appears in the XORed value of the inputs of the final round, the pair is called a **right pair**, otherwise it is called a **wrong pair**. Assume that a pair is Ω_p is a right pair with probability p.

Next, cryptanalyst guess a part of the secret key, say a **correct partial key** (c.p.k. for short) for each pair, applying a method introduced in [4]. For each guess, he has some candidates of the c.p.k.. Let ν be the average number of candidates par one guess. If a pair is a right pair, the c.p.k. is one of the candidates. Otherwise, these candidates are selected approximately randomly, so the c.p.k. may or may not contained in the candidates. On average, the c.p.k. is guessed pm times more often than each other candidate, say a **pseudo partial key** (p.p.k. for short), where m pairs are used. So, if pm is large enough, the cryptanalyst can find the c.p.k.. The S/N ratio introduced by Biham and Shamir[4] indicates that pm is large enough or not.

Now we observe the S/N ratio more precisely. Note that there are two cases in which the c.p.k. is in the candidates. They are the cases that

1. the key is guessed correctly with a right pair, and

2. the key is guessed unexpectedly with a wrong pair.

The *case 2* can also occur with almost the same probability to each p.p.k.. So the ratio of frequency of *case 1* to *case 2* indicates the ability of the attack to distinguish the c.p.k. from p.p.k.'s. This ratio is named S/N ratio by Biham and Shamir. It is easy to see that *case 1* is expected to occur pm times. Note that there are 2^k candidates of a c.p.k., where the length of the c.p.k. is k. So *case 2* expected to occur $\frac{m\nu}{2^k}$ times for each candidate. Now we have

$$S/N = \frac{2^k p}{\nu}.$$

The more the S/N ratio is, the better the efficiency of the attack is.

Although the S/N ratio indicates the efficiency, it does not suggest us how many pairs are necessary.

In the rest of this section, we discuss how to estimate the number of pairs.

Assume that we use m pairs to attack. The probability distribution of the number of times that the c.p.k. is guessed is a binary distribution with probability $p + \frac{\nu}{2^k}$, which we are going to approximate to normal distribution. And the distribution of the number of times that a p.p.k. is guessed is expected to be a binary distribution with probability $\frac{\nu}{2^k}$. We do not approximate the latter one to normal distribution because there are too many p.p.k.'s to achieve good approximation.

It is desired that the c.p.k. is guessed more often than any other candidates.

Let the probability that a particular p.p.k. is guessed exactly t times be $P_1(m,t)$. Then,

$$P_1(m,t) = \binom{m}{t}(\frac{\nu}{2^k})^t(1 - \frac{\nu}{2^k})^{m-t}.$$

For small t, we can approximate that

$$P_1(m,t) \simeq \frac{m^t}{t!}(\frac{\nu}{2^k})^t(1 - \frac{\nu}{2^k})^m.$$

Now, let the probability that at least one p.p.k. is guessed at least t times be $P_{all}(m,t)$. Assuming that the number of times that each p.p.k. is guessed is independet each other, we have

$$P_{all}(m,t) \simeq 1 - \left(1 - \sum_{i=t}^{m} P_1(m,i)\right)^{2^k}.$$

On the other hand, assume that the probability distribution of the number of times that the c.p.k. is guessed is the normal distribution, it is a binary distribution in reality however. Let π_r be the expectations of the numbers of times which the c.p.k. is guessed, and let σ_r be its standard deviations. Approximately,

$$\pi_r \simeq \frac{m\nu}{2^k} + pm$$

and

$$\sigma_r \simeq \sqrt{\frac{m\nu}{2^k} + pm}.$$

Let

$$t = \pi_r - \alpha\sigma_r,$$

where α should be chosen so that if you need 2.5\0.5% significant level of the attack, α should be 2\3 respectively for example.

To guarantee the attack to succeed, we have to choose m so that P_{all} is negligibly small. Thus,

$$\sum_{i=t}^{m} P_1(m,i) \ll \frac{1}{2^k}$$

must hold. Usually, the series of P_1's converge so rapidly that we can neglect latter terms. So, we have

$$P_1(m,t) \ll \frac{1}{2^k}.$$

As an example, we estimate the necessary number of pairs for 8-round-DES attack established by Biham and Shamir. In this case, $k = 30$, $p = \frac{1}{10486}$, and $\nu = 4^5$. Tentatively, we set $\alpha = 2$, then we have $m > 150000$ in a rough estimate. Similarly, when we set $\alpha = 3$, we have $m > 200000$ in a rough estimate. In other words, the differential attack using 150,000 ciphertext pairs succeed with 97.5% probability, and attack using 200,000 ciphertext pairs succeed with 99.5% probability.

We carry out a few experiments with computer to ensure that our estimation is adequate for 8-round-DES.

First, we count the times that the c.p.k. is guessed par 150,000\200,000 pairs. The result is shown in *Table 1*.

# of guess \m	150,000	200,000
2 or less	0.0%	0.0%
3	0.1	0.0
4	0.1	0.0
5	0.5	0.0
6	0.7	0.0
7	1.6	0.0
8	2.7	0.2
9	4.2	0.4
10	6.5	0.9
11	6.3	1.6
12	8.9	2.0
13	8.6	3.5
14 or more	59.9	91.5

Table 1 : Probability of the numbers of guesses of c.p.k.'s

Second, we count the times that each p.p.k. is guessed par 150,000 / 200,000 pairs. It is too hard to count for every p.p.k. because there are 2^{30} p.p.k.'s. We use a kind of narrowing down method to be shown in the following section to narrow the p.p.k.'s down to ones which are expected to be guessed many times. We show in *Table 2* and *Table 3* the average numbers of p.p.k.'s which are guessed each times par 150,000 / 200,000 pairs respectively.

m=150,000	Distance from c.p.k.				
# of guess	1	2	3	4	5
5 or less	very often				
6	1.71	.06	.01	.03	.13
7	.59	.01	0	0	0
8	.25	.01	0	0	0
9	.05	0	0	0	0
10	.02	0	0	0	0
11	.03	0	0	0	0

Table 2 : Average numbers of p.p.k.'s guessed each times par 150,000 pairs

m=200,000	Distance from c.p.k.				
# of guess	1	2	3	4	5
5 or less	very often				
6	5.22	.48	.02	.10	.63
7	2.70	.04	0	0	.02
8	1.24	.01	0	0	0
9	.46	0	0	0	0
10	.18	0	0	0	0
11	.08	0	0	0	0
12	.04	0	0	0	0

Table 3 : Average numbers of p.p.k.'s guessed each times par $200,000$ pairs

In the tables, 'Distance from c.p.k.' means extent of the difference from the c.p.k.. Since each S-box needs 6 key bits, our 30 bits of partial key can be divided into 5 blocks of 6 key bits. If a key is said to be of distance 2 from the c.p.k. for example, 2 blocks are different from and 3 blocks are the same as ones of the c.p.k.. In both cases of $m = 150,000$ and $m = 200,000$, it seems that a p.p.k. is often guessed more than 5 times, or sometimes more than 10 times. But in most of the cases, these p.p.k. is not so different from the correct one. The reason is that a p.p.k. which is not so different from the c.p.k. tends to be guessed more frequently for a right pair. So in such a case that these p.p.k.'s are guessed many times, it is expected that there are more right pairs which make the c.p.k. be guessed. So roughly speaking, we only need to be afraid of the p.p.k.'s far from c.p.k. (perhaps of distance 3 or more).

As a conclusion, our attack using $150,000$ pairs *must/can/does not* fail when the c.p.k. is guessed $5 - /6/7+$ times, respectively. And so, the empirical success ratio is 98.6% through 99.3%, which agree with our theoretical estimation. Similarly, our attack using $200,000$ pairs *can/does not* fail when the c.p.k. is guessed $7 - /8+$ times, respectively. So, empirically, our attack almost always succeeds, which roughly agree with our theoretical estimation also.

3 Cryptanalysis of DES with 8 rounds using less number of counters

In [4], Biham and Shamir's attack to 8-round-DES needs $150,000$ pairs for 95% success ratio and $250,000$ pairs for almost 100% success ratio. They do not achieve the level of our estimation shown in the last section. The reason of this is that the objective of their method is not only the efficiency of the attack but saving memories. In fact, 2^k counters are needed to achieve the simple differential attack to k bits' partial key, which is usually a heavy load. Of course, there is a trade off between efficiency and memories. However, we wish to keep efficiency even when saving memories.

In this section we show another device to save memories which do not enlarge the necessary number of pairs so much. Although the following algorithm is written for 8-round-DES, this method could be applied more generally.

Our algorithm is as follows.

ALGORITHM cryptanalysis_for_8-round-DES

input: a set of pairs of ciphertexts whose plaintexts' XOR is $405C000004000000_x$

output: a partial key

1. Setup a set of $2^6 \times 2^6$ counters, which correspond to the $2^6 \times 2^6$ values of key bits of K_8 entering S_2 and S_6.

2. For each ciphertext pair, guess the partial key for S_2 and S_6, and count the number of times each candidate is guessed.

3. Discard the candidates which is guessed less than τ_{26} times.

4. Let C_{26} be the set of candidates which are not discarded.

5. Setup $|C_{26}| \times 2^6$ counters. They correspond to the values of key bits of K_8 entering S_2, S_5 and S_6.

6. For each ciphertext pair do:

   (a) Let the subset C of C_{26} be the candidates of partial key which are guessed by the pair.

   (b) if C is empty, then Discard the pair,
   else guess the partial key for S_2, S_6 and S_5, and count the number of times each candidate is guessed.

7. Discard the candidates which is guessed less than τ_{256} times.

8. Let C_{256} be the set of candidates which are not discarded.

9. Setup $|C_{256}| \times 2^6$ counters. They correspond to the values of key bits of K_8 entering S_2, S_5, S_6a and S_7.

10. For each ciphertext pair do:

    (a) Let the subset C of C_{256} be the candidates of partial key which are guessed by the pair.

    (b) if C is empty, then Discard the pair,
    else guess the partial key for S_2, S_6, S_5 and S_7, and count the number of times each candidate is guessed.

11. Discard the candidates which is guessed less than τ_{2567} times.

12. Let C_{2567} be the set of candidates which are not discarded.

13. Setup $|C_{2567}| \times 2^6$ counters. They correspond to the values of key bits of K_8 entering S_2, S_5, S_6, S_7 and S_8.

14. For each ciphertext pair guess the partial key guess the partial key for S_2, S_6, S_5, S_7 and S_8, and count the number of times each candidate is guessed.

15. Pick up the candidate of maximal count. The value corresponds to the counter is the c.p.k..

Step 3, 7 and 11 in the algorithm above, we can restate "Discard all the candidates except ρ_{26} (or ρ_{256} or ρ_{2567}) candidates that are guessed most often" instead. Anyway, to apply this algorithm, we have to decide τ's or ρ's, the thresholds to discard candidates. There are no standard values of them. Generally speaking, there is a trade off between reliability and time or memory space. τ's should be smaller (or ρ's should be larger) if reliability takes priority and τ's should be larger (or ρ's should be smaller) if time or memory space takes priority.

In our experiments, we set $\tau_{26} = T/2^{12} + mp' - \alpha\sqrt{T/2^{12}}$, where T is the total number of guesses and α is the value defined in the last section. p' is the probability that the guess of partial keys for S_2 and S_6 of the final round can be done in the same manner of the guess using a right pair, which is equal to $4p$. The reason why we set τ_{26} as above is that the c.p.k. is expected to be guessed at least its expected value minus α times of its standard deviation. (The standard deviation can be approximated by $\sqrt{T/2^{12}}$ in this case.) In step 7 and 11 in our experiments, we discard candidates with setting ρ's instead of τ's. We set $\rho_{256} = \rho_{2567} = 1,000$. An advantage of the manner of setting ρ's is that the size of necessary counters is previously known. Anyway, the way to narrow candidates down is not so essential. In our case, $|C_{26}|$ is about $1,300$ when $m = 150,000$ and about $2,200$ when $m = 200,000$. Note that we can recycle the counters set in step 1, 5, 9 and 13. So the number of necessary counters is about $83,000$ for $m = 150,000$ and $140,000$ for $m = 200,000$ in our experiments. Our attack succeeded with probability 96.1% using $150,000$ ciphertext pairs, and 100.0% using $200,000$ ciphertext pairs. (We tried to attack 128 times in each case.)

4 Conclusion

We have introduced an adequate method to estimate the number of pairs of ciphertexts for differential cryptanalysis. It is useful not only for estimation of the number of necessary ciphertexts but for evaluation of an attack. Actually, it seems better than S/N ratio. It will be also useful in evaluation of security of DES-like cryptosystems.

We have also shown a scheme which achieves better performance using less memories.

Acknowledgment

The author wishes to thank Mr.Nakamura, Dr.Okamoto, and Mr.Ueno of C & C Information Technology Research Laboratories, NEC Corporation, for their useful advice and suggestion. The author also thanks Miss Tanaka, Mr.Masumoto, Mr.Okamura and the other members of the Laboratories for their eager discussion and friendship.

References

[1] National Bureau of Standards, *Data Encryption Standard, U.S. Department of Commerce*, FIPS pub. 46, January 1977.

[2] Shoji Miyaguchi, Akira Shiraishi, Akihiro Shimizu, *Fast data encryption algorithm Feal-8*, Review of electrical communications laboratories, Vol.36 No.4, 1988.

[3] Lawrence Brown, Josef Pieprzyk, Jennifer Seberry, *LOKI - A Cryptographic Primitive for Authentication and Secrecy Applications*, Advances in Cryptology - AUSCRYPT'90. Springer Verlag, Lecture Notes 453, pp.229-236, 1990.

[4] Eli Biham, Adi Shamir, *Differential Cryptanalysis of DES-like Cryptosystems*, proceedings of CRYPTO 90, 1990.

Construction of DES-like S-boxes Based on Boolean Functions Satisfying the SAC

Kwangjo Kim

Section 710, P.O.Box 8 Daedog Danji
Electronics and Telecommunications Research Institute
Daejeon,Chungnam, KOREA, 305-606

Abstract

In this paper, we present how to construct DES-like S-boxes based on Boolean functions satisfying the Strict Avalanche Criterion and compare their cryptographic properties with those of DES S-boxes in various points of view. We found that our designed DES-like S-boxes exhibit better cryptographical properties than those of DES S-boxes.

1 Introduction

Until now, DES[3] (Data Encryption Standard) is known as one of the most famous block cipher algorithms. In some published block cipher algorithm[9][16][14], half of the input block is nonlinearly mapped into a new half of input block by one important cryptographic primitive called f-function which is the core of the algorithm and determine their overall security. In DES f-function is implemented by 8 nonlinear look-up tables (each look-up table has 6-bit input and 4-bit output and is usually called S-box.) and 2 linear operations *i.e.*, expansion from 32-bit to 48-bit and 32-bit permutation.

For the good S-box design in the open literature, Webster and Tavares[0] proposed the notion of Strict Avalanche Criterion (SAC) in order to combine the concepts of the completeness[4] and the avalanche effect[1]. We[18][19] have already proposed the cryptographic properties of Boolean functions satisfying the SAC and a variety of construction methods of cryptographic functions satisfying the SAC. Forré[22] proposed the construction method of DES-like S-boxes by checking her defined statistical independence of 4 Boolean functions forming of a DES-like S-box. However, she did not consider that each row of DES-like S-boxes must be bijective.

Recently, Biham and Shamir[17] suggested the remarkable breaking method "differential cryptanalysis" of DES by utilizing the pairs XOR distribution of an DES S-box. They also insisted that their method can be applied for any DES-like cryptosystem.

In this paper, with our open design criteria including the possibility of differential cryptanalysis, we propose a design method of DES-like S-boxes based on Boolean functions satisfying the SAC and compare their cryptographic properties with those of DES S-boxes.

2 Design Criteria

We will state here some necessary notation and formal definitions in order to construct DES-like S-boxes.

Let Z denote the set of integers and Z_2^n denote the n dimensional vector space over the finite field $Z_2 = GF(2)$. Also, let $\oplus$ denote the addition over Z_2^n, or, the bit-wise exclusive-or. $|\cdot|$ denotes the cardinality of a set. $wt()$ denotes the Hamming weight function, and $c_i^{(n)}$ denotes an n dimensional vector with Hamming weight 1 at the i-th position. S denotes a DES-like S-box : $Z_2^6 \rightarrow Z_2^4$.

2.1 Strict Avalanche Criterion

Webster and Tavares[6] defined the SAC as follows:

Definition 1 (SAC) *We say that a function $f: Z_2^n \rightarrow Z_2^m$ satisfies the SAC, if for all i $(1 \leq i \leq n)$ there hold the following equations:*

$$\sum_{x \in Z_2^n} f(x) \oplus f(x \oplus c_i^{(n)}) = (2^{n-1}, 2^{n-1}, \dots, 2^{n-1}).$$

If a function satisfies the SAC, each of its output bits should change with a probability of one half whenever a single input bit is complemented. If some output bits depend on only a few input bits, then, by observing a significant number of input-output pairs such as chosen plaintext attack, a cryptanalyst might be able to detect these relations and use this information to aid in the search for the key.

Forré[10] extended this definition of the SAC into a higher order SAC which means every function obtained from an n-bit input function, f by keeping the k-th input bits constant and satisfies the SAC as well for every $k \in Z_2^k$. (this must be true for any choice of the positions and of the values of k constant bits). In this case, f is called to satisfy the k-th order SAC and if $k = n - 2$, f is called to satisfy the maximum order SAC. When we do not specify the order, we say that f at least satisfies the 0-th order SAC.

2.2 Nonlinearity

It has been widely accepted as a basic criterion that any cryptosystem must be nonlinear. Rueppel[8] suggested that the nonlinearity of a Boolean function can be measured by the Hamming distance to the set of affine functions and is related with the Walsh transform $\hat{F}$ of $\hat{f} \colon Z_2^n \to \{1, -1\}$ according to

$$\delta(\hat{f}) = 2^{n-1} - \frac{1}{2} max_{\mathbf{w}} |\hat{F}(\mathbf{w})|. \tag{1}$$

Meier and Staffelbach[13] suggested that nonlinearity criterion for Boolean functions can be classified in view of their suitability for cryptographic design. Their classification is set up in terms of the largest transformation group leaving a criterion invariant. One is the maximum distance to affine functions and the other is the maximum distance to linear structures. These two classifications are shown to be invariant under all affine transformations.

Pieprzyk[11] and Finkelstein[12] suggested the amount of nonlinearity and gave an equation to calculate the maximum achievable nonlinearity for an n-bit bijective function.

$$N_f = \begin{cases} \sum_{i=(n-2)/2}^{n-2} 2^i & \text{for } n = 2, 4, 6, \ldots \\ \sum_{i=(n-3)/2}^{n-3} 2^{i+1} & \text{for } n = 3, 5, 7, \ldots \end{cases} \tag{2}$$

When n is even, N_f equals to $2^{n-1} - 2^{n/2-1}$. Also they proposed one construction method of a row of DES-like S-boxes satisfying this nonlinearity.

By checking the nonlinearity of the 4 Boolean functions (*i.e.*, 4 functions : $Z_2^6 \to Z_2$) forming of one DES-like S-box by Eq. (1), we will evaluate the nonlinearity of a DES-like S-box later.

2.3 Bijection

We may or may not need this criterion depending on the structure of DES-like cryptosystems.

In order that an n-bit input function, f should be a bijection; that is, that every possible input vector maps to an unique output vector, Adams and Webster[15] suggested that a necessary and sufficient condition for f to be bijective is that any linear combination of a Boolean function has Hamming weight 2^{n-1}, *i.e.*

$$wt(\sum_{i=1}^{n} a_i f_i) = 2^{n-1} \tag{3}$$

for any $a_i \in \{0, 1\}$, $(a_1, a_2, \ldots a_n) \neq (0, 0, \ldots, 0)$, and $f = [f_1, f_2, \ldots, f_n]$. This condition gives us that each f_i $(1 \leq i \leq n)$ is basically required to be 0/1 balanced for f to be a bijection.

2.4   Cross Correlation of Avalanche Variables

Avalanche variables are the binary components of the so-called avalanche vectors, defined as the exclusive-or sums

$$\mathbf{V}_i \;=\; S(\mathbf{x}) \oplus S(\mathbf{x} \oplus \mathbf{c}_i^{(6)}). \tag{4}$$

It is stated[6] that, for a given set of avalanche vectors generated by the complementing of a single input bit, all the avalanche variables should be pairwise independent. Their degree of independence is measured by the correlation coefficient $\rho(A, B)$ which, for two random variables A and B, is given as

$$\rho(A, B) \;=\; \frac{cov(A, B)}{\sigma(A) \cdot \sigma(B)} \tag{5}$$

where

$$
\begin{aligned}
cov(A, B) \;&=\; \text{covariance of } A \text{ and } B \\
&=\; E[AB] - E[A] \cdot E[B], \\
\sigma^2[A] \;&=\; \text{standard deviation of } A \\
&=\; E[A^2] - \{E[A]\}^2 \\
\sigma^2[B] \;&=\; \text{standard deviation of } B \\
&=\; E[B^2] - \{E[B]\}^2 \\
E[A], E[B] \;&=\; \text{the expected value or mean of A, B} \\
E[AB] \;&=\; \text{the expected value of the product AB}
\end{aligned}
$$

Since the value of the avalanche variable has 0 or 1, the value of the correlation coefficient was proved[5] to range from -1 to 1.

The intuitive meanings of this correlation coefficient are:

1. when -1, the avalanche variables are always complements of one another.

2. when 1, the avalanche variables are always identical.

3. when 0, the avalanche variables are independent.

We denote $\rho_{i,j}(k)$ as the correlation coefficient of the avalanche variables i and j of an avalanche vector $\mathbf{V}_k$.

2.5   Pairs XOR Distribution

Biham and Shamir[17] proposed the differential cryptanalysis for DES-like cryptosystems. This method analyses the effect of particular differences in plaintext pairs on the differences of the resultant ciphertext pairs. These differences can be used to assign probabilities to the possible keys and locate the most probable key. Their method usually works on many pairs of plaintexts with the same particular difference using only the resultant ciphertext pairs. For DES-like cryptosystems, the difference is chosen as a fixed XORed value of the two plaintexts.

In DES any S-box has $64 \cdot 64$ possible input pairs, and each one of them has an input XOR and an output XOR. There are only $64 \cdot 16$ possible tuples of input and output XORs. Therefore, each tuple appears in average 4 pairs. However, not all the tuples exist as a result of a pair, and the existing ones do not have a uniform distribution. So, the important properties of S-boxes are derived from the analysis of the tables that show its particular distribution—called the pairs XOR distribution—defined as follows:

Definition 2 *A table that shows the distribution of the input XORs and output XORs of all possible pairs of an S-box is called the pairs XOR distribution table of the S-box. In this table, each row corresponds to a particular input XOR, each column corresponds to a particular output XOR and the entries in the table count the number of possible pairs with such an input XOR and an output XOR.*

Since differential cryptanalysis makes a direct use of the pairs XOR distribution in any S-box of DES-like cryptosystems, we must take the possibility of this attack of any block cipher algorithm into consideration.

We first consider how many entries appear in the pairs XOR distribution–% of entry.

$$\% \ of \ entry \ = \ \frac{nz}{64 \times 16} \times 100 \tag{6}$$

where nz denotes the number of non-zero entries in a pairs XOR distribution table.

Second, in order to compare the values of all entries with the ideal (uniform) value of entry, we check the standard deviation, σ of all entries by

$$\sigma \ = \ \sqrt{\sum_{i=0}^{63} \sum_{j=0}^{15} (e_{ij} - 4)^2 / 64 \times 16}. \tag{7}$$

where e_{ij} is a measured number of entry in pairs XOR distribution table.

Third, we check the nontrivial[1] maximum value of entries in DES S-boxes since the relatively higher valued entries are directly utilized in the differential attack.

We simply call these three parameters "differential characteristics of an S-box". We investigated the differential characteristics of DES S-boxes as shown in Table 1.

3 Construction of DES-like S-boxes

3.1 Boolean functions

In order to construct a 4-bit bijective function (one row of DES-like S-boxes), we need to choose 4 Boolean functions as a first step. We limit ourselves to consider

---

[1]It is trivial that the entry always has a value of 64 when the input XOR and output XOR of any DES-like S-box are zero.

Table 1: Differential characteristics of DES S-boxes

Box	% of entry	σ	max. entry
S1	79.49	3.76	16
S2	78.61	3.83	16
S3	79.69	3.78	16
S4	68.55	4.18	16
S5	76.56	3.86	16
S6	80.47	3.69	16
S7	77.25	3.95	16
S8	77.15	3.82	16

the following 4 cases for selecting a 4-bit input Boolean function to make the design criteria of DES-like S-boxes clear.

1. Boolean functions satisfying the 0-th order SAC

2. Boolean functions satisfying the 1-st order SAC

3. Boolean functions satisfying the 2-nd order SAC

4. Bent functions[2]

By computer search, we give the distribution of Boolean functions of satisfying any order SAC in Table 2. Since the function satisfying the 1-st order SAC always satisfies the 0-th order SAC, we did not count twice a function satisfying a higher order SAC in Table 2.

Table 2: Distribution of Boolean functions satisfying the SAC

n	2			3			4		
	B	U	S	B	U	S	B	U	S
No SAC	6	2	8	38	154	192	11502	49906	61408
0-th	0	8	8	24	24	48	1152	2656	3808
1-st	-	-	-	8	8	16	216	72	288
2-nd	-	-	-	-	-	-	0	32	32
Sum	6	10	16	70	186	256	12870	52666	65536

B : Balanced
U : Unbalanced
S : Sum

We have shown[20] that for odd number of input half of all functions satisfying the maximum order SAC are balanced and the other half ones are unbalanced.

But, for even number of input all functions satisfying the maximum order SAC are unbalanced. Also we[21] proved that there is an interesting relationship between Boolean functions satisfying the (maximum order) SAC and bent functions as follows:

Theorem 1 *Let $\mathcal{A}_n$ denote the set of all n-bit input Boolean functions, $\mathcal{B}_n$ denote the set of n-bit input bent functions, and $\mathcal{S}_n$ denote the set of n-bit input Boolean functions satisfying the SAC. In particular, we denote the set of n-bit input Boolean functions satisfying the maximum order SAC by $\mathcal{S}_n^{max}$. The relationship between these sets for even n can be stated as*

$$\mathcal{S}_n^{max} \subseteq \mathcal{B}_n \subseteq \mathcal{S}_n \subset \mathcal{A}_n.$$

Also we verified this relationship by checking their cardinality as shown in Table 4. We can consider that Boolean functions satisfying the maximum order SAC as a basic

Table 4: The cardinality of the set

n	2	3	4	5	6
$|\mathcal{A}_n|$	16	256	65,536	2^{32}	2^{64}
$|\mathcal{S}_n|$	8	64	4,128	?	?
$|\mathcal{B}_n|$	8	NE	896	NE	$2^{32.3}$*
$|\mathcal{S}_n^{max}|$	8	16	32	64	128

NE : Not Exist, * : 5,425,430,528

building block for a 4-bit bijective function. However, since all 4-bit input Boolean functions satisfying the 2-nd SAC are 0/1 unbalanced, we cannot construct a 4-bit input bijective function. By the same reason, we can not construct a 4-bit bijective function by using 4-bit input bent functions. (Note that unbalanced Boolean functions satisfying the maximum order SAC are bent for even number of input.) Therefore, we choose to use 4-bit input balanced Boolean functions satisfying the 0-th order or the 1-st order SAC.

3.2 DES-like S-boxes

The following criteria[7] on the DES S-boxes are well known.

1. No S-box is a linear or affine function.

2. Changing one bit of in the input of an S-box results in changing at least two output bits.

3. The S-boxes were chosen to minimize the difference between the number of 1's and 0's when any single input bit is held constant.

4. $S(\mathbf{x})$ and $S(\mathbf{x} \oplus (001100))$ differ in at least two bits.

5. $S(\mathbf{x}) \neq S(\mathbf{x} \oplus (11ef00))$ for any e and f.

If we think of an S-box as consisting of 2^r permutations on the set of all n-bit vectors (Note that a DES S-box has a value of n=4 and r=2), then a further possible criterion seems natural. In DES, key bits are used to select which permutation is used, and so it might be desirable to make permutations as different from one another as possible. This condition requires that two indexing entries in the same column of any two permutations disagree each other. We call this condition "column constraint".

Thus, we try to construct DES-like S-boxes based on the following criteria:

1. Neither linear nor affine.

2. Each row is bijective.

3. Using Boolean functions satisfying the 0-th (or 1-st) order SAC as described before.

4. Meets column constraint.

5. Resistance against differential attack.

Considering the differential characteristics of DES-like S-boxes, we choose two threshold parameters μ and λ. μ corresponds to % of entry and λ corresponds to max. of entry in our construction method. Intuitively, as many entries appear uniformly in the pairs XOR distribution table of a DES-like S-box, the differential cryptanalysis becomes more and more difficult. Also, the higher value of μ and the lower value of λ than DES are desirable, we set the threshold value $\mu_{DES} = 80$ and $\lambda_{DES} = 16$.

Afterwards, Boolean functions are restricted to have 4-bit number of input. We here summarize our method to construct DES-like S-boxes as follows:

step 1 Set m=4.

step 2 Initialize i=1 and j=1.

step 3 Choose randomly a candidate Boolean function satisfying the 0-th (or 1-st) order SAC.

step 4 Check weight constraint (*i.e.* Eq. (3)) of linear combinations of candidate Boolean functions. If satisfy then $i = i + 1$ else goto **step 3**.

step 5 If $i < m$ goto **step 3** else save a row of a DES-like S-box and $j = j + 1$.

step 6 If $j < m$, goto **step 2**.

step 7 Check column constraint. If not satisfy, goto **step 2**.

step 8 Compute μ and λ. If $\mu > \mu_{DES}$ and $\lambda \leq \lambda_{DES}$ then goto **step 2** else output a DES-like S-box.

If we choose Boolean functions satisfying the 1-st order SAC in the **step 3** of the above method, we found by experiment that $\mu_{1st-SAC}$ ranges from 73 to 80. Also we experimentally checked μ by using bent functions in the **step 3** too. But μ_{bent} ranges from 35 to 60 (Note that each row of an S-box is not bijective in these two cases.). These experiments give us that any function satisfying one of design criteria is not always good from other points of view. So we found again that it is better to use Boolean functions satisfying the 0-th order SAC.

In practice, after preparing with 10000 bijective functions by choosing 1152 balanced Boolean functions satisfying the 0-th order SAC randomly at **step 6** of this method, we have constructed 32 DES-like S-boxes with an engineering workstation (SONY) in a few minutes. 8 examples of DES-like S-boxes are listed in the **Appendix**.

4 Comparison

We denote s^2DES as a DES-like cryptosystem which all DES S-boxes are replaced by our designed DES-like S-boxes. For the sake of simplicity, we call our designed DES-like S-boxes "s^2DES S-boxes". Since 8 S-boxes are used in DES, we choose 8 typical s^2DES S-boxes constructed by our method for comparison and other constructed examples of DES-like S-boxes are listed in [23].

We compare the quantitative characteristics of s^2DES S-boxes with DES S-boxes in various points of view and evaluate the goodness-of-fit of them.

First we checked the nonlinearity of 4 Boolean functions: $Z_2^6 \rightarrow Z_2$ consisting of an S-box compared in Tables 5. In the output bit column of this Table, 4 denotes the most significant location of an output vector and 1 denotes the least significant location of an output vector.

Table 5: Comparison of nonlinearity of DES and s^2DES S-boxes

Box	DES				s^2DES			
	output bit				output bit			
	1	2	3	4	1	2	3	4
S1	18	20	22	18	22	20	20	22
S2	22	20	18	18	24	22	22	22
S3	18	22	20	18	20	24	22	22
S4	22	22	22	22	20	22	22	22
S5	22	20	18	20	22	24	22	24
S6	20	20	20	20	22	22	20	22
S7	18	22	14	20	22	20	22	18
S8	22	20	20	22	22	22	22	22

We can see that the nonlinearity of DES S-boxes ranges from 14 to 22, but the nonlinearity of s^2DES S-boxes ranges from 18 to 24. This can be said that s^2DES S-boxes are more nonlinear than DES S-boxes.

Second, we measured the differential characteristics of s^2DES S-boxes as shown in Table 6.

Table 6: Differential characteristics of s^2DES S-boxes

Box	% of entry	σ	max. entry
S1	81.38	3.11	14
S2	85.25	3.39	14
S3	84.38	3.34	14
S4	83.40	3.54	16
S5	82.91	3.57	16
S6	83.98	3.48	16
S7	81.93	3.62	16
S8	82.81	3.54	16

The differential characteristics of s^2DES S-boxes are found to exhibit better than those of DES S-boxes (Table 1). This suggests that s^2DES can resist better than DES against differential cryptanalysis.

Third, we checked the dependence matrix of a DES-like S-box. The dependence matrix $\mathbf{P} = (p_{i,j})$ of the S-box is defined as follows: The element $p_{i,j}$ of $\mathbf{P}$ is the probability that the output variable y_j of the S-box changes when the input variable x_i is complemented. The average values, *i.e.* $(p_{i,1} + p_{i,2} + p_{i,3} + p_{i,4})/4$ of $(p_{i,j})$ of DES S-boxes and s^2DES S-boxes are compared in Table 7. Whenever one bit of input

Table 7: Comparison of average $(p_{i,j})$ of DES and s^2DES S-boxes

Box	DES	s^2DES
S1	0.620	0.495
S2	0.633	0.510
S3	0.661	0.505
S4	0.615	0.521
S5	0.633	0.516
S6	0.651	0.516
S7	0.656	0.516
S7	0.625	0.508

in s^2DES S-boxes is complemented, every output bits can be said to change about probability $\frac{1}{2}$. But, the probability of dependence matrix in DES S-boxes is found to be greater than 0.6. This shows indirectly that Boolean functions consisting of DES S-boxes do not satisfy the SAC.

Finally we compare the average values of cross correlation of avalanche variables, $\rho_{i,j}(k)$ of DES S-boxes with those of s^2DES S-boxes in Table 8. See the definition of $\rho_{i,j}(k)$ in Section 2.4. In average, the cross correlations between the output bits of s^2DES S-boxes exhibit more independent than those of DES S-boxes.

Table 8: Comparison of average $\rho_{i,j}(k)$ of DES and s^2DES S-boxes

Box	DES	s^2DES
S1	-0.195	-0.051
S2	-0.188	-0.070
S3	-0.165	-0.053
S4	-0.232	-0.083
S5	-0.184	-0.074
S6	-0.183	-0.096
S7	-0.153	-0.105
S7	-0.176	-0.101

The detailed values of the dependence matrix and the cross correlation of avalanche variables of DES S-boxes and s^2DES S-boxes are given in [23].

5   Concluding Remarks

We proposed a simple method to construct DES-like S-boxes based on our design criteria. By using Boolean functions satisfying the 0-th order SAC, we can construct many DES-like S-boxes easily. Compared with DES S-boxes, s^2DES S-boxes exhibit better cryptographic properties from various points of view.

We[23] also checked the overall security of s^2DES and compared with other DES-like cryptosystems and verified that s^2DES exhibits good cryptographical properties. We can argue that s^2DES can be used for the private purpose instead of DES. Also, our approach can be easily generalized to construct any input size DES-like S-boxes. However, we think that the following topics require further research.

- Quantitative immunity of differential characteristics of a DES-like S-box to differential cryptanalysis.

- Existence of DES-like S-boxes having uniform pairs XOR distribution.

Acknowledgement

The author is very grateful to Prof. Hideki Imai and Prof. Tsutomu Matsumoto for their discussions and guidances.

References

[1] H. Feistel, "Cryptography and Computer Privacy", Scientific American, Vol.228, No.5, pp 15–23, 1973.

[2] O.S. Rothaus, "On "Bent" Functions", J. of Combinatorial Theory(A), Vol.20, pp.300–305, 1976.

[3] "Data Encryption Standard", National Bureau of Standards, Federal Information Processing Standard, Vol. 46, U.S.A., Jan., 1977.

[4] J.B. Kam and G.I. Davida, "Structured Design of Substitution Permutation Networks", IEEE Trans. on Comp., Vol. C-28, No.10, pp.747–753, Oct., 1979.

[5] A.F. Webster, "Plaintext/Ciphertext Dependences in Cryptographic Systems", Master's Thesis, Queen's Univ., CANADA, 1985.

[6] A.F. Webster and S.E. Tavares, "On the Design of S-boxes", Proc. of CRYPTO'85, Springer-Verlag, 1985.

[7] E.F. Brickell, J.H. Moore, and M.R. Purtill, "Structures in the S-boxes of the DES", Proc. of CRYPTO'86, Springer-Verlag, pp.3–8, 1986.

[8] R.A. Rueppel, *Analysis and Design of Stream Ciphers*, Springer-Verlag, Berlin, 1986.

[9] S. Miyaguchi, A. Shiraishi, and A. Shimizu, "Fast Data Encryption Algorithm FEAL-8", (*in Japanese*), Electr. Comm. Lab. Tech. J., NTT, Vol.37, No.4/5, pp.321–327, 1988.

[10] R. Forré, "The Strict Avalanche Criterion : Spectral Properties of Boolean Functions and an Extended Definition", Proc.of CRYPTO'88, Springer-Verlag, 1988.

[11] J. Pieprzyk, "Nonlinearity of Exponent Permutations", Proc. of EURO-CRYPT'89, Springer-Verlag, 1989.

[12] J. Pieprzyk and G. Finkelstein, "Towards Effective Nonlinear Cryptosystem Design", IEE, Pt.E, Vol.135, pp.325–335, 1988.

[13] W. Meier and O. Staffelbach, "Nonlinearity Criteria for Cryptographic Functions", Proc. of EUROCRYPT'89, Springer-Verlag, 1989.

[14] K. Takaraki, K. Sasaki, and F. Nakagawa, "Multi–Media Encryption Algorithm (*in Japanese*)", 89-MDP-40-5, 1989.1.19.

[15] C. Adams and S. Tavares, "The Use of Bent Sequences to Achieve Higher-Order Strict Avalanche Criterion in S-box Design", (Private Communication), 1990.

[16] L. Brown, J. Pieprzyk, and J. Seberry, "LOKI – a Cryptographic Primitive for Authentication and Secrecy", Proc. of AUSCRYPT'90, 1990.

[17] E. Biham and A. Shamir, "Differential Cryptanalysis of DES-like Cryptosystems", Proc. of CRYPTO'90, 1990.

[18] K. Kim, T. Matsumoto, and H. Imai, "On Generating Cryptographically Desirable Substitutions", Trans. IEICE, Vol. E73, No.7, Jul., 1990.

[19] K. Kim, T. Matsumoto, and H. Imai, "A Recursive Construction Method of S-boxes Satisfying Strict Avalanche Criterion", Proc. of CRYPTO'90, 1990.

[20] K. Kim, T. Matsumoto, and H. Imai, "Methods to Generate Functions Satisfying the Strict Avalanche Criterion", Technical Report on Information Security, ISEC90-30, Nov. 13, 1990.

[21] K. Kim, T. Matsumoto, and H. Imai, "On the Cryptographic Significance of Bent Functions", KSEAJ Letters, 1990.

[22] R. Forré, "Methods and Instruments for Designing S-boxes", J. of Cryptology, Vol.2, No.3, pp.115–130, 1990.

[23] K. Kim, "A Study on the Construction and Analysis of Substitution Boxes for Symmetric Cryptosystems", Ph.D Thesis, Yokohama National Univ., 1991.

Appendix: 8 s^2DES S-boxes

s^2DES S1-box

12	14	1	15	11	10	8	4	7	9	5	0	3	2	13	6
3	5	4	12	9	14	0	8	2	7	10	1	13	6	15	11
10	7	9	11	15	13	2	5	14	6	1	4	12	3	8	0
6	0	5	8	14	7	1	3	12	4	2	13	11	15	10	9

s^2DES S2-box

2	12	4	6	3	0	8	5	10	11	15	7	13	1	14	9
14	8	3	11	9	13	10	2	5	0	1	6	7	12	15	4
4	13	14	3	1	10	5	7	9	15	8	12	11	2	0	6
0	11	1	15	4	5	12	14	7	10	9	13	6	8	2	3

s^2DES S3-box

5	10	7	12	13	2	0	4	6	14	11	15	3	1	9	8
3	13	6	14	2	0	15	12	1	5	10	7	4	11	8	9
9	2	11	6	1	13	10	15	14	12	3	5	0	8	4	7
1	8	14	11	5	15	9	3	7	6	4	0	12	10	13	2

s^2DES S4-box

13	2	12	11	3	1	5	9	15	6	8	0	14	10	4	7
9	1	14	5	11	7	12	4	8	15	0	6	3	2	13	10
14	5	15	13	7	2	9	6	0	12	10	11	4	8	1	3
6	10	5	3	2	11	14	0	7	4	1	8	9	13	15	12

s^2DES S5-box

10	2	9	11	8	7	6	3	5	13	12	15	0	4	14	1
2	1	0	5	11	8	15	7	12	9	14	6	3	13	10	4
0	5	8	6	4	3	13	14	9	1	15	11	2	10	7	12
7	14	11	13	12	2	10	9	1	8	0	4	5	6	3	15

s^2DES S6-box

0	11	7	10	12	9	14	6	1	3	5	15	2	4	8	13
3	1	4	5	0	12	8	7	10	2	14	13	6	9	15	11
5	12	15	6	7	11	10	13	0	8	9	14	4	2	1	3
4	8	10	11	6	5	7	1	14	15	12	2	3	13	9	0

s^2DES S7-box

5	0	11	14	10	2	9	8	13	3	12	6	4	7	1	15
10	12	13	4	9	1	3	0	6	8	5	15	14	11	2	7
6	11	12	9	0	3	4	14	1	7	8	13	10	2	15	5
9	4	7	0	3	11	2	1	15	5	6	8	12	13	10	14

s^2DES S8-box

7	13	4	14	10	15	8	2	11	9	6	3	1	12	5	0
6	14	10	12	5	7	0	1	2	13	11	4	8	9	15	3
10	6	12	1	11	9	14	3	13	15	4	5	0	2	8	7
5	3	15	6	0	1	13	9	4	10	14	8	12	7	11	2

The Data Base of Selected Permutations

(Extented abstract)

Jun-Hui Yang

Computing Center, Academia Sinica, 100080, Beijing, China

Zong-Duo Dai*, Ken-Cheng Zeng

Graduate School, Academia Sinica, 100039-08, Beijing, China

Abstract

The so called selected permutations, which can be used to compose S-boxes in the DES-like ciphers, are classified under the action of a transformation group G^*. A method for building up a data base of the selected permutations is given. In fact, we have built up a data base which consists of one representative from each of the G^*-equivalent classes, it turns out that the data base is of size 17433 and can be easily memorized by the help of a small magnetic disc. The constructed data base together with the applications of the group G^* will produce totally $17433 \times 3 \times 2^9$ selected permutations, which can be used to provide ready raw material for composing S-boxes.

Key Words: DES, S-boxes, Permutations.

*This presentation was prepared during a visit to the Department of Mathematics, Royal Holloway and Bedford New College, University of London, and supported by SERC grant GR/F 72727.

1 Introduction

The Data Encryption Standard (DES) block cipher algorithm consists of 16 rounds of a key controlled function. The so called substitution boxes (S-boxes) are the only non-linear components of the function, and the strength of a DES-like block cipher relies heavily on the design of these S-boxes. In the DES algorithm, there are 8 S-boxes, each S-box consists of four permutations on the first 16 non-negative integers, and a cryptographically good design of S-boxes depends mainly on the careful selection of these permutations.

In this presentation, we are interested in the permutations (called "selected permutations") which satisfy some cryptographic desirable properties accordant to the "S-boxes design principles" released by IBM and NSA regarding DES [2, 3]. We classify these selected permutations under a transformation group G^*. A method for building up a data base of the selected permutations is given. In fact, we have built up a data base which consists of one representative from each of the G^*-equivalent classes, it turns out that the data base is of size 17433 and can be easily memorized by the help of a small magnetic disc. The constructed data base together with the applications of the group G^* will produce totally $17433 \times 3 \times 2^9$ selected permutations, which can be used to provide ready raw material for composing S-boxes.

2 The Selected Permutations

Let Z be the set of the first 16 non-negative integers $\{0, 1, 2, \cdots, 15\}$, and let $V_4(GF(2))$ be the vector space of dimension 4 over the binary field $GF(2)$. In this presentation, an element X in Z with $X = x_0 + x_1 2 + x_2 2^2 + x_3 2^3$ and $x_i \in \{0, 1\}$ will be identified to the vector (x_0, x_1, x_2, x_3) in $V_4(GF(2))$, and the addition of numbers in Z will be understood in the vectorial sense. Thus a permutation π on Z realize a surjective transformation from $V_4(GF(2))$ to $V_4(GF(2))$, sending the input vector $(x_0, x_1, x_2, x_3) \in V_4(GF(2))$ to the output vector $(y_0, y_1, y_2, y_3) \in V_4(GF(2))$. As is well-known, the functional relation between the output signal y_j and the input vector

(x_0, x_1, x_2, x_3) to the permutation π can be expressed by a uniquely determined polynomial $y_j = f_j(x_0, x_1, x_2, x_3)$ of degree at most 1 in each of the indeterminates $x_i, 0 \leq i \leq 3$, separately.

As the components of some cryptographic desirable substitution boxes (S-boxes), the permutations contained in S-boxes should satisfy some cryptographic requirements, the following "selection criteria" for permutions are accordant to the "S-boxes design principles" released by IBM and NSA regarding DES [2, 3].

1. No permutation is affine, i.e., $\deg f_j(x_0, x_1, x_2, x_3) > 1$ for at least one j, $0 \leq j \leq 3$.

2. A single bit of difference in the input vectors will lead to at least two bits of difference in the outputs, i.e., $W(\pi(X) + \pi(X + 2^i)) \geq 2$ for any number X in Z and any $i \in \{0, 1, 2, 3\}$, where by $W(X)$ with $X = x_0 + x_1 2 + x_2 2^2 + x_3 2^3$ and $x_i \in \{0, 1\}$ we mean the Hamming weight of the vector (x_0, x_1, x_2, x_3), i.e., the number of $1's$ in the vector (x_0, x_1, x_2, x_3).

3. For any X in Z the output vectors $\pi(X)$ and $\pi(X + 6)$ differ in at least two bits, i.e., $W(\pi(X) + \pi(X + 6)) \geq 2$.

4. For any given i and j in $\{0, 1, 2, 3\}$ and any fixed input signal x_i, the output signal y_j assumes the values "0" and "1" in approximately the same number of times, i.e., the conditional probability $Prob.(y_j = b | x_i = a)$ should be close to 1/2 for any given a and b, say $|Prob.(y_j = b | x_i = a) - 1/2| \leq 1/8$.

5. The indeterminate x_i affects the output polynomial $f_j(x_0, x_1, x_2, x_3)$ for any i and j in $\{0, 1, 2, 3\}$, i.e., there exists a number X in Z such that $f_j(X) \neq f_j(X + 2^i)$. In other words, if we denote by $C_{ij}(\pi)$ the number of integers $X = (x_0, x_1, x_2, x_3)$ in Z satisfying "$x_i = 0, f_j(X) \neq f_j(X + 2^i)$", then $C_{ij}(\pi) > 0$.

We call the permutations which satisfy the above requirements the **selected permutations**, of which the set will be denoted by $\mathcal{P}$. We are interested in producing all the selected permutations.

Note : The criteria 2 and 3 are the propagation properties. And the criterion 4 is the I/O-correlation immunity requirement. The values taken by $|Prob.(y_j = b \mid x_i = a) - 1/2|$ will be some integral multiples of 1/8, and for the permutations used in the S-boxes of the DES algorithm the probabilities $|Prob.(y_j = b \mid x_i = a)|$ are of the form $1/2 + b/8, b \in \{-1, 0, 1\}$, so we put a restriction that $|Prob.(x_i = a|y_j = b) - 1/2| \leq 1/8$ on the the criterion 4. And the criterion 5 is the completeness requirement. The property that $C_{ij}(\pi) = 4$ for all i, j is called "strong completeness" in [6], which is equivalent to the "strict avalanche" property studied in [1,4,5], but there does not exist any permutation satisfying the strong completeness as well as the requirement 2, in fact it is proved [6] that any permutation on the set Z satisfying the requirement 2 can not be strongly complete. The requirement 1 in fact is unnecessary because it is proved [6] that the requirement 1 is implied by the completeness requirement 5.

3 The Transformation Group G^*

We shall define a transformation group G^* which acts on S_Z and will be used in classifying the selected permutations, where S_Z is the symmetric group on Z.

First we consider the following subgroups of S_Z.

1. The group $T = \{t_{\vec{b}} | \vec{b} \in V_4(GF(2))\}$ of translations, where $t_{\vec{b}}(X) = X + \vec{b}$ for all $X \in Z$.

2. The group $I = \{\bar{\sigma} \mid \sigma \in S_4\}$, where S_4 is the symmetric group on the set $\{0, 1, 2, 3\}$, and $\bar{\sigma} \in S_Z$ is induced by σ such that $\bar{\sigma}(X) = x_0 2^{\sigma(0)} + x_1 2^{\sigma(1)} + x_2 2^{\sigma(2)} + x_3 2^{\sigma(3)}$ for any $X = x_0 + x_1 2 + x_2 2^2 + x_3 2^3$ in Z.

3. The group $J = \{\bar{\sigma} \mid \sigma \in \{(1), (12), (03), (12)(03)\}\}$, a subgroup of I.

4. The group $U = TI$ and the group $V = TJ$.

Let G be the group consisting of all transformations of the form $g = (u, v), u \in U, v \in V$, acting on S_Z and sending the permutation π ($\in S_Z$) to $u\pi v$ ($\in S_Z$). And let G^* be the subgroup of G consisting of all transformations of the form $g = (u, v), u \in U, v \in J$. We shall say two permutations π_1 and π_2 are G^*-equivalent if there exists an element $g \in G^*$ such that $g(\pi_1) = \pi_2$.

Theorem 1 *The set $\mathcal{P}$ of all the selected permutations is invariant under the action of the group G, hence $\mathcal{P}$ is also invariant under the action of the group G^*.* □

We see from Theorem 1 that G^* is a transformation group acting on the set $\mathcal{P}$, thus the selected permutations are divided into some G^*-equivalent classes. To determine all the selected permutations it is enough to build up a data base consisting of one representitave from each of the G*-equivalent classes, since then the constructed data base together with the applications of the group G^* would produce all the selected permutations. The basic idea that how to build up this kind of data base will be described in the next section.

Remark: The idea of classification of a certain kind of permutations (or S-boxes) under the action of a certain transformation group is of general significance.

4 Build up a Data Base

4.1 Permutation Skeletons

Let $\mathcal{P}_0$ be the set of all the selected permutations π with $\pi(0) = 0$. To determine the selected permutations upto G^*-equivalence we may consider only the elements in $\mathcal{P}_0$, since we have

Lemma 1 *Any selected permutation is G^*-equivalent to a permutation π with $\pi(0) = 0$.* □

Let H be the subgroup of G^* consisting of all elements in G^* under which the set $\mathcal{P}_0$ is invariant, i.e., $H = \{h \mid h \in G^*, h(\mathcal{P}_0) \subseteq \mathcal{P}_0\}$. H is called the stationary subgroup of $\mathcal{P}_0$ in G^*.

Lemma 2 *1. Two elements in $\mathcal{P}_0$ are G^*-equivalent if and only if they are H-equivalent.*

2. H consists of all transformations in G^* of the form $g = (\sigma, \tau), \sigma \in I, \tau \in J$.

| |

For any permutation $\pi \in S_Z$, consider the quadruple

$$q(\pi) = (\pi(2^0), \pi(2^1), \pi(2^2), \pi(2^3))$$

which is composed with the images of the four numbers $2^i, 0 \leq i \leq 3$, and will be called the skeleton of π; and meanwhile this π will be called a **desired extension** of the quadruple $(\pi(2^0), \pi(2^1), \pi(2^2), \pi(2^3))$ if $\pi \in \mathcal{P}_0$. We have

Lemma 3 *Let* $\pi \in \mathcal{P}_0$ *and* $q(\pi) = (a_0, a_1, a_2, a_3)$, *then*

1. *$a_i \neq a_j$, if $i \neq j$,*

2. *$W(a_i) \geq 2$ for all $i, 0 \leq i \leq 3$,*

3. *$W(a_1 + a_2) \geq 2$.*

$\square$

To facilitate the work of determining the selected permutations, we consider the set $\mathcal{A}$ which consists of all quadruples $\bar{a} = (a_0, a_1, a_2, a_3)$ satisfying the conditions listed in Lemma 3. And consider the action of the group H on the set $\mathcal{A}$ such that $h(\bar{a}) = (\sigma(a_{\tau(0)}), \sigma(a_{\tau(1)}), \sigma(a_{\tau(2)}), \sigma(a_{\tau(3)}))$ for any $h = (\sigma, \tau) \in H$ and $\bar{a} = (a_0, a_1, a_2, a_3) \in \mathcal{A}$. The set $\mathcal{A}$ is invariant under the action of H. We shall say two quadruples $\bar{a}$ and $\bar{a}'$ are H-equivalent if there exsists an element h in H such that $h(\bar{a}) = \bar{a}'$.

Lemma 4 1. *The skeleton of any element in $\mathcal{P}_0$ bolongs to $\mathcal{A}$, or in other words, any element in $\mathcal{P}_0$ is a desired extension of a quadruple in $\mathcal{A}$.*

2. *We have $q(h(\pi)) = h(q(\pi))$ for any $h \in H$ and any $\pi \in \mathcal{P}_0$, hence the skeletons of any two H-equivalent elements in $\mathcal{P}_0$ will be H-equivalent.*

$\square$

Lemma 5 *The set $\mathcal{A}$ consists of* 5616 *quadruples, which form 81 H-equivalent classes.* $\square$

Let $\tilde{A}$ be a set consisting of a representative from each of the H-equivalent classes in A, it follows from Lemma 5 that the set $\tilde{A}$ consists of 81 quadruples. It follows from the above Lemmas that every selected permutation is G^*-equivalent to a desired extension of a quadruple in $\tilde{A}$; and any two H-unequivalent desired extentions of a quadruple in $\tilde{A}$ are G^*-unequivalent; and any two desired extensions from different quadruples in $\tilde{A}$ are G^*-unequivalent.

Thus to determine the selected permutations upto G^*-equivalence, one needs only for each quadruple in $\tilde{A}$ produce all its desired extensions and classify them under the action of H.

4.2 Extentions

Producing all desired extensions of any given quadruple in $\tilde{A}$ is easy, here we omit it.

4.3 Classification

The classification is based on a test for the H-equivalence of any two permutations in $\mathcal{P}_0$. The test will be described below.

First we note that it is easy to decide whether a given permutation π belongs to the group I. To do this, we write $B = \{1, 2, 4, 8\}$, and behave in the following way.

1. Check the weights $W(\pi(X)), X \in B$. If one of them is greater than 1, then decide $\pi \in I$, otherwise go to 2;

2. Check whether $\pi(X + Y) = \pi(X) + \pi(Y)$ for all $X, Y \in B$. If this is untrue for one pair (X,Y), then decide $\pi \in I$, otherwise go to 3;

3. Check whether $\pi(X + Y + E) = \pi(X) + \pi(Y) + \pi(E)$ for all $X, Y, E \in B$. If this is untrue for one triple, decide $\pi \in I$, otherwise decide $\pi \in I$.

Now for any two permutations in $\mathcal{P}_0$ we formulate the test for their H-equivalence. Given any pair of permutations (π_1, π_2), compute the product $\pi_2 \tau \pi_1^{-1}$ for every $\tau \in J$. Check by help of the device just described whether such a product belongs to I. If this is the case for some

$\tau \in J$, then the permutations π_1 and π_2 will be H-equivalent, otherwise we conclude that they are H-unequivalent.

5 Permutations of Special Cryptographic Interest

As an application we sort out from the set $\mathcal{P}$ two kinds of permutations, which are of special cryptographic interest, namely, the so-called I/O-correlation immune permutations, and the the so-called improved completeness permutations to be defined below.

1. A permutation is said to be I/O-correlation immune if $Prob(y_j = b \mid x_i = a) = 1/2$ for all i, j, a and b. The I/O-correlation immune property is G-invariant and it turns out that among the 17433 G^*-equivalence classes, into which $\mathcal{P}$ decomposes, 46 classes consist of I/O-correlation immune permutations.

2. As the set $\mathcal{P}$ does not contain strongly complete (strict avalanche) permutations [6], it is natural to ask whether one can weaken in some manner the strong completeness requirement, so that it will still remain stronger than that of ordinary completeness, but will not result in a void subset of $\mathcal{P}$. We do this in the following way.

   Given a permutation π, improved completeness amounts to requiring that $2 \leq C_{ij}(\pi) \leq 6$ for all $0 \leq i, j \leq 3$. The improved completeness property is also G*-invariant, and so in searching for this kind of permutations we can again make use of the 17433 G*-equivalent class representatives in the data base. It turns out that all the improved completeness permutations in $\mathcal{P}$ form 2683 G*-equivalence classes .

References

[1] C. Adams and S. Tavares, "The Structured Design of Cryptographically Good S-Boxes", J. Cryptology, Vol.3, No. 1,1990.

[2] D. K. Bransted, J. Gait and S. Katzke, "Report of the Workshop on Cryptography in Support of Computer Security", NBS, Sept. 1977.

[3] E. F. Brickell, J. H. Moore and M. R. Purtill, "Structure in the S-boxe of the DES", in Advances in Cryptology: Proc. of CRYPTO'86.

[4] R. Forre, "The strict avalanche criterion:spectral properties of boolean functions and an extended definition", in Advances in Cryptology: Proc. of CRYPTO'88.

[5] A. F. Webster and S. E. Tavares, "On the design of S-boxes", in Advances in Cryptology: Proc. of CRYPTO'85.

[6] J. H. Yang, Z. D. Dai and K. C. Zeng, "A Cryptographic Study on S-boxes of DES Type (I). An integrated analysis of the Design Criteria for S-boxes", To appear at "System Science and Mathematics", 1991.

[7] K. C. Zeng, Z. D. Dai and J. H. Yang, "A Cryptographic Study on S-boxes of DES Type (II). An Entropy Leakage Analysis for the Key", To appear at "System Science and Mathematics", 1991.

A Framework for the Design of One-Way Hash Functions Including Cryptanalysis of Damgård's One-Way Function Based on a Cellular Automaton

Joan Daemen, René Govaerts and Joos Vandewalle

Katholieke Universiteit Leuven, Laboratorium ESAT,
Kardinaal Mercierlaan 94, B-3001 Heverlee, Belgium.

Abstract

At Crypto '89 Ivan Damgård [1] presented a method that allows one to construct a computationally collision free hash function that has provably the same level of security as the computationally collision free function with input of constant length that it is based upon. He also gave three examples of collision free functions to use in this construction. For two of these examples collisions have been found[2] [3], and the third one is attacked in this paper. Furthermore it is argued that his construction and proof, in spite of their theoretical importance, encourage inefficient designs in the case of *practical* hash functions. A framework is presented for the *direct* design of collision free hash functions. Finally a concrete proposal is presented named Cellhash.

1   Introduction

A *collision free* hash function h is an easily computable function that maps strings of arbitrary length to strings of some fixed length in such a way that finding two strings x and x', with $h(x) = h(x')$ is computationally infeasible. The pair x, x' is called a collision for h. A *collision free function* f from m bits to t bits with $m > t$ is an easily computable function that maps strings of length m to strings of length t, in such a way that finding a pair of strings x, x' of length m, with $f(x) = f(x')$ is computationally infeasible.

The presence of the term 'computationally infeasible' makes the previous definitions somewhat fuzzy. By interpreting this term in different ways, we get different definitions of collision free functions or collision free hash functions. At one end of the spectrum there is the pragmatic point of view and at the other the complexity theoretic framework.

For readability a collision free hash function will be denoted by 'hash function' and a collision free function with an output shorter than its input by 'CF function' in the sequel. A function that takes an input of fixed length to an output of fixed length that is shorter than the input will be denoted by 'FI function'. A CF function is a special case of a FI function.

In [1] a construction is given where finding a collision for the hash function yields a collision for the underlying CF function. Hence the hash function is provably as secure as the CF function. This reduces the design to that of the CF function. It is widely believed that this is much easier. In our opinion the direct design of a hash function is not harder and leads to more efficient, i.e. faster hash functions.

With the breaking of the Damgård function based on cellular automata in the following section, all three CF functions proposed in [1] are broken. This does not necessarily mean that collisions have been found for the corresponding hash functions built on these CF functions. However, because the underlying FI functions are shown not to be computationally collision free, the proof of Damgård does not apply. The hash functions are robbed of their *provable* collision free-ness. For the hash function based on cellular automata we show that the probability of the hash value being unaffected by the complementation of certain messagebits is large. Hence collisions for the hash function can be created ad libitum.

We give a general framework that allows the direct design of hash functions. Finally a concrete proposal for a hardware oriented high speed hash function is described. It is named Cellhash and its design follows this framework. Like Damgård's CF function it makes use of cellular automata, but this time in a well-considered way.

2   Cryptanalysis of Damgård's CF Function

Damgård's CF function $g_c()$ computes a 128-bit string h starting from a 256-bit string x. We have $h = g_c(x)$. In the following we will explain in an informal way how it operates and how to create collisions. The function is based on a cellular automaton thoroughly investigated by Wolfram in [4]. When using results of Wolfram the reader is referred to [4] for a proof or further explanation.

A constant 256-bit string z is appended to x resulting in a 512-bit string. This is taken to be the initial state $a^0 = a_0^0 a_1^0 \ldots a_{511}^0$ of a 512-bit binary cellular automaton(CA) with periodic boundary conditions[4]. When the CA is iterated the bits of the state are modified according to the local updating rule

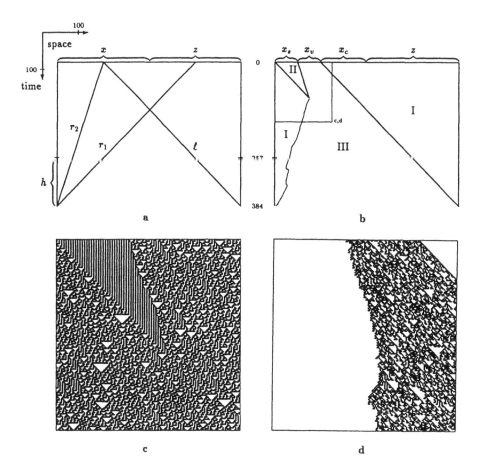

Figure 1: Space-time diagrams of the cellular automaton operation of the CF function.   a: Propagation of information.   b: Scheme for creating collisions.   c: Upper left corner of b, black denotes 1, white denotes 0.   d: bitwise XOR of two space-time diagrams for arguments of the function with the same output h.

($\oplus$ denotes XOR, $\vee$ denotes OR) :

$$a_i^{t+1} = a_{i-1}^t \oplus (a_i^t \vee a_{i+1}^t)$$

where a_i^t denotes the i-th bit of the state of the CA after t iterations. The cellular automaton performs 384 iterations and the output h consists of the concatenation of the first bits of the strings a^{257} to a^{384}, hence $h = h_0 h_1 \ldots h_{127}$ with $h_i = a_0^{257+i}$. h can be seen as the concatenation of subsequent values in time of one bit of the state, i.e. a temporal sequence. The state of the CA at a certain moment can be seen as an array of values in space, i.e. a spatial sequence. Figure 1.a gives a space-time diagram [4] of the computation. Bit a_i^t is at the position i units to the right and t units downwards from the upper left corner of the diagram. The top line of the diagram is the initial state a^0 containing x and z. Because there are periodic boundary conditions the right and left edge are in fact adjacent and the diagram can be seen as an unfolded cylinder. Mathematically this implies that the lower indices should be taken modulo 512.

Because of the local nature of the CA updating each bit a_i^t can only affect bits a_{i-1}^{t+1} to a_{i+1}^{t+1} of the next state. Hence each bit a_i^t can only depend on bits a_{i-1}^{t-1} to a_{i+1}^{t-1} of the previous state. Applying this recursively implies that a_i^t depends only on a_{i-k}^{t-k} to a_{i+k}^{t-k} of the state k iterations ago . The union of these sets for all k is called the dependence area of a_i^t, denoted by $\mathcal{A}_i^t$. In words: if all bits in $\mathcal{A}_i^t$ are left unchanged, a_i^t cannot change, whatever happens outside of $\mathcal{A}_i^t$. In Figure 1.a the borders of the dependence area of the last bit of h are indicated. The left border is line ℓ with slope $+1$ and the right border is r_1 with slope -1. This comprises the dependence areas of all other bits of h that are mere translations upwards of the indicated area. Hence the indicated area can be seen as the dependence area of all h and will be denoted by $\mathcal{H}$.

However, the updating rule is asymmetric, and so is the propagation of information [4]. It turns out that for *almost all* initial states a^0, $\mathcal{H}$ is bounded at the right by a border that is much steeper than r_1. In Figure 1.a the actual border is approximated by a diagonal r_2 with slope -3 such that all of a_0 lies in $\mathcal{H}$. Of course we can't just choose this slope. In reality it is fixed by the initial state a_0. For most a_0 however r_2 can be fairly well approximated by a straight line with slope -4, hence we've been rather pessimistic. This implies that for these states some bits of a_0 lie outside of $\mathcal{H}$, providing a first type of collisions. For most inputs the output is independent of the bits a_{126-k}^0 to a_{126}^0 for some i. Hence *for the majority of inputs x there exists an integer $k > 0$ such that all x' that only differ from x in bit positions $126 - k$ to 126 have the same output h, leading to 2^k simultaneous collisions.*

But even *larger sets of simultaneous collisions can be found*. If the initial state contains a substring that exhibits the alternating pattern (010101...), a remarkable observation can be made. As can be seen in Figure 1.c the left edge of the pattern shifts one bit the right each timestep and the right edge follows an irregular path that moves on the average one bit to the *right* per three iterations[4]. This pattern is stable with respect to the updating rule, in that it expands, albeit slowly, to the right. With respect to information propagation, this pattern acts as a buffer, as can be seen in Figure 1.d. Here the bitwise XOR of Figure 1.c with another space-time diagram is shown. The initial state of the second space-time diagram differs only from that in Figure 1.c in the 64 bits on the right from the alternating pattern. It can be seen that the differences in the subsequent states propagate in the cone bounded at the left by the alternating area. This can be used to construct collisions as illustrated in Figure 1.b.

The description in [1] does not specify the constant z. However our attack works independently of the particular value of z. x is the concatenation of x_s(64 bits), x_v(62 bits) and x_c(130 bits). x_s contains the alternating pattern. This results in the division of Figure 1.b. Area II contains the alternating pattern that 'buffers' changes in area III from area I. The dependence area for this configuration is formed by area I. Figure 1.c and 1.d are close ups of the upper left corner of Figure 1.b. Figure 1.c clearly shows area II and in Figure 1.d the difference pattern originating from two initial states that differ only in x_v is seen to be limited to area III. In conclusion: all x that have the same x_c and for which x_s contains the alternating pattern have the same output. This yields 2^{130} different sets, each containing 2^{62} inputs with the same output h. Finding these sets requires no computations whatsoever.

3 Collisions for the Hash Function Itself

The input to the hash function is a message m that has been padded in a one-to-one way to a multiple of 128 bits. Suppose $m = m_0 \parallel m_1 \ldots \parallel m_n$, with every m_i a 128 bit string. The hash value h is the result of applying the hash function $h_c()$ to m. We have $h = h_c(m)$. This function is defined in terms of the CF function g_c.

$$h^i = g_c(h^{i-1} \parallel m_{i-1}) \quad \text{for} \quad i = 1, \ldots, n$$

h^0 is equal to IV, a fixed initial value that is part of the specification of $h_c()$. The hashvalue h is given by h^n. Suppose we have an intermediate hash value h^i. h^{i+1} is then computed by $h^{i+1} = g_c(h^i, m_i)$. In figure 2.a it can be seen that changes in the first k bits of m_i (denoted by $\leftrightarrow$) influence only the $k + 1$ last bits

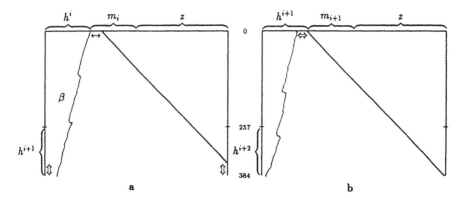

Figure 2: Space-time diagrams of two subsequent applications of $g_c()$. a: The first k bits of m_i influence only the last $k + 1$ bits of h^{i+1} b: h^{i+2} is independent of the $k + 1$ last bits of h^{i+1}

of h^{i+1} (denoted by $\Leftrightarrow$) if the border β is steep enough. This is the case for almost all initial configurations. Now suppose the last bit of h^{i+1} is not affected. This happens with probability $1/2$. In the next step h^{i+2} is computed from h^{i+1} and m_{i+1}. In figure 2.b it can be seen that, for small k there is a large probability that h^{i+2} will be independent of the k bits of h^{i+1} that were affected. From this it can be concluded that changes in bits that are situated at the beginning of the blocks m_i have a fair probability not to influence the hash value h.

4 Provably secure hashing vs. hashing in practice

Suppose $g()$ is a CF function for which it can be proven that finding a collision is equivalent to solving a hard problem. With Damgård's method a hash function $h()$ can now be constructed based on $g()$. Finding a collision for $h()$ is equivalent to solving the hard problem. However it cannot be proven that a certain problem is hard, i.e. that there is no efficient (e.g. polynomial) algorithm to solve it. Hence the security of $h()$ ultimately depends on the unproven assumption of the hardness of the problem. If the assumption is widely accepted, the hash function is widely considered to be secure. Complexity theory provides a fundament to a variety of widespread assumptions. Furthermore, there is emerging a vast literature treating hash functions in a complexity theoretic setting.

In practice, very fast hash functions are required. However, CF functions that are based on some widespread complexity assumption require a lot of calculation and are generally slow. Hence for the sake of speed practical hash function have been proposed that are based on functions for which finding collisions *seems* hard. These functions often incorporate elements that *remind of* hard problems. Damgård's[1] three CF functions are of this type. They are 'based' respectively on modular squaring, knapsack and cellular automata. The inventors of this type of functions assume that the use of the calculations that are related to some hard problem will provide the needed security. The fact that most of these functions have been broken, suggests that this is not a sound argument.

A more rational approach is the construction of a hash function using a FI function that is *designed to be collision free*. If the correct construction is used the Damgård proof applies. Observe that the hash function is not computationally collision free, but that finding a collision for the hash function is at least as hard as finding a collision for the underlying FI function. The distinction between these functions and the ones described in the previous paragraph lies in the design of the presumed CF functions. The ones meant in this paragraph are those designed using sound principles of diffusion and confusion. The ones in the previous paragraph are not really designed but merely 'based' on hard problems.

Another class of hash functions makes use of an FI function that is not especially designed to be collision free. The security of the hash function is based on the fact that collisions for the FI function are of no use in the construction of a collision for the hash function. Damgård's proof does not apply in this case.

The last two classes comprise most of the practical hash function designs. There are numerous designs where the FI function is realized by a block cipher. Because a block cipher is designed for another purpose, a dedicated function is likely to provide a more efficient solution. Two recent examples of dedicated hash functions are Snefru[6] and MD4[7]. The design of the underlying functions is based on considerations of diffusion and confusion.

To our knowledge all proposed hash functions can be described by the following general scheme:

- Segmentation: The input is padded in a prescribed way and subdivided in a number of blocks of equal length.

- Initialization: The initial value of the *intermediate hash result* is fixed by the specification.

- Iteration: For all blocks: the FI is applied to block i and the intermediate hash result. the output is the new intermediate hash result. Observe that each block is used only once.

- Result: The hash result is the output of the last iteration. Eventually a specified compression may take place, e.g. not all bits may be used.

In the following section we will present a framework that is even more general. It allows the design of hash functions that don't have certain restrictions imposed by this description. Because all proposed hash functions can be modeled to fit in our framework, it can also be considered as an alternative viewpoint. The framework is partly inspired by the design of MD4[7].

5 A Framework for the Direct Design of Hash Functions

The security of the hash function should be guaranteed by the diffusion and confusion of the messagebits in a hash result. In our framework this is realized in a way similar to the diffusion and confusion of the key in a symmetric block cipher, e.g. DES. The security of the overall scheme is based on executing a number of rounds each performing easily implementable functions. This is exactly what happens in the FI function in MD4. It is however different in that the number of rounds in our framework is not constant but depends on the length of the input, i.e. the message. In practical designs the number of rounds will be proportional to the length of the message, with a lower bound for short messages.

The scheme of the framework is given in Figure 2. It can be seen as an 'encryption' of the initial value IV that is a constant for a particular scheme. The number of rounds is $k(n)$. Each round consists of the application of a function $f_{j,n}$ on the intermediate hash result H^{j-1}. Each $f_{j,n}$ is 'keyed' by a string m_j, consisting of messagebits specified by a selection window $s_{j,n}$. The output after $k(n)$ rounds is the hash result. A concrete design has to define the padding, the function $k(n)$, the functions $f_{j,n}$ and the selection windows $s_{j,n}$ for all n and $1 \leq j < k(n)$.

In using this framework to design a hash function, a number of problems arise because of the variability of the number of rounds. It is for instance not clear how to devise a general selection schedule s for a variable length input. This problem *can* be tackled by subdividing the (possibly padded) message M in a number of equal blocks and designing the scheme for one such block. This comes down to designing an IF function that inputs one block and the intermediate hash result. This is what happens in *all* proposed hash functions thus far.

This is however not *the* solution, but just *one of the many*. Instead of being a design *principle*, this blocking is a design *option*. When blocking is applied and the designer wants to apply the Damgård proof, he has to construct the FI functions to be collision free. This restricts these FI functions in an unnecessary way. Seen within our framework, the advantage of having a FI function that is designed to be collision free is no longer self evident. What is relevant here is the contribution of the FI function to the purpose of the hash function related to its

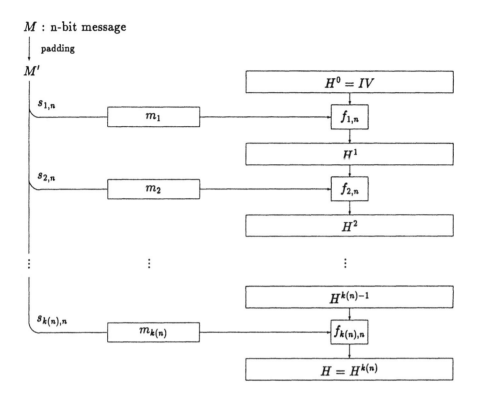

Figure 3: Schematic overview of the framework.

'computational cost', i.e. its effectiveness and efficiency.

A global approach can lead to more elegant solutions and a better use of the available computing resources, both in a hardware as in software oriented designs. Although we believe that the Damgård proof is of no value here, it keeps on being a valuable and necessary step in the construction of computationally collision free hash functions based on CF functions for which finding a collision is equivalent to solving some (generally assumed) hard problem.

It is impossible to give a complete list of design criteria for hash functions in our framework. However, in the rest of this section we will informally describe some general criteria for the components of the scheme. This will be illustrated by the design of our concrete proposal Cellhash. A more theoretical treatment is in preparation.

- Size: To prevent a birthday attack the hash result has to be at least 128 bits.

- Diffusion: The functions $f_{j,n}$ have to guarantee diffusion of information. With equal keys, two inputs H^{j-1} and $\tilde{H}^{j-1}$ that differ in only a *few* bits must give rise to two outputs H^j and $\tilde{H}^j$ differing in *substantially more* bits even in worst case. If the $f_{j,n}$ are invertible for a fixed key m_j it is guaranteed that no two inputs give the same output. The key-dependence of the functions has to maximize the diffusion of the messagebits into the hash result. Therefore it is required that each $f_{j,n}$, with the input fixed and seen as a function from key to output, is an injection for all possible inputs.

- Confusion: The hash result has to depend on the messagebits in an involved and complicated way. Therefore it is necessary that the $f_{j,n}$ provide the indispensable nonlinearity (in relation to all thinkable algebras). To promote confusion the selection schedule has to use each and every messagebit several times. Preferably a messagebit would be selected in a number of rounds that are *mutually distant in time*. The rounds that precede and succeed these would preferably use as many different messagebits as possible.

- Feasibility: The goal is a hash function actually working at *high speed*. A succinct and clear description is an advantage both in software, leading to a short program, and in hardware, possibly leading to a simple and compact device. Moreover, a long or complicated description enlarges the probability of implementation errors.

    - Software: Specific widespread instructions of processors can be advantageously used in the $f_{j,n}$. Restrictions on the selection schedule are imposed by the *relatively* slow memory access in contrast to the access of on-chip registers or cache.

- Hardware: For speed, the implementation should be on one chip, with the traffic onto and off the chip minimized. The $f_{j,n}$ must be decomposable in a limited number of primitives, which are actually implemented. By parallel computation optimal use can be made of available storage and time.

6 Cellhash: A Hardware oriented Hash Function

Cellhash is a hash function capable of very high speeds if implemented in hardware. In applications where speed is important and very large amounts of data have to be authenticated (e.g. electronic financial transactions, software integrity) hardware implementations are the natural solution. The core of Cellhash is formed by two cellular automata operations and a permutation. These guarantee the confusion and diffusion properties and allow for an efficient one chip hardware implementation by well established methods[11]. By adoption of Cellhash (or a similar algorithm) as a standard, costs would be reduced even further because of the high quantities of chips produced. This low cost opens up the possibility of the installation of these IC's in a standard way on computers.

6.1 The Global Scheme

The hash result of a message M of length n is computed in two phases :

- Preparation of the message
  The message is extended with the minimum number of 0's so that its length in bits is at least 248 and congruent to 24 modulo 32. The number of bits added is represented in a byte that is subsequently appended, most significant bit first. At this point the resulting message has a length that is an exact multiple of 32 bits. The resulting message can be written as $M_0 M_1 \ldots M_{N-1}$, i.e. the concatenation of N (32-bit) words M_i.

- Application of the hash function
  $F_c(H, A)$ is a function with argument H a bitstring of length 257 and A a bitstring of length 256. It returns a bitstring of length 257. IV is the all-zero bitstring of length 257.
  $H^0 = IV$
  $H^j = F_c(H^{j-1}, M_{j-1} M_{j \bmod N} \ldots M_{j+6 \bmod N}), j = 1 \ldots N$
  H^N is the hash result

Hence the number of rounds $k(n) = \max(8, \lceil \frac{n+8}{32} \rceil)$, All functions $f_{j,n}$ are the same, namely F_c, and each selection window s_j selects messagewords M_{j-1} to M_{j+6}. By reaching the end of the message the first 224 bits can be seen as repeated (e.g. if j is $N - 3$, s_j selects $M_{N-4}M_{N-3}M_{N-2}M_{N-1}M_0M_1M_2M_3$). In this way every bit of the message is used exactly 8 times in the hashing process.

6.2 Description of $F_c(H, A)$

The computation can be considered as a 5-step transformation of H. The calculations in each step are done simultaneously on all bits of H. Let $h_0h_1 \ldots h_{256}$ denote the bits of H and $a_0a_1 \ldots a_{255}$ the bits of A. All indices should be taken modulo 257, $\vee$ means OR and $\oplus$ means XOR.

Step 1 : $\quad h_i = h_i \oplus (h_{i+1} \vee \bar{h}_{i+2}), \quad 0 \leq i < 257$

Step 2 : $\quad h_0 = \bar{h}_0$

Step 3 : $\quad h_i = h_{i-3} \oplus h_i \oplus h_{i+3}, \quad 0 \leq i < 257$

Step 4 : $\quad h_i = h_i \oplus a_{i-1}, \quad 1 \leq i < 257$

Step 5 : $\quad h_i = h_{10*i}, \quad 0 \leq i < 257$

The five steps are further clarified in Figure 3. Step 1 is a nonlinear cellular automaton (CA) operation where each bitvalue h_i is updated according to the bitvalues in its neighborhood (in this step and step 3 periodic boundary conditions apply). The nonlinearity of the updating rule has to guarantee the needed confusion. This particular CA operation is invertible if the length of H is odd [9]. Step 2 consists merely of complementing 1 bit to eliminate circular symmetry in case bitstring A consists of only 0's. Step 3 is a linear CA operation that greatly increases the diffusion. This step is invertible if the length of H is no multiple of 9 [9]. In step 4 the actual messagebits are injected in H, to be diffused and confused during subsequent rounds. Step 5 is a bit permutation where bits are placed away from their previous neighbors. This step is needed to maintain a permanent diffusion during the hashing process. The length of H is 257 (a prime) to make step 1 and 3 invertible and to avoid circular symmetric patterns in H.

6.3 Properties

- **Size**

  The hash result is 257 bits long. If it is more convenient to use only 256 bits of the hash result, its first or last bit can easily be omitted without jeopardizing the security significantly.

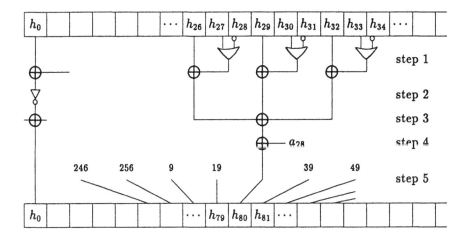

Figure 4: Schematic overview of the calculation of one output bit using the F_c function. It clearly shows that each output bit depends on 9 input bits and that nearby output bits depend on non-overlapping sets of input bits.

- **Diffusion**

  Each bit of H^i depends on 9 bits of H^{i-1}. This dependence is linear in 3 bits. Because of the bit permutation in step 5, the neighborhood is 'refreshed' each iteration. By this we mean that nearby bits of H^i depend on non-overlapping sets of bits of H^{i-1}. This property assures that *every* bit of an intermediate hash result H^i depends on *all* bits of H^{i-3}, the intermediate hash result three rounds before, in a nonlinear way. Moreover, F_c is invertible if the key A is regarded as a constant.

- **Confusion**

  This paragraph uses ideas of differential cryptanalysis as described in [8]. With the same m_i, a difference pattern in H^i (input XOR's) gives rise to a number Λ of possible difference patterns in H^{i+1} (output XOR's) *each appearing with the same probability* $1/\Lambda$. We have

  $$\Lambda = 2^{k+p}$$

  where k is the number of 1's and p is the number of 001 patterns in the input XOR. So only input XOR's with few 1's give rise to high probability characteristics. These phenomena are caused by the nonlinearity and the regular structure of the cellular automaton mechanism in step 1. Because of the diffusion caused by step 3 and 5, output XOR's of these characteristics consist of isolated 1's on a zero background. On the input of the next F_c iteration these will give rise to characteristics with high Λ. Hence simple difference patterns in H^i give rise to a vast amount of possible difference patterns, each with very small probability, only a few rounds later. Moreover every bit of the message is selected 8 times in the hashing process. Because of the diffusion mechanism in F_c described above, the influence of the first injection of a messagebit has already spread out over all bits of the intermediate hash result by the time of the last injection.

- **Hardware Feasibility**
  The description allows a straightforward chip implementation[11]. One round comprises one F_c calculation and the parallel input of 32 new messagebits and can be performed in one clockcycle. The total hashing process takes one round per 32 bits and 8 supplementary initial rounds for the installation of the first 256 messagebits. A very conservative guess of about 10 Mhz clockfrequency[11] gives a hashing speed of 0.3 Gbit/sec. for long messages.

- **Provable Security ?**
  The designers of Cellhash do not claim that finding a collision for Cellhash is equivalent to any hard problem other than finding a collision for Cellhash.

7 Conclusions

Collisions have been found for Damgård's CF function based on cellular automata. A general framework is presented that allows the direct design of hash functions. A concrete proposal is described named Cellhash. Unlike Snefru and MD4, Cellhash is designed for hardware implementation, making extremely high speed possible.

8 Acknowledgements

The authors would like to thank Ivan Damgård[10]. There has been an extensive revision of this paper in consequence of his comments.

References

[1] I. Damgård, Design Principles for Hash Functions, in *Advances in Cryptology: Proceedings of Crypto '89*, 416–427, Springer-Verlag,1990.

[2] P. Camion, J. Patarin, The Knapsack Hash Function proposed at Crypto '89 can be broken, *Advances in Cryptology–Eurocrypt '91*, Proceedings, Springer-Verlag, to appear.

[3] Bert Den Boer, *Internal Report RIPE*.

[4] S. Wolfram, Random Sequence Generation by Cellular Automata, *Advances in Applied Mathematics*, **7** (1986) 123–169.

[5] W. Meier, O. Staffelbach, Analysis of Pseudo Random Sequences Generated by Cellular Automata, *Advances in Cryptology–Eurocrypt '91*, Proceedings, Springer-Verlag, to appear.

[6] R. Merkle, A Fast Software One-Way Hash Function, *Journal of Cryptology*, **3** (1990) 43–58.

[7] R. Rivest, The MD4 Message Digest Algorithm, *Abstracts of Crypto '90*, 281–291.

[8] E. Biham, A. Shamir, Differential Cryptanalysis of DES-like Cryptosystems, *Abstracts of Crypto '90*, 1–32 .

[9] J. Daemen, R. Govaerts, J. Vandewalle, Properties of Partially Linear Cellular Automata, *Internal Report ESAT*.

[10] Ivan Damgard, *personal communication*.

[11] I. Verbauwhede and Z. Sahraoui of IMEC (Interuniversitair Micro Electronica Centrum), *personal communication*.

How to Construct
A Family of Strong One Way Permutations *

Babak Sadeghiyan

Yuliang Zheng

Josef Pieprzyk

Department of Computer Science,

University College,

University of New South Wales,

Australian Defence Force Academy,

Canberra, A.C.T. 2600, Australia.

Abstract

Much effort has been spent to identify the hard bits of one way functions, such as RSA and Rabin encryption functions. These efforts have been restricted to $O(\log n)$ hard bits. In this paper, we propose practical solutions for constructing a family of strong one way permutations such that when a member is chosen uniformly at random, with a high probability we get a one way permutation m, with $t < n - O(\log n)$, the maximum number of simultaneous hard bits. We propose two schemes. In the first scheme m is constructed with $O(\log n)$ fold iteration of $f \circ g$, where f is any one way permutation, $g \in_r G$ and G is a strongly $universal_2$ family of polynomials in Galois field. In the second scheme $m = f \circ g \circ h$, where h is a hiding permutation. We suggest a practical solution based on this scheme. The strong one way permutations can be applied as an efficient tool to build pseudorandom bit generators and universal one way hash functions.

1 Introduction

One way functions are those functions which are, roughly speaking easy to compute but hard to invert. One way functions are primitive tools for many cryptographic

---

*Support for this project was provided in part by TELECOM Australia under the contract number 7027 and by ARC Grant under reference number A48830241.

applications such as pseudorandom bit generation, hashing schemes and so on. In [7], it is proved that the existence of any one way function is a necessary and sufficient condition for the existence of pseudorandom bit generators. And in [9], it is shown that the existence of one way functions is also a necessary and sufficient condition for the existence of one way hash functions and secure signature schemes.

On the other hand, Blum and Micali [2] discovered the hard predicates b of one way functions f and applied it for the construction of pseudorandom bit generators (PBG). Such a $b(x)$, is hard to evaluate, when the value of $f(x)$ is given. They applied the hard core predicate b to extract some bits from each iteration of f. Some one way functions have the interesting feature of existence of *hard bits* in their argument. These bits cannot be computed by any family of probabilistic polynomial time algorithms with a probability of success better than by flipping a coin. In other words, hard bits are those bits of the argument of a one way function that distinguishing them from random bits is as difficult as inverting the one way function itself. If such a one way function is applied in the Blum-Micali pseudorandom bit generator scheme, a function which extracts the hard bit of the one way function could be used as a hard core predicate. The efficiency of Blum-Micali pseudorandom bit generator can be improved if a longer string of bits could be extracted with b.

In [14], a duality between pseudorandom bit generators and one way hash functions was revealed. Applying the revealed duality, a universal one way hash function scheme, which is the dual of the Blum-Micali pseudorandom bit generator is proposed. This scheme is based on a insertion function which substitutes a bit of the position of a hard bit with a bit of the string of the message. The efficiency of the proposed scheme also can be improved if the applied one way permutation has more hard bits. So if there was a one way permutation which had more hard bits, a more efficient pseudorandom bit generator and also a more efficient universal one way hash function could be constructed.

Much effort has been spent to identify hard bits of some specific number theoretic one way functions. In [1], it is shown that the $O(\log n)$ least siginficant bits of the RSA and Rabin encryption functions are individually hard, and that those $O(\log n)$ bits are also simultaneously hard. In addition, it is shown in [8] that the exponentiation function, i.e. $f(x) = g^x(\bmod P)$, where P is a prime and g is a generator of Z_P^*, has also $O(\log n)$ hard bits. Both these works take advantage of complicated techniques based on number theoretic approaches.

A breakthrough in this area is due to Goldreich and Levin [5] who have shown how to build a hard core predicate for all one way functions. They have extended the construction to show that $O(\log n)$ pseudorandom bits could be extracted from any one way function. Their result cannot be improved without imposing additional assumptions on the one way function [5], leaving the problem of constructing a function with $O(n)$ simultaneous hard bits open. However, if a one way function is proven to have a better security, then more number of pseudorandom bits could be extracted with the same method.

In [3], [6] a construction for a probabilistic encryption function has been presented, for which all the bits of the presented one way function are simultaneously hard. This construction is based on the composition of hard bits from many one way functions.

Another significant work in this area is due to Schrift and Shamir. They have shown in [11] that half of the input bits of exponentiation function modulo a Blum integer, i.e. $f(x) = g^x \pmod{N}$ where N is a Blum integer, are simultaneously hard and almost all bits are individually hard to evaluate. This work also take advantage of complicated techniques. As a result exponentiation modulo a Blum integer is yet the only natural function with $O(n)$ proven simultaneous hard bits.

In [10], a n-bit one way permutation such that each input bits is individually hard and any $t < n - O(\log n)$ input bits are simultaneously hard is called a strong one-way permutation [1].

In this paper, we show how to construct a family of strong one way permutations, such that when a member is chosen uniformly at random, a one way function that all its input bits are hard and any $t < n - O(\log n)$ input bits is indistinguishable from a random string is obtained, with a high probability. In contrast with [6], which composes hard bits of many one way functions, we compose the one way permutations to get a strong one way permutation. Two practical schemes has been proposed. Both schemes take avdantage of the family of polynomials in Galois field. The first scheme is based on the existence of any one way function and is constructed with $O(\frac{n}{\log n})$ fold composition of the one way function and a randomly chosen element of the family of polynomials in Galois field. The second scheme is based on the existence of a hiding one way permutation, and is constructed with a three layer structure applying a hiding one way permutation, a randomly chosen element of the family of polynomials in Galois field and any one way permutation.

Section 2 provides the notation used in the paper. Section 3 gives preliminary definitions and represents the concept of hard bits. In section 4, first the definition for strong one way permutations are given, then some properties of polynomials in Galois field have been investigated, and finally the proposed constructions have been studied.

2 Notations

The notation we use here is similar to [14]. The set of all integers is denoted by N. Let $\Sigma = \{0, 1\}$ be the alphabet we consider. For $n \in N$, Σ^n is the set of all binary strings of length n. The concatenation of two binary strings x, y is denoted by $x \parallel y$. The bit-by-bit exclusive-or of x, y is denoted by $x \oplus y$. The length of a string x is denoted by $| x |$. The number of elements in a set S is denoted by $\#S$.

---

[1]Note that the term of *strongly* one-way permutation has been used in [4], with a different meaning.

Let l be a monotone increasing function from N to N and f a function from D to R, where $D = \bigcup_n D_n$, $D_n \subseteq \Sigma^n$ and $R = \bigcup_n R_n$, $R_n \subseteq \Sigma^{l(n)}$. D is called the domain and R the range of f. Denote by f_n the restriction of f on Σ^n. f is a permutation if each f_n is a 1 to 1 and onto function. f is polynomial time computable if there is a polynomial time algorithm computing $f(x)$ for all $x \in D$. The composition of two functions f and g is defined as $f \circ g(x) = f(g(x))$. The i-fold composition of f is denoted by $f^{(i)}$.

A (probability) ensemble E, with length $l(n)$, is a family of probability distributions $[E_n \mid E_n : \Sigma^{l(n)} \to [0,1], n \subset N]$. The uniform ensemble U with length $l(n)$ is the family of uniform probability distributions U_n, where each U_n is defined as $U_n(x) = \frac{1}{2^{l(n)}}$, for all $x \in \Sigma^{l(n)}$. By $x \in_E \Sigma^{l(n)}$ we mean that x is randomly selected from $\Sigma^{l(n)}$ according to E_n, and in particular by $x \in_r S$ we mean that x is chosen from the set S uniformly at random.

3 Preliminaries

Definition 1 *A __statistical test__ is a probabilistic algorithm T that on an input x, where x is an n-bit string, halts in $O(n^t)$ and outputs a bit $0/1$, where t is some fixed positive integer.*

Definition 2 *Let l be a polynomial, and E^1 and E^2 be ensembles both with length $l(n)$. E^1 and E^2 are called __indistinguishable__ from each other, if for each statistical test T, for each polynomial Q, for all sufficiently large n,*

$$\mid Prob\{T(x_1) = 1\} - Prob\{T(x_2) = 1\} \mid < \frac{1}{Q(n)}$$

where $x_1 \in_{E^1} \Sigma^{l(n)}$, $x_2 \in_{E^2} \Sigma^{l(n)}$.

Definition 3 *Let $f : D \to R$, where $D = \bigcup_n \Sigma^n$ and $R = \bigcup_n \Sigma^{l(n)}$, be a polynomial time computable function. We say that f is __one-way__ if for each probabilistic polynomial time algorithm M, for each polynomial Q and for all sufficiently large n,*

$$Prob\{f_n(M(f_n(x))) = f_n(x)\} < \frac{1}{Q(n)}$$

where $x \in_U D_n$.

Note that the one-way property of a function is relative to a specific model of computation with a specific amount of computing resources. For the remaining of this paper, we assume that we have a computing resource for at most $2^{k(n)}$ operations, where $k(n) = O(\log n)$ bits. We also assume $k^+(n)$ is a function with a growth rate slightly more than $k(n)$ such that $n - k^+ > k$. As an example, consider $k^+(n) = O(\log^{1.01}(n))$.

If a function f is one-way then given $f(x)$ the argument x must be unpredictable. If every bit of the argument x were easily computable from $f(x)$, then f would not be a one-way function. Hard bits are some specific bits of the argument which are unpredictable and cannot also be guessed with a probability better than by flipping a coin [2], [14].

Definition 4 *Let $f : D \to R$ be a one-way function, where $R = \bigcup_n \Sigma^n$ and $D = \bigcup_n \Sigma^{l(n)}$. Let $i(n)$ be a function from N to N with $1 \leq i(n) \leq n$. If for each probabilistic polynomial time algorithm M, for each Q and for all sufficiently large n,*

$$Prob\{M(f_n(x)) = x'_{i(n)}\} < \frac{1}{2} + \frac{1}{Q(n)}$$

where $x \in_r \Sigma^n$ and $x'_{i(n)}$ is the $i(n)$-th bit of an $x' \in \Sigma^n$ satisfying $f(x) = f(x')$, then $i(n)$-th bit is a <u>hard bit</u> of f.

It can be shown that all the hard bits are independent of one another, and any string of $t \leq k$ hard bits is indistinguishable from a random string, when only $f(x)$ is given. Such strings of bits which are indistinguishable from random strings bits are called simultaneous hard bits.

Note that, there may exist some one way functions which do not have any hard bit. Note also that, any one-way function should have more than $k(n) = O(\log n)$ bits which are unpredictable, although they can be biased.

Blum and Micali discovered the hard core predicates of functions and applied it to construct pseudorandom bit generators (PBG) [2]. If a one way permutation f had $t = n - k^+$ (the maximum possible) known simultaneous hard bits, it could be used for the Blum-Micali pseudorandom bit generator scheme to get maximum number of bits per iteration of f. We call such a one way permutation a strong one way permutation or simply a strong permutation.

4 Strong One Way Permutations

In this section we construct a strong one-way permutation. First, we give a formal definition for strong one way permutations, which achieve the maximum number of simultaneous hard bits.

Definition 5 *Assume that $m : D \to D$ is a one-way permutation, where $D = \bigcup_n \Sigma^n$. Also assume that $I = \{i_1, i_2, \ldots, i_t\}$ where $i_1, i_2, \ldots, i_t$ are functions from N to N, with $1 \leq i_j(n) \leq n$ for each $1 \leq j < n - k$. Denote by E_n^I and E_n^R the probability distributions defined by the random variables $x_{i_t(n)} \cdots x_{i_2(n)} \, x_{i_1(n)} \parallel m(x)$ and $r_t \ldots r_2 \, r_1 \parallel m(x)$ respectively, where $x \in_r \Sigma^n$, $x_{i_j(n)}$ is the $i_j(n)$-th bit of x and $r_j \in_r \Sigma$. Let $E^I = \{E_n^I \mid n \in N\}$ and $E^R = \{E_n^R \mid n \in N\}$. We call m a <u>strong one way permutation</u> if E^I and E^R are indistinguishable from each other, for any I.*

In other words, given $m(x)$, any string of $t < n - k$ bits of x is indistinguishable from a random string.

In the proposed schemes for the construction of strong one way permutations, we take advantage of polynomials over Galois field $GF(2^n)$. In the following, some properties of these polynomials which are of our interest are presented. First, we consider the notion of strongly $universal_2$ hash functions, presented by Carter and Wegman in [13].

Definition 6 *Suppose G is a set of functions and each element of G being a function from A to B. G is strongly universal$_2$ if given any two distinct elements a_1, a_2 of A and any two elements b_1, b_2 of B, then $\frac{(\#G)}{(\#B)^2}$ functions take a_1 to b_1 and a_2 to b_2.*

In other words, the values of $g(x)$ and $g(y)$ are independently and uniformly distributed in B for every $x, y \in A$, when $g \in G$ is chosen uniformly at random. Strongly $universal_2$ sets of functions can be created using polynomials over finite fields. As the simplest example consider $G = \{g \mid g(x) = px + q; p, q \in GF(2^n)\}$ in the finite field $GF(2^n)$ [13].

Here, we give the definition of complete permutations, which is a design criteria for good encrypting functions.

Definition 7 *We say a permutation v is complete if each output bit depends on all input bits and vice-versa. In other words, the Boolean expression for each output bit contains all the input bits, and vice-versa.*

Since the operation in Galois field $GF(2^n)$ is done modulo an irreducible polynomial the resultant permutation is also complete, due to the properties of operations in Galois fields.

Example: We investigate the case for $GF(2^3)$. Here, x is a string of three bits, i.e. x_2, x_1, x_0, and is presented as the polynomial $x_2 z^2 + x_1 z + x_0$ and p is presented as $p_2 z^2 + p_1 z + p_0$ and q as $q_2 z^2 + q_1 z + q_0$. There are two irreducible polynomial in $GF(2^3)$, which are $z^3 = z + 1$ and $z^3 = z^2 + 1$.

For $z^3 = z + 1$:

$$
\begin{aligned}
g(x) = px + q = \quad & (q_2 + p_2 x_0 + p_1 x_1 + p_0 x_2 + p_2 x_2)z^2 \\
+ \quad & (q_1 + p_1 x_0 + p_0 x_1 + p_2 x_2 + p_2 x_1 + p_1 x_2)z \\
+ \quad & (q_0 + p_0 x_0 + p_2 x_1 + p_1 x_2)
\end{aligned}
$$

and for $z^3 = z^2 + 1$:

$$
\begin{aligned}
g(x) = px + q = \quad & (q_2 + p_2 x_0 + p_1 x_1 + p_0 x_2 + p_2 x_2 + p_2 x_1)z^2 \\
+ \quad & (q_1 + p_1 x_0 + p_0 x_1 + p_2 x_2)z \\
+ \quad & (q_0 + p_0 x_0 + p_2 x_1 + p_1 x_2 + p_2 x_2)
\end{aligned}
$$

Either irreducible polynomial has produced some common terms, which are functions of all the input bits, and some different terms in the coefficients of $g(x)$. So, it is clear that polynomials in $GF(2^n)$ result in a complete permutation. This would happen if the operation performed in any $GF(2^n)$. Notice that when p, q are chosen at random, for every x, y the outputs would be uniformly and independently distributed.

When we are operating in $GF(2)$ the multiplication is equivalent to 'AND' operation and addition is equivalent to 'XOR' operation. So each coefficient in $g(x)$ is the inner product of x with a different string obtained from p. Note that, if we present x and $g(x)$ as two vectors, they can be related to each other with a system of n linear equations. In the above example when the irreducible polynomial is $z^3 = z + 1$ the above equations can be represented as

$$g(x) = px + q = \left(\begin{bmatrix} p_0 + p_2 & p_1 & p_2 \\ p_1 + p_2 & p_0 + p_2 & p_1 \\ p_1 & p_2 & p_0 \end{bmatrix} \begin{bmatrix} x_2 \\ x_1 \\ x_0 \end{bmatrix} + \begin{bmatrix} q_2 \\ q_1 \\ q_0 \end{bmatrix} \right) \begin{bmatrix} z^2 & z^1 & z^0 \end{bmatrix}$$

If we had applied polynomials of higher degrees, such as $g(x) = \alpha x^2 + \beta x + \gamma$, a similar result would be obtained. For the remaining of the paper, we use the simplest case $g(x) = px + q$, although all the following lemmas and theorems are also true when g is of a higher degree. Notice that the above relation can be stated in a vector representation as

$$\underline{g}(\underline{x}) = \underline{p}\,\underline{x} + \underline{q}$$

Moreover, $\underline{p}$ and $\underline{q}$ can be modified in a way such that $\underline{g}$ and $\underline{x}$ become related to each other through a Toeplitz matrix, where a Toeplitz matrix is a matrix M such that $M_{i,j} = M_{i+1,j+1}$ for all i, j. For the above example, we may write

$$\underline{g} = \begin{bmatrix} p_0 + p_2 + r_2 & p_1 & p_2 \\ p_1 + p_2 + r_1 & p_0 + p_2 + r_2 & p_1 \\ p_1 + r_0 & p_1 + p_2 + r_1 & p_0 + p_2 + r_2 \end{bmatrix} \begin{bmatrix} x_2 \\ x_1 \\ x_0 \end{bmatrix} + \begin{bmatrix} q_2' \\ q_1' \\ q_0' \end{bmatrix}$$

where $\underline{r} = \begin{bmatrix} r_2 \\ r_1 \\ r_0 \end{bmatrix}$ is a randomly chosen vector, and $q_2' = q_2 + r_2 x_2$ and $q_1' = q_1 + r_1 x_2 + r_2 x_1$ and $q_0' = q_0 + r_0 x_2 + (p_1 + r_1)x_1 + (p_2 + r_2)x_0$.

Lemma 1 *If $g(x) = px + q$, where $p, q \in GF(2^n)$ are chosen randomly and are known, and $n - k^+$ bits of the $g(x)$ is known (or its k^+ bits are unknown), then the overall probability to guess each bit of x is equal to $\frac{1}{2^{k^+}}$.*

Proof : Assume that there is an algorithm L that given p, q and some bits of $g(x)$ lists all possible values of x. Since g is a permutation, if one bit of $g(x)$ is given, L would list all possible values of x which would be 2^{n-1} elements on average. In general, given i bits of $g(x)$, L lists 2^{n-i} possible values of x. If $n - k^+$ bits of $g(x)$ is given, L would list 2^{k^+} possible elements for x. One can guess the correct value of x with a probability of $\frac{1}{2^{k^+}}$. Since g is a complete permutation and p and q are chosen randomly, the probability of guessing any bit of x is equal to the probability of guessing the value of x. □

Note that, for some specific values of p, q some bits of x could be guessed efficiently, but when we consider the probability of guessing any bit of x over all values of p, q, it is equal to $\frac{1}{2^{k^+}}$.

Lemma 2 *Let $m : D \to D$ be a one-way permutation where $D = \bigcup_n \Sigma^n$ and $m = f \circ g$, where f is a one-way permutation, and $g = px + q$ where $p, q \in_r GF(2^n)$. Also assume that $i_1, i_2, \ldots, i_k$ are functions from N to N, with $1 \le i_j(n) \le n$ for each $1 \le j \le k$ and $k = O(\log n)$. Denote by E_n^1 and E_n^2 the probability distributions defined by the random variables $x_{i_k(n)} \ldots x_{i_2(n)} \, x_{i_1(n)} \parallel m(x)$ and $r_k \ldots r_2 \, r_1 \parallel m(x)$ respectively, where $x \in_r \Sigma^n$, $x_{i_j(n)}$ is the $i_j(n)$-th bit of x and $r_j \in_r \Sigma$. Let $E^1 = \{E_n^1 \mid n \in N\}$ and $E^2 = \{E_n^2 \mid n \in N\}$, then E^1 and E^2 are indistinguishable from each other.*

In other words, given $m(x)$, the probability of distinguishing any k-bit string of x from random strings is less than $\frac{1}{2^k}$, where $k = O(\log n)$ when the probability is calculated over all values of p, q.

Note that Lemma 2 virtually says that, *given $f(x')$ and $p', q' \in_r GF(2^n)$ it is hard to guess any $O(\log n)$ bits of x, where $f(x') = f \circ g(x) = m(x)$.* As $x = p'x' + q'$ is the inverse of $g(x) = x' = px + q$, $p, q \in_r GF(2^n)$. So x is actually the concatenation of the inner products of x' with n different strings obtained from p'.

Proof : Goldreich and Levin [5] showed that:

given $f(x')$ and p', where f is any one way function and $\mid p' \mid = \mid x' \mid$ and p' is an arbitrary string, the inner product of x' and p' is an hard-core predicate of f, and cannot be guessed with a probability better than $\frac{1}{2} + \frac{1}{Q(|x'|)} = \frac{1}{2} + \frac{1}{Q(n)}$ for each probabilistic polynomial time algorithm and for each Q.

They also extended their result and showed that $O(\log n)$ hard bits can be obtained from any one way function, where the simultaneous hard bits are the inner product of $O(\log n)$ different n-bit strings with x'. According to [5], the set of strings may also form a Toeplitz matrix, where a Toeplitz matrix is a matrix M such that for all i, j $M_{i,j} = M_{i+1,j+1}$. As mentioned earlier, the matrix which relates x to $g(x)$ can be rearranged into a Toeplitz matrix, so the same sort of proof that has been given in [5] could similarly be presented here to show that any k bit of x is indistinguishable from a random string when $m(x)$ is given.

As a simple and informal justification, assume that f is a one way permutation which acts on k^+ bits and keeps the other bits unchanged. So, given $m(x) = f \circ g(x)$, it is hard to guess k^+ bits of $g(x)$, but $n - k^+$ bits of $g(x)$ can be guessed efficiently. As it was suggested earlier, x and $g(x)$ are related to each other with a system of n equations with n variables. When $n - k^+$ bits of $g(x)$ are known, the system of equations would be reduced to a system of k^+ variables. However, if any k^+-bit string of x is given, the system of equations can be solved and the values of k^+ unknown bits of $g(x)$ would be revealed. Since there is no algorithm which can invert f with a probability better than $\frac{1}{2^{k^+}}$, any bit of x cannot be guessed with an overall probability better than $\frac{1}{2} + \frac{1}{2^{k^+}}$. Moreover, any probabilistic algorithm M that could distinguish any $t \leq k$ bit string of x from a random string with a probability better than $\frac{1}{2^{k^+}}$, would be able to invert f with an overall probability better than $\frac{2^t}{2^{k^+}}$ which contradicts our assumption that f is a one way permutation. $\qquad \square$

The result of Lemma 2 can be compared with the results of Vazirani and Vazirani [12], where it is shown that the XOR of any non-empty subset of hard bits is also hard to guess. Altogether, it can be concluded that all bits of x are individually hard, and any $k = O(\log n)$ bits of x are simultaneously hard bits of $f \circ g$ and cannot be distinguished from a random string with a probability of success better than $\frac{1}{2^k}$, when the probability is computed over all values of p, q.

4.1 A scheme for the construction of strong permutations

With Goldreich-Levin method, only $O(\log n)$ pseudorandom bits could be extracted from any one way function. This number of pseudorandom bits cannot be improved without additional assumptions. The reason is that a one way function which cannot be inverted with a probability better than $\frac{1}{Q(n)}$, may act only on $\log Q(n)$ of the bits of x and leave the rest unchanged [5]. In Lemma 2, we constructed a one way permutation $f \circ g$ such that any k input bits cannot be distinguished from a random string. If we apply $f \circ g$ as a one way permutation in the Blum-Micali pseudorandom bit generator, any k bits can be extracted per iteration of $f \circ g$. We take advantage of such a one way permutation to construct a family of strong permutations.

In the following, we suggest two schemes to obtain strong permutations and present the theorems behind them. The first scheme is based on the t fold composition of the $f \circ g$.

Theorem 1 *Let* $m : D \rightarrow D$ *be a one-way permutation where* $D = \bigcup_n \Sigma^n$ *and* $m = (f \circ g)^s = \underbrace{(f \circ g) \circ \ldots \circ (f \circ g)}_{s \text{ times}}$, *where* f *is a one-way permutation, and* $s = O(\frac{n}{\log n})$, *and* $g = px + q$ *where* $p, q \in_r GF(2^n)$. *Then* m *is a strong one way permutation.*

Proof: First, we show that $(f \circ g)^2 = f \circ g \circ f \circ g$ has $2k$ hard bits. Let's notate the first k bits string of x as $x_{\leftarrow 1}$, its second k bits as $x_{\leftarrow 2}$ and so on, and consider $y = f \circ g(x)$. Given $(f \circ g)^2(x)$, $x_{\leftarrow 2} \parallel y_{\leftarrow 1}$ is indistinguishable from a random string, since according to Blum-Micali pseudorandom bit generator [2] the concatenation of hard bits from each iteration would be indistinguishable from a random string, and according to Lemma 2 any k bit of $f \circ g$ is indistinguishable from a random string. As $x_{-2} \parallel y_{-1}$ forms a $2k$-bit string, then $x \oplus y$ has $2k$ bits which cannot be guessed efficiently. Since any k bit of $f \circ g$ is indistinguishable from a random string, any $2k$-bit string of $x \oplus y$ is indistinguishable from a random string. So, for each probabilistic polynomial time algorithm M

$$\text{Prob}\{M[(f \circ g)^2(x)] = x \oplus y\} < \frac{1}{2^{2k+}}$$

In the following, it is shown that the above relation implies that:

$$\text{Prob}\{M[(f \circ g)^2(x)] = x\} < \frac{1}{2^{2k+}}$$

To justify the above claim, by contradiction assume that there is an probabilistic algorithm M' that can compute x with a probability better than $\frac{1}{2^{2k}}$, we apply M' in another probabilistic algorithm M'' to compute the value of $x \oplus y$ with a probability better than $\frac{1}{2^{2k}}$. M'' first runs M' on $(f \circ g)^2(x)$ to get the value of x. As the description of f and g is known, then M'' runs $f \circ g$ on x to find the value of y. M'' gives $x \oplus y$ as its output. If the value of x is correct, the value of y would be correct with probability 1. Hence, M'' outputs the correct value for $x \oplus y$ with a probability better than $\frac{1}{2^{2k}}$. This contradicts our assumtion that $x \oplus y$ cannot be guessed with a probability better than $\frac{1}{2^{2k}}$. So, it can be concluded that for each probabilistic polynomial time algorithm M

$$\text{Prob}\{M[(f \circ g)^2(x)] = x\} < \frac{1}{2^{2k+}}$$

As the number of pseudorandom bits extracted from a one way function with Goldreich-Levin method is confined to the security parameter of the one way function, then $2k$ simultaneous hard bits could be extracted from $(f \circ g)^2$. This can be done by choosing a random $2k \times n$ Toeplitz matrix, and multiplying it by x. Notice that, since p, q are chosen randomly and independently, the matrix which relates $g(x)$ to x can be arranged in a Toeplitz matrix form. In addition, the matrix could be arranged in a way that for any determined $2k$ bits of x, the corresponding rows form a Toeplitz matrix. Hence, given $(f \circ g)^2(x)$, any $2k$ bit of x cannot be distinguished from a random string. This completes the proof that two fold iteration of $f \circ g$ produces a one way function such that every $2k$ input bits are indistinguishable from a random string.

By induction method and performing a proof similar as above, it can be shown that for $(f \circ g)^i(x)$, each input bit is individually a hard bit and any ik input bits are simultaneous hard bits. Therefore, to obtain a one way permutation with any $n-k$ input bits simultaneously hard, it is enough to construct an $\frac{n}{k}$ fold of $f \circ g$. With $k = O(\log n)$, a construction of $O\left(\frac{n}{\log n}\right)$ fold of $f \circ g$ is needed, which would be performed in polynomial time anyhow. □

4.2 A three-layer construction for strong permutations

In Lemma 2, we constructed a one way permutation $f \circ g$ such that any k input bits cannot be distinguished from a random string. If there existed another transformation (permutation) h, such that given any t bit string of its input, x, where $t < n-k$, it would be difficult to guess any k bits of its output, $h(x)$, then we could apply this function before g and get $f \circ g \circ h(x)$ as a one way permutation such that, when its output is given the overall probability of distinguishing any $t < n - k$ bit string of x from random strings would be less than $\frac{1}{2^{k^+}}$, as it will be proved in the next subsection. Since h would be able to hide any k bits of its output, we call it a hiding permutation.

First, we introduce a definition for hiding permutations.

Definition 8 *Assume that $h : D \rightarrow D$ is a permutation. Also assume that $i_1, \ldots, i_t$ and $j_1, \ldots, j_k$ are functions from N to N, where $1 \leq i_l(n), j_l(n) \leq n$ for each $1 \leq l \leq n$. We call h a hiding permutation if for each probabilistic polynomial time algorithm M, for each $t < n - k$ and for each polynomial Q and for all sufficiently large n,*

$$\mid Prob\{M(x_{i_t}, \ldots, x_{i_1} \parallel y_{j_n}, \ldots, y_{j_{k+}}) = y_{j_k}, \ldots, y_{j_1}\} - \frac{1}{2^k} \mid < \frac{1}{Q(n)}$$

where $x \in_r \Sigma^n$ and x_i denotes i-th bit of x, and y_j denotes j-th bit of $h(x)$ and $k = O(\log n)$.

The following theorem shows how to make a strong permutation from a hiding permutation.

Theorem 2 *If h is a hiding permutation and $g = px + q$, where $p, q \in_r GF(2^n)$, and f is any one way permutation, then $m = f \circ g \circ h$ is a strong one way permutation.*

Proof Sketch: To prove that m is a strong one way permutation we should show that any bit of x is hard to guess, and any $n - k^+$ bits of x are simultaneously hard, given only $m(x)$.

Assume that, by contradiction, that a probabilistic polynomial time algorithm M could guess x_i, given $m(x)$. Let $x' = h(x)$. Since any x_i is a function of some bits of x', and according to Lemma 2 any input bit of $f \circ g$ is individually a hard bit, then any algorithm which can guess x_i could a guess hard bit of $f \circ g$. This is contradictory to the assumption that f is a one way permutation, since computing any hard bit of $f \circ g$ is equivalent to reversing f. Hence, every x_i is a hard bit of m.

Moreover, having $t < n - k$ bits of x does not reveal any k bits of $h(x)$, since h is a hiding permutation. Then, having $t < n - k$ bits of x would not help in inverting m, given $m(x) = f \circ g \circ h(x)$. So, for each polynomial Q and for large enough n, any probabilistic polynomial time algorithm M cannot distinguish any string of $t < n - k$ bits of x from a random string with a probability better than $\frac{1}{Q(n)}$, when $m(x)$ is given. $\square$

A method for hiding x is based on the application of a one way permutation which acts on all bits, and serves as a hiding permutation due to the following lemma.

Lemma 3 *Any one way permutation h which is complete, is a hiding permutation.*

Proof : By contradiction assume that a one way permutation which is complete is not a hiding permutation. Then, there is a probabilistic polynomial time algorithm M that can obtain $y_{j_k}, \ldots, y_{j_1}$, given $x_{i_t}, \ldots, x_{i_1}$, with the available computing resource. On the other hand, since $n - t > k$ bits of x are not given, then the obtained k bits of the output do not depend on at least $n - t - k$ bits of the input. This is equivalent to say that h is formed of the two functions, namely h_1, h_2, i.e. $h(x) = h_1(x_{i_{t+k}}, \ldots, x_{i_1}) \parallel h_2(.)$. Obviously h_1 would not be a function of x, which contradicts our assumption that h is a complete permutation. □

Then, a concrete example for construction of a family of strong one way permutation, based on using a complete one way permutation as the hiding permutation, is $m(x) = f \circ g \circ h(x)$, where f is any one way permutation, $g = px + q$ where $p, q \in_r GF(2^n)$, and h is a complete one way permutation which acts on all its bits.

5 Conclusions

There are many functions that is considered to be one way, so if someone knows the value of $f(x)$, he can find the value of x for less than a fraction of $\frac{1}{Q(n)}$ of x's. This does not necessarily mean that any bit of x cannot be guessed efficiently. On the other hand, it is shown that $O(\log n)$ bits of RSA and Rabin encryption schemes are hard to guess [1]. Also, it was shown that $O(\log n)$ bits of the exponentiation function is also hard to guess [8]. This does not mean that the remaining bits are easy to guess and it simply means that we do not have yet any proof about remaining bits. As, recently it is shown that $\frac{n}{2}$ bits of exponentiation function are simultaneously hard to guess, when the operation is done modulo a Blum integer [11]. In this paper, we showed how to make a family of strong one way permutation, such that when a member is chosen uniformly at random, we get a one way permutation such that all its input bits are hard and any $t < n - k$ bit of input bits is indistinguishable from a random string, with a high probability. Two schemes for this purpose was suggested. The second scheme relies on the existence of a hiding permutation. An open problem is to show that a one way permutation is complete, or cannot be splitted into two parts. We also took advantage of the simplest family of polynomials in Galois field and showed that the family is also a family of complete permutations, where it had already been shown that it is also a family of strongly *universal*$_2$ functions. The proposed schemes for the construction of a family of strong one way permutations can be shown to work with families of polynomials of higher degrees in Galois field as well, where such polynomials form a family of strongly *universal*$_n$ functions. As it is shown [10], a strong one way permutation is an effective tool for the construction of efficient pseudorandom bit generators and universal one way hash functions.

ACKNOWLEDGMENT

We would like to thank Cathy Newberry and other members of CCSR for their help and assistance during the preparation of this work. The first author would also like to thank the organising committee of ASIACRYPT '91 for their generous support.

References

[1] W. Alexi, B. Chor, O. Goldreich, and C. P. Schnorr. RSA and Rabin functions: Certain parts are as hard as the whole. *SIAM Journal on Computing*, 17(2):194–209, 1988.

[2] M. Blum and S. Micali. How to generate cryptographically strong sequences of pseudo-random bits. *SIAM Journal on Computing*, 13(4):850–864, 1984.

[3] Manuel Blum and Shafi Goldwasser. An efficient probabilistic public-key encryption scheme which hides all partial information. In *Advances in Cryptology - CRYPTO '84*, volume 196 of *Lecture Notes in Computer Science*, pages 289–299. Springer-Verlag, 1985.

[4] O. Goldreich, H. Krawczyk, and M. Luby. On the existence of pseudorandom generators. In *Proceedings of the 29th IEEE Symposium on the Foundations of Computer Science*, pages 12–24, 1988.

[5] O. Goldreich and L. A. Levin. A hard-core predicate for all one-way functions. In *the 21st ACM Symposium on Theory of Computing*, pages 25–32, 1989.

[6] Shafi Goldwasser and Silvio Micali. Probabilistic encryption. *Journal of Computer and System Sciences*, 28:270–299, 1984.

[7] R. Impagliazzo, L. A. Levin, and M. Luby. Pseudo-random generation from one-way functions. In *the 21st ACM Symposium on Theory of Computing*, pages 12–24, 1989.

[8] Douglas L. Long and Avi Wigderson. The Discrete Logarithm Hides $O(\log n)$ Bits. *SIAM Journal on Computing*, 17(2):363–372, 1988.

[9] J. Rompel. One-way functions are necessary and sufficient for secure signatures. In *the 22nd ACM Symposium on Theory of Computing*, pages 387–394, 1990.

[10] B. Sadeghiyan and J. Pieprzyk. A construction for one way hash functions and pseudorandom bit generators. Technical Report CS 91/2, University College, The University of New South Wales, 1991. Also in the Abstracts of EUROCRYPT '91.

[11] A. Scherift and A. Shamir. Discrete logarithm is very discreet. In *Proceedings of the ACM Symposium on Theory of Computing*, pages 405–415, 1990.

[12] U. V. Vazirani and V. V. Vazirani. Efficient and Secure Pseudo-random Number Generation. In *Proceedings of the IEEE Symposium on Foundations of Computer Science*, pages 458–463, 1984.

[13] M. N. Wegman and J. L. Carter. New hash functions and their use in authentication and set equality. *Journal of Computer and System Sciences*, 22:265–279, 1981.

[14] Y. Zheng, T. Matsumoto, and H. Imai. Duality between Two Cryptographic Primitives. In *the 8-th International Conference on Applied Algebra, Algebraic Algorithms and Error Correcting Codes*, page 15, 1990.

On claw free families

Wakaha OGATA   Kaoru KUROSAWA

1 Introduction

Claw free families [1] are useful for collision free hash functions [2] and secure digital signatures [1]. However, only a few number theoretic examples are known.

This paper first shows that there are two types of claw free families and defines them as weak claw free families and strong claw free families. No attention has been paid for this important difference so far. Second, we show how to construct weak claw free families and strong claw free families from homomorphic one-way permutations, respectively. Some new examples are also presented.

2 Known claw free families[1][2][3]

Informally speaking, a set of one-way permutaions $\{f_0, f_1, \ldots, f_{r_1}\}$ is a claw free family if it is hard to find (x, y) such that

$$f_i(x) = f_j(y),$$

for any $i \neq j$. If the set is $\{f_0, f_1\}$, it is a claw free pair.

There are only a few known examples.

$< A >$ Let $n = pq$, where p, q are k-bit primes and equivalent to 3 mod 4. Define $QR(n)$ as

$$QR(n) = \{x \mid x = z^2 \bmod n \text{ for some } z\}.$$

The claw free family is

$$\mathcal{D} = QR(n),$$

$$f_i(x) = a_i x^2 \bmod n,$$

where $a_i \in QR(n)$ $(i = 0, 1, \ldots, r - 1)$

$< B >$ Let p be a k-bit prime and equivalent to 3 mod 4. Let g is a generator of Z_p^*.

$$\mathcal{D} = Z_p^*,$$

$$f_i(x) = a_i g^x \bmod p,$$

where $a_i \in Z_p^*$ $(i = 0, 1, \ldots, r - 1)$.

$< C >$ Let

$$n = \prod_{i=1}^{t} p_i$$

where $p_i (i = 1, 2, \ldots, t)$ are primes such that

$$p_i = 3 \bmod 4$$

(n is a Blum number).
   Define $J(a)$ as

$$J(a) = \left(\left(\frac{a}{p_1} \right), \left(\frac{a}{p_2} \right), \ldots, \left(\frac{a}{p_t} \right) \right).$$

The claw free family is

$$\mathcal{D} = QR(n),$$

$$f_i(x) = (a_i x)^2 \bmod n,$$

where $a_i \in Z_p^*$ $(i = 0, 1, \ldots, r - 1)$, $J(a_i) \neq J(a_j)$ $(i \neq j)$.

   Example $<$A$>$ and example $<$C$>$ are based on the difficulty of factorization. Example $<$B$>$ is based on the difficulty of the discrete log problem.

3 Weak claw free and strong claw free

3.1 What is the difference?

The example $<$A$>$ and the example $<$C$>$ in section 2 are both factoring type. However, they have the following difference.
   In example $<$A$>$, the claw is

$$\begin{aligned} f_i(x) &= a_i x^2 \bmod n \\ &= a_j y^2 \bmod n = f_j(y). \end{aligned}$$

It is hard to find a claw pair, (x, y), on input (n, a_i, a_j). However, on input n, it is easy to find $((a_0, a_1), (x, y))$ which makes a claw. (First, choose x, y and a_0 randomly. Next, let $a_1 = a_0 x^2 / y^2 \bmod n$). This situation is no good for some applications. We call such a claw free family "a weak claw free".

   In example $<$C$>$, it is hard to find a claw pair, (x, y), on input (n, a_i, a_j). On input n, it is also hard to find $((a_0, a_1), (x, y))$, which makes a claw. This claw free family is stronger than example $<$A$>$. We call such a claw free family "a strong claw free".

3.2 Definitions

We give the formal definitions of weak claw free family and strong claw free family.

Definition 1 *Let G be a probabilistic polynomial time algorithm. We say that G is a one-way function generator if there is a polynomial p such that*

- *G's input is 1^k and its output is an ordered set (D, F) of algorithms.*

- *Algorithm D always halts within $p(k)$ steps and defines a uniform probability distribution over the finite set $\mathcal{D} = [D()]$. (That is, running D with no inputs uniformly selects an element from $\mathcal{D}$.)*

- *F halts within $p(k)$ steps on any input $x \in \mathcal{D}$.*

- *For all (inverting) probabilistic polynomial time algorithms $I(\cdot, \cdot, \cdot, \cdot)$, for all c and sufficiently large k:*

$$Prob[x \in F^{-1}(F(z)) \mid (D, F) \leftarrow G(1^k); z \leftarrow D(); x \leftarrow I(1^k, D, F, F(z))] \leq k^{-c}.$$

Definition 2 *Let G be a one-way function generator. We say that G is a one-way permutation generator if the following conditions are satisfied.*

- *G outputs (D, F).*

- *$|D()|$ is constant.*

- *F is a permutation in $\mathcal{D} = [D()]$.*

Now we give formal definitions of weak claw free and strong claw free.

Definition 3 *Let G be a one-way function generator, and H be a probabilistic polynomial time algorithm. We say H is a weak claw free generator on G, if the following condition is satisfied.*

- *H's input is (D, F) generated by G.*

- *H outputs (d, f_0, f_1).*

- *d is a poly-time algorithm which outputs an element of $\mathcal{D}$ randomly and uniformly.*

- *f_0 and f_1 are permutations.*

- For all (weak claw making) probabilistic polynomial time algorithms $A_0(\cdot, \cdot, \cdot)$, for all c and sufficiently large k:

$$Prob[f_0(x) = f_1(y) \mid (D, F) \leftarrow G(1^k); (d, f_0, f_1) \leftarrow H(D, F);$$
$$(x, y) \leftarrow A_0(d, f_0, f_1)] \leq k^{-c}.$$

Definition 4 *Let G be a one-way function generator, and H be a probabilistic time algorithm. We say H is a strong claw free generator on G, if the following condition is satisfied.*

- *H's input is (D, F) generated by G.*

- *H outputs (d, f_0, f_1).*

- *d is a poly-time algorithm which outputs an element of $\mathcal{D}$ randomly and uniformly.*

- *f_0 and f_1 are permutations.*

- *For all (strong claw making) probabilistic polynomial time algorithms $A_1(\cdot, \cdot, \cdot)$, for all c and sufficiently large k:*

$$Prob[f_0(x) = f_1(y) \mid (D, F) \leftarrow G(1^k); ((d, f_0, f_1), (x, y)) \leftarrow A_1(D, F)] \leq k^{-c}.$$

In this section, we only give definitions of claw free generator which generates a pair of permutations. We can define them which generates $r \geq 2$ permutations similarly.

We also give a definition of self claw free function which is used later.

Definition 5 *Let G be a one-way function generator. We say G is a self claw free function generator, if for all (self claw making) probabilistic polynomial time algorithms $A(\cdot, \cdot)$, for all c and sufficiently large k :*

$$Prob[F(x) = F(y); \ x \neq y \mid (D, F) \leftarrow G(1^k); (x, y) \leftarrow A(D, F)] \leq k^{-c}.$$

4 Sufficient conditions for claw free families

4.1 Weak claw free pair

Theorem 1 *Suppose that a one-way permutation generator G exists such that*

- $(D, F) \leftarrow G(1^k).$

- $F(x \oplus y) = F(x) \odot F(y)$ $(x, y \in \mathcal{D} = [D()])$ *where $\odot$ and $\oplus$ are commutative operators.*

Define H as follows :

- *H's input is (D, F) generated by G.*

- *Choose randomly $a_0, a_1 \leftarrow D()$.*

- *Construct two permutations as*

$$f_0(x) = a_0 \odot F(x)$$

$$f_1(x) = a_1 \odot F(x)$$

- *Define $d = D$.*

- *Output (d, f_0, f_1).*

Then, H is a weak claw free generator on G.

proof : First, we show an informal sketch of the proof. Suppose that we can find a pair (x, y) such that $f_0(x) = f_1(y)$. Since $f_i(x) = a_i \odot F(x)$,

$$a_0 \odot F(x) = a_1 \odot F(y)$$
$$a_0 \odot^{-1} a_1 = F(x) \odot^{-1} F(y).$$

Let $F(z) = F(x) \odot^{-1} F(y)$. Then

$$F(x) = F(z) \odot F(y)$$
$$= F(z \oplus y)$$
$$x = z \oplus y$$
$$z = x \oplus^{-1} y.$$

So, $a_0 \odot^{-1} a_1 = F(x \oplus^{-1} y)$. Thus, we can compute the preimage of $a_0 \odot^{-1} a_1$.

We now show a formal proof. Suppose that there exists a (weak claw making) probabilistic polynomial time algorithm $A(\cdot, \cdot, \cdot)$ and a constant c such that for infinitely large k :

$$Prob[f_0(x) = f_1(y) \mid (D, F) \leftarrow G(1^k); (d, f_0, f_1) \leftarrow H(D, F);$$
$$(x, y) \leftarrow A(d, f_0, f_1)] \geq k^{-c}.$$

Then, we can obtain a (inverting G) probabilistic polynomial-time algorithm I as follows.

- Input of I is $(1^k, D, F, F(z))$.

- Choose $a_0 \in \mathcal{D} = [D()]$ at random and let $a_1 = F(z) \odot a_0$.

- Run A on input $(d, a_0 \odot F, a_1 \odot F)$ and get (x, y).

- Let $\tilde{z} = x \oplus^{-1} y$.

- output $\tilde{z}$.

Note that $F(\tilde{z}) = F(z)$. Then, I is a probabilistic polynomial time algorithm and for infinitely large k :

$$Prob[F(\tilde{z}) = F(z) \mid (D, F) \leftarrow G(1^k); z \leftarrow D(); \tilde{z} \leftarrow I(1^k, D, F, F(z))] \geq k^{-c}.$$

This is a contradiction. ¶

We can make a generator which outputs $r = O(k)$ permutations, and can prove similarly.

<A> and <B> in section 2 are examples of theorem 1.

4.2   Strong claw free families

Theorem 2 *Suppose that there exists a one-way function generator G as follows.*

C0) $(D, F) \leftarrow G(1^k), \mathcal{D} = [D()]$.

C1) G *is a self claw free function generator.*

C2) F *has a homomorphic property, such that $F(x \oplus y) = F(x) \odot f(y)$.*

C3) $F(\mathcal{D}) = \{y \mid y = F(x); x \in \mathcal{D}\}$ *is a subgroup of $\mathcal{D}$. Let $a_0, a_1, \ldots$ be the coset leaders.*

C4) $F(F(\mathcal{D})) = F(\mathcal{D})$.

Define H as follows :

- *H's input is (D, F) generated by G.*

- *Choose coset leaders $a_0, \ldots, a_{r-1}$.*

- *Construct r permutations as*

$$f_i(x) = F(a_0 \odot x) \quad (i = 0, \ldots, r - 1).$$

- *Define $d() = F(D())$.*

- *Output $(d, f_0, \ldots, f_{r-1})$.*

Then, H is a strong claw free generator on G.

proof:

1. Clearly, f_i's are permutations.

2. Suppose that there exists a (strong claw making) probabilistic polynomial-time algorithm A and a constant c such that for sufficient large k :

$$Prob[f_i(x) = f_j(y); \; i \neq j \mid (D, F) \leftarrow G(1^k);$$
$$((d, f_0, \ldots, f_{r-1}), (x, y)) \leftarrow A(D, F)] \; \geq \; k^{-c}.$$

Since $f_i(x) = F(a_i \odot x)$ $(i = 0, \ldots, r-1)$, we get

$$F(a_i \odot x) = F(a_j \odot y) \quad (a_i \odot x \neq a_j \odot y).$$

So, we can obtain a (self claw making) probabilistic polynomial time algorithm A' as follows.

- Input of A' is (D, F).

- Run A on input (D, F) and get $((d, f_0, \ldots, f_{r-1}), (x, y))$.

- Compute $f_l(x), f_l(y)$ $(l = 0, \ldots, r-1)$ and compare them. If $f_i(x) = f_j$ then let $X = a_i \odot x$, $Y = a_j \odot y$.

- Output (X, Y).

The probability A' succeeds is

$$Prob[F(X) = F(Y); X \neq Y \mid (D, F) \leftarrow G(1^k); (X, Y) \leftarrow A'(D, F)] \geq k^{-c}.$$

This is a contradiction. ¶

$<C>$ is an example of theorem 2. We will give another example in section 5.

5 New claw free families

5.1 Weak claw free pair

We give a new weak claw free pair based on the difficulty of the certified discrete log problem. This claw free pair is not included in theorem 1. This fact shows that theorem 1 is a sufficient condition, but not a necessary condition.

< Proposed-1 > Let G be a one-way function generator which output (D, F') :

$$\mathcal{D} = [D()] = Z_p^*,$$

$$F' = F \circ C_p,$$

$$F(x) = g^x \bmod p,$$

where p a prime number, g is a generator on $(\bmod\ p)$ and C_p is a factorization of $p - 1$. Define a generator H as follows :

- H's input is (D, F).

- Define d be an algorithm which chooses $x \in \mathcal{D} = Z_{p-1}^*$.

- Let $g_0 = g$ and choose generators g_1 randomly. (Using C_p, H can check whether g_0 is a generator.)

- Construct two permutations as follows :

    $$f_0(x) = g_0^x \bmod p$$

    $$f_1(x) = g_1^x \bmod p$$

- Output (d, f_0, f_1).

Theorem 3 *The above H is a weak claw free generator on G, if the certified discrete log problem is hard.*

Sketch of proof : Suppose that we can find a pair (x, y) that satisfies $f_0(x) = f_1(y)$. Since $f_i(x) = g_i^x \bmod p$,

$$g_0^x = g_1^y \bmod p,$$
$$g_0^{\left(\frac{x}{y}\right)} = g_1 \bmod p.$$

Thus, we can compute a discrete log of g_0 when generator is g_1. This is a contradiction.

proof : We assume that discrete log problem is difficult, that is, for all probabilistic polynomial time algorithms I and for all c and sufficiently large k :

$$Prob[x = g^y \bmod p \mid p \text{ is prime}; |p| = k; g \text{ is a generator };$$
$$x \leftarrow Z_p^*; y \leftarrow I(p, g, x)] \leq k^{-c}.$$

We assume that H is not a weak claw free generator. Then there is a weak claw making probabilistic polynomial time algorithms $A(\cdot, \cdot, \cdot)$, such that for any c and sufficiently large k:

$$Prob[f_0(x) = f_1(y); x, y \in Z_{p-1}^* \mid (D, F) \leftarrow G(1^k);$$
$$(d, f_0, f_1) \leftarrow H(D, F); (x, y) \leftarrow A(d, f_0, f_1)] \geq k^{-c}. \tag{1}$$

Now, using A, we construct a probabilistic polynomial time algorithm I which outputs a discrete log, y on input (p, g, x).

- I's input is (p, g, x).

- Let $\mathcal{D} = Z_{p-1}^*$.

- Construct two permutations as

$$f_0(z) = g^z \bmod p,$$
$$f_1(z) = x^z \bmod p.$$

- Run A on input (d, f_0, f_1) and get (x', y').

- Let $s = x'/y'$.

- output s.

From (1), we get

$$Prob[g^s = x \bmod p \mid s \leftarrow I(p, g, x)]$$
$$\geq Prob[g^{x'} = x^{y'} \bmod p \mid (D, F) \leftarrow G(1^k); (d, f_0, f_1) \leftarrow H(D, F);$$
$$(x', y') \leftarrow A(d, f_0, f_1)] \times Prob[x \text{ is a generator}]$$
$$\geq k^{-c} \times Prob[x \text{ is a generator}]$$

Since

$$Prob[x \text{ is a generator}] = \frac{\varphi(p-1)}{p-1}$$
$$= \frac{p'-1}{2p'} \approx 1/2,$$

there exists a c' such that

$$Prob[g' = x \bmod p \mid s \leftarrow I(p,g,x)] \geq k^{-c'}.$$

This is a contradiction. ¶

5.2 Strong claw free family

We give a new example of theorem 2. n is a Blum number in $<C>$ whereas N does not need to be a Blum number in $<$ Proposed-2 $>$,

$<$ Proposed-2 $>$ Define $B(N, s)$ as follows :

$$B(N, s) = \{x \mid x = y^s \bmod N \text{ for some } y\}$$

Let B_i be a coset of $B(N, s)$ and a_i be a coset leader of B_i. Let G be a one-way function generator which outputs (D, F) :

$$\mathcal{D} = [D()] = B(N, s),$$

$$F(x) = x^s \bmod N,$$

where $N = pq$, p and q are primes of k-bits, and s is an odd number, and

$$gcd(p - 1, s) = gcd(p - 1, s^2) = e_1, \tag{2}$$
$$gcd(q - 1, s) = gcd(q - 1, s^2) = e_2, \tag{3}$$
$$gcd(e_1, e_2) = 1. \tag{4}$$

Define a generator H as follows :

- H's input is (D, F).

- Choose coset leaders, $a_0, \ldots, a_{r-1}$.

- Construct r permutations as follows :

$$f_i(x) = (a_i x)^s \bmod N \quad (i = 0, \ldots, r - 1)$$

- Define $d() = F(D())$.

- Output $(d, f_0, \ldots, f_{r-1})$.

Theorem 4 *The above H is a strong claw free generator on G, if G is a one-way function generator, that is, the factorization is hard.*

Before the proof, we give some lemmas which are used in the proof. In the following, we assume that p, q, s, e_1, e_2 satisfy the conditions (2), (3), (4).

Lemma 1

$$\gcd(p-1, e_2) = 1$$
$$\gcd(q-1, e_1) = 1$$

Proof : From eq.(2), (3) and (4) we get

$$s = le_1 e_2, \tag{5}$$
$$gcd(p-1, l) = 1, \tag{6}$$
$$gcd(q-1, l) = 1, \tag{7}$$

where l is an integer. Suppose that

$$gcd(p-1, e_2) = d\ (> 1). \tag{8}$$

Eq.(8) means that $p-1$ and $s(= le_1 e_2)$ must be divided by d. Then, from eq.(2), $d|e_1$. On the other hand, from eq. (8), $d|e_2$. This is against eq.(4). The other part is proven in the same way. $\qquad$ (Lemma 1) ¶

Lemma 2 *G is a self claw free function generator, if the factorization is hard.*

Proof :   Suppose that we find a self claw pair (a, b). From eq.(5), $a^{le_1 e_2} = b^{le_1 e_2}$ (mod N). From eq.(6) and (7), we obtain $a^{e_1 e_2} = b^{e_1 e_2}$ (mod N). Let

$$a'' = a^{e_2} \bmod N,$$
$$b'' = b^{e_2} \bmod N.$$

Then, $a''^{e_1} = b''^{e_1}$ (mod N). From lemma 1, we get $a'' = b''$ (mod q).
First, we assume $a'' \neq b''$ (mod p). Then, $\gcd(a'' - b'', N) = q$.
Next, we assume $a'' = b''$ (mod p). Then, $a^{e_2} = b^{e_2} \bmod N$. From lemma 1, we get $a = b$ (mod p). Since $a \neq b$ (mod N), $a \neq b$ (mod q). Then, $\gcd(a - b, N) = p$. Therefore we can factor N. $\qquad$ (Lemma 2) ¶

Lemma 3 *Define functions* F, F' *as*

$$F(x) = x^s \bmod N,$$
$$F'(x) = x^{s^2} \bmod N.$$

Then, F *and* F' *are* $e_1 e_2 : 1$ *if the domain is* Z_N^*.

Proof : In $GF(p)$, the number of s-th roots of unity is $\gcd(p-1, s)$, where p is a prime [3]. Then, equation : $x^s = 1 \pmod{p}$ has e_1 solutions, $w_1, w_2, \ldots, w_{e_1}$, since eq.(2). For $x_0 \in Z_p^*$, equation : $x^s = x_0^s \pmod{p}$ has e_1 solutions, because

$$\frac{x^s}{x_0^s} = 1 \pmod{p},$$

$$\left(\frac{x}{x_0}\right)^s = 1 \pmod{p},$$

$$\frac{x}{x_0} = w_i \pmod{p} \ (i = 1, 2, \ldots, e_1),$$

$$x = x_0 w_i \pmod{p} \ (i = 1, 2, \ldots, e_1).$$

Similarly, equation : $x^s = x_0^s \pmod{q}$ has e_2 solutions. So, equation : $x^s = x_0^s \pmod{N}$ has $e_1 e_2$ solutions. Namely, F is an $e_1 e_2 : 1$ function. Similarly, F' is $e_1 e_2 : 1$. (Lemma 3) ¶

Lemma 4 F *is a permutation if the domain is* $B(N, s)$.

Proof : For any $x \in B(N, s)$, it is clear that $F(x) \in B(N, s)$. Suppose that there exists a pair (x_1, x_2) such that $x_1, x_2 \in B(N, s)$ and $f(x_1) = f(x_2)$ $(x_1 \neq x_2)$. Since F is $e_1 e_2 : 1$ (lemma 3) and $x_1, x_2 \in B(N, s)$, there exist $a_1, a_2, \ldots, a_{e_1 e_2}, b_1, b_2, \ldots, b_{e_1 e_2} \in Z_N^*$ such that

$$x_1 = a_1^s = a_2^s = \cdots = a_{e_1 e_2}^s \qquad (\bmod\ N),$$
$$x_2 = b_1^s = b_2^s = \cdots = b_{e_1 e_2}^s \qquad (\bmod\ N).$$

Because $F(x_1) = F(x_2)$,

$$a_1^{s^2} = a_2^{s^2} = \cdots = a_{e_1 e_2}^{s^2} = b_1^{s^2} = \cdots = b_{e_1 e_2}^{s^2} \pmod{N}.$$

Namely, F' has $2 e_1 e_2$ solutions in Z_N^*. It contradicts to lemma 3. Therefore there is no such pair (x_1, x_2). (Lemma 4) ¶

Proof of theorem 4 : We show that the conditions of theorem 2 are satisfied.

0) G's output is (D, F), where

$$\mathcal{D} = [D()] = Z_N^*,$$

$$F(x) = x^s \bmod N.$$

1) From lemma 2, G is a self claw free function generator.

2) F has a homomorphic property, because

$$
\begin{aligned}
F(xy) &= (xy)^s \bmod N \\
&= x^s y^s \bmod N \\
&= F(x)F(y) \bmod N.
\end{aligned}
$$

3) $F(Z_N^*) = B(N, s)$. $B(N, s)$ is a subgroup of Z_N^*.

4) From lemma 4, $F(F(Z_N^*)) = B(N, s) = F(Z_N^*)$.

From above 0) $\sim$ 4) and theorem 3, G is a strong claw free generator. ¶

6   Conclusion

In this paper, first, we have defined the weak claw free family and strong claw free family. Next we have given sufficient conditions for the existence of weak and strong claw free families.

References

[1] Goldwasser, Micali and Rivest: *A "paradoxical" solution to the signature problem*, Proc. of 25th FOCS, 1984, pp.441-448

[2] Damgard: *Collision free hash functions and public key signature schemes*, Eurocrypt '87

[3] Goldwasser, Micali and Rivest: *A digital signature scheme secure against adaptive chosen massage attacks*, SIAM J. COMPUTE. Vol.17, No.2, 1988

[4] Koblitz: *A course in number theory and cryptography*, Springer-Verlag, 1987

Sibling Intractable Function Families and Their Applications [1]

(Extended Abstract)

YULIANG ZHENG, THOMAS HARDJONO AND JOSEF PIEPRZYK

Centre for Computer Security Research
Department of Computer Science
University of Wollongong
Wollongong, NSW 2500, Australia
E-mail: yuliang@cs.uow.edu.au

Abstract

This paper presents a new concept in cryptography called the *sibling intractable function family* (SIFF) which has the property that given a set of initial strings colliding with one another, it is computationally infeasible to find another string that would collide with the initial strings. The various concepts behind SIFF are presented together with a construction of SIFF from any one-way function. Applications of SIFF to many practical problems are also discussed. These include *the hierarchical access control problem* which is a long-standing open problem induced by a paper of Akl and Taylor about ten years ago, *the shared mail box problem*, *access control in distributed systems* and *the multiple message authentication problem*.

1 Introduction

This paper presents a new concept in cryptography called the *sibling intractable function family* (SIFF). SIFF is a generalization of the concept of the universal one-way hash function family introduced in [13], and it has the property that given a *set* of initial strings colliding with one another, it is computationally infeasible to find another string that would collide with the initial strings. We also present a simple method for transforming any universal one-way hash function family into a SIFF. As Rompel has proved that universal one-way hash function family can be constructed from any one-way function, we obtain the theoretically optimal result that SIFF also can be constructed from any one-way function.

SIFF has many nice features, and can be applied to a number of cryptographic problems. We will describe in detail a solution to the hierarchical access control problem, which includes a way to generate and update keys for a hierarchical organization. In a hierarchical organization it is assumed that authority is distributed in a hierarchical manner, where higher level members of the organization have access to resources and

---

[1]Supported in part by Telecom Australia under the contract number 7027 and by the Australian Research Council under the reference number A48830241.

data classified at a lower level. In this way we solve, under the weakest assumption of the existence of any one-way function, a long-standing open problem induced by a paper of Akl and Taylor about ten years ago [1]. Applications of SIFF to other three problems, namely the shared mail box problem, access control in distributed systems and the multiple message authentication problem, will also be discussed.

The rest of the paper is organized as follows: In Section 2, we give basic definitions for one-way functions, pseudo-random function families and universal hash function families. In the same section we also introduce the new notion of SIFF. In Section 3, we show a method for transforming any universal one-way hash function family into a SIFF. As a corollary, we obtain the theoretically optimal result that SIFF can be constructed from any one-way function. In Section 4, we give a formal definition for the security of a key generation scheme for hierarchical organizations, and present a solution to the hierarchical access control problem by the use of SIFF and pseudo-random function families. We describe in detail the following three aspects of the solution: key generation, key updating and proof of security. In Section 5, we suggest three other applications of SIFF to show its usefulness. The first application is to the shared mail box problem, the second to the access control in distributed systems and the third to the multiple message authentication problem. Section 6 closes the paper with a summary of the results and a suggestion for further research.

2 Basic Definitions

In this section we introduce the definitions for one-way functions, pseudo-random function families, universal hash function families and sibling intractable function families.

2.1 Pseudo-random Function Families

Denote by $\mathcal{N}$ the set of all positive integers, n the security parameter, Σ the alphabet $\{0,1\}$ and $\#S$ the number of elements in a set S. By $x \in_R S$ we mean that x is chosen randomly and uniformly from the set S. The composition of two functions f and g is defined as $f \circ g(x) = f(g(x))$. Throughout the paper ℓ and m will be used to denote polynomials from $\mathcal{N}$ to $\mathcal{N}$. First we give our formal definition of one-way functions.

Definition 1 *Let* $f : D \to R$ *be a polynomial time computable function, where* $D = \bigcup_n \Sigma^{\ell(n)}$ *and* $R = \bigcup_n \Sigma^{m(n)}$. f *is a one-way* function *if for each probabilistic polynomial time algorithm* M, *for each polynomial* Q *and for all sufficiently large* n, $\Pr\{f_n(x) = f_n(M(f_n(x)))\} < 1/Q(n)$, *where* $x \in_R D_n$ *and* f_n *denotes the restriction of* f *on* $\Sigma^{\ell(n)}$.

Let $F = \{F_n | n \in \mathcal{N}\}$ be an infinite family of functions, where $F_n = \{f | f : \Sigma^{\ell(n)} \to \Sigma^{m(n)}\}$. Call F a function family mapping $\ell(n)$-bit input to $m(n)$-bit output strings. F is *polynomial time computable* if there is a polynomial time algorithm (in n) computing all $f \in F$, and *samplable* if there is a probabilistic polynomial time algorithm that on input $n \in \mathcal{N}$ outputs uniformly at random a description of $f \in F_n$. In addition, we call F a *one-way* family of functions if the function g defined by $g_n \in_R F_n$ is a one-way function.

Now we introduce the definition of pseudo-random function families [5] which will be applied in Section 4.2. Intuitively, $F = \{F_n | n \in \mathcal{N}\}$ is a pseudo-random function family if to a probabilistic polynomial time Turing machine (algorithm), the output of a function f chosen randomly and uniformly from F_n, whose description is unknown to the Turing machine, appears to be totally uncorrelated to the input of f, even if the algorithm can choose input for f. The formal definition is described in terms of *(uniform) statistical tests for functions*. A (uniform) statistical test for functions is a probabilistic polynomial time Turing machine T that, given n as input and access to an oracle O_f for a function $f : \Sigma^{\ell(n)} \to \Sigma^{m(n)}$, outputs a bit 0 or 1. T can query the oracle only by writing on a special tape some $x \in \Sigma^{\ell(n)}$ and will read the oracle answer $f(x)$ on a separate answer-tape. The oracle prints its answer in one step.

Definition 2 *Let $F = \{F_n | n \in \mathcal{N}\}$ be an infinite family of functions, where $F_n = \{f | f : \Sigma^{\ell(n)} \to \Sigma^{m(n)}\}$. Assume that F is both polynomial time computable and samplable. F is a pseudo-random function family iff for any statistical test T, for any polynomial Q, and for all sufficiently large n,*

$$|p_n^f - p_n^r| < 1/Q(n),$$

where p_n^f denotes the probability that T outputs 1 on input n and access to an oracle O_f for $f \in_R F_n$ and p_n^r the probability that T outputs 1 on input n and access to an oracle O_r for a function r chosen randomly and uniformly from the set of all functions from $\Sigma^{\ell(n)}$ to $\Sigma^{m(n)}$. The probabilities are computed over all the possible choices of f, r and the internal coin tosses of T.

In [5], it has been shown that pseudo-random function families can be constructed from pseudo-random string generators. By the result of [11,10], the existence of one-way functions is sufficient for the construction of pseudo-random function families.

2.2 Universal Hash Function Families

Universal hash function families, first introduced in [3] and then developed in [18], play an essential role in many recent major results in cryptography and theoretical computer science. (See for example [10,11,16].) Let $U = \bigcup_n U_n$ be a family of functions mapping $\ell(n)$-bit input into $m(n)$-bit output strings. For two strings $x, y \in \Sigma^{\ell(n)}$ with $x \neq y$, we say that x and y collide with each other under $u \in U_n$ or x and y are siblings under $u \in U_n$, if $u(x) = u(y)$.

Definition 3 *Let $U = \bigcup_n U_n$ be a family of functions that is polynomial time computable, samplable and maps $\ell(n)$-bit input into $m(n)$-bit output strings. Let $D_n = \{x \in \Sigma^{\ell(n)} | \exists u \in U_n, \exists y \in \Sigma^{m(n)} \text{ such that } u(x) = y\}$ and $R_n = \{y \in \Sigma^{m(n)} | \exists u \in U_n, \exists x \in \Sigma^{\ell(n)} \text{ such that } y = u(x)\}$. Let $k \geq 2$ be a positive integer. U is a (strongly) k-universal hash function family if for all n, for all k (distinct) strings $x_1, x_2, \ldots, x_k \in D_n$ and all k strings $y_1, y_2, \ldots, y_k \in R_n$, there are $\#U_n/(\#R_n)^k$ functions in U_n that map x_1 to y_1, x_2 to y_2, ..., and x_k to y_k.*

An equivalent definition for the (strongly) k-universal hash function family is that for all k distinct strings $x_1, x_2, \ldots, x_k \in D_n$, when h is chosen uniformly at random from U_n, the concatenation of the k resultant strings $y_1 = h(x_1), y_2 = h(x_2), \ldots, y_k = h(x_k)$ is distributed randomly and uniformly over the k-fold Cartesian product R_n^k of R_n. The following *collision accessibility property* is a useful one.

Definition 4 *Let $U = \bigcup_n U_n$ be a family of functions that is polynomial time computable, samplable and maps $\ell(n)$-bit input into $m(n)$-bit output strings. Let $k \geq 1$ be a positive integer. U has the k-collision accessibility property, or simply the collision accessibility property, if for all n and for all $1 \leq i \leq k$, given any set $X = \{x_1, x_2, \ldots, x_i\}$ of i initial strings in $\Sigma^{\ell(n)}$, it is possible in probabilistic polynomial time to select randomly and uniformly functions from U_n^X, where $U_n^X \subset U_n$ is the set of all functions in U_n that map $x_1, x_2, \ldots,$ and x_i to the same strings in $\Sigma^{m(n)}$.*

k-universal hash function families with the collision accessibility property can be obtained from polynomials over finite fields [3,18]. Denote by P_n the collection of all polynomials over $GF(2^{\ell(n)})$ with degrees less than k, i.e.,

$$P_n = \{a_0 + a_1 x + \cdots + a_{k-1}x^{k-1} | a_0, a_1, \ldots, a_{k-1} \in GF(2^{\ell(n)})\}.$$

For each $p \in P_n$, let u_p be the function obtained from p by chopping the first $\ell(n) - m(n)$ bits of the output of p whenever $\ell(n) \geq m(n)$, or by appending a fixed $m(n) - \ell(n)$-bit string to the output of p whenever $\ell(n) < m(n)$. Let $U_n = \{u_p | p \in P_n\}$, and $U = \bigcup_n U_n$. Then U is a (strongly) k-universal hash function family, which maps $\ell(n)$-bit input into $m(n)$-bit output strings and has the collision accessibility property.

2.3 Sibling Intractable Function Families

Let $k = k(n)$ be a polynomial with $k \geq 1$. Let $H = \{H_n | n \in \mathcal{N}\}$, where $H_n = \{h | h : \Sigma^{\ell(n)} \to \Sigma^{m(n)}\}$, be an infinite family of functions that is one-way, polynomial time computable and samplable, and that has the collision accessibility property. Also let $X = \{x_1, x_2, \ldots, x_i\}$ be a set of i initial strings in $\Sigma^{\ell(n)}$, where $1 \leq i \leq k$, and h be a function in H_n that maps $x_1, x_2, \ldots, x_i$ to the same string. Let F, called a *sibling finder*, be a probabilistic polynomial time algorithm that on input X and h, outputs either "?" ("I cannot find") or a string $x' \in \Sigma^{\ell(n)}$ such that $x' \notin X$ and $h(x') = h(x_1) = h(x_2) = \cdots = h(x_i)$. Informally, H is a k-*sibling intractable function family*, or k-SIFF for short, if for any $1 \leq i \leq k$, for any sibling finder F, the probability that F outputs an x' is negligible. More precisely:

Definition 5 *Let $k = k(n)$ be a polynomial with $k \geq 1$. Let $H = \{H_n | n \in \mathcal{N}\}$, where $H_n = \{h | h : \Sigma^{\ell(n)} \to \Sigma^{m(n)}\}$, be a family of functions that is one-way, polynomial time computable and samplable, and that has the collision accessibility property. Also let $X = \{x_1, x_2, \ldots, x_i\}$ be any set of i initial strings, where $1 \leq i \leq k$. H is a k-sibling intractable function family, or simply k-SIFF, if for each $1 \leq i \leq k$, for each sibling finder F, for each polynomial Q, and for all sufficiently large n,*

$$\Pr\{F(X, h) \neq ?\} < 1/Q(n),$$

where h is chosen randomly and uniformly from $H_n^X \subset H_n$, the set of all functions in H_n that map x_1, x_2, ..., and x_i to the same strings in $\Sigma^{m(n)}$, and the probability $\Pr\{F(X,h) \neq ?\}$ is computed over H_n^X and the sample space of all finite strings of coin flips that F could have tossed.

Here are several remarks on SIFF which follow directly from the definition of SIFF:

1. If $H = \{H_n | n \in \mathcal{N}\}$ is a k-SIFF for some $k \geq 1$, then the function f defined by $f_n \in_R H_n$ is a one-way function.

2. A one-way one-to-one function is a 1-SIFF.

3. A universal one-way hash function family introduced in [13] is a 1-SIFF.

4. If $H = \{H_n | n \in \mathcal{N}\}$ is a k-SIFF, then it is also an i-SIFF for any $1 \leq i < k$.

In the next section we give an explicit construction of SIFF from any one-way function.

3   Construction of SIFF

In [16], Rompel showed that universal one-way hash function families, that is, 1-SIFF, can be constructed from any one-way function. Rompel's result is the starting point of our construction of k-SIFF. The following theorem shows that 1-SIFF can be transformed into 2^s-SIFF for any $s = O(\log n)$. This result is general enough owing to the fact that a k-SIFF is also an i-SIFF for any $1 \leq i < k$.

Theorem 1 Let ℓ, m' and m be polynomials with $m'(n) - m(n) = O(\log n)$. Let $k = 2^{m'(n)-m(n)}$. Assume that $H' = \{H_n' | n \in \mathcal{N}\}$ is a 1-SIFF mapping $\ell(n)$-bit input to $m'(n)$-bit output strings, and $U = \{U_n | n \in \mathcal{N}\}$ a k-universal hash function family that has the collision accessibility property and maps $m'(n)$-bit input to $m(n)$-bit output strings. Let

$$H_n = \{u \circ h' | h' \in H_n', u \in U_n\}$$

and $H = \{H_n | n \in \mathcal{N}\}$. Then H is a k-SIFF mapping $\ell(n)$-bit input into $m(n)$-bit output strings.

Proof: First we observe that H has the collision accessibility property, simply because that H' is samplable and that U has the collision accessibility property.

Now assume for contradiction that there exists a sibling finder F that, for infinitely many n, on input some $X = \{x_1, x_2, \ldots, x_i\}$ and $h \in_R H_n^X$ where $1 \leq i \leq k$, outputs with probability at least $1/Q(n)$ a string $x' \in \Sigma^{\ell(n)}$ such that $x' \notin X$ collides with all strings in X, where Q is a polynomial and H_n^X is the set of all functions in H_n that map x_1, x_2, ..., and x_i to the same strings in $\Sigma^{m(n)}$. We show a contradiction to the assumption that H' is a 1-SIFF. More specifically, we construct a probabilistic polynomial time algorithm M that uses F as an oracle and finds, with probability $1/(2kQ(n))$, either a string colliding with x_j for some $1 \leq j \leq i$, or (when $i < k$) the inverse of some of the

$k - i$ strings $y_{i+1}, y_{i+2}, \ldots, y_k$, where $h' \in_R H'_n$ and each y_j, $i+1 \leq j \leq k$, is generated by first picking randomly an element from $\Sigma^{\ell(n)}$ and then evaluating the function h' at the random point.

Given F, X, $\{y_{i+1}, y_{i+2}, \ldots, y_k\}$ and h', M runs according to the following steps:

1. Choose $z \in_R \Sigma^{m(n)}$.

2. Choose randomly $u \in U_n$ such that $u(y_1) = u(y_2) = \cdots = u(y_i) = u(y_{i+1}) = \cdots = u(y_k) = z$, where $y_j = h'(x_j)$ for all $1 \leq j \leq i$.

3. Call F with X and $h = u \circ h'$ as input. Let the output of F be x'.

Note that for $h' \in_R H'_n$, the probability that $y_{j_1} = y_{j_2}$ for some $1 \leq j_1 \neq j_2 \leq k$ is negligible. Otherwise $H' = \{H'_n | n \in \mathcal{N}\}$ would not be a 1-SIFF. In the following discussion, we will assume that $y_1, y_2, \ldots, y_k$ are all distinct.

The function $h = u \circ h'$ is clearly a random element of H_n^X, $\{y_{i+1}, y_{i+2}, \ldots, y_k\}$, as U is a k-universal hash function family with the collision accessibility property, and h', z and u are all chosen randomly. Denote by S_1 the set of all the siblings of $x_1, x_2, \ldots, x_i$ and by S_2 the set of the inverses of $y_{i+1}, y_{i+2}, \ldots, y_k$, both with respect to h'. Note that S_1 and S_2 are disjoint sets when $y_1, y_2, \ldots, y_k$ are all distinct, and that $x' \neq ?$ iff $x' \in S_1$ or $x' \in S_2$. Therefore, we have

$$\Pr\{x' \neq ?\} = \Pr\{x' \in S_1\} + \Pr\{x' \in S_2\}.$$

By assumption we have

$$\Pr\{x' \neq ?\} \geq 1/Q(n).$$

This implies that either

$$\Pr\{x' \in S_1\} \geq 1/2Q(n)$$

or (when $i < k$)

$$\Pr\{x' \in S_2\} \geq 1/2Q(n).$$

$\Pr\{x' \in S_1\} \geq 1/2Q(n)$ implies that x' collides, with probability at least $1/(2iQ(n)) \geq 1/(2kQ(n))$, with x_j for some $1 \leq j \leq i$ under the randomly chosen function h'. This contradicts our assumption that H' is a 1-SIFF. On the other hand, when $i < k$, $\Pr\{x' \in W\} \geq 1/2Q(n)$ implies that with probability at least $1/(2(k-i)Q(n)) \geq 1/(2kQ(n))$, x' is the inverse of y_j for some $i+1 \leq j \leq k$ with respect to $h' \in_R H'_n$, which contradicts the fact that if H' is a 1-SIFF then the function defined by choosing $h' \in_R H'_n$ is a one-way function. In summary, $\Pr\{x' \neq ?\} \geq 1/Q(n)$ is a contradiction to the assumption that H' is a 1-SIFF. This completes the proof. $\qquad\square$

Combining Theorem 1 with Rompel's result that universal one-way hash function families, i.e., 1-SIFF, can be obtained from any one-way function, and with the fact that a 2^s-SIFF is also an i-SIFF for all $1 \leq i < 2^s$, we have:

Theorem 2 *k-SIFF can be constructed from any one-way function.*

In the following section we will apply the results of this section to the hierarchical access control problem and provide a solution based on SIFF.

4   The Hierarchical Access Control Problem

In today's modern society, hierarchical structures exist in various forms, from business corporations to government departments, each resembling a directed graph (such as a tree) in its figurative shape, with a particular node of the graph the highest point of command. In mathematical terms, such a hierarchy usually take the form of a *partially ordered set* with the highest point of command being the *maximal node*. The various positions throughout the hierarchical structure are then represented as internal nodes, each being the point of command over its underlying sub-graph, consisting also of nodes.

In a hierarchical organization which deals with some amount of sensitive information, the security of certain pieces of information must often be maintained at a certain level which corresponds to a particular depth in the hierarchical organization represented by the graph. A typical case would be that of a banking corporation where the manager deals with private and sensitive data. Such data should not be accessible to company members with positions and authority lower than the manager. However, the opposite condition is often required to be fulfilled. The manager should be able to access data belonging to employees working underneath him/her in the hierarchical organization.

In past years cryptography has often been used to ensure the security of sensitive data. Of more recent interest, however, is the problem of organizing cryptographic keys in a hierarchical manner to mirror the structure of the organization employing the cryptographic techniques for security. The problem of generating and updating keys for a hierarchical organization, called the *hierarchical access control problem*, was first posed by Akl and Taylor in 1982 [1]. Since then many solutions or partial solutions to the problem have been proposed [2,12,4,6,8,9,14]. A common drawback with these schemes is that all of them are based on a single cryptographic assumption, that is the (supposed) difficulty of breaking the RSA cryptosystem [15], and make heavy use of the underlying algebraic properties of the crypto-function.

In [17], Sandhu gave a solution to the *special* case when an organization has a tree structure, using a set of one-way functions. However, the problem of solving the general case of partially ordered sets under the weakest assumption of the existence of one-way functions, remains an interesting open problem. In this section we give a simple solution to the open problem. Incorporated into our solution are the idea of Sandhu for the tree structure and an elegant use of SIFF. A remarkable feature of our solution is that each node in the hierarchical structure needs to keep only one secret key.

The problem of access control in hierarchical organizations is described more formally in Section 4.1. A definition for security of key generation schemes is introduced in the same section. This is followed by a detailed description of the key generation scheme and a proof of its security in Section 4.2 and Section 4.3 respectively. An improvement of the scheme is presented in Section 4.4 and several issues on updating keys are briefly discussed in Section 4.5.

4.1   Preliminaries

Usually, an organization G consists of a set of $P(n)$ members together with a hierarchical relation among the members, here P is a polynomial and the computational power of all members in the organization is bounded by probabilistic polynomial time. Such a hierarchical organization can be well modeled by an algebraic system called a *partially ordered set*. Let $S = \{N_1, N_2, \ldots, N_{P(n)}\}$ be a set of $P(n)$ nodes, each of which represents a member of the organization. Denote by $\geq$ the hierarchical relation within the organization. Then G is determined by the pair of S and $\geq$. In mathematical terms, the organization G is called a partially ordered set or *poset* for short. For convenience, in the following discussions we will sometimes interchange the terms *(hierarchical) organization* and *poset*, and the terms *member* and *node*.

Every poset has some nodes called *maximal nodes*. Each maximal node N_i has the property that there is no node $N_j \in S$ such that $N_j \geq N_i$ and $N_j \neq N_i$. In this paper, we will only be concerned with such a hierarchical organization that has only one *maximal node* N_0. Results in this paper can be readily generalized to the case where a hierarchical organization has multiple maximal nodes. Assume that N_i and N_j are two different nodes in S. N_i is called an *ancestor* of N_j (or equivalently, N_j is a *descendant* of N_i) if $N_i \geq N_j$. N_i is called a *parent* of N_j (or equivalently, N_j is a *child* of N_i) if N_i is an ancestor of N_j and there is no other node $N_k \in S$ with $N_i \geq N_k \geq N_j$. Now assume that $S' \subset S$ is a subset of S. S' induces a sub-poset that consists of the set $\Theta(S')$ and the partial order relation $\geq$, where $\Theta(S')$ consists of both the nodes in S' and the nodes which are descendants of nodes in S'.

A *Hasse diagram* of an organization is a figure consisting of nodes, with an arrow directed downwards from N_i to N_j whenever N_i is a parent of N_j. As the correspondence between an organization and its Hasse diagram is obvious, in the following discussions we will not distinguish between an organization and its Hasse diagram. In particular, we will not distinguish between a node in S and the member of the organization represented by the node.

The *hierarchical access control problem* for an organization G essentially reduces to that of generating a key K_i for each node N_i in such a way that for any nodes N_i and N_j, the node N_i is able to derive from K_i the key K_j of N_j iff $N_i \geq N_j$. Related to this is the problem of key updates of the nodes. Key updating is required when the structure of the organization is modified. Typical changes to the structure includes the deletion and addition of nodes. Key updating is also required when some keys are lost or when the duration of validity of the keys has expired.

Next we discuss the definition of security of a key generation scheme for a hierarchical organization. Any key generation scheme should at least fulfill the requirement that it is computationally difficult for members of the organization, represented by a subset S' of S, to *find* by collaboration the key K_i of a node N_i not in $\Theta(S')$, where $\Theta(S')$ consists of both the nodes in S' and the nodes which are descendants of nodes in S'. When N_i is an internal node or the maximal node N_0, which implies that N_i has at least one child, the following more general requirement should be fulfilled. That is, it is computationally difficult for S' to *simulate* N_i's procedure for generating the key of a child of N_i. Note that S' may or may not be able to find the key of N_i and that the child of N_i may or may not be in $\Theta(S')$. The reason for considering the general requirement is that N_i's procedure for generating the key of the child, even if the child is in $\Theta(S')$, is a privilege of N_i, and the privilege should not be shared by any other node that is not an ancestor of N_i. The following is a formal definition of security of a key generation scheme.

Definition 6 *Let G be a hierarchical organization with P(n) nodes (members). Denote by S the set of the P(n) nodes. A key generation scheme for a hierarchical organization is secure if for any $S' \subset S$, for any node $N_i \notin \Theta(S')$, for any polynomial Q and for all sufficiently large n, the probability that the nodes in S' are able to find by collaboration the key K_i of the node N_i whenever N_i has no child, or to simulate N_i's procedure for generating the key of a child of N_i whenever N_i is an internal node or the maximal node N_0, is less than $1/Q(n)$.*

4.2 Key Generation

Denote by ID_i the identity of the node N_i. Assume that every ID_i can be described by an $\ell(n)$-bit string, where ℓ is a polynomial. Let $F = \{F_n | n \in \mathcal{N}\}$ be a pseudo-random function family, where $F_n = \{f_K | f_K : \Sigma^{\ell(n)} \to \Sigma^n, K \in \Sigma^n\}$ and each function $f_K \in F_n$ is specified by an n-bit string K. Let $H = \{H_n | n \in \mathcal{N}\}$ be a k-SIFF mapping n-bit input to n-bit output strings. Also assume that k is sufficiently large so that no nodes could have more than k parents. The following key generation procedure can be done either by a trusted third party or by the maximal node itself.

First a random string $K_0 \in_R \Sigma^n$ is chosen for the maximal node N_0. For the nodes without a key, either with one or more parents, the following two steps should be completed until all the nodes in S have been assigned keys.

1. *Nodes with one parent*
   Given a node N_i with its parent N_j which has already been assigned a key K_j, the key to be assigned to N_i is the n-bit string $K_i = f_{K_j}(ID_i)$.

2. *Nodes with two or more parents*
   Given a node N_i with all its p parents $N_{j_1}, N_{j_2}, \ldots, N_{j_p}$ having been assigned keys $K_{j_1}, K_{j_2}, \ldots, K_{j_p}$, the key K_i for N_i is chosen as a random string $K_i \in_R \Sigma^n$. From H_n a function h_i is chosen randomly and uniformly such that $f_{K_{j_1}}(ID_i)$, $f_{K_{j_2}}(ID_i), \ldots, f_{K_{j_p}}(ID_i)$ are mapped to K_i, i.e.,

$$h_i(f_{K_{j_1}}(ID_i)) = h_i(f_{K_{j_2}}(ID_i)) = \cdots = h_i(f_{K_{j_p}}(ID_i)) = K_i.$$

   The function h_i is then made public, making all the ancestors of N_i aware of h_i.

It is clear that if $N_j \geq N_i$, then K_i can be derived from K_j. When N_j is an immediate parent of N_i, K_i can be computed via either $K_i = f_{K_j}(ID_i)$ if N_j is the single parent of N_i, or $K_i = h_i(f_{K_j}(ID_i))$ if N_i has other immediate parents. When N_j is not an immediate parent of N_i, all keys in the path from N_j to N_i are computed downwards and K_i is obtained in the final stage.

4.3 Security of the Key Generation Scheme

This section proves that the key generation scheme is secure. A corollary of the result is that secure key generation schemes for hierarchical organizations can be obtained from any one-way function.

Theorem 3 *The key generation scheme for a hierarchical organization is secure.*

Proof (Sketch): Assume that $S' \subset S$ can find the key K_i of some node $N_i \notin \Theta(S')$ that has no child. This implies that S' is able to predict the output of the pseudo-random function family, which is a contradiction.

Now assume that $N_i \notin \Theta(S')$ is an internal node or the maximal node N_0, and that S' can simulate N_i's procedure for generating the key K_j of some child N_j of N_i. Note that for our key generation scheme, being able to simulate N_i's procedure for generating the key K_j of the child N_j of N_i implies being able to get either K_i when N_i is the single parent of N_j, or $f_{K_i}(ID_j)$ when N_j has other parents than N_i. Also note that getting K_i or $f_{K_i}(ID_j)$ means getting the keys of all the descendants of N_j besides the key K_j of N_j. Thus there are only two situations to be considered when S' is able to get K_i or $f_{K_i}(ID_j)$ but fails to mimic any of the parents of N_i. These two cases are:

Case-1: N_i is an ancestor of some node(s) in $\Theta(S')$.

Case-2: N_i is not the ancestor of any node in $\Theta(S')$.

In Case-1, being able to get K_i or $f_{K_i}(ID_j)$ implies being able to do one of the following three actions: invert the pseudo-random function family, find a collision string for the sibling intractable function family, or invert the sibling intractable function family. The success of any of these actions with a high probability is a contradiction. In Case-2, being able to get K_i or $f_{K_i}(ID_j)$ implies being able to predict the output of the pseudo-random function family. This is also a contradiction. $\qquad\square$

4.4 Improvement of the Key Generation Scheme

A problem with the above key generation scheme is that a node must pass through a number of intermediate descendants in order to arrive at a given distant (non-child) descendant node may be an inconvenience to the members of the hierarchical organization. This traversal down the structure requires the use of the sibling intractable function family and the pseudo-random function associated with each node along the traversed path in order to find the keys of the intermediate nodes.

A solution to the problem consists of a modification to the key generation phase. For a given node N_i with q ancestors (*including the parents*) $N_{j_1}, N_{j_2}, \ldots,$ and N_{j_q}, the generation of the key of N_i involves the selection of a random string $K_i \in_R \Sigma^n$ and the selection of $h_i \in H_n$ randomly and uniformly such that $f_{K_{j_1}}(ID_i)$, $f_{K_{j_2}}(ID_i)$, $\ldots,$ and $f_{K_{j_q}}(ID_i)$ are all mapped to K_i, i.e.

$$h_i(f_{K_{j_1}}(ID_i)) = h_i(f_{K_{j_2}}(ID_i)) = \cdots = h_i(f_{K_{j_q}}(ID_i)) = K_i.$$

In this way any ancestor of N_i which was involved in the generation of K_i can access N_i directly without the need to pass any intermediate nodes. An extensive treatment of this issue together with other solutions to the hierarchical access control problem is presented in [19].

4.5 Key Updating

It is natural to expect that the structure of an organization (i.e., the shape of the corresponding Hasse diagram) will change throughout time, and thus the keys of the nodes will also need to be updated. Some typical changes include the addition and deletion of nodes, and the establishment and removal of links between nodes. Another reason for the renewal of the keys is the replacement of one member of the organization with another, without involving any change to the structure (Hasse diagram) itself. The arrival of a new member to the organization implies the creation of a new identity information for that member. Key updating is also required when some keys are lost or when the duration of validity of some keys has expired. In this section we will consider briefly three of the most typical problems related to the maintenance of the internal nodes of the hierarchical structure. These are the addition and deletion of nodes, and the replacement of the identification information of nodes. The case of leaf nodes is trivial and will not be discussed.

4.5.1 Addition and Deletion of Nodes

When nodes are added or deleted there are a number of possibilities as to how the sub-posets affected by the change should be maintained. When a new node N_k is added between node N_j and its immediate parent node N_i, N_k becomes the new parent of N_j, and N_i the parent of N_k. The descendants of N_k which includes N_j are effectively shifted one level down in the organization (Hasse diagram). It is clear that the addition of a new node followed by a shift down of all its descendants requires the generation of new keys for that node and its descendants. Only the new node requires a new identity information.

In the case of the deletion of a node, its descendants becomes the descendants of its parent(s) and new keys must be generated for the descendants. This corresponds to an upward shift by one level of these descendants in the organization.

4.5.2 Replacement of an Identity

The replacement of the identity information of a node can be due to a number of changes in the organization. A member at a node can be replaced by another current member or by a new member from outside the organization. Often, the identity of a node simply needs to be changed following some organizational decision. In all these cases the keys for all the descendants of that node need to be generated again. The node itself is assigned a new key which is used to generate and assign the keys of its immediate children nodes.

5   Other Applications of SIFF

There are numerous ways that SIFF can be applied. In the following, three other applications of SIFF are briefly discussed. The first application is to the shared mail box problem, the second to the access control in distributed systems, and the third to the multiple message authentication problem. A further application of SIFF to database authentication is presented in [7].

5.1 The Shared Mail Box Problem

Suppose that there is a group consisting of k users $U_1, U_2, \ldots, U_k$. Each user U_i has a private mail box B_i. Assume also that there is a shared mail box B_s. The shared mail box problem consists of the design of a cryptographic system for the group that has the following features:

1. Each user U_i holds just one secret key K_i.

2. For each $1 \leq i \leq k$, the mail box B_i can only be opened by the user U_i who possesses the secret key K_i.

3. The shared mail box B_s can be opened by every user in the group, but not by any outsider.

4. Even when $k - 1$ users conspire together, it is computationally difficult for the $k - 1$ users to open the other user's private mail box.

This problem represents an abstraction of many practical applications where the determination of access rights to various resources is required. The traditional way for solving the access problem with the private and shared mail boxes is to let each user hold two keys, one for his or her private mail box and the other for the shared mail box. The traditional solution becomes very impractical when the user is a member of a number of different groups, and hence has to hold as many keys for shared mail boxes as the total number of groups he or she belongs to. We present a simple solution to the problem by the use of a SIFF and a secure secret-key block cipher, both of which can be constructed from any one-way function. First of all, we, a trusted third party, choose a secure secret-key block cipher for the purpose of locking the private and shared mail boxes. Then we choose a $(k - 1)$-SIFF $H = \{H_n | n \in \mathcal{N}\}$ that maps n-bit input to n-bit output strings. The following is the key generation procedure for the group.

- Choose for each user U_i a random n-bit string K_i as his or her secret key.

- Choose a random n-bit string K_s for the shared mail box B_s.

- Select from H_n a random function h such that all private keys $K_1, K_2, \ldots, K_k$ are mapped to K_s, i.e., $h(K_1) = h(K_2) = \cdots = h(K_k) = K_s$. Then make h public.

It is easy to see that the scheme fulfills all the four requirements. In particular, each user U_i can apply the secure block cipher with K_i as a key to open or lock his or her private mail box B_i, and with $h(K_i)$, which is mapped to K_s, as a key to open or lock the shared mail box B_s.

5.2 Access Control in Distributed Systems

SIFF can also be used to control access of data in a distributed system which are geographically dispersed. Each site in the system would have two levels of access, the first being applied to the site as a whole, while the second to control access to resources and data stored at that site.

In the first access level, the access by one site to another is determined using SIFF. Hence, a given site can determine which other sites that may have access to it. This, in effect, classifies sites according to their sensitivity, and may be governed by various performance and practical necessities. In the second access level, which assumes the granting of access in the first level, transactions that access multilevel resources and data can be controlled also using SIFF.

5.3 The Multiple Message Authentication Problem

Message authentication is an important part of information security. A common method is to append to a message to be authenticated a short tag such as a checksum, by using a modification detection code. In some cases, we have many (independent) messages to be authenticated. Two usual methods are for each message to be given a tag independent of one another, and for the concatenation of all the messages to be given a single common tag. In the first method the resulting number of tags may prove too impractical to be maintained, while in the second method the validation of one message requires the use of all other (unrelated) messages in the re-calculation of the tag.

A preferred method would be one that employs a single common tag for all the messages in such a way that a message can be verified individually without involving other messages. As we will see, SIFF can be used as an ideal tool for the above purpose. In order to explain it more clearly, we will use the software protection problem as an illustrative example.

Consider a group of software companies that produce popular softwares. For obvious reasons, it is important for the software companies to protect their software products from being infected by computer viruses and from being modified illegally. Suppose that all software products of the software companies can be represented by strings of $\ell(n)$-bit long (short softwares can be extended by padding some fixed pattern.) Also suppose that the length of the tags is $m(n)$ bits and k is an integer that is larger than the total number of softwares the companies can produce in the foreseeable future. To protect softwares, the companies first choose a k-SIFF $H = \{H_n | n \in \mathcal{N}\}$ that maps $\ell(n)$-bit input to $m(n)$-bit output strings. Then the companies execute the following procedure.

1. Choose a random $m(n)$-bit tag t.

2. Choose randomly and uniformly from H_n a function h that maps all softwares of the company to the same tag t.

3. Publish in some mass media, such as popular newspapers or computer magazines, the function h and the tag t, together with a list of the names of the softwares.

When some new products come out, the companies can either generate a tag for the new products, or re-generate a tag for all the products including the new ones, by using exactly the same procedure described above.

A user can now easily detect any modification to a software (from one) of the companies, simply by computing the output t' of the public function h, with the string representing the software as input of h, and comparing the output t' with the public tag t. If t' and t are the same, the user is confident that the software has not been tampered with illegally. Otherwise, if t' and t differ, the user should not proceed to use the software.

From these short examples it is clear that SIFF promises its usefulness in a wide area of applications.

6 Conclusion and Further Work

We have introduced the notion of SIFF which includes as a special case the notion of the universal one-way hash function family defined Naor and Yung in [13]. We have also shown a simple method for transforming any universal one-way hash function family into a SIFF. Putting together this and Rompel's results, we have obtained the theoretically optimal result that SIFF can be constructed from any one-way function. As applications of SIFF, we have presented a key generation scheme for hierarchical organizations, and suggested solutions to the shared mail box problem, access control in distributed systems and the multiple message authentication problem.

The applicability of a solution based on SIFF to a practical problem is largely determined by the compactness of SIFF. An improvement in the compactness of SIFF results directly in the improvement in the efficiency of a solution based on SIFF. For the construction of k-SIFF presented in Theorem 1, the length of a description of a function in H_n is of order $O(L_1(n) + L_2(n))$, where $L_1(n)$ is the length of a description of a function in U_n and $L_2(n)$ that of a function in H'_n respectively. Searching for more compact constructions of SIFF from one-way functions, together with other applications of SIFF, is an interesting subject for further research.

Acknowledgements The authors are grateful to Prof. Jennifer Seberry for continuous interest and support, and to Babak Sadeghiyan and other members of the Centre for Communications and Computer Security Research for helpful discussions and comments. Special thanks go to Toshiya Itoh for pointing out crucial errors in early versions of this paper.

References

[1] AKL, S. G., AND TAYLOR, P. D. Cryptographic solution to a multilevel security problem. In *Advances in Cryptology - Proceedings of Crypto'82* (Santa Barbara, August 1982), D. Chaum, R. L. Rivest, and A. T. Sherman, Eds., Plenum Press, NY, pp. 237–250.

[2] AKL, S. G., AND TAYLOR, P. D. Cryptographic solution to a problem of access control in a hierarchy. *ACM Transactions on Computer Systems 1*, 3 (1983), 239–248.

[3] CARTER, J., AND WEGMAN, M. Universal classes of hash functions. *Journal of Computer and System Sciences 18* (1979), 143–154.

[4] CHICK, G. C., AND TAVARES, S. E. Flexible access control with master keys. In *Advances in Cryptology - Proceedings of Crypto'89*, Lecture Notes in Computer Science, Vol. 435 (1990), G. Brassard, Ed., Springer-Verlag, pp. 316–322.

[5] GOLDREICH, O., GOLDWASSER, S., AND MICALI, S. How to construct random functions. *Journal of ACM 33*, 4 (1986), 792–807.

[6] HARDJONO, T., AND SEBERRY, J. A multilevel encryption scheme for database security. In *Proceedings of the 12th Australian Computer Science Conference* (1989), pp. 209–218.

[7] HARDJONO, T., ZHENG, Y., AND SEBERRY, J. A new approach to database authentication. In *Proceedings of the Third Australian Database Conference (Database'92)* (1992).

[8] HARN, L., AND KIESLER, T. Authentication group key distribution scheme for a large distributed network. In *Proceedings of the 1989 IEEE Symposium on Security and Privacy* (1989), pp. 300–309.

[9] HARN, L., AND LIN, H.-Y. A cryptographic key generation scheme for multi-level data security. *Computer & Security 9*, 6 (1990), 539–546.

[10] HÅSTAD, J. Pseudo-random generation under uniform assumptions. In *Proceedings of the 22-nd ACM Symposium on Theory of Computing* (1990), pp. 395–404.

[11] IMPAGLIAZZO, R., LEVIN, L., AND LUBY, M. Pseudo-random generation from one-way functions. In *Proceedings of the 21-st ACM Symposium on Theory of Computing* (1989), pp. 12–24.

[12] MACKINNON, S. J., TAYLOR, P. D., MEIJER, H., AND AKL, S. G. An optimal algorithm for assigning cryptographic keys to access control in a hierarchy. *IEEE Transactions on Computers C-34*, 9 (1985), 797–802.

[13] NAOR, M., AND YUNG, M. Universal one-way hash functions and their cryptographic applications. In *Proceedings of the 21-st ACM Symposium on Theory of Computing* (1989), pp. 33–43.

[14] OHTA, K., OKAMOTO, T., AND KOYAMA, K. Membership authentication for hierarchical multigroup using the extended Fiat-Shamir scheme. In *Advances in Cryptology - Proceedings of EuroCrypt'90*, Lecture Notes in Computer Science, Vol. 473 (1991), I. B. Damgård, Ed., Springer-Verlag.

[15] RIVEST, R. L., SHAMIR, A., AND ADLEMAN, L. A method for obtaining digital signatures and public-key cryptosystems. *Communications of the ACM 21*, 2 (1978), 120–128.

[16] ROMPEL, J. One-way functions are necessary and sufficient for secure signatures. In *Proceedings of the 22-nd ACM Symposium on Theory of Computing* (1990), pp. 387–394.

[17] SANDHU, R. S. Cryptographic implementation of a tree hierarchy for access control. *Information Processing Letters 27*, 2 (1988), 95–98.

[18] WEGMAN, M., AND CARTER, J. New hash functions and their use in authentication and set equality. *Journal of Computer and System Sciences 22* (1981), 265–279.

[19] ZHENG, Y., HARDJONO, T., AND SEBERRY, J. New solutions to access control in a hierarchy, September 1991. (Submitted for publication).

A Digital Multisignature Scheme
Based on the Fiat-Shamir Scheme

Kazuo Ohta　　　　　*Tatsuaki Okamoto*

NTT Laboratories
Nippon Telegraph and Telephone Corporation
1-2356, Take, Yokosuka-shi, Kanagawa-ken, 238-03, Japan

Abstract: We show the sequential multisignature scheme based on the Fiat-Shamir scheme which is a slight variant of simultaneous multisignature scheme, and discuss the security of a digital multisignature scheme. The following properties are proven;

(1) The difficulty of deriving secret information from public information in a multisignature scheme with already used signatures is equivalent to that of deriving it in a single signature scheme; and

(2) The difficulty of forging a partial multisignature so that the total multisignature is valid is equivalent to that of deriving a single signature in the Fiat-Shamir scheme.

1. Introduction

Computer-based message systems have become the principal carriers of business correspondence in many offices. In such sytems, the integrity and authenticity of the digital message must be ensured by digital information instead of handwritten signatures. This can usually achieved through the use of a digital signature technique [D].

In addition to the originator's signature, supervisors are often required to sign office documents for verifying and approving an originator's message. In such a case several persons may sign the same message. This is referred to as *multisignature*. There are two kinds of *multisignature scheme*; one is the sequential version where persons sign the same messages sequentially, and the other is the simultaneous version where persons sign them simultaneously.

A number of digital signature schemes have been developed in the last ten years [RSA, R, OS]. These schemes, which were originally developed for *single signatures*, are also applicable to multisignature by iterating them directly. However, because of the increase in signature length, they are not satisfactory. Though an extended RSA method has been proposed as a scheme for solving this problem [IN], the order of signing with this method is fixed and determined by the signer's key, based on his or her position in the office. In addition, Okamoto proposed a multisignature scheme that overcomes this problem, combining a one-way function and any bijective public-key cryptosystem, such as the RSA scheme [O]. However, all these multisignature schemes are not efficient, because they are based on the RSA scheme, which is very slow in creating digital signatures.

In 1986, Fiat and Shamir proposed a promising signature scheme [FS], which is efficient and proven under the difficulty of factoring a large composite number that the probability of forgery against any known or chosen message attack is $1/2^{kt}$. Here, k denotes the number of secret information integers stored in a signer and the signature size is proportional to t. For example, in order to attain the security level of 2^{-72}, i.e., $tk = 72$, when a signer chooses the minimum signature size as $t = 1$ and stores seventy two $(k = 72)$ secret integers, the speed of their scheme's typical implementation is about twenty times faster than that of the RSA signature scheme. Moreover, when a trusted center exists, the Fiat-Shamir scheme is one of identity-based cryptosystems. Identity-based cryptosystems, in which each user's public key is his identification information, such as his name, address, etc., do not require any key directories and can simplify key management [S, OSK].

In this paper, we show the sequential multisignature scheme based on the Fiat-Shamir scheme which is a slight variant of simultaneous multisignature scheme [BLY, GQ2], and discuss the security of a digital multisignature scheme. All Fiat-Shamir based scheme has the following features:

(1) the speed is about twenty times faster than that of the RSA based multisignature schemes;

(2) it is one of the identity-based signature schemes;

(3) the order of signing is not restricted; and

(4) the redundancy of signed message is restricted by the security level, for example to 72 bits.

2. The Sequential Multisignature Scheme

A multisignature scheme will be described, in which m users joining the signature system sign the same messages sequentially, and convince a verifier that the checked message is signed by each user and has not been modified by any intruder. Hereafter we denote a message as P, and identification information of user i as ID_i.

2.1 Key generation

A trusted center publishes a modulus n which is the product of two secret large primes p and q. For user i whose identification information is ID_i, the center calculates integers S_{ij} $(1 \le j \le k)$:

$$S_{ij} = \frac{1}{\sqrt{f(ID_i, j)}} \mod n, \tag{1}$$

where f is a public one-way function.

Finally, the center issues a smart card to user i after properly checking his physical identity. This smart card includes the set of $(n, f, h, S_{i1}, \cdots, S_{ik})$, where h is another public one-way function. In practice, we can use f instead of h.

2.2 Multisignature generation

To sign a message P, the m users execute the following procedure:

Step 1) Repeat while $i = 1, \cdots, m$

Step 1-i) Signer i receives X_{i-1}, where $X_0 = 1$ holds, generates a random integer $R_i \in Z_n$, where Z_n denotes $\{0, \cdots, n-1\}$, calculates

$$X_i = R_i^2 \cdot X_{i-1} \mod n, \tag{2}$$

and sends it to the next signer $(i + 1)$, where signer $(m + 1)$ is considered as signer 1.

Step 2) Repeat while $i = 1, \cdots, m-1$

Step 2-i) Signer i receives (P, I_m, X_m, Y_{i-1}) from signer $(i-1)$, where $Y_0 = 1$ holds, calculates

$$(e_1, \cdots, e_k) = h(P, I_m, X_m), \tag{3}$$

and

$$Y_i = Y_{i-1} \cdot R_i \cdot \prod_{e_j=1} S_{ij} \mod n, \tag{4}$$

and sends (P, I_m, X_m, Y_i) to the next signer $(i + 1)$, where $I_m = ID_1 * \cdots * ID_m$. Here $*$ denotes concatenation (e.g., "010" $*$ "110" = "010110").

Step 3) Signer m receives (P, I_m, X_m, Y_{m-1}) from signer $(m-1)$, calculates equations (3) and (4), and sends $(P, I_m, (e_1, \cdots, e_k), Y_m)$ to a verifier.

Note: A variant of the above-mentioned scheme is: in Step 2-i), signer i checks the validity of the multisignature (P, I_m, X_m, Y_{i-1}) generated by the preceeding $(i - 1)$ signers (signer 1 through $i - 1$). If it is valid, signer i generates Y_i and sends (P, I_m, X_m, Y_i) to the next signer. Otherwise, (s)he notifies all signers of this fact and halts the procedure. This variant seems to be more secure against some active attacks.

2.3 Multisignature verification

A verfier verifies the message with the multisignature $(P, I_m, (e_1, \cdots, e_k), Y_m)$ from signer m using the public modulus n and one-way functions f and h as follows:

Step 1) The verifier calculates V_{ij} with I_m as follows,

$$V_{ij} = f(ID_i, j) \quad (1 \le i \le m, 1 \le j \le k). \tag{5}$$

Step 2) The verifier calculates Z_m with V_{ij}, $(e_1, \cdots, e_k)$ and Y_m as follows,

$$Z_m = Y_m^2 \cdot \prod_{i=1}^{m} \prod_{e_j=1} V_{ij} \mod n. \tag{6}$$

Step 3) The verifier calculates $h(P, I_m, Z_m)$ and checks whether the following equation holds;

$$(e_1, \cdots, e_k) = h(P, I_m, Z_m). \tag{7}$$

If the above equation holds, the multisignature message is considered to be valid.

Note: If all m signers follow this procedure, a verifier will accept the multisignature as valid. By definition, $S_{ij}^2 = 1/V_{ij} \bmod n$ $(1 \leq i \leq m, 1 \leq j \leq k)$,

$$Z_m = (Y_{m-1}^2 \cdot R_m^2 \cdot \prod_{e_j=1} 1/V_{mj}) \cdot \prod_{i=1}^m \prod_{e_j=1} V_{ij} \bmod n$$
$$= (Y_{m-1}^2 \cdot \prod_{i=1}^{m-1} \prod_{e_j=1} V_{ij}) \cdot R_m^2 \bmod n$$
$$\vdots$$
$$= R_1^2 \cdots R_m^2 = X_m$$

and thus $(e_1, \cdots, e_k) = h(P, I_m, X_m) = h(P, I_m, Z_m)$ holds.

Remark: In this section, we described a simple case, where $(e_1, \cdots, e_k) = h(P, I_m, X_m)$. Our scheme is easily extended to a more general case, $((e_{11}, \cdots, e_{1k}), \cdots, (e_{m1}, \cdots, e_{mk})) = h(P, I_m, X_m)$. Moreover, our technique is applicable to both the more general version $(t \geq 2)$ and the higher degree version of the Fiat-Shamir scheme [GQ1, OO], where t is one of the security parameters of the Fiat-Shamir scheme and $t = 1$ in the above-mentioned multisignature scheme. Note that parameter t to be used in Problem (P2) in Section 3 is different from this security paprameter.

3. Security Consideration

In this section, it is shown that for passive attack our multisignature scheme based on the Fiat-Shamir scheme is considered to be as secure as the Fiat-Shamir scheme. Active attack is not treated here.

The security of any scheme against passive attack depends on two factors. The first is the difficulty of deriving secret information S_{ij} $(1 \leq j \leq k)$ from all signers' identification information ID_i, messages, and signatures (security condition 1). The second is the difficulty of forging a partial multisignature message (P, I_m, X_i, Y_i) so that the total multisignature message $(P, I_m, (e_1, \cdots, e_k), Y_m)$ satisfies verfication equations (5)-(7) (security condition 2).

In computational complexity theory, the difficulty of solving a problem P1 is considered to be equivalent to that of solving another problem P2, when we can construct an expected polynomial time algorithm for solving P2 with an oracle, O1, which solves P1 with non-negligible probability and the reverse also holds [GJ, O, R]. Hereafter, the expression "equivalent" and "as hard as", and so on, are used in reference to this definition.

It is shown in Theorem 1 and 2 that the problem of breaking security conditions 1 and 2 of our multisignature scheme is equivalent to the problem of breaking the Fiat-Shamir signature scheme by passive attack. These theorems are proven under two conditions that 1) the one-way function f is omitted, that is, we assume the following equation: $S_{ij}^2 = 1/V_{ij} \bmod n$ $(1 \leq i \leq m, 1 \leq j \leq k)$ instead of equation (1), and 2) the component I_i of h's input is omitted, that is, we assume the following equation: $(e_1, \cdots, e_k) = h(P, X_m)$ instead of equation (3). The former condition means that not just any one-way function can be used. A scheme with one-way function for key generation is more intractable than one without any one-way function. Thus, the former condition is not considered essential in our security considerations. Since a scheme with the component I_i in verification is more intractable than one without I_i, the latter condition is not considered essential, either.

Therefore, Theorem 1 and 2 indicate that the Fiat-Shamir based multisignature scheme is considered to be at least as secure as the Fiat-Shamir signature scheme against passive attack.

Here, let P1, P2, P3, and P4 refer to the problems presented below and let O1, O2, O3, and O4 be the oracles that solve non-negligible fractions of these problems.

(P1) Given $V_{ij}, X_i^{(\alpha)}, P^{(\alpha)}, Y_i^{(\alpha)}$ $(1 \le i \le m, \; 1 \le j \le k, \; 1 \le \alpha \le \Gamma)$, derive S_{ij} $(1 \le i \le m, 1 \le j \le k)$ which satisfies

$$S_{ij} = \frac{1}{\sqrt{V_{ij}}} \mod n,$$

where $(P^{(\alpha)}, X_m^{(\alpha)}, Y_i^{(\alpha)})$ is the partial multisignature message generated by i signers from the first to the ith signer for message $P^{(\alpha)}$, where identification information of signer i is $(1/V_{i1}, \cdots, 1/V_{ik})$.

(P2) Given V_{ij} $(1 \le i \le T, \; 1 \le j \le k)$, X_i, P, Y_i $(i = 1, \cdots, t, \; 1 \le t < T)$, and X_T, forge a partial multisignature message $(P, X_T, \tilde{Y}_{t+l})$ $(t, \; l \in \{0, 1, 2, \cdots, T\}, \; t + l \le T)$ for (P, X_T, Y_i) $(i = 1, \cdots, t)$ so that the total multisignature message satisfies verfication equations (5)-(7), where identification information of signer i is $(1/V_{i1}, \cdots, 1/V_{ik})$. Here, $(P, X_T, \tilde{Y}_{t+l})$ indicates the partial multisignature message generated by l signers, from the $(t + 1)$th to the $(t + l)$th signer.

(P3) Given V_{0j} $(1 \le j \le k)$, derive S_{0j} $(1 \le j \le k)$ which satisifies

$$S_{0j} = \frac{1}{\sqrt{V_{0j}}} \mod n.$$

(P4) Given V_{0j} $(1 \le j \le k)$ and P_0, derive $(P_0, (e_{01}, \cdots, e_{0k}), Y_0)$ which is a signature of P_0 in the Fiat-Shamir scheme by a signer, whose identification information is $(1/V_{01}, \cdots, 1/V_{0k})$, that is, $(P_0, (e_{01}, \cdots, e_{0k}), Y_0)$ satisfies

$$Z_0 = Y_0^2 \cdot \prod_{e_{0j}=1} V_{0j} \mod n,$$

and

$$(e_{01}, \cdots, e_{0k}) = h(P_0, Z_0).$$

Theorem 1. *The difficulty of solving problem P1 is equivalent to that of solving problem P3, when $k = O(1)$.*

Proof. It is obvious that we can construct an expected polynomial time algorithm with O3 for solving problem P1.

It is proven below that we can construct an expected polynomial time algorithm with O1 for solving problem P3. The algorithm for solving P3 can be constructed as follows using O1:

Do while $1 \le j_0 \le k$.

Step 1) Set $V_{i_0 j_0} = V_{0 j_0}$.

Step 2) Generate $(m-1) \cdot k + (k-1)$ pairs (V_{ij}, S_{ij}) for $1 \leq i \leq m$ and $1 \leq j \leq k$ (if $i = i_0$ then $j \neq j_0$), where $V_{ij} = 1/S_{ij}^2 \bmod n$ holds.

Step 3) Do while $1 \leq \alpha \leq \Gamma$, do while $1 \leq i \leq m$,

if $i \neq i_0$, then follow the procedure of signer,

otherwise, generate a random number $Y^{(\alpha)} \in Z_n$ and a random k bits $(\tilde{e}_1^{(\alpha)}, \cdots, \tilde{e}_k^{(\alpha)})$, calculate

$$Z^{(\alpha)} = Y^{(\alpha)^2} \prod_{\tilde{e}_j^{(\alpha)} = 1} V_{i_0 j} \bmod n,$$

and send $Z^{(\alpha)}$ as $X_i^{(\alpha)}$ to signer $(i_0 + 1)$.

When signer i_0 receives $(P^{(\alpha)}, X_m^{(\alpha)}, Y_{i_0 - 1}^{(\alpha)})$, calculates $(\tilde{e}_1'^{(\alpha)}, \cdots, \tilde{e}_k'^{(\alpha)})$

$= h(P^{(\alpha)}, X_m^{(\alpha)})$.

If $\tilde{e}_{j_0}' = \tilde{e}_{j_0}$ holds, put

$$Y_{i_0}^{(\alpha)} = Y_{i_0 - 1}^{(\alpha)} \cdot Y^{(\alpha)} \cdot \prod_{j \neq j_0} S_{i_0 j}^{\tilde{e}_j' - \tilde{e}_j}.$$

Otherwise, return to the beginning of Step 3).

Step 4) Query oracle O1 with input V_{ij}, $X_i^{(\alpha)}$, $P^{(\alpha)}$, and $Y_i^{(\alpha)}$ $(1 \leq i \leq m, 1 \leq j \leq k, 1 \leq \alpha \leq \Gamma)$. Then check whether the output of O1 includes a correct answer $S_{i_0 j_0}$ such that

$$S_{i_0 j_0} = \frac{1}{\sqrt{V_{0 j_0}}} \bmod n.$$

If it is correct, output the value as $S_{i_0 j_0}$. Otherwise return to Step 1.

It is obvious that the distribution of $X_i^{(\alpha)}$, $P^{(\alpha)}, Y_i^{(\alpha)}$ generated in the above steps is equivalent to the valid input of O1.

The probability that $\tilde{e}_{j_0}' = \tilde{e}_{j_0}$ holds in Step 3 is $1/2$, and the probability that a correct answer is output in Step 4 is non-negligible from the assumption. Therefore, the expected running time of the above algorithm is polynomial in the size of $|V_{0j}|$. Q.E.D.

Theorem 2. *The difficulty of solving problem P2 is equivalent to that of solving problem P4.*

Proof. It is obvious that we can construct an expected polynomial time algorithm with O4 for solving problem P2.

It is proven below that we can construct an expected polynomial time algorithm with O2 for solving problem P4 for any t $(t < T)$ and any l $(t+l \leq T)$. The algorithm for solving P4 can be constructed as follows using O2:

Step1) Divide V_{0j} into l integers V_{ij} $(t + 1 \leq i \leq t + l)$, that is, $V_{0j} = \prod_{i=t+1}^{t+l} V_{ij}$ $(1 \leq j \leq k)$.

Step 2) Generate $(T - l) \cdot k$ pairs (V_{ij}, S_{ij}) $(1 \le i \le t$ or $t + l + 1 \le i \le T,\ 1 \le j \le k)$, where $V_{ij} = 1/S_{ij}^2 \bmod n$ holds.

Step 3) For i $(1 \le i \le t)$, calculates X_i and (P_0, X_T, Y_i) using equations (2)-(4).

Step 4) Query oracle O2 with input V_{ij} $(1 \le i \le T,\ 1 \le j \le k)$, X_i, P_0, Y_i $(1 \le i \le t)$ and X_T. Then obtain $\tilde{Y}_{t+l}$.

Step 5) Calculate Y_0 as follows:

$$Y_0 = \tilde{Y}_{t+l} \cdot R_{t+l+1} \cdots R_T / \prod_{i=1}^{t} \prod_{e_j=1} S_{ij} \bmod n,$$

and set $E_0 = (e_1, \cdots, e_k)$, where $(e_1, \cdots, e_k) = h(P_0, X_T)$. Check whether Y_0 is a corect answer. If it is correct, output the value. Otherwise, return to Step 1.

It is obvious that the distribution of X_i, P_0, Y_i $(1 \le i \le t)$ and X_T generated in the above steps is equivalent to the distribution of the valid input of O2, because these values are calculated in the same manner as regular signers, who know the pair (V_{ij}, S_{ij}), where $V_{ij} = 1/S_{ij}^2 \bmod n$ holds.

It is clear that (P_0, E_0, Y_0) is a valid signature by a signer, whose identification information is $(1/V_{01}, \cdots, 1/V_{0k})$. This is because: since the total multisignature message $(P_0, (e_1, \cdots, e_k), Y_T)$ is valid, the following two equations hold:

$$X_T = Y_T^2 \cdot \prod_{i=1}^{T} \prod_{e_j=1} V_{ij} \bmod n,$$

$$(e_1, \cdots, e_k) = h(P_0, X_T).$$

Thus

$$Y_0^2 = X_T / \prod_{i=t+1}^{t+l} \prod_{e_j=1} V_{ij} = X_T / \prod_{e_j=1} V_{0j} \bmod n$$

holds, since

$$Y_T = \tilde{Y}_{t+l} \cdot \prod_{i=t+l+1}^{T} (R_i \cdot \prod_{e_j=1} S_{ij}) = Y_0 (\prod_{i=1}^{t} \prod_{e_j=1} S_{ij})(\prod_{i=t+l+1}^{T} \prod_{e_j=1} S_{ij}) \bmod n.$$

Finally,

$$X_T = Y_0^2 \cdot \prod_{e_j=1} V_{0j} \bmod n,$$

and

$$E_0 = h(P_0, X_T).$$

The probability that Y_0 is a correct answer in Step 5 is non-negligible from the assumption. Therefore, the expected running time of the above algorithm is polynomial in the size of $|V_{0j}|$. \hfill Q.E.D.

4. Efficiency

In this section, we compare the multisignature scheme based on the Fiat-Shamir scheme versus the previous multisignature schemes based on RSA scheme [O] and a direct iteration scheme for multisignature with the Fiat-Shamir scheme.

Processing speed   The amount of processing needed for our scheme is compared using the average number of modular multiplications required to generate signature, because f and h are much faster than modular multiplications.

The RSA based scheme requires $(3|n|)/2$ steps, the direct iteration scheme requires $t(k + 2)/2$ steps, and the our scheme requires $t(k + 6)/2$ steps, where $|n|$ means the bit length of n.

For example, when $tk = 72$, our scheme requires 39 steps (where $k = 72, t = 1$), while the direct iteration scheme requires 37 steps (where $k = 72, t = 1$), and the RSA based scheme requires 768 steps where $|n| = 512$.

Redundancy   We compare the three schemes by the information size stored as a signature in verifier's memory when m users join the signature system and $|n| = 512$. $\{|ID| \cdot m + k \cdot t + 512\}$ bits are stored as a signature in the Fiat-Shamir based scheme, while $\{|ID| \cdot m + 512\}$ bits are stored in the RSA based scheme and $\{(|ID| + 512 \cdot t + k \cdot t) \cdot m\}$ bits are stored in the iteration scheme of the Fiat-Shamir scheme.

Here, the value of $k \cdot t$ indicates the security level and must be at least 72, because the security level is comparable to the difficulty of factoring [FS]. The redundancy of signed message is restricted by security level, for example to 72 bits.

Numbers of Transmission   While the total numbers of message transmissions among m signers in the RSA based scheme and the iteration scheme are $(m - 1)$, the Fiat-Shamir based scheme requires $(2m - 1)$. If a bridge node exists in the multisignature scheme, the simultaneous version requires 3 steps for transmissions for any m, because massages can be broadcasted simultaneously [BLY].

5. Conclusion and Open problems

It has been shown that a digital multisignature scheme based on the Fiat-Shamir's signature scheme overcomes the problems inherent in previous schemes. Moreover, the security of Fiat-Shamir's multisignature schemes has been discussed. Our discussion focuses on the condition of passive attack. Therefore, the discussion about active attack is a remaining problem (e.g., see the note of subsection 2.2). In addition, the security of simultaneous multisignature scheme [BLY, GQ2] is a remaining problem against both passive and active attacks.

Finally, we conclude with an open problem relating to the efficiency: when a bridge node does not exist, is there any sequential multisignature schemes which can reduce the number of transmission to $(m - 1)$ with the same order of redundancy as a single signature scheme?

References

[BLY] Brickell, E., Lee, P. and Yacobi, Y.: "Secure Audio Teleconference," Advances in Cryptology - Crypto'87, Lecture Notes in Computer Science 293, 1988, pp.429-433

[D] Davies, D. W.: "Applying the RSA digital signature to electric mail," IEEE Computer (Feb. 1983), pp.55-62

[FFS] Feige, U., Fiat, A. and Shamir, A.: "Zero Knowledge Proofs of Identity," Proceedings of the 19th Annual ACM Symposium on Theory of Computing, 1987, pp.210-217

[FS] Fiat, A. and Shamir, A.: "How to Prove Yourself: Practical Solution to Identification and Signature Problems," Advances in Cryptology - Crypto'86, Lecture Notes in Computer Science 263, 1987, pp.186-199

[GQ1] Guillou, L.C., and Quisquater, J.J.: "A Practical Zero-Knowledge Protocol Fitted to Security Microprocessor Minimizing Both Tranamission and Memory," Eurocrypt'88 Abstracts, 1988, pp.71-75

[GQ2] Guillou,L.C., and Quisquater,J.J.:"A Paradoxical Identity-Based Signature Scheme Resulting from Zero-Knowledge," Proceedings of Crypto'88,Lecture Notes in Computer Science 403, 1988, pp.216-231

[GJ] Garey, M. R., and Johnson, D. S.: "Computers and Intractability - A Guide to the Theory of NP-Completeness," W. H. Greeman, San Francisco, 1979

[IN] Itakura, K., and Nakamura, K.: "A public-key cryptosystem suitable for digital multisignature," NEC J. Res. Dev. 71 (Oct. 1983)

[O] Okamoto, T.: "A digital Multisignature Scheme Using Bijective Public-Key Cryptosystems," ACM Trans. on Comp. Systems, Vol. 6, No. 8, 1988, pp.432-441

[OO] Ohta, K. and Okamoto, T.: "Practical Extension of Fiat-Shamir Scheme," Electron.Lett., 24, No. 15, 1988, pp.955-956 (Revised version: Proceedings of Crypto'88, 1988, pp.232-243)

[OS] Okamoto, T., and Shiraishi, A.: "A fast signature scheme based on quadratic inequalities," Proceedings of the IEEE Symposium and Provacy (Oakland, Calif., April, 1979), IEEE, New York, 1985, pp.123-132 (Revised version: IEEE Trans. Information Theory, Vol.IT-36, No.1, 1990, pp.47-53)

[OSK] Okamoto, T., Shiraishi, A., and Kawaoka, T.: "A Single Public-Key Authentication Scheme for Multiple Users," Technical report of IECE Japan, IN83-92 (January 1984) (Revised version: Systems and Computers in Japan, 18, 10, pp.14-24 (1987); translated from IECE Japan Transactions, J69-D, 10, pp.1481-1489 (1986)).

[R] Rabin, M. O.: "Digitalized signatures and public-key functions as intractable as factorization," Tech. Rep. MIT/LCS/TR-212, MIT, Cambridge, Mass., 1979

[RSA] Rivest, R.L., Shamir, A. and Adleman, L.: "A Method for Obtaining Digital Signatures and Public-Key Cryptosystems," Communication of the ACM, Vol. 21, No. 2, 1978, pp.120-126

[S] Shamir, A. : "Identity-based cryptosystems and signature schemes", Proceedings of Crypto'84, Lecture Notes in Computer Science 196, 1985, pp.47-53.

A Generalized Secret Sharing Scheme With Cheater Detection

Hung-Yu Lin, Lein Harn
Computer Science Telecommunications Program
University of Missouri-Kansas City
Kansas City, MO 64110

<Abstract>

A new secret sharing scheme is presented in this paper to realize the generalized secret sharing policy. Different from most of previous works, it is computationally secure and each participant holds only one single shadow. Any honest participant in this scheme can detect and identify who is cheating even when all of the other participants corrupt together. An extended algorithm is also proposed to protect the secret form dishonest participant without the assumption of simultaneous release of the shadows. With $(\times,\times)$-homomorphism property, it can also be used to protect individual secrets while revealing the product of these secrets.

I. Introduction

A secret sharing scheme is a method of hiding a secret among several shadows such that the secret can be retrieved by some subset(s) of these shadows but not by the others according to a given sharing policy. In real applications, the shadows are held by a group of participants with designated privileges and no one in the group is fully trusted by the others. Only if all of the participants of a "qualified" group agree to pool their shadows can this secret be reconstructed.

Since the introduction of threshold schemes in 1979 [4, 9], many secret sharing schemes have been proposed and extensively discussed in literature. However, most all of them pursue the unconditional security and deal with only special sharing policies. For example, Shamir's well-known (l,m)-threshold scheme can only realize the secret sharing policy in which any l or more than l participants can reconstruct the secret. This sharing policy is far too simple in many applications because, implicitly, it assumes that every participant in the group has equal privilege to the secret or every participant is equally trusted. Although this assumption can be removed by assigning different number of shadows to the participants with different privileges, it is still not robust enough to realize any sharing policy and the management of the multiple shadows held by each participant becomes a new problem.

In 1987, Saito, and Nishizeki [7] proposed a secret sharing scheme for the general access structure(explained in Section II). They tried to realize the general access structure by using the multiple shadows assignment approach. Later Benaloh and Leichter [3] proposed a simpler method of developing a secret sharing scheme by translating the access structure into a monotone formula. They also stated that "there exists access structures for which any generalized secret sharing must give some trustee shares which are from a domain larger than that of the secret". However, this conclusion may only applied to those secret sharing schemes without cryptographic assumption.

Different from most of previous works, we propose a computationally secure scheme, based on RSA assumption, to realize the generalized secret sharing policy with each participant keeping only one shadow and having the ability to detect and identify who is cheating. Most importantly, a dishonest participant can obtain the secret with only small probability even when the simultaneous release of shadows is not available. With $(\times,\times)$-homomorphism property, it can also be used to protect individual secrets while revealing the product of these secrets.

II. Basic Idea and Preliminaries

PROPOSITION 1. Given k, e and n, where $k=\alpha^e \bmod n$, and n is the product of two large primes, p and q, it is computationally infeasible to find α.

This is the well known RSA assumption.

THEOREM 1. Given k_1, k_2, e_1, and e_2 such that $k_1 = k^{e_1}$ mod n and $k_2 = k^{e_2}$ mod n, k^r mod n can be easily computed if $\gcd(e_1, e_2) = r$.

<Proof>: Since $\gcd(e_1, e_2) = r$, we can find one pair of (s_1, s_2) from Euclid algorithm such that $s_1 * e_1 + s_2 * e_2 = r$, so

$$(k_1)^{s_1} * (k_2)^{s_2} = (k)^{s_1 * e_1 + s_2 * e_2} = k^r \pmod{n}.$$

If k is the secret and k_1 and k_2 are the shadows assigned to participants u_1 and u_2, respectively, then by choosing e_1 and e_2 with $\gcd(e_1, e_2) = 1$, u_1 and u_2 together can reconstruct the secret. On the contrary, if $\gcd(e_1, e_2) \neq 1$, then u_1 and u_2 together cannot reconstruct the secret by PROPOSITION 1. This simple idea will be followed throughout the development of the generalized secret sharing scheme.

Now, with the RSA assumption in mind, let's give the generalized secret sharing scheme a more concise definition. Suppose a secret k is to be shared according to a given sharing policy by a group of m participants $U = \{u_1, u_2, \ldots, u_m\}$. Each participant may be designated with a different privilege. A generalized secret sharing scheme is a method of breaking k into m pieces $k_1, k_2, \ldots, k_m$, with k_i secretly distributed to u_i such that

(1) if $A \subseteq U$ is a qualified subset of participants, called positive access instance, according to the sharing policy, then k can be reconstructed from shadows $\{k_i \mid u_i \in A\}$.

(2) if $A \subseteq U$ is not a qualified subset of participants, called negative access instance, according to the sharing policy, then k cannot be reconstructed from $\{k_i \mid u_i \in A\}$.

The set F of all positive access instances is called the positive access structure of the sharing policy and the set N of all negative access instances is called the negative access structure of the sharing policy.

PROPOSITION 2. Every combination of the participants is either a positive access instance or a negative access instance, but not both. That is,

$F \cup N = 2^\mathcal{U}$ and $F \cap N = \phi$,

where $2^\mathcal{U}$ is the power set of $\mathcal{U}$.

PROPOSITION 3. If $\mathcal{A}$ is a negative access instance, then any subset $\mathcal{B}$, $\mathcal{B} \subseteq \mathcal{A}$, is also a negative access instance.

DEFINITION 1. The maximum set of A, $A \subseteq 2^\mathcal{U}$, is given as
$\mathfrak{M}(A) = \{\mathcal{A} \in A \mid \forall \mathcal{B} \in A, \mathcal{A} \not\subset \mathcal{B} \}$.

Let $U_i = \{\mathcal{A} \in 2^\mathcal{U} \mid u_i \in \mathcal{A} \}$ denote the set of all instances in which u_i is present, and $U'_i = \{\mathcal{A} \in 2^\mathcal{U} \mid u_i \notin \mathcal{A} \}$ denote the set of instances in which u_i is absent. Therefore, $U_i \cap U'_j = \{\mathcal{A} \in 2^\mathcal{U} \mid u_i \in \mathcal{A} \wedge u_j \notin \mathcal{A} \}$ denotes the set of all instances in which u_i is present and u_j is absent in all of them. For simplicity, "$\cap$" is omitted in the rest of the paper. According to the propositions and definitions described above, we have the following lemmas:

LEMMA 1. If $\mathcal{A}$ is a negative access instance and F is the negative access structure of the same sharing policy, then there exists at least one element $\mathcal{B}$ in $\mathfrak{M}(F)$ such that $\mathcal{A} \subseteq \mathcal{B}$.

LEMMA 2. $\mathfrak{M}(U'_{i_1} U'_{i_2} \ldots U'_{i_s}) = \{ \mathcal{U} - \{u_{i_1}, u_{i_2}, ., u_{i_s} \}\}$.

LEMMA 3. $\mathfrak{M}(U_{i_1} U_{i_2} \ldots U_{i_h} U'_{j_1} U'_{j_2} \ldots U'_{j_s}) = \mathfrak{M}(U'_{j_1} U'_{j_2} \ldots U'_{j_s})$, if $U_{i_x} \neq U'_{j_y}$ for $x \in [1, h]$ and $y \in [1, s]$.

LEMMA 4. If A_i's $\subseteq 2^\mathcal{U}$, for i = 1, 2, ..., s, then
$\mathfrak{M}(A_1 \cup A_2 \cup \ldots \cup A_s) = \mathfrak{M}((\mathfrak{M}(A_1) \cup \mathfrak{M}(A_2) \cup \ldots \cup \mathfrak{M}(A_s))$.

LEMMA 5. Let $\mathfrak{M}(A_i) = \{\mathcal{A}_i\}$, where $\mathcal{A}_i$ is a subset of $\mathcal{U}$, $i = 1$ to $|2^{\mathcal{U}}|$, then

$$\mathfrak{M}(\bigcup_{i=1}^{s} A_i) = \bigcup_{i=1}^{s} \mathfrak{M}(A_i) \quad (s \leq |2^{\mathcal{U}}|)$$

if $\mathcal{A}_i \not\subset \mathcal{A}_j$ and $\mathcal{A}_j \not\subset \mathcal{A}_i$.

III. Realization of the Generalized Secret Sharing Scheme

All of the existing secret sharing schemes proposed so far try to realize secret sharing policies from the viewpoint of "who can reconstruct the secret". Can the same problem be solved by another approach? Instead of realizing the generalized secret sharing scheme by analyzing the positive access structure, this paper will propose a different approach to realize the scheme by analyzing the negative access structure, especially the maximum set of the negative access structure. The reason will become clear in the next section.

Suppose the positive access structure of a given sharing policy is F. The corresponding negative access structure is therefore $N = 2^{\mathcal{U}} - F$. As we know, this negative access structure can be represented in the sum_of_product disjunctive normal form(DNF). The maximum set of the negative access structure, $\mathfrak{M}(N)$, can be easily found by applying lemmas 1-5.

Now the central authority assigns a distinct prime p_j to each negative access instance $\mathcal{N}_j$ of $\mathfrak{M}(N)$ and computes the tag t_i associated with participant u_i as

$$t_i = \prod_{u_i \in \mathcal{N}_j} p_j.$$

The shadows assigned to the participants are therefore computed as

$$k_i = k^{t_i} \mod n, \text{ for } i = 1, 2, \ldots, m.$$

Each shadow k_i is then secretly distributed to participant u_i.

Quite interestingly, we can find the similarity in assigning keys in a partial-order hierarchy proposed by Akl and Taylor [1] and assigning shadows in this generalized secret sharing scheme by analyzing the negative access structure. The fact that one's shadow may be derived from another shadow does not conflict the two secret sharing requirements described in section II and it gives us the flexibility in sharing secret among participants whose designated privileges have a partial-order relationship.

IV. Analysis

Now we can start to prove that this secret sharing scheme can satisfy the two requirements described in section II. The size of the tags associated with participants will also be discussed in this section.

THEOREM 2. No negative access instance can reconstruct the secret if shadows are created according to the above scheme.
<proof>: Any negative access instance is a subset of some element(s) of the maximum set of the negative access structure(by LEMMA 1), so the greatest common divisor of the tags associated with participants in it is greater than 1 and the secret cannot be reconstructed(by THEOREM 1).

THEOREM 3. Any positive access instance can reconstruct the secret if shadows are created according to the above scheme.
<proof>: Assume that the positive access instance $\mathcal{A}$ cannot reconstruct the secret and the greatest common divisor of the tags associated with participants in $\mathcal{A}$ is greater than 1. That is, $\mathcal{A}$ is a subset of some element(s) of the maximum set of the negative access structure and is therefore a negative access instance. This contradicts PROPOSITION 2, $F \cap N = \phi$. So, any positive access instance can reconstruct the secret if shadows are created according to the above scheme.

One potential problem for this generalized secret sharing scheme is the size of tags associated with each participant. From the algorithm described in the previous section we know that the number of primes required in this scheme is exactly the cardinality of the maximum set of the negative access structure N (that is, the number of products in the DNF). From Spernre's Theorem [9], the worst case happens when the cardinality of $\mathfrak{M}(N)$ is maximum, which is $C^m_{int(m/2)}$, and is exactly the cardinality of the maximum set of the negative access structure of the $(int(m/2) +1, m)$ -threshold scheme. This result is not what we woluld like to see since, in the worst case, the number of primes required grows exponential in m. But fortunately, the tags associated with participants are public information and therefore processing and storing these tags requires no security mechanism.

Furtheremore, for some secret sharing policies, with only several primes we can implement a generalized secret sharing scheme for a large number of participants. In fact, in the best case, with only m primes a secret can be shared by up to $C_1^m + C_2^m + \ldots + C_{m-1}^m = 2^m - 2$ participants. That is, for m participants it needs only $O(lg\ m)$ primes in this scheme. So, at least, this new scheme, provides an alternative when realizing a generalized secret sharing policy. In consideration of the cheater problem to be discussed in the next section, this new scheme shows its robutness in dealing with dishonest participants.

V. How to Deal with Cheaters

The cheating problem in the information-theoretic (l,m)-threshold scheme has been discussed in [5, 6, 10]. However, it has never been discussed for the generalized secret sharing schemes. Here a modified scheme, which idea is similar to that proposed by Tompa and Woll [10], to keep the secret from the cheater will be discussed.

THEOREM 4. Any honest participant can detect and determine who is cheating.

<proof> Suppose Alice is the honest participant with shadow, k_{Alice}, and Bob tries to cheat her by presenting a fake shadow, $k'_{Bob} = (k)^{t'_{Bob}}$, instead of his genuine shadow, k_{Bob}. From public information t_{Alice} and t_{Bob} with $\gcd(t_{Alice}, t_{Bob}) = g$, Alice can find a and b such that

$$a * t_{Alice} + b * t_{Bob} = g ,$$

then Allice will compute

$$(k_{Alice})^a (k'_{Bob})^b = (k)^{a * t_{Alice} + b * t'_{Bob}} = k^{g'} \neq k^g \pmod{n},$$

where k is the secret of the system. Since $(k^{g'})^{t_{Alice}/g}$ (mod n) is not equal to the value of $k^{t_{Alice}}$, Alice can easily detect and determine that Bob is cheating.

Even when Alice is the only honest participant, her ability to detect and determine who is cheating is not influenced by the corruption of the other participants.

However, the ability to detect and determine who is cheating cannot prevent the cheater from gaining more information than the honest participant. Tompa and Wool proposed a clever method to solve this problem. Here, this generalized secret sharing scheme can also solve the same problem.

Suppose the secret is k. The dealer first chooses a series of dummy secrets, $k_1, k_2,...,$ and k_j such that

$k_1 < k_2 < ......< k_{i-1}$ and $k_{i-1} > k$

then creates the shadows of $k_1, k_2,..., k_{i-1}, k, k_{i+1},...,$ and k_j according to the generalized secret sharing scheme and distributes to each participant their shadows of $k_1 ,......,$ and k_j (in fact, the shadows of $k_{i+1}, k_{i+2},...,$ and k_j can be randomly chosen without following the origional scheme). When participants of any access instance agree to reconstruct the secret k, they have to reconstruct $k_1, k_2,...,$and k_{i-1} and k one at a time until they get $k, k < k_{i-1}$, and can then be sure that k is the secret. To keep the secret from cheaters, every time participants have to present their shadows simultaneously. Without knowing the position of k in this series, the probability for the cheater to obtain the secret by presenting a fake shadow is about $1/j$, where j is the number of the shadows distributed to each participant. Even if the simultaneous release of shadows is not supported, the scheme can also be slightly modified to fullfill the requirements due to the cheater detecting ability discussed above. For example, the participants can reveal their own shadows one by one according to some predefined order which may be different for each candidate secret reconstructed. For such arrangement, a cheater can obtain the secret only in the round in which the candidate secret to be reconreconstructed is the real secret and he is the last participant that releases the shadow.

VI. Secret Sharing Homomorphism

The property of secret sharing homomorphism was first introduced by Benaloh in 1986 [2]. Again, this property was first found in the traditional (l,m)-threshold scheme. The secret sharing homomorphism can be stated as follows:

Suppose k_1 and k_2 are two secrets and $k_{i,1}$'s and $k_{i,2}$'s are the shadows of k_1 and k_2, respectively, for $i =1, 2,...,l$. $\mathfrak{f}$ is the function to reconstruct the secret from the shadows and $\oplus$ and $\otimes$ are the binary operations on elements of the secret domain and of the shadow domain, respectively. A (l,m)-threshold scheme has the $(\oplus,\otimes)$-homomorphism property if

$$k_1 = \mathfrak{f}(k_{1,1}, k_{2,1}, ..., k_{l,1}) \text{ and } k_2 = \mathfrak{f}(k_{1,2}, k_{2,2}, ..., k_{l,2})$$

then

$$k_1 \oplus k_2 = \mathfrak{f}(k_{1,1} \otimes k_{1,2}, ..., k_{l,1} \otimes k_{l,2}).$$

That is, $k_{i,1} \otimes k_{i,2}$'s are the shadows of $k_1 \oplus k_2$.

Shamir's (l,m)-threshold scheme is (+,+)-homomorphic and it can be transformed to be (+,×)-homomorphic. Obviously, this new generalized secret sharing scheme has the (×, ×)-homomorphism property. From this property, participants can work together to reconstruct the product of some secrets without revealing the individual secrets.

VII. Conclusion

A computationally secure secret sharing scheme is presented in this paper which can realize the generalized secret sharing policy. Each participant holds only one shadow and any honest participant can detect and determine who is cheating in the process of secret reconstruction. The protection of the secret from cheaters can also be achieved through the same scheme by assigning extra dummy shadows. This generalized secret sharing scheme has also the (×,×)-homomorphism property which can be applied in the release of the product of secrets without revealing the individual secrets.

Reference

[1] Akl, G.S. and P.D. Taylor, "Cryptographic Solution to A Multilevel Security Problem", Proc. Crypto '82, Plenum Press, 237-250.

[2] Benaloh, J., "Secret Sharing Homomorphisms: Keeping Shares of a secret secret", Proc. Crypto '86, Springer-Verlag, 251-260.

[3] Benaloh, J. and J. Leichter, "Generalized Secret Sharing and Monotone Functions", Proc. Crypto '88, 27-35.

[4] Blackly, G., "Safeguarding Cryptographic Keys", Proc. NCC, Vol. 48, AFIP Press, 1979, 313-317.

[5] Brickle, E.f., and D.R. Stinson, "The Detection of Cheaters in Threshold Schemes", Proc. Crypto '88, Springer-Verlag, 564-577.

[6] Chor, B., S. Goldwasser, S. Micali,, and B. Awerbuch, "Verifiable Secret Sharing and Achieving Simultaneity in the Presence of Faults", Proc. 26th IEEE Symp. on Fooundations of Computer Science, 1985, 383-395.

[7] Ito, M, A. Saito, and T. Nishizeki, "Secret Sharing Scheme Realizing General Access Structure", Proc. Glob. Com(1987).

[8] Lubell, D., J. Comb. Th.1(1966), 299.

[9] Shamir, A. "How to Share a Secret", Comm. of the ACM 22, 11, Nov. 1979, 612-613.

[10] Tompa, M., and H. Woll, "How to share a secret with cheaters", J. of Cryptology 1(1988), 133-138.

Generalized Threshold Cryptosystems

Chi Sung Laih

Department of Electrical Engineering

National Cheng Kung University

Tainan, Taiwan, ROC

Lein Harn

Computer Science Telecommunications Program

University of Missouri - Kansas City

Kansas City, MO 64110, USA

<Abstract>

In a threshold cryptosystem, one can send an encrypted message to a group without knowing the internal secret sharing policy of the group. The encrypted ciphertext can only be deciphered by some users of the group according to the secret sharing policy. In this paper, we propose solutions to handle the generalized secret sharing policy. In addition, we investigate two different models for the group: one with a mutually trusted party in the group and the other one without.

I Introduction

In a group oriented cryptosystem, one can send an encrypted message to this group without knowing the internal secret sharing policy. By doing so, each group (organization), rather than each individual user, has a single encryption key known to the outside. However, within each group, the encrypted ciphertext can only be deciphered by some subset(s) of the users but not by others according to a given secret sharing policy. The cryptosystem must be designed in such a way that the sender needs only one encryption key to encrypt a message to the group. On the other hand, internal users can decrypt the message only when they form a qualified subset of users according to the secret sharing policy.

Some earlier solutions [1,2] to the above problem require interactive protocols and are very impractical. Frankel [3] proposed a solution employing a trusted clerk or tamperfree modulars to distribute the encrypted message. A noninteractive solution which incorporates the concept of a (t,n) threshold scheme and ElGamal's cryptosystem [4] was proposed by Desmedt and Frankel [5] in 1989. In their system, each organization has a single public key and the outside sender can use this key to encrypt the message and broadcast the ciphertext to the organization. Each user in the organization will first calculate a modified shadow in Z_{q-1}, where $q-1$ is a prime, and then compute their "partial result" in $GF(2^m = q)$ and transmit this result to a designated individual. The designated individual will be able to decipher the ciphertext if there are t partial results available. Since the (t,n) threshold scheme has been used, the RSA encryption scheme [6] cannot be incorporated in the design [5].

As shown by Benaloh and Leichter [7], a (t,n) threshold scheme can only handle a small fraction of the secret sharing policy which people may wish to follow. Their paper proves that there are many access structures in which the conventional (t,n) threshold scheme cannot work efficiently.

In this paper we propose a generalized threshold cryptosystem to accommodate generalized access structures for two kinds of group. In the first group, there is a mutually trusted party to determine the public key for the group and the secret keys for each user according its internal access policy. In the second group, there is no mutually trusted party. The public key for the group is determined by all users in the group and users can choose their own private keys individually. In our generalized threshold cryptosystem with mutually trusted party, we also show how the RSA scheme and the Diffie-Hellman public key distribution scheme [8] can be incorporated in the design.

II. Generalized threshold cryptosystem with mutually trusted party

In this model, we assume that there is a mutually trusted center in the group. The job of the center is to determine the public key of the group and generate the private keys for all members according to the access structure. Our scheme has two phases:

(A) key generation

We assume that the access structure within the group can be represented by the disjunctive normal form (DNF):
$$F = f_1 + f_2 + ... + f_t,$$
where each f_i is an access instance. For example, for a group of four users A, B, C and D, the access structure allows the ciphertext to be decrypted by either A and B together or by C and D together. We can represent this access structure as
$$F = AB + CD,$$
where AB and CD are the access instances and $AB = f_1$, $CD = f_2$. Since A belongs to access instance f_1, we can represent this fact by $A \in f_1$.

The trusted center will construct an RSA scheme for the group. Let (e, N) and (d, p, q) be the public key and private key of the group respectively, where $e*d = 1 \pmod{\phi(N)}$ and N is the product of two large primes p and q. The center chooses a generator α over Z_N. Let s be the total number of users in the group.

<u>Step 1</u> For each user i, the center randomly chooses a secret key $x_i \in [1, \phi(N)]$
$(1 \leq i \leq s)$ such that

$$d = \sum_{i=1}^{s} x_i \pmod{\phi(N)}.$$

Step 2 For each access instance f_j, the center also chooses a random secret key $v_j \in [1, \phi(N)]$ $(1 \le j \le t)$.

Step 3 Based on the Diffie-Hellman's public key algorithm, the center calculates the public keys for each user i and access instance f_j as

$$y_i = \alpha^{x_i} \pmod{N}, \ 1 \le i \le s,$$

$$z_j = \alpha^{v_j} \pmod{N}, \ 1 \le j \le t.$$

Step 4 The center calculates the common secret keys $K_{i,j}$, for $1 \le i \le s$ and $1 \le j \le t$, shared between each user i and access instance f_j as

$$K_{i,j} = \alpha^{x_i v_j} \pmod{N}$$

$$= (y_i)^{v_j} \pmod{N}$$

$$= (z_j)^{x_i} \pmod{N}$$

Step 5 The center calculates a public value T_j for each access instance f_j such

that
$$\sum_{i \notin f_j} x_i = \sum_{i \in f_j} K_{i,j} + T_j \pmod{\phi(N)}. \qquad (1)$$

The center distributes the secret keys x_i to the users secretly and publishes the public keys y_i, z_j and T_j. The secret key for each access instance can be discarded and the center also can be closed completely.

(B) Encryption and Decryption

Encryption: Any outside sender can use the single public key (e, N) of the group to encrypt the message m into ciphertext C as

$$C = m^e \pmod{N},$$

and then broadcast the ciphertext C to all users in the group.

Decryption: We assume that f_j is one of the access instances and each user $i \in f_j$ is willing to work together to decrypt C. Then each user i needs to calculate his/her partial results as

$$C^{x_i + K_{i,j}} \pmod{N},$$

where $K_{i,j}$ is the common secret key between user i and access instance f_j. Each user will send this partial result to one designated individual. Once all partial results have been received by the designated individual, the ciphertext C can be decrypted as

$$(\prod_{i \in f_j} C^{x_i + K_{i,j}}) C^{T_j} \pmod{N}$$

$$= C^{\sum_{i \in f_j} x_i} C^{(\sum_{i \in f_j} K_{i,j}) + T_j} \pmod{N}$$

$$= C^{\sum\limits_{i \in f_j} x_i} C^{\sum\limits_{i \notin f_j} x_i} \quad (\text{mod } N)$$

$$= C^{\sum\limits_{i=1}^{s} x_i} \quad (\text{mod } N)$$

$$= C^d = m \quad (\text{mod } N).$$

Example 1 Assume that the group has four users A, B, C and D. The access structure of the group can be described as $F = f_1 + f_2 = AB + CD$. In Figure 1 we indicate the private and public keys associated with each user and access instance.

When users A and B want to decrypt the ciphertext $C = m^e$ (mod N), they need to calculate the partial results $C^{x_A + K_{A,1}}$ (mod N) and $C^{x_B + K_{B,1}}$ (mod N) separately, and send their partial results to the designated individual. The designated individual then decrypts the ciphertext as

$$C^{x_A + K_{A,1}} C^{x_B + K_{B,1}} C^{T_1} \quad (\text{mod } N)$$

$$= C^{x_A + x_B + x_C + x_D} \quad (\text{mod } N)$$

$$= C^d \quad (\text{mod } N)$$

$$= m.$$

Discussions: It is obvious that our scheme is noninteractive and has one single public key for each group. The sender does not need to know any internal information of the group (e.g. the secret sharing policy, public key for each user). Only if all users of an access instance work together can the ciphertext be decrypted correctly. Our scheme incorporates the RSA and Diffie-Hellman public key schemes and all operations are accomplished over Z_N. The security of our scheme heavily relies on solving Eq.(1). That is, given a public value T_j for each access instance f_j, can we find the sum of secrets x_i for user i when $i \notin f_j$? It is obvious that solving Eq.(1) for the sum of the secrets x_i, $i \notin f_j$, needs to determine the sum of the common keys K_{ij} for all users i and access instance f_j, $i \in f_j$. However, it is equavalent to solving the Diffie-Hellman public key distribution system. And it is believed that the problem is equivalent to solving discret logarithm over Z_N [8]. On the other hand, the security of the proposed encryption scheme also relies on the factorization probelm since it uses RSA as the basic encryption and decryption functions. Note that this system can also work successfully by using the ElGamal's encryption scheme [4] over GF(P). In that case, the security of the overall system only relies on the discret logarithm problem.

III. Generalized threshold cryptosystem without mutually trusted party

As pointed out by Ingemarsson and Simmons [9], in most applications there does not exist any trusted party in the group. This situation becomes common in some commercial and/or international applications. Under this condition, the scheme presented in the previous section cannot work properly. Here, we propose a modified scheme which allows all users in the group to determine the group key and each user can select their own secret key individually.

(A) key generation

We assume that the access structure in the group can be represented as
$$F = f_1 + f_2 + \ldots + f_t,$$
where each f_i is an access instance. Since there is no trusted party, the public and private keys of the group are determined by all users in the group. Assume all users agree to use a prime number P as the modulo and a primitive number α over GF(P) as the generator. Then each user randomly selects his/her private key $x_i \in [1, P-1]$ $(1 \le i \le s)$ and publishes his/her public key $y_i = \alpha^{x_i} \pmod P$. The public key for the group can be computed as

$$y = \prod_{i=1}^{s} y_i \pmod P.$$

The secret decryption key of the group is $x = \sum_{i=1}^{s} x_i \pmod{P-1}$ and it is not known by any individual user. Any user i who does not belong to the access instance f_j needs to publish a corresponding public key $T_{i,j}$ that satisfies

$$x_i = \sum_{h \in f_j} K_{i,h} + T_{i,j} \pmod{P-1}. \tag{2}$$

(B) Encryption and Decryption

Encryption: Any outside sender can use the ElGamal scheme to encrypt the message m. The ciphertext (C_1, C_2) can be computed as

$$C_1 = \alpha^r \pmod P,$$

$$C_2 = my^r \pmod P,$$

where r is a random number within [1, P-1] and broadcast (C_1, C_2) to all users in the group.

Decryption: We assume that f_j is one of the access instances and each user $i \in f_j$ is willing to work together to decrypt (C_1, C_2). Then each user i needs to calculate his/her partial result as

$$C_1^{x_i} \prod_{h \notin f_j} C_1^{K_{i,h}} \pmod P$$

where $K_{i,h}$ is the common secret key between users i and $h (h \notin f_j)$. Each user will send this partial result to one designated individual. Once all partial results have been received by the designated individual, the ciphertext (C_1, C_2) can be decrypted as

$$C_2 (\prod_{i \in f_j} (C_1^{x_i} \prod_{h \notin f_j} C_1^{K_{i,h}})(\prod_{h \notin f_j} C_1^{T_{h,j}}))^{-1} \pmod P$$

$$= m y^r ((\prod_{i \in f_j} C_1^{x_i})(\prod_{i \in f_j} (\prod_{h \notin f_j} C_1^{K_{i,h}} C_1^{T_{h,j}})))^{-1} \pmod P$$

$$= m y^r (\prod_{i \in f_j} C_1^{x_i} \prod_{h \notin f_j} C_1^{\sum_{i \in f_j} K_{i,h} + T_{h,j}})^{-1} \pmod P$$

$$= m y^r (\prod_{i \in f_j} C_1^{x_i} \prod_{h \notin f_j} C_1^{x_h})^{-1} \pmod P$$

$$= m y^r (C_1^{\sum_{i=1}^{s} x_i})^{-1} \pmod P$$

$$= m y^r (\alpha^{x^r})^{-1} \pmod P$$

$$= m \pmod P$$

<u>Example 2</u>   Assume that the group has four users A, B, C and D. The access structure of the group can be described as $F = f_1 + f_2 = AB + CD$. In Figure 2, we indicate the private and public keys associated with each user.

When users A and B want to decrypt the ciphertext (C_1, C_2), they need to calculate the partial results $C_1^{x_A + K_{A,C} + K_{A,D}} \pmod P$, $C_1^{x_B + K_{B,C} + K_{B,D}} \pmod P$ separately, and send their partial results to the designated individual. The designated individual then decrypts the ciphertext as

$$C_2 ((C_1^{x_A + K_{A,C} + K_{A,D}})(C_1^{x_B + K_{B,C} + K_{B,D}})(C_1^{T_{C,1} + T_{D,1}}))^{-1} \pmod P$$

$$= m y^r ((\alpha^x)^r)^{-1} \pmod P$$

$$= m \pmod P.$$

Discussions: The security of this modified scheme is very similar to the previous one.   The only difference is that there is no mutually trusted party in the group.   For any user i who does not belong to access instance f_j, he/she needs to calculate and publish a corresponding public key $T_{i,j}$. The security of this modified scheme heavily relies on solving Eq.(2). It is very similar to solving Eq.(1). Based on the discussions in previous section, it is obvious that the security of our scheme is equivalent to the difficulty of solving the discrete logarithm problem.

IV. Conclusion

Two threshold cryptosystems have been proposed: one with a mutually trusted party in the group and the other one without.   Our scheme can also handle a generalized secret sharing policy.   All computations can be accomplished either over Z_N or GF(P).

References

[1] O. Goldreich, S. Micali, and A Wigderson. How to play any mental game. In Proceedings of the Nineteenth ACM Symp. Theory of Computing, STOC, pp. 218-229, May 25-27, 1987.

[2] Y. Desmedt. Society and group oriented cryptography: a new concept.   In Advances in Cryptology, Proc. of Crypto '87, pp. 120-127, August 16-20, 1988.

[3] Y Frankel.   A practical protocol for large group oriented networks.   To appear in : Advances in Cryptology, Proc. of Eurocrypt '89, April, 1989.

[4] T. ElGamel.   A public key cryptosystem and a signature scheme based on discrete logarithms.   In IEEE Trans. Inform. Theory, Vol. IT-31, pp. 469-472, July, 1985.

[5] Y. Desmedt, and Y. Frankel.   Threshold cryptosystem. In Advances in Cryptology, Proc. of Crypto '89, pp. 307-315, August 20-24, 1989.

[6] R. L. Rivest, A. Shamir, and L. Adelman.   A method for obtaining digital signatures and public-key cryptosystem. In Commun. of ACM, Vol. 21, No. 2, pp. 120-126, Feb. 1978.

[7] J. Benaloh, and J. Leichter.   Generalized secret sharing and monotone functions.   In Advances in Cryptology, Proc. of Crypto '88, pp. 27 35, August 21-25, 1988.

[8] W. Diffie, and M. E. Hellman.   New directions in cryptography.   In IEEE Trans. Inform. Theory, Vol. IT-22, pp. 644-654, Nov. 1976.

[9] I. Ingemarsson, and G. J. Simmons.   A protocol to set up shared secret schemes without the assistance of a mutually trusted party.   To appear in : Advances in Cryptology, Proc. of Eurocrypt '90, May 21-24, 1990.

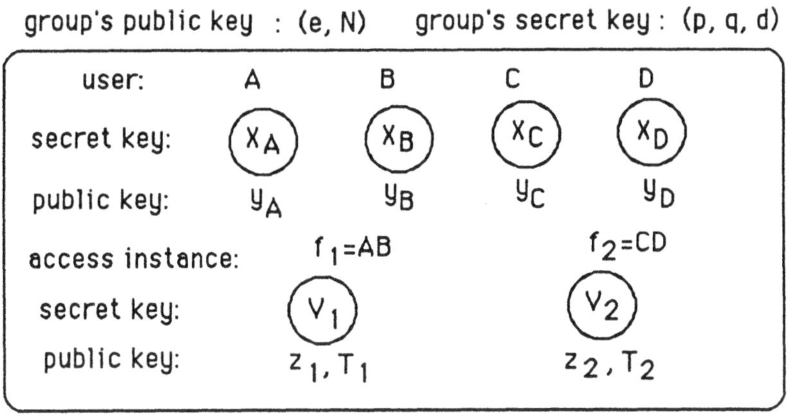

group's public key : (e, N) group's secret key : (p, q, d)

user:	A	B	C	D
secret key:	x_A	x_B	x_C	x_D
public key:	y_A	y_B	y_C	y_D
access instance:	$f_1 = AB$		$f_2 = CD$	
secret key:	v_1		v_2	
public key:	z_1, T_1		z_2, T_2	

where $d = x_A + x_B + x_C + x_D \pmod{\phi(N)}$

$$y_i = \alpha^{x_i} \pmod{N}, \ i = A, B, C \text{ and } D$$

$$z_j = \alpha^{v_j} \pmod{N}, \ j = 1 \text{ and } 2$$
$$x_C + x_D = K_{A,1} + K_{B,1} + T_1 \pmod{\phi(N)}$$
$$x_A + x_B = K_{C,2} + K_{D,2} + T_2 \pmod{\phi(N)}$$

Figure 1. The public and private keys of example 1.

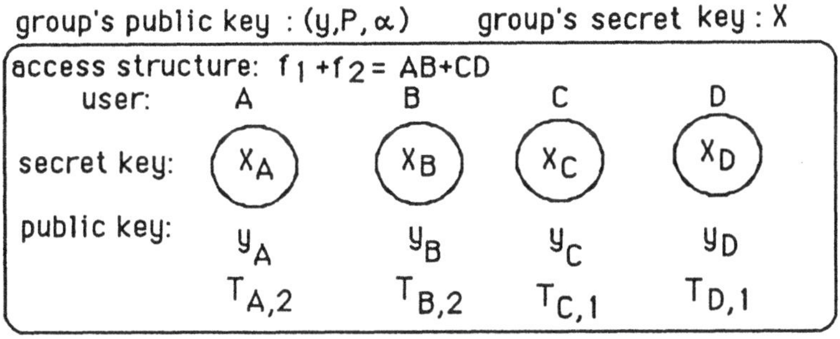

group's public key : (y, P, α) group's secret key : X

access structure: $f_1 + f_2 = AB + CD$

user:	A	B	C	D
secret key:	x_A	x_B	x_C	x_D
public key:	y_A	y_B	y_C	y_D
	$T_{A,2}$	$T_{B,2}$	$T_{C,1}$	$T_{D,1}$

where $X = x_A + x_B + x_C + x_D \pmod{P-1}$, $y = y_A y_B y_C y_D \pmod{P}$

$$y_i = \alpha^{x_i} \pmod{P}, \ i = A, B, C \text{ and } D$$

$x_A = K_{A,C} + K_{A,D} + T_{A,2} \pmod{P-1}$, $x_B = K_{B,C} + K_{B,D} + T_{B,2} \pmod{P-1}$

$x_C = K_{C,A} + K_{C,B} + T_{C,1} \pmod{P-1}$, $x_D = K_{D,A} + K_{D,B} + T_{D,1} \pmod{P-1}$

Figure 2. The public and private keys of example 2.

Feistel Type Authentication Codes

Reihaneh Safavi-Naini *

Department of Maths., Stats. and Computing Science
University of New England
Armidale, NSW 2351, AUSTRALIA

December 19, 1991

Abstract

In this paper we generalise Luby-Rackoff construction of pseudorandom permutation generators to generalised invertible function generators and prove that
if there exits a generalised pseudorandom function generator then there exist a
generalised pseudorandom invertible generator. This construction is then used for
a pseudorandom authentication code which offers provable security against T-fold
chosen plaintext/ ciphertext attack and provable perfect protection against strong
spoofing of order T. The performance of the code is compared with that of a code
obtained from a Feistel type permutation generator. The code, called Feistel type
A-code, provides a new approach to the design of practically good A-codes and
hence is of high practical significance.

1   Introduction

Theoretical study of authentication systems has followed a path similar to secure systems,
i.e., initially the theoretical limits of the authentication channel are established [1] and
then the codes that reach these limits were sought. The computation model was the ideal
case when no bound on the computational power of the enemy was set.

In [2] Simmons suggested that only classical combinatorial constructions have sufficient symmetry required for "good" authentication codes. The constructions proposed
by Stinson [3], [4] and De Soete [5], [6] can be seen as the realisation of this suggestion.
However these constructions suffer from either a large amount of key to be exchanged or
the small size of message space which ultimately render them impractical.

---

*Support for this project was provided in part by Australian Research Council grant A49030136 and
TELECOM Australia under the contract number 7027

The next natural and important step is to seek for "good" codes in a finite computational model. In this paper we study the authentication scenario of Simmons [2] when the computational model corresponds to polynomially bounded resources.

Traditionally authentication is achieved by appending a key dependent hash value to the message or concatenating the message with a known bit pattern and then applying a block cipher algorithm to spread the added redundancy over the whole block [7]. The assessment of such systems is usually limited to estimating the probability of success of an intruder if he/she randomly chooses a key or a cryptogram from the remaining cryptograms as it is deemed that the cryptographic algorithm is secure and doesn't reveal any information about the key or the rest of cryptogram space.

In [8] a construction of an A-code that offers perfect protection against impersonation and substitution when the enemy has polynomially bounded reources is given. This construction is then extended to A-codes that can *provably* provide protection against strong spoofing of order T [9]. However the major shortcoming of this latter construction is that the amount of redundancy is high and fixed, i.e., the size of cryptogram is twice the size of the message. This is a serious obstacle in most cases when less redundancy is affordable. The code uses at least four rounds, and up to ten rounds depending on the type of added redundancy, of Feistel type cipher to achieve provable protection.

We introduce a class of A-codes with secrecy that is provably secure against chosen T-fold plaintext attack and also can be proved to limit the best strategy of the enemy in a strong spoofing attack of order T to that of impersonation in which case the best strategy of the enemy is random selection from the cryptogram space. Hence the code provides provable perfect protection against strong spoofing of order T, also. The construction is similar to Feistel type block ciphers and consists of a number of identical rounds. The input message is divided into two parts (not necessarily equal) and in each round an expansion function is applied to provide confusion and add redundancy. The two parts of the message are then interchanged to provide diffusion.

It is proved that only three rounds is required to achieve provable perfect protection and provable security. For these codes the probability of deception of the enemy can be made arbitrarily low by using proper expansion functions and large n. The construction is particularly important as it suggests a new approach to the design of A-codes that are 'good' in a practical sense. It is also instructive to note the parallelism of the results to those of the block ciphers which can be interpreted as the extension of Shannon's principle of confusion/diffusion for good block cipher systems to confusion/expansion/diffusion for good authentication systems.

Firstly we give all the preliminary definitions and then proceed to define A-code generators. The assessment of these codes in terms of the amount of security and protection offered by them is discussed and pseudorandom A-codes (PA-codes) which are secure against T-fold plaintext/ciphertext attack and can effectively limit the probability of success of an intruder in a strong spoofing attack of order T, to that of an impersonation attack are introduced. Next the construction of PA-codes that can be considered as Feistel type construction is given and the superior performance of them compared to the traditional A-codes obtained from provably secure block ciphers is established. This comparison is further extended in the concluding section which also discusses the interpretation of a Feistel type construction for future research on practically 'good' A-codes.

2 Preliminaries

Let $V_n = \{0,1\}^n$ be the set of all binary strings of length n. For $x, y \in V_n$, denote by $x \oplus y$, the n-tuple $z \in V_n$ obtained from bitwise modulo two addition of x and y. The set of all functions from V_n to V_m is denoted by $F_{n,m}$, i.e.,

$$F_{n,m} = \{f | f : V_n \rightarrow V_m\}$$

and $|F_{n,m}| = 2^{m2^n}$. If $n = m$ then F_n is the set of functions from V_n to V_n and $|F_n| = 2^{n2^n}$.

For $m \geq n$, $f \in F_{n,m}$ has an inverse $\bar{f}$ if $\bar{f}(f(x)) = x$ for all $x \in V_n$.

Definition 2.1 *Let $l(n)$ and $m(n)$ be polynomials in n. A generalised function generator with key length $l(n)$ is a family of functions $f = \{f^{n,m}\}$ where, for each key k of length $l(n)$, $f_k^{n,m} \in F_{n,m}$ and, for all $x \in X$, $f_k^{n,m}$ is computable in polynomial time.*

A generalised function generator is invertible if every function $f_k^{n,m}$ is invertible. This implies $m \geq n$.

An *expansion generator* is a generalised invertible function generator with $m > n$.

The similarity of a generalised function generator to a family $F = \{F_{n,m}^R\}$ of random functions, $|F_{n,m}^R| = 2^{l(n)}$, can be determined by introducing the concept of distinguishing circuit family.

Definition 2.2 *An oracle circuit $C_{n,m}$ for $f^{n,m}$ is an acyclic circuit consisting of AND, OR, NOT and constant gates together with t oracle gates. Each oracle gate can realise a function of $F_{n,m}$, i.e., takes an input of length n and produces an output of length m using the selected member of $F_{n,m}$. The output of the circuit is one bit.*

Definition 2.3 *A distinguishing circuit family $\{C_{n,m}\}$ for f is an infinite class of oracle circuits, $C_{n_1,m_1}, C_{n_2,m_2}, \cdots, n_1 < n_2 < n_3 \cdots$ such that there exists two constants s and c such that*

- *the size of $C_{n,m}$ is smaller than n^c (size is defined as the number of all connections between gates);*

- *if $P[C_{n,m}(f^{n,m})]$ and $P[C_{n,m}(F_{n,m}^R)]$ denote the probability of an output bit of one when the oracle gates are evaluated using a function randomly selected from $f^{n,m}$*

(randomly choosing index) and $F_{n,m}^R$ respectively, the probability of distinguishing the two function generators satisfy

$$|P[C_{n,m}(f^{n,m})] - P[C_{n,m}(F_{n,m}^R)]| > \frac{1}{n^s}.$$

We define f to be a *pseudorandom function generator* if there is no distinguishing circuit family for f.

The above definitions are extensions of Luby-Rackoff's definitions [10] to the case when $n \neq m$.

In GGM [11] it is shown that it is possible to construct a generalised pseudorandom function generator from cryptographically secure bit generators(CSB). According to this construction to construct $f^{n,m}$ with key length $l(n)$ two CSB generators G and G' are required. G maps $l(n)$ input bits to $2l(n)$ bits while G' maps an input of length $l(n)$ bits to an output of length $m(n)$ bits. The existence of CSB generators is ultimately connected to the fundamental question of existence of one-way-functions [12].

A *permutation generator* $p = \{p^n\}$ is an invertible function generator with $n = m$ and $p^n \in P_n$, P_n is the set of permutation functions of F_n. Pseudorandom permutation generators can be used for symmetric block encryption with provable security. Luby-Rackoff constructed pseudorandom permutation generators such that p^{2n} uses $f^n \in F_n$. Their construction consists of a number of consecutive rounds similar to a Feistel cipher. They proved that after three rounds, for large enough n, the output is not distinguishable from a truly random block of length $2n$. This result was later generalised to the case when two functions from f^n are available and finally to the case of only one function [13],[14]. Pseudorandom permutation generators result in provably secure block symmetric algorithms under the assumption that a pseudorandom bit generator exists.

3 Pseudorandom Authentication Code

3.1 Authentication Code

We consider an authentication scenario in which a transmitter wants to send the state of a source (message) to a distant receiver over a publicly exposed channel. An enemy tries to deceive the receiver in making it accept a fraudulent message originated by him/her as a genuine one sent by the transmitter. He/she uses an *impersonation attack* by injecting a false message, *substitution attack* in which a received message from the transmitter is replaced by the one devised by the enemy and finally a *spoofing attack of order T* in which the knowledge of T intercepted messages is used to devise a message to substitute the $T+1^{th}$ message (impersonation is spoofing of order zero and substitution is of order one). We assume each message is used by the transmitter only once and hence the T received messages are all distinct.

An *authentication code* is a family of invertible mappings, indexed by key, from the set X of messages to the set Y of possible codewords (valid cryptograms). Each mapping

A_i specifies a subset $A_i(x) \subset Y$ of authentic messages such that it is easy for an insider (who knows the key) to generate $A_i(x)$, for all $x \in X$, and also, for all $y \in Y$, to verify if $y \in A_i(x)$ holds. However both of these are hard for an outsider (who does not have the key).

An A-codes can be with secrecy or without secrecy. In an A-code with secrecy, a codeword is in fact a cryptogram and its corresponding message remains secret while in a code without secrecy there is an easy algorithm to derive the message from the codeword even without the knowledge of the key. In the rest of this paper we restrict ourselves to A-codes with secrecy.

For an A-code with secrecy we define a *strong spoofing of order T* as an attack in which the enemy can choose a set X_T of T messages and the set Y_T of the corresponding cryptograms will be given to him/her. The enemy's goal is to devise a cryptogram $y \notin Y_T$ that is acceptable by the receiver under the secret key. This is in fact the counterpart of chosen plaintext attack for authentication systems. Assessment of the authentication capability of the system is by calculating the probability of success of an enemy in an attack which has the maximum probability of success, i.e., when the best strategy is used. An *explicit authentication code* is an A-code for which the redundancy added to the code is explicit and independant of the key. For example in a traditional A-code a fixed bit string is added to the message and then a cryptographic algorithm is used to spread the redundancy over the whole block. In an *implicit authentication code* redundancy is added during the encoding process and depends on the key.

An A-code can offer protection only if $|Y| > |X|$, i.e., the code adds redundancy to the messages. Let $|X| = 2^k$ and $|Y| = 2^n$.

Definition 3.1 *(Massey, [15]) An A-code provides perfect type (n,k) protection against spoofing of order T if the enemy's best strategy in such attack is random selection from $Y^T = Y \backslash Y_T$ where Y_T is the set of intercepted cryptograms.*

3.2 Pseudorandom A-codes

Definition 3.2 *An A-code generator $A = \{A^{n,m}\}$ with key length function $l(n)$ and block size function $m(n) > n$ is a family of invertible generalised functions where $A^{n,m}$ is an authentication code with message set V_n and cryptogram set V_m.*

This definition corresponds to an expansion generator.

Assessment of an A-code generator is by studying the probability of success of an enemy in a strong spoofing attack. It is known [10] that pseudorandomness can provide for provable security. We will show that it is also sufficient for provable perfect protection.

To assess protection offered by an A-code generator we note that prior to the knowledge of T message/cryptogram pairs the enemy's best strategy is the best impersonation strategy. When a set X_T of messages is chosen and the corresponding set Y_T of cryptograms is given, the first stage is to be able to distinguish Y_T from the set Y_T^R of T randomly chosen cryptograms. Only if this distinguishing is possible is a review of the strategy necessary. Otherwise the best strategy will remain the same as in the impersonation case (the effective size of cryptogram space is reduced). Hence effectively the assessment is performed in two stages. In the first stage the enemy tries to devise a distinguishing circuit family for the expansion generator and only proceeds to the second stage if the first stage is successful. Hence an oracle circuit with T oracle gates constitutes the first stage of a *strong spoofing attack of order T*.

An A-code generator $A = \{A^{n,m}\}$ is pseudorandom (PA-code) if A is a pseudorandom expansion generator.

Lemma 3.1 *In a PA-code the best strategy of the enemy in a strong spoofing of order T is effectively the same as impersonation on a cryptogram set whose size is reduced by T and the pobability of having a best impersonation strategy other than random selection from the cryptogram space can be made arbitrarily small by choosing large block size.*

Proof(sketch):

Because of the pseudorandomness, for an arbitrary ϵ, there exists an n large enough such that

$$|P[C_{n,m}(f^{n,m})] - P[C_{n,m}(F^R_{n,m})]| < \epsilon,$$

i.e., the set Y_T is not distinguishable from Y_T^R and the knowledge of T chosen message/cryptogram pairs does not allow the enemy to derive any information about the rest of the cryptogram space or key space and so the best strategy is the same as that of impersonation (on the set $Y \backslash Y_T$).

To prove that the best impersonation strategy is random selection we have to show that there does not exist an infinite family $\{S_{n_1}, S_{n_2}..\}$ of strategies better than random selection (which is equivalent to having a uniform distribution on V_m). Consider the $2^l \times 2^m$ incidence matrix I of the code $A^{n,m}$ whose rows correspond to individual keys and columns labelled by the elements of Y such that

$$
\begin{aligned}
I(k,y) \;=\; & 1 \text{ if } A^{n,m}_k(x) = y \text{ for some } x \in V_n \\
=\; & 0 \text{ otherwise.}
\end{aligned}
$$

Also let I^x denote a $2^l \times 2^m$ matrix such that

$$
\begin{aligned}
I^x(u,v) \;=\; & 1 \text{ if } A^{n,m}_u(x) = v; \\
=\; & 0 \text{ otherwise.}
\end{aligned}
$$

Since A is pseudorandom for a given x and a randomly chosen key, $p(y|x)$ is 'almost' uniform, i.e., there does not exist a constant s such that

$$|p(y|x) - 2^{-m}| > n^{-s}$$

(otherwise there exists a distinguishing circuit family for A).

But

$$p(y|x) = 2^{-l(n)}(\Sigma_u I^x(u,y)).$$

Hence $\Sigma_u I^x(u,y) = N_x$ is 'almost' independent of y and we have

$$\Sigma_k I(k,y) = \Sigma_k \Sigma_x I^x(k,y) = \Sigma_x(\Sigma_k I^x(k,y)) = \Sigma_x N_x$$

which is 'almost' uniform on Y.

$\square$

4 Construction of PA-code

In this section first we give a construction of an implicit PA-code which is called Feistel-type becuase of its similarity to Feistel construction of block codes. Next we mention another construction for explicit PA-codes which is based on traditional construction of A-codes.

4.1 Invertible Function Generators from Generalised Function Generators

Let k, m and p denote strings of length $l(n)$, $m(n)$ and $p(n)$, where $p < m$ and all the polynomials are integer valued. $f = \{f^{n,m}\}$ is a generalised function generator with input $I = L \bullet R \in V_{n+p}$ and $L \in V_p$, $R \in V_n$.

We define an index set $K_{i,j} = \{i+1, \cdots, i+j\}$ which specifies a restriction of a vector $v \in V_n$ to $v_{K_{i,j}} \in V_j$

$$v_{K_{i,j}} = v_{i+1}, \cdots v_{i+j}.$$

Given a generalised function generator f, we define a generalised invertible function generator $g = \{g^{n+p,m+n}\}$ as follows:

$$g^{n+p,n+m}(L \bullet R) = L' \bullet R', \qquad\qquad L' \in V_n,\ R' \in V_m$$
$$L' = R$$
$$R' = [f_k^{n,m}(R)]_{K_{0,m-p}} \bullet [[(f_k^{n,m}(R)]_{K_{m-p,p}} \oplus L].$$

The inverse function $\bar{g}^{n+p,n+m}$

$$\bar{g}^{n+p,m+n}(\alpha \bullet \beta) = [f_k^{n,m}(\alpha) \oplus \beta]_{K_{m-p,p}} \bullet \alpha, \tag{1}$$

takes an input $I' \in V_{n+m}$, $I' = \alpha \bullet \beta$, $\alpha \in V_n$, $\beta \in V_m$, and returns the corresponding output. This can be seen as a two step inverse mapping from V_{n+m} into V_{n+m} and then a restriction to V_{n+p}. It is easy to verify

$$\bar{g}^{n+p,m+n}(g^{n+p,n+m}(L \bullet R)) = L \bullet R.$$

$g^{n+p,n+m}$ expands its $n + p$ input bits to $n + m$ output bits and defines an invertible mapping $V_{n+p} \rightarrow V_{n+m}$ for all $p < m$. The amount of added redundancy (expansion) is $m - p$ (Figure 1).

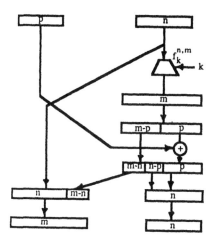

Figure 1

We call g an *elementary expansion generator (EEG(m))* and define a cryptographic composition of two such generators.

Let g and h denote two EEG(m)s obtained from two generalised function generators of key length functions $l_1(n)$ and $l_2(n)$ with the same expansion polynomial $m(n)$.

The composition $h \circ g$ of g and h is an expansion generator t with the following properties:

- the key length function for t is $l(n) = l_1(n) + l_2(n)$;

- $t_{k_2 \bullet k_1}^{n+p,n+m} = h_{k_2}^{n+m,n+m} \circ g_{k_1}^{n+p,n+m}$.

$t^{n+p,n+m}$ takes an input $L \bullet R$, $L \in V_p$, $R \in V_n$ of size $n + p$ and produces $n + m$ bits as output:

$$
\begin{aligned}
t_{k_2 \bullet k_1}^{n+p,n+m}(L \bullet R) &= h_{k_2}^{n+m,n+m}(g_{k_1}^{n+p,n+m}(L \bullet R)) \\
&= h_{k_2}^{n+m,n+m}[(L' \bullet R'_{K_0,m-n}) \bullet R'_{K_{m-n,n}}]
\end{aligned}
$$

where $L' \bullet R'$, $L' \in V_n$, $R' \in V_m$ is the output of g, i.e., $g_{k_1}^{n+p,n+m}(L \bullet R) = L' \bullet R'$.

The inverse function $\bar{t}$ is obtained using composition of inverses of g and h. Let $\alpha \bullet \beta$ denote the input to $\bar{t}^{n+p,n+m}$ where $\alpha \in V_n$, $\beta \in V_m$ and $\alpha' \in V_n$, $\beta' \in V_m$, $\alpha' \bullet \beta' = \bar{h}_{k_2}^{n+m,n+m}(\alpha \bullet \beta)$, we have:

$$\bar{t}_{k_2 \bullet k_1}^{n+p,n+m}(\alpha \bullet \beta) \;=\; \bar{g}_{k_1}^{n+p,n+m}(\bar{h}_{k_2}^{n+m,n+m}(\alpha \bullet \beta)) \;=\; \bar{g}_{k_1}^{n+p,n+m}(\alpha' \bullet \beta')$$

and $\bar{g}^{n+p,n+m}$ and $\bar{h}^{n+m,n+m}$ are defined as in equation (1), i.e., $\bar{t}^{n+p,m+n}$ consists of an inverse mapping into V_{n+m} followed by a restriction to V_{n+p}.

4.2 Feistel Type PA-Code

In the previous section we showed how to construct a class of expansion generators from a generalised function generator and defined a composition for these expansion generators. This can be considered as a Feistel type construction for A-code generators where the main difference from the Feistel type permutation generator generators (used for secure block encryption) is the replacement of the function generator with a generalised function generator.

The main result of this section is the counterpart of Luby-Rackoff theorem in this generalised case and states that a pseudorandom expansion generator can be obtained by composition of three EEG(m), obtained from a pseudorandom generalised function generator f.

Theorem 4.1 *Let g be an EEG(m). Then*

- *there exists a distinguishing circuit family for $g \circ g$;*

- *$g \circ g \circ g$ is a pseudorandom expansion generator.*

Proof: The proof is parallel to Luby-Rackoff [10] and is based on a main lemma. We use the following definitions to simplify the presentation of the lemma.

The operator
$$H : F_{n,m} \times F_{n,m} \times F_{n,m} \to F_{n+p,n+m}$$

can be defined using composition of EEG(m). Let f_0, f_1, $f_2 \in F_{n,m}$ and $L^0 \in V_n$, $R^0 \in V_m$. Define

$$
\begin{aligned}
\alpha &= f_0(R^0) \\
R^1 &= [\alpha_{K_{0,m-p}} \bullet [\alpha_{K_{m-p,p}} \oplus L^0]]_{K_{m-n,n}}, \\
L^1 &= R^0 \bullet [\alpha_{K_{0,m-p}} \bullet [\alpha_{K_{m-p,p}} \oplus L^0]]_{K_{0,m-n}}, \\
\beta &= f_1(R^1), \\
R^2 &= [\beta \oplus L^1]_{K_{m-n,n}}, \\
L^2 &= R^1 \bullet [\beta \oplus L^1]_{K_{0,m-n}} \\
\gamma &= f_2(R^2),
\end{aligned}
$$

and

$$H(f_2, f_1, f_0)(L^0 \bullet R^0) = R^2 \bullet (\gamma \oplus L^2).$$

Lemma 4.1 *(Main Lemma)*

Let $C_{n+p,n+m}$ be an oracle circuit with distinct inputs to the oracle gates. Then

$$|P[C_{n+p,n+m}(F^R_{n+p,n+m})] - P[C_{n+p,n+m}H(F^R_{n,m}, F^R_{n,m}, F^R_{n,m})]| < \frac{t^2}{2^n}$$

where t is the number of oracle gate, m and p are polynomials in n and $p < n$.

$\square$

The theorem gives a construction of implicit authentication codes with secrecy that offer provable protection against strong spoofing of order T and are provably secure against T-fold plaintext/ciphertext attack.

The generalised pseudorandom function $f^{n,m}$ adds $d_f(n) = m - n$ reduandant bits to its input block and the Feistel type expansion generator $h^{n,m}$ constructed from it has a redundancy $d_h(n) = m - p$ which depends on the choice of p. From Lemma 3.1 the probability of success of an intruder in a strong spoofing of order T is

$$\frac{2^{n+p} - T}{2^{m+n} - T} \approx 2^{-d_h(n)}.$$

This result holds for all $p(n) < m(n)$ which implies that it is possible to obtain different levels of protection with a given generalised function generator f by varying the block size of the message.

The code can also be used if the message size is less than n simply by padding the message with extra bits to reach the minimum acceptable size n (explicit authenticatio). In fact this will increase the amount of protection offered by the code. The highest protection is obtained when the input block is of size one, i.e., when one bit is padded by $n-1$ bits and subsequently encoded. The probability of deception in this case is 2^{-m-n+1}.

4.3 PA-codes From Pseudorandom Permuation Generators

It is possible to construct an explicit PA-code from pseudorandom permutation generators. This is in fact the extension of the traditional approach to A-code generators.

Let $f = \{f^n\}$ be a pseudorandom function generator and let $g = \{g^{2n}\}$ denote the Feistel type pseudorandom permutation generator constructed from it. If $p(n) < 2n$ is an integer valued polynomial , we can define an authentication code $A = \{A^{p,2n}\}$ as

$$A^{p,2n}_k(x) = g^{2n}_k(0_{2n-p} \bullet x), \ x \in V_p$$

where 0_{2n-p} is the all zero vector of length $2n-p$, $x \in V_p$ is the message to be authenticated and $A^{p,2n}_k(x) \in V_{2n}$ is the corresponding codeword. It is not hard to verify that A is a pseudorandom A-code as the existence of a distinguishing circuit family for A implies the existence of such family for g. The maximum protection offered by the code is when a one bit message is padded by $2n-1$ bits and then enciphered in which case the probability of deception in impersonation attack is $2^{-(2n-1)}$. This is much higher than the one obtained by a Feistel type A-code with expansion polynomial $m > n$ where the maximum protection is 2^{-m-n+1}.

5 Concluding Remarks

Pseudorandom expansion generators correspond to A-code generators with provable perfect protection and provable security. They can be realised by using pseudorandom permutation generators (traditional approach) or Feistel type A-codes.

In both cases the amount of redundancy can be adjusted to the protection required from the code. In fact for both codes the probability of deception is directly related to the amount of redundancy d as 2^{-d}. However the Feistel type A-code can provide more redundant bits for the same input block, hence provides more protection. amount of redundancy it can introduce which is the In both codes the maximum probability of deception corresponds to an input message of length $2n - 1$ (in fact Feistel A-code can also have input message of lenght $2n$) which is 2^{-m+n-1} for Feistel A-codes and 2^{-1} for traditional A-codes. The minimum probability of deception corresponds to an input of length one and the probability of deception in the two cases are 2^{-m-n+1} and 2^{-2n+1} respectively.

Another important aspect of this construction is that it suggests a new approach to the design of practically 'good' authentication code by using elementary rounds where in each round the input is subjected to confusion/expansion/diffusion. Consecutive application of the same elementary round will result in a 'good' authentication code. This is essentially the approach taken in the design of practically good block ciphers (e.g. DES, FEAL), however, in a good A-codes the non-linear functions should also provide for expansion.

Acknowledgement:

I am grateful of Peter Pleasants for his careful reading of an earlier version of this paper.

References:

1. G.J. Simmons, *Authentication theory/coding theory*, in Advances in Cryptology, Proceedings of Crypto 84, Springer-Verlag (Berlin), 1985, pp. 411-431.

2. G.J. Simmons, *A game theory model of digital message authentication*, Congressus Numerantium, 34(1982), pp. 413-424.

3. D.R. Stinson, *A construction for authentication/secrecy codes from certain combinatorial designs*, in Advances in Cryptology: Proceedings of Crypto 87, Springer-Verlag (Berlin), 1988, pp. 355-366.

4. D.R. Stinson, *Some constructions and bounds for authentication codes*, Journal of Cryptology 1 (1988), pp. 37-51.

5. M. De Soete, *Some constructions for authentication-secrecy codes*, in Advances in Cryptology: Eurocrypt '88, Springer-Verlag (Berlin), pp. 57-76.

6. M. De Soete, *Bounds and constructions for authentication-secrecy codes*, in Advances in Cryptology: Crypto '88, Springer-Verlag (Berlin), pp. 311-317.

7. G.J. Simmons, *A survey of information authentication*, Proceedings of IEEE, pp. 603-620, 1988, Vol. 76, No. 5.

8. R.S. Safavi-Naini and J.R. Seberry, *Error correcting codes for authentication and subliminal channel*, IEEE Transaction on Information Theory, pp. 13-17, Vol. 37, No.1, 1990.

9. J. Pieprzyk and R. S. Safavi-Naini, *Pseudorandom authentication systems*, Abstarcts of Eurocrypt '91, Brighton.

10. M. Luby and C. Rackoff, *How to construct pseudorandom permutations from pseudorandom functions*, SIAM J. Comput., 17(1988), pp.373-386.

11. O. Goldreich, S. Goldwasser and S. Micalli, *How to construct random functions*, in Proceedings of the 25th Annual Symposium on Foundation of Computer Science, October 24-26, 1984.

12. L.A. Levin, *One-way functions and pseudorandom generators*, in Proceedings of the 17th ACM Symposium on Theory of Computing, Providence, RI, 1985, pp. 33-365.

13. Y. Zheng, T. Matsumoto and H. Imai, *Impossibility and optimality results on constructing permutations*, in Abstracts of Eurocrypt '89, Houthalen, Belgium, April 1989.

14. J. Pieprzyk, *How to construct pseudorandom permutations from single pseudorandom functions*, in Abstarcts of Eurocrypt '90, Aarhus, Denmark, May 1990.

15. J. L. Massey, *Cryptography-a selective survey*, Digital Communications, Elsvier Science Publishers, 1986, pp. 3-21.

Research Activities on Cryptology in Korea

Man Y. Rhee

Hanyang University
Dept. of Electronic Communication Eng.
17 Haengdang Dong, Seongdong Ku
Seoul, Korea 133-791

Abstract
In this paper, we introduce the recent development of cryptologic research in Korea since 1989. Two workshops on information security and cryptology and a symposium on data security have been held. The Korea Institute of Information Security and Cryptology (KIISC) was organized in December 1990. The KIISC is playing a leading role for promoting research activities on cryptology in Korea. Since Japan is very active in the area of cryptologic research and development since 1984, it is worth to include Japanese research activities in this paper. Also, the recent cryptologic research activities in the Republic of China (Taiwan) are introduced.

With regard to the research activities from other nations in Asia, it is unfortunate not to compile their activities in this issue due to the lack of information available at the present time.

1   Introduction

The widespread applications of data storage and transmission have given rise to a need for some types of cryptosystems. Data teleprocessing makes sensitive or valuable information vulnerable to unauthorized access while in storage and transmission. Cryptography embraces both the privacy problem and authentication problem. The privacy problem is to prevent an opponent from extracting the message information from the channel ; whereas the authentication problem is to prevent an opponent from injecting false data into the channel or altering messages so that their meaning would change.

Until the late 1970s, cryptographic technology was exclusively used for the military and diplomatic sectors, but business community and private individuals are recently recognizing to protect their valuable information in computer communication networks against the unauthorized interception.

With the advent of modern computer and communication technology, the need for cryptographic protection becomes widely recognized and the direction of researches has been changed drastically. At the beginning of the 1970s only symmetric (one-key) ciphersystems were known, and the classified researches in cryptography discouraged cryptologists who seek to discover new cryptosystems. Nevertheless the explosive growth in unclassified research in all aspects of cryptology has been progressed since 1976.

In 1977, the Data Encryption Standard (DES) [5] has been adopted by the National Bureau of Standard (NBS) as the standard algorithm for the information security of the U.S. federal government. Since then research activities on cryptology have started to expand to open academic environments beyond the traditional realm of special government agencies. The introduction to the public key cryptosystem by Diffie and Hellman [4], where the key for the crypto-algorithm can be made public, has had a tremendous impact on the development in the field of cryptology.

This trend for open research has influenced academic activities across the world. Since 1982, two conferences, the CRYPTO and the EUROCRYPT, have been held annually at the University of California, Santa Barbara in the U.S.A. and an European country, respectively, under the sponsorship of the International Association for Cryptologic Research (IACR) within the IEEE. In Australia, the AUSCRYPT conference has been held since 1990. Most importantly, the first international conference on cryptology in Asia was held in Fujiyoshida, Japan on November 11 to 14, 1991. In Japan, open academic researches on cryptology started in as early as 1984 and are expected to be mature with annual workshops and symposia in this area by now.

In Korea, recently there has been a rapid increase in numbers of telephone lines and computers and, accordingly, reports of computer crimes such as fixing the apartment lottery and the stoppage of several dozens of Personal Computers on the important network due to the virus were announced. Moreover, the recent government-led construction of the national backbone networks for communications has stimulated academic research activities on the communication and information security by a number of scholars in the mid 1980s. In 1989, the first domestic conference on cryptology, coded the WISC89, was held, and have had a strong lasting influence on research activities of cryptology in Korea.

In this paper, we will discuss on a brief historical review of cryptosystem, the outline of cryptologic researches in Korea, some important results from research works in Japan, and the recent cryptologic researches in the Republic of China (Taiwan).

2 Brief Historical Review of Cryptosystems

A short survey on cryptologic evolution is presented in this section. We introduce classical ciphers formerly in use, followed by modern cryptosystems presently in use. There are three important types of ciphers, that is, the transposition, substitution, and product ciphers. Since the transposition or the substitution cipher needs the simplest encryption technique, they are of little use now, but are important components in more complex product ciphers. The transposition ciphers are block ciphers that change the position of characters or bits of the message plaintext. The oldest known transposition cipher is the Scytale cipher used by the ancient Greeks as early as 400 B.C.

There are four types of substitution ciphers: monoalphabetic substitution, homophonic substitution, polyalphabetic substitution, and polygram substitution. A monoalphabetic cipher is the simplest substitution cipher because there is a one-to-one correspondence between a character and its substitute. The so-called Ceaser cipher is an example of the monoalphabetic cipher with the ith letter replaced by $(i+3)$th letter modulo 26. The homophonic substitution cipher maps each character x of the plaintext alphabet into a set of ciphertext elements $f(x)$ called homophones. The Beale cipher (1880s) is an example of the homophonic cipher. The polyalphabetic ciphers are multiple substitution ciphers using different keys. Most polyalphabetic ciphers are periodic substitution ciphers based on a period p. A popular form of the periodic substitution ciphers based on shifted alphabets is the Vigenere cipher and Beaufort cipher.

The Kasiski method, introduced in 1863 by a Prussian officer, Fredrich W. Kasiski, analyzes repetitions in the ciphertext to determine the exact period. Hence, using this method, the period can be found first and then a set of the monoalphabetic ciphers are solved. In an effort to remove the weakness of the periodic polyalphabetic ciphers, cryptologists turned to running key ciphers which are aperiodic polyalphabetics. A running-key cipher, designed for complete concealment of information, has the key sequence which is as long as the message plaintext, possibly foiling a Kasiski attack. A Kasiski solution is no longer possible due to the aperiodic selection of the 26 alphabets used. But Bazeries proposed to solve this problem facing with the running key ciphers in the late 1890s. If the key has redundancy, the cipher may be breakable using the Friedman's method. The Hagelin machine (M-209) and all Rotor ciphers are the polyalphabetic substitution ciphers generating a stream with a long period.

The American Sigaba (M-134), the British Typex, the German Enigma, the Japanese Purple were all rotor ciphers, and undoubtedly many are still in use today. The Hagelin and all Rotor machines are polyalphabetic ciphers generating a stream with a large period.

Another important substitution cipher is the one-time pad in which the ciphertext is the bit-by-bit modulo 2 sum of the plaintext and a nonrepeating keystream of the same length. However, the one time pad is inappropriate to use because the large amount of nonrepeating key makes it impractical for most applications. The first implementation of the one-time pad cipher was the Vernam cipher, designed by Gilbert Vernam in 1917.

Up to this point, all substitution ciphers encipher a single character of plaintext at a time. But, in polygram ciphers, several characters are enciphered at the same time using one key. The Playfair cipher is the 2-gram substitution cipher, whereas the Hill cipher is the more generalized N-gram substitution cipher employing linear transformations.

A product cipher is a composition of two or more ciphers such that the ciphertext space of one cipher becomes the message plaintext space of the next. The combination of ciphers is called cascaded superencipherment in such a way that the final product is superior to either its components. In fact, the principle of product ciphers is evident in all the rotor ciphers which use several steps of encryption. The product cipher involves the steps of both substitution and transposition. An early product cipher is the German ADFGVX cipher used in World War I.

The DES, which belongs to a block cipher, is the most widely known encryption algorithm. In block encryption, the plaintext is divided into blocks of fixed length which are then enciphered using the same key. The DES is the algorithm with which the 64-bit block of plaintext is enciphered into the 64-bit ciphertext under the control of the 56-bit internal key. The iterated process consists of 16 rounds of encipherment, while the deciphering process is exactly the same as the process of encryption except that the keys are used in the reverse order. Also, NTT Japan had developed the DES-like cryptosystem called the Fast Data Encipherment Algorithm (FEAL-8) in 1986. Currently, various DES-like cryptosystems like MULTI2, LOKI, etc. are publicly reported.

There are two types of stream ciphers, one for which the key-bit stream is independent of the plaintext and the other for which the key-bit stream is a function of either the plaintext or the ciphertext. The former is called the synchronous cipher and the latter is often called the self-synchronizing cipher. Stream ciphers operate on a bit-by-bit bases, combining the plaintext with a key stream. The key-bit stream in a synchronous stream cipher is generated independently of the plaintext. The loss of a bit in ciphertext causes loss of synchronization and all ciphertext following the loss will be decrypted incorrectly. We also know that synchronous stream ciphers can be broken when the key-bit stream is repeated periodically. For this reason, synchronous stream ciphers have limited applicability to file and database encryption. However, synchronous stream ciphers protect the ciphertext to some extend against opponent's searching, because identical blocks in the plaintext are enciphered under a different part of the key stream. They also protect against injection of false ciphertext, replay, and ciphertext deletion, because insertion or deletion in the ciphertext causes loss of synchronization. Therefore, all the subsequent ciphertext will decipher incorrectly. However, key-autokey stream ciphers have the advantage of no error propagation.

A ciphertext-autokey cipher is the one in which the key is derived from the ciphertext and is called a self-synchronizing stream cipher with ciphertext feedback. Since each key bit is computed from the preceding ciphertext bits, it is functionally dependent on all preceding bits in the plaintext plus the priming key. This nature will make cryptanalysis more difficult because the statistical properties of the plaintext are diffused across the ciphertext.

The invention of public-key cryptosystems in the late 1970s drastically changed the direction of cryptologic researches. A solution to the key-distribution problem was suggested by Diffie and Hellman in 1976. Since then, asymmetric cryptosystems were proposed which used two different keys. One key used for enciphering can be made public, while the other using for deciphering is kept secret. Thus, the two keys are generated such that it is computationally infeasible to find the secret key from the public key. The famous RSA public-key cryptosystem was invented by Rivest, Shamir, and Adleman in 1978, and McEliece proposed the algebraic codes cryptosystem in 1978.

Integer factorization has advanced significantly in the last decade and the largest hard integer up to almost 90 decimal integers in length had been factored. If a machine is built that could factor 150 digit integers, such a machine could then break the RSA cryptosystem. There are several integer-factoring algorithms proposed by Lenstra and Lenstra (1980), Gordon (1985), and Pomerance (1987).

It was known till 1982 that the public-key algorithms for solving knapsack problem are of exponential computational complexity because the knapsack problem is NP-complete. In particular, the iterated Merkle-Hellman scheme was claimed to be secure. However, the basic Merkle-Hellman additive knapsack system was shown by Shamir (1982) to be easy to crack. Subsequently, Adleman (1983) proposed attacks on the iterated Merkle-Hellman cryptosystem by cracking the enciphering key obtained using the Graham-Shamir knapsack, but Adleman's attack on the multiple iterated knapsack systems seems not quite to succeed. Furthermore, Brickell, Lagarias, and Odlyzko (1984) have evaluated the Adleman attacks on multiple iterated knapsack system. In addition, Desmedt, Vandewalle, and Govaerts (1982,1984) claimed that the iterated Merkle-Hellman systems are insecure, since an enciphering key can be cracked. A multiple iterated knapsack scheme is the cryptosystem relying on modular multiplications which are used for disguising an easy knapsack. Lagaries (1984) and Brickell (1985) have given an intensive study on these systems by using simultaneous diophantine approximation. Goodman-McAuley (1985) proposed another knapsack cryptosystem using modular multiplication to disguise an easy knapsack, but that system is substantially different from these mentioned above. However, this cryptosystem can still be broken using lattice basis reduction. Pieprzyk (1985) designed a knapsack system which is similar to the Goodman-McAuley knapsack except that integers have been replaced by polynomials over $GF(2)$. Brickell (1984) and Lagarias and Odlyzko (1985) proposed the algorithms for solving knapsacks of low density. Schnorr (1988) developed the more efficient lattice basis reduction algorithm which could be successful on knapsacks of much higher density.

The 1980s were the first full decade of public key cryptography. Many systems were proposed and many systems were broken. But no one can deny asymmetric two-key cryptosystems in full bloom over this period.

The validity of business contracts and agreements is guaranteed by signatures. Business applications such as bank transactions, military command and control orders, and contract negotiations using computer communication networks, will require digital signatures.

Let us review several authentication scheme built using public-key cryptosystems as well as conventional symmetric cryptosystems. The ElGamal's algorithm (1985) for authentication relies on the difficulty of computing discrete logarithm over the finite field $GF(p)$, where p is a prime. Ong, Schnorr, and Shamir (1984) proposed a signature scheme based on polynomial equation modulo 2. However, Pollard (1987) broke the quadratic and cubic OSS schemes. The Okamoto-Shiraishi signature scheme (1985) is based on the difficulty of finding approximate kth roots modulo n. Their original scheme was for $k = 2$. Not only the $k = 2$ case, but also $k = 3$ case were quickly broken by Brickell and DeLawrentis in 1986. The cryptanalysis of the Shamir's fast signature scheme (1978) is somewhat similar to the attacks on the multiplicative knapsack system. In 1984, Odlyzko has shown a method of attack to break the Shamir's scheme which was designed for authentication. Next, the Rabin signature scheme (1978) is based on symmetric crypto-operations. Cryptographic operations used in the Diffie-Lamport scheme (1979) are based on a symmetric encryption and decryption. The drawback of Diffie-Lamport scheme is the length of signature which are too long. The signature scheme invented by Matyas and Meyer in 1981 is based on the DES algorithm.

In 1984, Shamir proposed an innovative type of cryptographic scheme what is called ID-based cryptosystems and signature schemes. This system enables two communicators to exchange their messages securely and to verify each other's signatures without exchanging secret or public keys. Consequently this scheme is not required to keep the key directory, and to use any service from a third party. But it is required to provide an individual smart card for each user, because this card enables the user to sign and encrypt the messages to be transmitted and to decrypt and verify the messages received. Even though the scheme is based on a public-key cryptosystem, the user can use any combination of his name, social security number, telephone number, or network address as his public key without the generation of a random pair of public and secret keys. The secret key is computed by a key generation center and issued to the user in the form of smart card when he first joins the network.

Since Shamir introduced ID-based cryptosystems in 1984, several cryptologists have been investigated and promoted this particular subject through their researches. In 1986, Desmedt and Quisquater proposed a public-key cryptosystem based on the difficulty of tampering. Tsujii and Itoh (1989) presented their paper on an ID-based cryptosystems based on the discrete logarithm problem. Okamoto and Tanaka (1989) introduced ID-based information security management system for personal computer networks. Matsumoto and Imai (1990) discussed on the security of some key sharing schemes. Akiyama, Torii, and Hasabe (1990) presented a paper on ID-based key management system using discrete exponential calculation. Recently, Youm and Rhee published a paper titled on ID-based cryptosystem and digital signature scheme using discrete logarithm problem in 1991.

Since 1985, interest has been shown by the new idea on zero-knowledge proofs. This new area provides the cryptologists with the fascinating protocols for obtaining provably secure communications. A basic question in complexity theory is how much knowledge should yield in order to convince a polynomial-time verifier for validity of some theorem. In 1985, issues in complexity theory and cryptography motivated Goldwasser, Micali, and Rackoff to introduce the concept of an interactive proof system. Since then, zero-knowledge interactive proofs (ZKIP) have been the focus of much attention in recent years. Goldreich, Micali, and

Widgerson (1986) showed that any language in NP has a computational zero-knowledge proof, and presented a methodology of cryptographic protocol design. In 1988, Ben-Or, Goldreich, Goldwasser, Hasted, Killian, Micali, and Rogaway extended the result of Goldreich, Micali and Widgerson that, under the same assumption, all of NP admit zero-knowledge interactive proofs and they also showed that every language that admits an interactive proof admits a perfect zero knowledge interactive proof. Ben-Or, Goldwasser, Killian, and Widgerson (1988) introduced the idea of multi-prover interactive proofs, in order to show how to achieve perfect zero-knowledge interactive proofs without using any intractability assumption. Crepeau (1987) described a zero-knowledge poker protocol that achieves confidentiality of the player's strategy or how to achieve an electronic poker face. Chaum, Evertse, Van der Graaf, and Peralta (1987) showed how to demonstrate possession of a discrete logarithm without revealing it. Several practical protocols have been proposed for this type of knowledge proof without revealing any actual information related to the secret. Tompa and Woll (1987) showed that any random self-reducible problem has a zero-knowledge interactive proof. They proved that the computation of both square roots modulo n (a composite) and discrete logarithms modulo p (a prime) are in random self-reducible class. Shizuya, Koyama, and Itoh (1990) presented an interactive protocol demonstrating (1) the prover actually knows two factors a and b (not necessarily primes) of a composite n and showing that (2) it is really ZKIP.

In the 1990s, it is needless to say that the concept of differential cryptanalysis proposed by Biham and Shamir (1990) is the hot research topic in cryptologic society. They demonstrated that many symmetric cryptosystems can be breakable by their method.

3 Cryptologic Research in Korea

In this section, we introduce cryptologic research activities in Korea.

3.1 WISC89

In July 1989, the first domestic conference on information security and cryptology, called the WISC89, was held. About 80 attendees working in the cryptology and related area have participated in the WISC89 and 22 papers were presented. In this workshop, six sessions were organized as follows:

1. Cryptology(I)

2. Cryptology(II)

3. Information and Computer Security(I)

4. Information and Computer Security(II)

5. Application and Implementation (I)

6. Application and Implementation (II)

There also was a panel discussion on the direction of the cryptology and related area in Korea. This workshop not only provided an opportunity to form special research groups on cryptology in the existing academic associations and institutes, but also helped to form the theoretical research group on the computers within the Korea Institute of Information Engineering and the subgroup on cryptology within the Korea Institute of Communications. Besides these developments, the need for the special institute for the sole research on cryptology has grown due to the rapid expansion in the number of researchers in this field, and lack of organized efforts and guidance in cryptologic research was recognized.

3.2 WISC90

The second annual workshop (WISC90) was held in September 1990. This time, the workshop had 110 participants, showing a steady increase in size and 23 papers were presented. Among the contributed papers, the following ones were noted for their excellence.

- Hyun S. Park *et al.* (KNDA), "A Study on Crypt-complexity of the Algebraic Coded Cryptosystems using the Reed-Solomon Codes".

- Yoon Y. Oh *et al.* (KIT), "Factorization of 17^n for $n \leq 48$".

- Jong I. Lim (Korea Univ.), "Proof of Lenstra's Conjecture Related to Elliptic Curve Method".

- Jae H. Lee (Seoul Univ.), "Frequency Distribution and Entropy of a Korean Syllable".

In particular, Pil J. Lee who has with the Bellcore U.S.A. before his return to Korea as a professor in Pohang Institute of Science and Technology, presented an invited talk on "Secure Remote Verification of Personal Identity".

3.3 Present Activities

Since two workshops have been held in the past two years, it became imperative that researchers in this area felt of necessity to establish an institute which would lead to the direction of organized efforts in research activities on cryptology. On December 12, 1990, the Korea Institute of Information Security and Cryptology (KIISC) was founded in Seoul in cooperation with a number of academic institutes for higher learning, private industries, and the government. Man Y. Rhee was elected as the first president of KIISC.

The purposes of the KIISC are to advance the theory and practice of cryptology and related fields, to contribute to the developments of the relevant area, and to promote the common interests among its members in cryptologic research. To achieve these objectives, the KIISC intends to pursue the following activities.

(1) To organize and sponsor an annual workshop and appropriate meetings on communication and information security.

(2) To investigate the academic and technological researches on cryptology and to publish journals and KIISC reviews.

(3) To research on the communication and data security standard.

(4) Training and education for engineers and scientists in the field of cryptology.

(5) To pursue subsidiary projects related to the above ones and other projects as its Board of Directors deems appropriate.

The major plans of the KIISC in 1991 are as follows:

(1) To publish KIISC journals, reviews, and proceedings of workshop.

(2) To organize the annual conference and workshop. The General Assembly was held on November 23, 1991 at the Korea National Defense Academy (KNDA).

(3) To pursue government project on "Information Society and the Role of Cryptology".

(4) Setting up frameworks for the communication and information security evaluation according to the international trends of the cryptologic standard.

(5) Publication of KIISC Membership Directory.

In compliance with the above plans, the two KIISC Reviews were published in April and August 1991, respectively. The first issue contains 2 congratulatory speeches, a special contribution, and 15 papers. (See **Appendix** for details). The production schedule for the 3rd KIISC Review and the first KIISC journal are in progress at present.

As of September 1991, the KIISC board of directors consists of one president, two vice presidents, 17 directors, and 2 auditors as shown in Table 1. There are about 200 members including 173 regular and 22 special members representing various private and government organizations.

Table 1: Board of Directors of KIISC

President	Man Y. Rhee (Hanyang Univ.)
Vice President	Byung S. Ahn (ETRI) and In H. Cho (Korea Univ.)
Directors	In S. Back (NCA), Sung Y. Cho (DACOM), Dae H. Kim (ETRI), Dong K. Kim (Ajou Univ.), Jae K. Kim (Dongkuk Univ.), Se H. Kim (KAIST), Young J. Kim (KT), Sang J. Moon (Kyungbuk National Univ.), Hong J. No (ADD), Jeong W. Oh (Yonsei Univ.), Kil H. Nam (KNDA), Sang H. Park (OPC), Sung A. Park (Sogang Univ.), Tae K. Sung (KMTC), Dong H. Won (SKK Univ.), Jong Y. Woo (Samsung Ele. Co.), Yoo J. Yoon (Daeyoung Ele. Co.),
Auditors	Youn H. No (SPPO), Un H. Song (KIDA)

Moreover, the symposium on Data Security sponsored by the KIISC was held in Seoul with 150 participants on August 23, 1991. The following 19 papers were presented and brought the consensus of opinion about the importance of cryptologic research :

1. Special Lecture : Man Y. Rhee (Hanyang Univ.), "Cryptology and Information Society".

2. Invited Lecture I : Sang J. Moon (Kyungbuk National Univ.), "An Application of Cryptography to Satellite Communications".

3. Invited Lecture II : In H. Cho and Jong I. Lim (Korea Univ.), "Role of Mathematics in Cryptology".

4. Jong I. Lim and Chang H. Kim (Korea Univ.), "A Study of Integer Factorization Algorithms".

5. Young J. Choie (Pohang Institute of Science and Technology), "Discrete Logarithm and Cryptography".

6. Seung A. Park, Min S. Lee, Jae H. Lee, and Hyun Y. Shin (Sogang Univ.), "Random Self-Reducibility, Hard Bits, and Pseudo-Random Generators".

7. Sang K. Hahn (KAIST), "Elliptic Curve Cryptosystem and Discrete Logarithm over Finite Fields".

8. You J. Sim, Jae K. Ryu, Jeong H. Ahn, Sun Y. Jung, and Jin W. Chung (Sungkyunkwan Univ.), "Cryptographic Protocols".

9. Hyung K. Yang, Yun H. Lee, Ki W. Sohn, Chang Y. Kwon, and Dong H. Won (Sungkyunkwan Univ.), "A Study on the Zero Knowledge Interactive Proof".

10. Pil J. Lee, Hee C. Moon, and Chae H. Lim (Pohang Institute of Science and Technology), "On ID-based Cryptosystems".

11. Chang S. Park (Dankook Univ.), "On the Cryptosystems Based on Error-Correcting Codes".

12. Se H. Kim and Bong S. Um (KAIST), "A Cryptanalysis on Knapsack Cryptosystems and an Improved Chor-Rivest Cryptosystem".

13. Bong D. Choi (KAIST) and Yang W. Shin (Changwon National Univ.), "Randomness and Statistical Tests".

14. Suk G. Hwang (Kyungbuk National Univ.) and Han H. Cho (Seoul National Univ.), "On Safe Hash Functions".

15. Heung Y. Youm (Sooncheonhang Univ.) and Man Y. Rhee (Hanyang Univ.), "ID-based Cryptosystems and Digital Signature Scheme Using Discrete Logarithm Problem".

16. Yoon H. Choi and Chang S. Jeong (Pohang Institute of Science and Technology), "Parallel Gaussian Elimination on SMM with Application to Cryptography".

17. Kil H. Nam (KNDA), "Research Trends of Information Security in Advanced Countries".

18. Young J. Lim, Yun H. Cheong, Kyun D. Cha and Dong K. Kim (Ajou Univ.), "A Study on the Development of Data Protection Mechanisms and Model in a Single Network".

19. Young H. Ahn (Kangnam Univ.) and Dong H. Won (Sungkyunkwan Univ.), "A Study on the Implementation of the Product Cryptosystem for Finite Field".

As a result of the foundation of the KIISC, the research on cryptology has become a part of open academic activities in Korea.

4 Survey of Some Important Research Results in Japan

In Japan, research activities on cryptology already started in 1984. Since then, eight symposia and four workshops have been held in Japan so far. The research activities in Japan are mainly led by CIS (Cryptography and Information Security) group of IEICE (Institute of Electronics, Information and Communication Engineers).

We introduce the brief history of the development of cryptologic research in Japan since 1984. For more information, refer to Prof. Imai's paper[7].

4.1 SCIS84

The first symposium on cryptography and information security (SCIS84) sponsored by the CIS group was held with 66 participants from February 9 to 11, 1984. Eleven papers including a proposal of a new public key cryptosystem, a study on secret-sharing communication systems *etc.* were presented and a panel discussion on research subjects of cryptography and information security was held.

4.2 SCIS85

The second symposium was held from January 31 to February 2, 1985 with 74 participants and 11 papers were presented. The Solovay-Strassen method for finding prime numbers was one of the hottest topic and the joint group of Yokohama National University (YNU) and University of Tokyo presented a new concept called "obscure representation".

4.3 SCIS86 and WCIS86

The third symposium sponsored by the CIS group was held from February 6 to 8, 1986. The 98 researchers attended and 23 papers were presented. Six out of 23 papers treated new constructions of public key cryptosystems and three of them were based on the obscure representation.

In August 1986, the first workshop on cryptography and information security (WCIS86) was held. Twelve papers were presented in front of 156 attendees. Two good presentations were given by drawing a great attention : the one was on KPS (Key Predistribution System) by a group of YNU and the other was on FEAL by NTT.

4.4 SCIS87 and WCIS87

The fourth symposium (SCIS87) was held from February 5 to 7, 1987 with 100 participants and 28 papers. In this symposium, several papers treated new topics and attracted attention by the audience. Presentations on the zero-knowledge proof and ID-based cryptosystem were firstly treated.

The second WCIS (WCIS87) was held in July 1987. 150 participants were gathered and 17 papers were presented. The hot dispute on ID-based cryptosystem was resumed in a panel discussion and theory of computational complexity was another big topic.

4.5 SCIS88 and WCIS88

The fifth symposium (SCIS88) was held from February 22 to 24 1988, which had 115 attendees and 41 papers. The ID-based cryptosystems was still the most popular topic. There was a panel discussion on the hardware implementation of cryptosystems. Several devices realizing FEAL-8 and many experimental implementations of RSA cryptosystem were reported.

The third WCIS (WCIS88) was held in July 1988 with 91 participants and 18 papers. Zero knowledge interactive proofs were the main theme in this workshop.

4.6 SCIS89 and WCIS89

The sixth symposium (SCIS89) was held from February 2 to 4, 1989. The number of participants was 118. The symposium had 35 papers in eleven sessions and one panel discussion on verifiable secret asking to insecure servers. Studies on ZKIP had increased remarkably and a new topic on computer virus was introduced.

The fourth WCIS (WCIS89) was held in August 1989 and organized with 110 participants and 11 papers. ZKIP had still great popularity and four invited papers on this topic were presented. In this workshop, NTT announced call for attack of breaking FEAL-8, which Prof. Shamir disclosed the chosen plaintext attack last year.

4.7 SCIS90 and WCIS90

The seventh symposium (SCIS90) was held from January 31 to February 2, 1990. It had 127 participants and 43 papers. There were also demonstrations of cryptosystems for FAX communications and extended FEAL family.

The fifth workshop (WCIS90) was held on August 1990. The whole program was composed of invited papers (mostly on ZKIP).

4.8 SCIS91

This year the eighth symposium (SCIS91) was held in Fujiyoshida from January 31 to February 2, 1991. 54 papers were presented and 133 participants attended. It was still a hot issue on ZKIP and ID-based cryptosystem.

Besides these symposia and workshops, the CIS group holds a meeting once every two months and there are small group discussion activities by Akaruiangoukai (bright CIS meeting) in Kwando area and by the Kansai Crypto group in Kansai area.

5 Cryptologic Research in Republic of China

5.1 Background

This section is simply transprinted from the information provided by C.C.Chang at the National Chung Cheng Univ., Republic of China (Taiwan). In Taiwan, there are only a few people working in the area of cryptology and a few cryptologic research projects granted by National Science Council (NSC) in recent years. Fortunately, in July 1990, the NSC founded a group to draw up several cryptologic research topics in order to integrate researches in this field as well as to attract the attention of researches in other fields. This group has been investigating the security related requirements and events in both the industry, academy and government units of Taiwan; formulating several cryptologic research topics as well as important references; proposing ways to emphasize this research area. They aim at training more students in this area, working out more results and keeping up with the modern cryptologic research level.

5.2 Current Situations

To understand the requirements of security problems faced in Taiwan, the above-mentioned group contacted with many computer centers of both industry, academy, and government units by ways of questionnaires and interviews. Based on the results of thirty four received questionnaires and ten interviews, they concluded as follows:

1. Current situation

   (a) 96 % of them adopts cryptologic approaches to obtain information security, mainly the password authentication.

   (b) Almost all units use DES as a basic encryption algorithm.

   (c) Only a few centers have the experience of security violations.

   (d) Almost all centers solve security problems by themselves or with the aid of equipment suppliers. They did not consult with domestic experts for some reasons.

2. Security requirements from the interviewed centers

   (a) Enforce users' sense of security.

   (b) Protect systems from computer virus.

   (c) Establish computer audit trails.

   (d) Provide computer network security.

   (e) Develop cost-effective encryption algorithms.

   (f) Provide authorization controls.

   (g) Design encryption/decryption algorithms for IC-card.

At this moment, only ten professors offer graduate courses on data security and cryptology within eight universities. Five professors have cryptologic research projects supported by the NSC since 1989.

5.3 Topics for Future Research

Since cryptologic research is a long term and promising research area, the following fifteen important research topics for further research have been proposed :

1. Private Key Cryptography

2. Public Key Cryptography

3. Cryptanalysis

4. Key Management

5. System Security

6. Data Base Security

7. Network Security

8. Computer Virus

9. Digital Signature and Authentication

10. Zero-Knowledge Proofs

11. Group-Oriented Cryptography

12. Implementation

13. Applications

14. Data Compression

15. Fundamental Theories

As computer networks become more popular, data communications over the networks increase. This also gives outsider a good chance to attack the security of computer systems at the same time. To assure information security, the followings are emphasized :

- Methods of computer audit trails.

- Establishment of computer security policy based on applications.

- Legislation for criminal attacks.

- Vaccines for computer virus.

Furthermore, in order to increase users' sense of security, the followings are proposed :

- To hold local information security conferences.

- To publish basic security concepts on the popular computer journals.

- To offer more information security courses.

- To schedule programs on public TV to introduce the basic concept on information security.

- To emphasize the importance of data security and encourage the utilization of data security

- To publish information security books (or events).

- To found security research committee and publish journals.

6 Concluding Remarks

Judging from the current trends, it is no doubt that the world of information society is gradually open. Now it becomes even more important to bring together our domestic technologies in protecting the valuable communication information through vigorous academic activities in cryptology.

Currently researchers in Korea have just started to pursue their active researches in cryptology, but we feel that Korea still lags far behind the world-leading nations in this field. In the near future, we need to accumulate and expand our research capability by training and educating new experts and stimulating academic research activities. At the same time in conjunction with industry and research labs, practical applications of cryptologic research have to be carried out. In the government level, the regulations, systems on communication and data security should be reorganized to reflect the ever-changing new environments.

To achieve these objectives, we need to make continuous efforts to push the domestic level of research on cryptology to the international level. The KIISC is expected to play a leading role by providing an open forum for communicating ideas and results, and stimulating active discussions and academic research on cryptology. The Korean government is also moving ahead by preparing to enact and enforce the law for the protection of the private information.

As a final remark, considering all these developments, time is ripe for hosting the ASIACRYPT conference in Korea in a near future, which will surely help to elevate current domestic research on cryptology to the international level.

Please contact the editor of the KIISC journal for any comments and suggestions if any :

KIISC
Section 0700,
P.O.Box 12, Daedog Science Town,
Daejeon, Chungnam,
KOREA, 305-606.

References

[1] D. Kahn, The Codebreakers, Macmillan Co., New York, 1967.

[2] The Beale Ciphers, The Beale Cypher Association, Nedfield, MA, 1978.

[3] D.E. Denning, Cryptography and Data Security, Addison Wesley, Reading, Mass., 1982.

[4] W. Diffie and M. Hellman, "New Directions in Cryptography", IEEE Trans. on IT, Vol.22, pp.644–654, Nov., 1976.

[5] "Data Encryption Standard", National Bureau of Standards, Federal Information Processing Standard, Vol. 46, U.S.A., Jan., 1977.

[6] C.H. Meyer and E.M. Matyas, Cryptography : A New Dimension in Computer Data Security, John Wiley and Sons, New York, 1982.

[7] H. Imai, "Recent Development of Cryptology in Japan", Trans. of IEICE, Vol.E 73, No.7, pp.1026–1030, Jul.,1990.

Appendix :   The contents of the first issue of KIISC Review

- Speech on Publication : The President, Man Y. Rhee.

- Congratulatory Speech : The Minister of Communications, Eon J. Song.

- Congratulatory Speech : The Vice-Minister of Science and Technology, Jeong W. Seo.

- Special Contribution : The Director of the Electric Wave Management Office of the Ministry of Communications, Seong D. Park.

- Man Y. Rhee (Hanyang Univ.), *invited paper*, "Historical Reviews on Cryptography".

- Byung S. Ahn (ETRI), "About the Policy of Information Security for Information Society".

- Jin W. Chung (Sungkyunkwan Univ.), "An Outlook of Cryptology".

- Man Y. Rhee (Hanyang Univ.),*invited paper*, "A Study on the Stream Cipher System(I)".

- Sang J. Moon and Hoon J. Lee (Kyeongbuk Univ.), "Analysis of Intersymbol Dependence in Lucifer Cipher System".

- Kil H. Nam (KNDA), "The Digital Signature System using Cryptosystem".

- Dong H. Won (Sungkyunkwan Univ.), "The Cryptographic Methods and Key Distribution".

- Heung R. Youm (Sooncheonhang Univ.), "The Cryptographic Key Generation, Distribution and Management Method in Computer Network".

- Man Y. Rhee (Hanyang Univ.), *invited paper*, "A Study on the Public-key Cryptosystem(I)".

- In H. Cho (Korea Univ.), "Continued Fraction and Cryptology".

- Jong I. Lim and Chang H. Kim (Korea Univ.), "Prime Factorization and Cryptology".

- Pil J. Lee and Hee C. Moon (Pohang Institute of Science and Technology), "On Security of Password System".

- Dong K. Kim (Ajou Univ.), "An Outline and Trends on Information Security System".

- Se H. Kim (KAIST), "On the Classification of the Confidential Level of Information".

- Kyoung S. Lee (KDI), "Technology of Cryptosystem and Analysis of its Trends".

On Necessary and Sufficient Conditions for the Construction of Super Pseudorandom Permutations

Babak Sadeghiyan
Josef Pieprzyk *

Department of Computer Science,
University College,
University of New South Wales,
Australian Defence Force Academy,
Canberra, A.C.T. 2600, Australia.

Abstract

In this paper, we present the necessary and sufficient conditions for super pseudorandomness of DES-like permutations. We show that four rounds of such permutations with a single random function is not super psuedorandom and we present a distinguishing circuit for $\psi(f^2, f, f, f)$ and another circuit for $\psi(f^l, f^k, f^j, f^i)$. Then, we investigate the necessary and sufficient conditions for super pseudorandomness of type-1 Feistel type transformations, and we show that k^2 rounds of this transformation is super pseudorandom.

1 Introduction

Pseudorandom bit generators have many cryptographic applications. Classical pseudorandom bit generators are deterministic algorithms which generate binary strings that have statistical properties similar to a truly random one. However, even if a generator passes all known statistical tests, it is sometimes possible to predict the next bit knowing some previous ones. The notion of unpredictability of the next bit was first introduced

---

*Support for this project was provided in part by TELECOM Australia under the contract number 7027 and by ARC Grant under the reference number A48830241.

by Yao [5]. He showed that a generated string is indistinguishable from a random one, if and only if, knowing the first s bits of the string, predicting the $s + 1$ bit is difficult, in other words the next bit is unpredictable. Goldreich, Goldwasser and Micali introduced the notion of pseudorandom function generators [1] and showed how to construct a pseudorandom function generator from a pseudorandom bit generator adopting Yao's unpredictability criteria for the next bit.

Luby and Rackoff showed how to construct a pseudorandom invertible permutation generator with three pseudorandom function generators and application of three rounds of DES-like permutations [2] (this structure is notated as $\psi(h, g, f)$). A practical implication of their work is that a private key block cipher which can be proven to be secure against chosen plaintext attack can be constructed. Their result is quite astonishing since it is not based on any unproven hypothesis. Later Pieprzyk [4] showed that four rounds of DES-like permutations with a single pseudorandom function generator $\psi(f^2, f, f, f)$ is also pseudorandom and it is secure against a chosen plaintext attack. Luby and Rackoff also introduced the notion of super pseudorandomness, where the block cryptosystem is secure against a chosen plaintext/ciphertext attack. They suggested that $\psi(h, g, f, e)$ is super pseudorandom. It is a question of how to construct super pseudorandom permutations and whether $\psi(f^2, f, f, f)$ is super pseudorandom.

In this paper, we present the necessary and sufficient conditions for super pseudorandomness of DES-like permutations. Then we show that four rounds of such permutations with a single random function is not super psuedorandom. We present a distinguishing circuit for $\psi(f^2, f, f, f)$ and another one for some cases of $\psi(f^l, f^k, f^j, f^i)$.

Then, we investigate the necessary and sufficient conditions for super pseudorandomness of type-1 Feistel type transformations, and we show that k^2 rounds of this transformation is super pseudorandom. We also show that $k^2 - k + 1$ rounds of the inverse of this type of transformations are pseudorandom.

2 Notations

The notation we use, is similar to [4]. The set of all integers is denoted by N. Let $\Sigma = \{0, 1\}$ be the alphabet we consider. For $n \in N$, Σ^n is the set of all 2^n binary strings of length n. The concatenation of two binary strings x, y is denoted by $x \parallel y$. The bit by bit exclusive-OR of x and y is denoted by $x \oplus y$. By $x \in_r S$ we mean that x is chosen from a set S uniformly at random. By f a function we mean a trasformation from Σ^n to Σ^n. The set of all functions on Σ^n is denoted by H_n, i.e., $H_n = \{f \mid f : \Sigma^n \to \Sigma^n\}$ and it consists of 2^{n2^n} elements. The composition of two functions f and g is defined as $f \circ g(x) = f(g(x))$. The i-fold composition of f is denoted by f^i. A function f is a permutation if it is a 1 to 1 and onto function. The set of all permutations on Σ^n is defined by P_n and it consists of $2^n!$ elements.

3 Preliminaries

This section provides the preliminary definitions and notions which are used throughout the paper.

Definition 1 *We associate with a function $f \in H_n$ the DES-like permutation $D_{2n,f} \in P_{2n}$ as,*

$$D_{2n,f}(L \parallel R) = (R \oplus f(L) \parallel L)$$

where R and L are n-bit strings, i.e., $R, L \in \Sigma^n$.

Definition 2 *Having a sequence of functions $f_1, f_2, \ldots, f_i \in H_n$, we define the composition of their DES-like permutations as $\psi \in P_{2n}$, where*

$$\psi(f_i, \ldots, f_2, f_1) = D_{2n,f_i} \circ D_{2n,f_{i-1}} \circ \ldots \circ D_{2n,f_1}$$

Definition 3 *Let $l(n)$ be a polynomial in n, a function generator $F = \{F_n : n \in N\}$ is a collection of functions with the following properties:*

- *Indexing: Each F_n specifies for each k of length $l(n)$ a function $f_{n,k} \in H_n$.*

- *Poly-time evaluation: Given a key $k \in \Sigma^{l(n)}$, and a string $x \in \Sigma^n$, $f_{n,k}(x)$ can be computed in polynomial time in n.*

Definition 4 *An oracle circuit C_n is an acyclic circuit which contains Boolean gates of type AND, OR and NOT, and constant gates of type zero and one, and a particular kind of gates named oracle gates. Each oracle gate has an n-bit input and an n-bit output and it is evaluated using some function from H_n. The oracle circuit C_n has a single bit output.*

Definition 5 *The size of an oracle circuit C_n is the total number of connections between gates, Boolean gates, constant gates and oracle gates.*

Definition 6 *A distinguishing circuit family for a function generator F is an infinite family of circuits $\{C_{n_1}, C_{n_2}, \ldots\}$, where $n_1 < n_2 < \ldots$, such that for some pair of constants c_1 and c_2 and for each n there exist a circuit C_n such that:*

- *The size of C_n is less than or equal to n^{c_1}.*

- *Let $\mathrm{Prob}\{C_n[H_n] = 1\}$ be the probability that the output bit of C_n is one when a function is randomly selected from H_n and used to evaluate oracle gates. Let $\mathrm{Prob}\{C_n[F_n] = 1\}$ be the probability that the output bit of C_n is one when a key k of length $l(n)$ is randomly chosen and $f_{n,k}$ is used to evaluate the oracle gates.*

The distinguishing probability for C_n is greater than or equal to $\frac{1}{n^{c_2}}$, that is,

$$\mid \mathrm{Prob}\{C_n[H_n] = 1\} - \mathrm{Prob}\{C_n[F_n] = 1\} \mid \geq \frac{1}{n^{c_2}}$$

Definition 7 *A function generator F is pseudorandom if there is no distinguishing circuit family for F.*

4 Pseudorandom and Super Pseudorandom Permutations

How to construct pseudorandom permutation from pseudorandom functions was first presented by Luby and Rackoff applying a DES-like structure. The following lemma is due to Luby and Rackoff and has been stated in [2] as the main lemma.

Lemma 1 *Let $f_1, f_2, f_3 \in_r H_n$ be indepedent random functions and C_{2n} be an oracle circuit with $m < 2^n$ oracle gates, then*

$$| \text{Prob}\{C_{2n}[P_{2n}] = 1\} - \text{Prob}\{C_{2n}[\psi(f_3, f_2, f_1)] = 1\} | \leq \frac{m^2}{2^n}$$

In other words, a block cryptosystem with three rounds of DES-like permutation and three different random functions is secure against chosen plaintext attack, when a cryptanalyst can ask for only a polynomial number of plaintexts. This result is quite astonishing since it does not depend on any unproved hypothesis and can be used as a justification for the design rule of DES. Luby and Rackoff also showed that the above Lemma remains valid even when functions are selected from three pseudorandom function generators.

When the block cryptosystem is secure against chosen plaintext/ciphertext attack, it is called super pseudorandom. This notion only applies for invertible permutations and is stated formally in three following definitions.

Definition 8 *A permutation generator F is a function generator such that each function $f_{n,k}$ is 1 to 1 and onto. Let $\overline{F} = \{\overline{F_n} : n \in N\}$, where $\overline{F_n} = \{\overline{f}_{n,k} : k \in \Sigma^{l(n)}\}$, where $\overline{f}_{n,k}$ is the inverse function of $f_{n,k}$. F is called invertible if $\overline{F}$ is also a permutation generator.*

Definition 9 *A super distinguishing family of circuits for an invertible permutation generator F is an infinite family of circuits $\{SC_{n_1}, SC_{n_2}, \ldots\}$, where $n_1 < n_2 < \ldots$, where each circuit is an oracle circuit containing two types of oracle gates, normal and inverse, such that for some pair of constants c_1 and c_2 and for each n there exist a circuit SC_n such that:*

- *The size of SC_n is less than or equal to n^{c_1}.*

- *Let $\text{Prob}\{SC_n[P_n] = 1\}$ be the probability that the output bit of SC_n is one when a permutation p is randomly selected from P_n and p and $\overline{p}$ are used to evaluate normal and inverse oracle gates. Let $\text{Prob}\{SC_n[F_n] = 1\}$ be the probability that the output bit of SC_n is one when a key k of length $l(n)$ is randomly chosen and $f_{n,k}$ and $\overline{f}_{n,k}$ is used to evaluate the normal and inverse oracle gates respectively.*

The distinguishing probability for SC_n is greater than or equal to $\frac{1}{n^{c_2}}$, that is,

$$| \operatorname{Prob}\{SC_n[P_n] = 1\} - \operatorname{Prob}\{SC_n[F_n] = 1\} | \geq \frac{1}{n^{c_2}}$$

Definition 10 *A permutation generator F is super pseudorandom if there is no super distinguishing circuit family for F.*

If F is a super pseudorandom permutation generator, it is secure against a chosen plaintext/ciphertext attack where a cryptanalyst can interactively choose plain blocks and see their encryptions and choose encryptions and see their corresponding plaintext blocks.

5 Necessary and Sufficient Conditions for Super Pseudorandomness

Luby and Rackoff [2] also suggested that, it is possible to make a super pseudorandom permutation with four independent random functions, i.e. if $f_1, f_2, f_3, f_4 \in H_n$ are independent random functions then $\psi(f_4, f_3, f_2, f_1)$ is a super pseudorandom permutation. It is a matter of question whether it is possible to build a super pseudorandom permutation with less number of random functions. In this section, we present necessary and sufficient conditions for construction of super pseudorandom permutations in Theorem 1. Later, these conditions are applied to construct a super pseudorandom permutation with a less number of random functions. Then, they are also applied to show that with a single random function and four rounds of DES-like permutations, a super pseudorandom permutation cannot be constructed.

First a definition for independent permutations is given, which would be used in the proof of Theorem 1.

Let π_1, π_2 be two pseudorandom permutation generators, and C_n be an oracle circuit with two types of oracles. Let $\operatorname{Prob}\{C_n[\pi_1, \pi_2] = 1\}$ be the probability that the output bit of the oracle circuit C_n is 1, when a key k of length $l(n)$ is randomly chosen and $p_{1,k} \in \pi_1$, $p_{2,k} \in \pi_2$ are used to evaluate each type of the oracle gates respectively. Let $\operatorname{Prob}\{C_n[P_n, P_n] = 1\}$ be the probability that the output bit of the oracle circuit C_n is 1, when two permutations p_1, p_2 are chosen independently and randomly from P_n and are used to evaluate the oracle gates of C_n. If π_2 is the inverse of π_1, then let $\operatorname{Prob}\{C_n[P_n, P_n] = 1\}$ be the probability that the output bit of the oracle circuit C_n is 1, when a permutation p is chosen randomly from P_n and p and $\bar{p}$ are used to evaluate the oracle gates of C_n. We say π_1 and π_2 are two dependent permutation generators if there is a distinguishing circuit with two types of oracles C_n such that

$$| \operatorname{Prob}\{C_n[\pi_1, \pi_2] = 1\} - \operatorname{Prob}\{C_n[P_n, P_n] = 1\} | \geq \frac{1}{n^{c_2}}$$

for some constant c_2.

Definition 11 *We say π_1 and π_2 are two independent permutation generators, if there is no distinguishing oracle circuit family with two types of oracle gates for (π_1, π_2).*

The following lemma suggests how to construct two independent permutations, applying DES-like structures.

Lemma 2 *Let $f_1, f_2, \ldots, f_i \in_r F_n$, and $G_2 = \psi(f_i, \ldots, f_2)$ and $G_3 = \psi(f_1, \ldots, f_{i-1})$ are two independent permutations, if and only if they are pseudorandom.*

Proof Sketch: We first show if $G_2 = \psi(f_i, \ldots, f_2)$ and $G_3 = \psi(f_1, \ldots, f_{i-1})$, are pseudorandom, they are independent permutations, where $f_1, \ldots, f_i \in_r F_n$. For simplicity of the proof, consider that $f_1, \ldots, f_i \in_r H_n$, but it is allowed to examine only a polynomial m number of oracle gates. In other words, it is assumed that the probability of distinguishing G_2 or G_3 from a random permutation is less than $\frac{Q(m)}{2^n}$, for any polynomial Q. Since, both G_2 and G_3 would not be distinguishable from a random permutation when polynomial number of oracles are examined, any distinguisher circuit for dependency of G_2 and G_3 would virtually be a distinguisher for dependency of at least a branch of of G_2 from a branch of G_3, for both G_2 and G_3 have two branches. Two situations may arise:

- i is even:

  When i is even, each branch of G_2 and G_3 is fed with a different set of random functions. So each of the four branches is statistically independent from other branches. Hence, there would be no distinguishing circuit for dependency of G_2 and G_3.

- i is odd:

  In this structure, one branch of G_2 is fed with the same set of random functions which feeds a branch of G_3, but the other branches are independent from the others. Since the former branches are fed with a reverse order of the same set of random functions, they are also independent from each other.

When random functions are substituted with pseudorandom ones, the probability of dependency between G_2 and G_3 would still remain less than 1 over any polynomial (see [2]).

Although, in the above proof, we considered that pseudorandom functions $f_1, \ldots, f_i$ are chosen independently from F_n, but in the general case, it is sufficient that only f_{i-1} and f_2 be chosen independently to make G_2 and G_3 two indepedent permutations, if G_2 and G_3 are pseudorandom, (no matter whether the other pseudorandom functions are chosen independently). ·

In addition, it can be shown if G_2 and G_3 are independent, they would be pseudorandom. Since using one type of oracles reduces the possibility of distinguishing, then

$$| \operatorname{Prob}\{C_{2n}[G_2] = 1\} - \operatorname{Prob}\{C_{2n}[P_{2n}] = 1\} | \ <$$
$$| \operatorname{Prob}\{C_{2n}[G_2, G_3] = 1\} - \operatorname{Prob}\{C_{2n}[P_{2n}, P_{2n}] = 1\} | \ < \ \frac{1}{n^{c_2}}$$

and also,

$$| \operatorname{Prob}\{C_{2n}[G_3] = 1\} - \operatorname{Prob}\{C_{2n}[P_{2n}] = 1\} | \ <$$
$$| \operatorname{Prob}\{C_{2n}[G_2, G_3] = 1\} - \operatorname{Prob}\{C_{2n}[P_{2n}, P_{2n}] = 1\} | \ < \ \frac{1}{n^{c_2}}$$

the above inequalities shows that both G_2 and G_3 are pseudorandom if they are independent. This completes the proof of lemma 2. □

Theorem 1 *Let $f_1, f_2, \ldots, f_i \in F_n$ such that $G_1 = \psi(f_i, \ldots, f_1)$ be a pseudorandom permutation. G_1 is super pseudorandom if and only if $G_2 = \psi(f_i, \ldots, f_2)$ and $G_3 = \psi(f_1, \ldots, f_{i-1})$ are two independent permutations.*

Proof :

To prove the validity of Theorem 1, we first should show *if G_2 and G_3 are independent, then G_1 is super pseudorandom*. First, it is necessary to show that $G_3 = \psi(f_1, \ldots, f_{i-1})$ and $\overline{G}_1 = \psi(f_1, \ldots, f_{i-1}, f_i)$ are independent of each other. In order to prove this claim, by contradiction assume that they are not independent, so

$$| \operatorname{Prob}\{C_{2n}[G_3, \overline{G}_1] = 1\} - \operatorname{Prob}\{C_{2n}[P_{2n}, P_{2n}] = 1\} | \geq \frac{1}{n^{c_2}}$$

Without changing the inequality relation, we have,

$$| \operatorname{Prob}\{C_{2n}[G_3, \overline{G}_1] = 1\} - \operatorname{Prob}\{C_{2n}[G_3, G_3] = 1\} +$$
$$\operatorname{Prob}\{C_{2n}[G_3, G_3] = 1\} - \operatorname{Prob}\{C_{2n}[P_{2n}, P_{2n}] = 1\} | \geq \frac{1}{n^{c_2}}$$

then,

$$| \operatorname{Prob}\{C_{2n}[G_3, \overline{G}_1] = 1\} - \operatorname{Prob}\{C_{2n}[G_3, G_3] = 1\} | \ +$$
$$| \operatorname{Prob}\{C_{2n}[G_3, G_3] = 1\} - \operatorname{Prob}\{C_{2n}[P_{2n}, P_{2n}] = 1\} | \ \geq \ \frac{1}{n^{c_2}}$$

If $| \operatorname{Prob}\{C_{2n}[G_3, G_3] = 1\} - \operatorname{Prob}\{C_{2n}[P_{2n}, P_{2n}] = 1\} | \geq \frac{1}{n^{c_2}}$, then G_3 is not pseudorandom, which contradicts our assumption. If $| \operatorname{Prob}\{C_{2n}[G_3, \overline{G}_1] = 1\} - \operatorname{Prob}\{C_{2n}[G_3, G_3] = 1\} | \geq \frac{1}{n^{c_2}}$, then the oracle circuit virtually distinguishes f_i from a randomly chosen function. This also contradicts our assumption that f_i is a pseudorandom function. Since both cases conclude to contradictions to our assumptions, it can be concluded G_3 and $\overline{G}_1$ should be independent of each other.

Note that, in order that $\overline{G}_1$ and G_3 to be two independent permutations, there is no need that the pseudorandom function f_i be chosen independently from $f_1, \ldots, f_{i-1}$.

Considering the claim, when it is given that G_2 and G_3 are independent, then

$$| \operatorname{Prob}\{C_{2n}[G_2, G_3] = 1\} - \operatorname{Prob}\{C_{2n}[P_{2n}, P_{2n}] = 1\} | < \frac{1}{n^{c_2}}$$

Without changing the sign of inequality, we may expand the above relation as,

$$
\begin{aligned}
| \operatorname{Prob}\{C_{2n}[G_1, \overline{G}_1] = 1\} - \operatorname{Prob}\{C_{2n}[G_1, \overline{G}_1] = 1\} \; + \\
\operatorname{Prob}\{C_{2n}[G_3, \overline{G}_1] = 1\} - \operatorname{Prob}\{C_{2n}[G_3, \overline{G}_1] = 1\} \; + \\
\operatorname{Prob}\{C_{2n}[G_2, G_3] = 1\} - \operatorname{Prob}\{C_{2n}[P_{2n}, P_{2n}] = 1\} | \; < \; \frac{1}{n^{c_2}}
\end{aligned}
$$

With reordering and separation of absolute values, we can get:

$$
\begin{aligned}
|| \operatorname{Prob}\{C_{2n}[G_2, G_3] = 1\} - \operatorname{Prob}\{C_{2n}[G_3, \overline{G}_1] = 1\} | \; - \\
| \operatorname{Prob}\{C_{2n}[G_3, \overline{G}_1] = 1\} - \operatorname{Prob}\{C_{2n}[G_1, \overline{G}_1] = 1\} | \; - \\
| \operatorname{Prob}\{C_{2n}[G_1, \overline{G}_1] = 1\} - \operatorname{Prob}\{C_{2n}[P_{2n}, P_{2n}] = 1\} || \; < \; \frac{1}{n^{c_2}}
\end{aligned}
$$

Since, it was assumed that G_2 and G_3 are two independent permutations, and so are G_3 and $\overline{G}_1$, then $| \operatorname{Prob}\{C_{2n}[G_2, G_3] = 1\} - \operatorname{Prob}\{C_{2n}[G_3, \overline{G}_1] = 1\} |$ would be less than $\frac{1}{n^{c_2}}$, since

$$
\begin{aligned}
| \operatorname{Prob}\{C_{2n}[G_2, G_3] = 1\} - \operatorname{Prob}\{C_{2n}[G_3, \overline{G}_1] = 1\} | \; < \\
| \operatorname{Prob}\{C_{2n}[G_2, G_3] = 1\} - \operatorname{Prob}\{C_{2n}[P_{2n}, P_{2n}] = 1\} | \; + \\
| \operatorname{Prob}\{C_{2n}[G_3, \overline{G}_1] = 1\} - \operatorname{Prob}\{C_{2n}[P_{2n}, P_{2n}] = 1\} | \; < \; \frac{1}{n^{c_2}}
\end{aligned}
$$

Hence

$$
\begin{aligned}
| \operatorname{Prob}\{C_{2n}[G_1, \overline{G}_1] = 1\} - \operatorname{Prob}\{C_{2n}[P_{2n}, P_{2n}] = 1\} | \; + \\
| \operatorname{Prob}\{C_{2n}[G_3, \overline{G}_1] = 1\} - \operatorname{Prob}\{C_{2n}[G_1, \overline{G}_1] = 1\} | \; < \; \frac{1}{n^{c_2}}
\end{aligned}
$$

So, each of the above absolute values would be less than $\frac{1}{n^{c_2}}$. In other words

$$| \operatorname{Prob}\{C_{2n}[G_1, \overline{G}_1] = 1\} - \operatorname{Prob}\{C_{2n}[P_{2n}, P_{2n}] = 1\} | < \frac{1}{n^{c_2}}$$

Hence G_1 and $\overline{G}_1$ are independent of each other, and G_1 is a super pseudorandom permutation, as it is pseudorandom.

To conclude Theorem 1, we should also show that *if G_1 is a super pseudorandom permutation, then G_2 and G_3 are two independent permutations.* Hence, it is given that

$$| \operatorname{Prob}\{C_{2n}[G_1,\overline{G}_1] = 1\} - \operatorname{Prob}\{C_{2n}[P_{2n},P_{2n}] = 1\} | < \frac{1}{n^{c_2}}$$

Without changing the sign of inequality,

$$| \operatorname{Prob}\{C_{2n}[G_1,\overline{G}_1] = 1\} - \operatorname{Prob}\{C_{2n}[G_3,\overline{G}_1] = 1\} \ +$$
$$\operatorname{Prob}\{C_{2n}[G_3,\overline{G}_1] = 1\} - \operatorname{Prob}\{C_{2n}[G_3,G_2] = 1\} \ +$$
$$\operatorname{Prob}\{C_{2n}[G_3,G_2] = 1\} - \operatorname{Prob}\{C_{2n}[P_{2n},P_{2n}] = 1\} | \ < \ \frac{1}{n^{c_2}}$$

With reordering and separation of absolute values, we get:

$$|| \operatorname{Prob}\{C_{2n}[G_1,\overline{G}_1] = 1\} - \operatorname{Prob}\{C_{2n}[G_3,\overline{G}_1] = 1\} | \ -$$
$$| \operatorname{Prob}\{C_{2n}[G_3,\overline{G}_1] = 1\} - \operatorname{Prob}\{C_{2n}[G_3,G_2] = 1\} | \ -$$
$$| \operatorname{Prob}\{C_{2n}[G_3,G_2] = 1\} - \operatorname{Prob}\{C_{2n}[P_{2n},P_{2n}] = 1\} || \ < \ \frac{1}{n^{c_2}}$$

If either G_2 or G_3 is not pseudorandom, it can be shown that G_1 is not super pseudorandom. As a justification, consider there is a distinguishing circuit with m oracle gates such that distinguishes G_3 from a random permutation with a probability better than $\frac{1}{n^{c_2}}$ when m random $2n$ bit strings are input to its oracle gates, then there is a super distinguishing circuit with at most $2m$ normal oracles and $2m$ inverse oracles for G_1 with the same probability.

Consider it can be proven that G_3 is pseudorandom. It can be shown that

$$| \operatorname{Prob}\{C_{2n}[G_1,\overline{G}_1] = 1\} - \operatorname{Prob}\{C_{2n}[G_3,\overline{G}_1] = 1\} | < \frac{1}{n^{c_2}}$$

Since G_3 and $\overline{G}_1$ would be independent of each other, and G_1 is assumed to be super pseudorandom. Hence, the following inequality would also be valid,

$$| \operatorname{Prob}\{C_{2n}[G_3,\overline{G}_1] = 1\} - \operatorname{Prob}\{C_{2n}[G_3,G_2] = 1\} | \ +$$
$$| \operatorname{Prob}\{C_{2n}[G_3,G_2] = 1\} - \operatorname{Prob}\{C_{2n}[P_{2n},P_{2n}] = 1\} | \ < \ \frac{1}{n^{c_2}}$$

It can be concluded that

$$| \operatorname{Prob}\{C_{2n}[G_3,G_2] = 1\} - \operatorname{Prob}\{C_{2n}[P_{2n},P_{2n}] = 1\} | < \frac{1}{n^{c_2}}$$

In other words, G_2 and G_3 are independent of each other. This completes the proof of Theorem 1. $\square$

Corollary 1 *Let $f_1, f_2, \ldots, f_i \in_r F_n$ such that $G_1 = \psi(f_i, \ldots, f_1)$ be a pseudorandom permutation. G_1 is super pseudorandom if and only if $G_2 = \psi(f_i, \ldots, f_2)$ and $G_3 = \psi(f_1, \ldots, f_{i-1})$ are pseudorandom permutations.*

Proof: It was shown in lemma 2, given f_2 and f_{i-1} are two independent pseudorandom functions, if G_2 and G_3 are pseudorandom, they would be independent too. As it was shown in Theorem 1, when G_2 and G_3 are independent, G_1 is super pseudorandom. Moreover, It was also shown that if G_1 is super pseudorandom then G_2 and G_3 are independent of each other, and when they are independent both should be pseudorandom and satisfy the conditions given in the axiom of Theorem 1. This finishes the proof. $\square$

Ohnishi [3] showed it is possible to use two independent pseudorandom functions, instead of three, in a three round DES-like structure to obtain a pseudorandom permutation, i.e., $\psi(f_2, f_2, f_1)$ is a pseudorandom permutation generator. Applying his results and result of Theorem 1 the following corollary can be shown to be true.

Corollary 2 Let $f_1, f_2 \in_r F_n$, then $G_1 = \psi(f_2, f_2, f_1, f_1)$ is a super pseudorandom permutation.

Although Theorem 1 states the necessary and sufficient conditions for super pseudorandomness in probabilistic terms, in practice we can construct super distinguishing circuits with a distinguishing probability near 1, for some structures. We investigate the validity of the above theorem and we construct super distinguishing circuits for some special cases.

Lemma 3 Let $f \in_r H_n$ then $\psi(f^2, f, f, f)$ is not super pseudorandom and there is a super distinguishing circuit SC_{2n} with 4 normal and inverse oracle gates.

Proof : Considering Theorem 1, $\psi(f^2, f, f, f)$ would be super pseudorandom if $\psi(f^2, f, f)$ and $\psi(f, f, f)$ are independent. Zheng, Matsumoto and Imai [6] showed that it is impossible to get pseudorandom permutation with three rounds of DES-like permutations and a single random functions, so neither $\psi(f^2, f, f)$ nor $\psi(f, f, f)$ are pseudorandom, and they are not independent. Therefore $\psi(f^2, f, f, f)$ cannot be a super pseudorandom permutation, due to Theorem 1. The structure of a super distinguishing circuit SC_{2n} is as follows:

Let $\hat{O}_0, \hat{O}_1, \hat{O}_3$ be normal oracle gates and $\check{O}_2$ be an inverse oracle gate. Denote by $(L_u \parallel R_u)$ and $(S_u \parallel T_u)$ the input to and the output of the u-th oracle gate respectively, and by $0^n \in \Sigma^n$ a n-bit string of all 0.

1. The input to $\hat{O}_0$ is $(L_0 \parallel R_0) = (0^n \parallel 0^n)$

2. The input to $\hat{O}_1$ is $(L_1 \parallel R_1) = (0^n \parallel T_0)$

3. The input to $\check{O}_2$ is $(L_2 \parallel R_2) = (0^n \parallel 0^n)$

4. The input to $\hat{O}_3$ is $(L_3 \parallel \acute{R}_3) = (S_2 \parallel T_2 \oplus T_0)$

5. SC_{2n} outputs a bit 1 if and only if $T_3 = T_0 \oplus T_1$

When a function $\psi(f^2, f, f, f)$ is used to evaluate the oracle gates the probability that SC_{2n} outputs a bit 1 is equal to 1, and when a function is drawn randomly and uniformly from P_{2n} the probability that SC_{2n} outputs 1 is equal to $\frac{1}{2^n}$. Thus SC_{2n} is a super distinguishing circuit for $\psi(f^2, f, f, f)$. $\qquad\square$

Lemma 3 can be generalised in the following theorem.

Theorem 2 *Let $f \in_r H_n$, there is a super distinguishing circuit for $\psi(f^l, f^k, f^j, f^i)$ with $p + q + 4$ normal and inverse oracle gates, where p, q satisfy*

$$i + j - l = (q - p)(j + k)$$

Proof : Due to the impossibility results of Zheng et al. [6], neither $\psi(f^l, f^k, f^j)$ nor $\psi(f^i, f^j, f^k)$ is pseudorandom. Thus, due to Theorem 1, $\psi(f^l, f^k, f^j, f^i)$ is not super pseudorandom .

We managed to construct a distinguisher for the following case:

Let $\hat{O}_0, \hat{O}_1, \ldots, \hat{O}_{p+1}$ and $\hat{O}_{p+q+3}$ be normal oracle gates and $\check{O}_{p+2}, \ldots, \check{O}_{p+q+2}$ be inverse oracle gates, where p, q satisfy: $i + j + k + p(j + k) = l + k + q(j + k)$ or,

$$i + j - l = (q - p)(j + k)$$

The structure of SC_{2n} is as follows:

1. The input to $\hat{O}_0$ is $(L_0 \parallel R_0) = (0^n \parallel 0^n)$

2. The input to $\hat{O}_1$ is $(L_1 \parallel R_1) = (0^n \parallel T_0)$ and the input for $\hat{O}_2$ to $\hat{O}_{p+1}$ is $(L_u \parallel R_u) = (0^n \parallel T_{u-1} \oplus R_{u-1})$

3. The input to $\check{O}_{p+2}$ is $(L_{p+2} \parallel R_{p+2}) = (0^n \parallel 0^n)$

4. The input to $\check{O}_{p+3}$ is $(L_{p+3} \parallel R_{p+3}) = (S_{p+3} \parallel 0^n)$ and the input for $\check{O}_{p+4}$ to $\check{O}_{p+q+2}$ is $(L_u \parallel R_u) = (L_{u-1} \oplus T_{u-1} \parallel 0^n)$

5. The input to $\hat{O}_{p+q+3}$ is $(L_{p+q+3} \parallel R_{p+q+3}) = (S_{p+q+2} \parallel T_p \oplus T_{p+q+2})$

6. SC_{2n} outputs a bit 1 if and only if $T_{p+q+3} = T_p \oplus T_{p+1}$

When a function $\psi(f^l, f^k, f^j, f^i)$ is used to evaluate the oracle gates and $i + j + k + p(j + k) = l + k + q(j + k)$, the probability that C_{2n} outputs a bit 1 is equal to 1, and when a function is drawn randomly and uniformly from P_{2n} the probability that C_{2n} outputs 1 is equal to $\frac{1}{2^n}$. $\qquad\square$

6 Super Pseudorandomness in Generalised DES-like Permutations

A DES-like permutation is a permutation in P_{2n} by applying functions in H_n. Zheng et al. [7] made three types of permutations in P_{kn} by generalisations of the construction of the DES-like permutation and application of functions in H_n, and called them type-1, type-2 and type-3 Feistel type transformations. They suggested that a permutation consisting of $k + 2$ rounds of type-2 or type-3 transformations is a super pseudorandom permutation, where each round is associated with a random function tuple. In this section, we show that k^2 rounds of type-1 transformation are required to get a super pseudorandom permutation.

First, we give the definition of type-1 transformations, according to [7]. Then, the necessary and sufficient conditions for super pseudorandomness of this type of transformations would be presented, and accordingly, some cases which are pseudorandom but cannot be super pseudorandom would be given. Finally, it is shown that k^2 rounds of type-1 transformations makes a super pseudorandom permutation.

6.1 Type-1 Transformations

Let $g_{1,i} \in H_{kn}$ be a function associated with an $f_i \in H_n$ and be defined by

$$g_{1,i}(B_1 \| B_2 \| \ldots \| B_k) = (B_2 \oplus f_i(B_1) \| B_3 \| \ldots \| B_k \| B_1)$$

where $B_j \in \Sigma^n$ for $1 \le j \le k$ and $k \in N$. Functions defined in such a way are called type-1 transformations. $g_{1,i}$ can be decomposed into $g_{1,i} = L_{rot} \circ \pi_{1,i}$, where

$$\pi_{1,i}(B_1 \| B_2 \| \ldots \| B_k) = (B_1 \| B_2 \oplus f_i(B_1) \| B_3 \| \ldots \| B_k)$$
$$L_{rot}(B_1 \| B_2 \| \ldots \| B_k) = (B_2 \| B_3 \| \ldots \| B_k \| B_1)$$

$g_{1,i}$ is an invertible permutation on Σ^{kn}, and its inverse, denoted by $\overline{g}_{1,i}$ is given by $\overline{g}_{1,i} = \pi_{1,i} \circ R_{rot}$ where

$$R_{rot}(B_1 \| B_2 \| \ldots \| B_k) = (B_k \| B_1 \| B_2 \| \ldots \| B_{k-1})$$

For $f_1, f_2, \ldots, f_s \in H_n$, define $\psi_1(f_s, \ldots, f_2, f_1) = g_{1,s} \circ \ldots \circ g_{1,2} \circ g_{1,1}$. Note that ψ_1 is an invertible permutation on Σ^{kn}, and its inverse $\overline{\psi}_1$ is defined by

$$\overline{\psi}_1(f_1, f_2, \ldots, f_s) = \overline{g}_{1,1} \circ \overline{g}_{1,2} \circ \ldots \circ \overline{g}_{1,s}$$

6.2 The Necessary and Sufficient Conditions for Super Pseudorandomness of Type-1 Transformations

It can be shown that $2k - 1$ rounds of type-1 transformations where, each round is associated with a randomly and independently chosen function from F_n, is a pseudorandom permutation.

The following lemma is from [7] and formally represents the above statement.

Lemma 4 *Let Q be a polynomial in n and C_{kn} be an oracle circuit with $Q(n) < 2^n$ oracle gates, then*

$$| \operatorname{Prob}\{C_{kn}[P_{kn}] = 1\} - \operatorname{Prob}\{C_{kn}[\psi_1(f_{2k-1}, \dots, f_2, f_1)] = 1\} | \le \frac{(k-1)Q^2(n)}{2^n}$$

where $f_1, f_2, \dots, f_{2k-1} \in_r F_n$.

Although the above lemma states that $\psi_1(f_{2k-1}, \dots, f_2, f_1)$ is pseudorandom, it is interesting to note that it is not super pseudorandom. This has been presented in the following lemma, where a super distinguishing circuit has been presented in the proof.

Lemma 5 *For any $f_{2k-1}, \dots, f_2, f_1 \in_r F_n$, there is super distinguishing circuit SC_{kn} for $\psi_1(f_{2k-1}, \dots, f_2, f_1)$.*

Proof : Let $B_1, B_2, \dots, B_k$ be strings of length n. The super distinguishing circuit has two oracles, i.e., a normal oracle and an inverse oracle. The input to the normal oracle is $B_1 \parallel B_2 \parallel \dots \parallel B_k$. Let $S_1 \parallel S_2 \parallel \dots \parallel S_k$ be the output of this oracle. Let the input to the inverse oracle gate be $S_1 + \alpha \parallel S_2 \parallel \dots \parallel S_k$ where α is an arbitrary n-bit string. The output of SC_{2n} is 1 if and only if the last n bits of the output from the inverse oracle gate is equal to $B_k + \alpha$. It can be verified that the output of SC_{2n} is always 1 when the normal and inverse oracle gates are computed by using ψ_1 for normal oracle gates and $\overline{\psi}_1$ for inverse oracle gates. On the other hand, if the oracle gates are computed using a permutation randomly chosen from P_{kn}, the output of SC_{kn} is 1 with probability $\frac{1}{2^n}$. □

It can be easily verified that by using an inverse oracle, together with a normal oracle, the effect of f_{2k-1} is virtually removed. In other words, the super distinguisher actually evaluates the inverse oracle with $\overline{\psi}_1(f_1, f_2, \dots, f_{2k-2})$, which is not pseudorandom by any means.

It can also be shown that the effect of $f_{2k-2}, \dots, f_{k+2}$ and f_k can also be removed (individually) by taking procedures similar to the procedure which is given in the proof of Theorem 1. If there existed a construction with type-1 transformations G_1 such that removing the last k random functions in G_1 and the first k random functions in $\overline{G}_1$, the remaining structures would remain pseudorandom, then G_1 would be a super pseudorandom permutation.

Based on the obove observation, we give the necessary and sufficient conditions for super pseudorandomness of i rounds of type-1 transformations.

Theorem 3 *Let $G_1 = \psi_1(f_i, \dots, f_1)$ be a pseudorandom permutation where $G_1 \in P_{kn}$ and consists of i rounds of type-1 transformations and $f_1, f_2, \dots, f_i \in_r H_n$. G_1 is a super* pseudorandom permutation if and only if $G_{2,j} = \psi_1(f_i, \ldots, f_{j+1}) \circ L_{rot} \circ \psi_1(f_{j-1}, \ldots f_1)$ and $G_{3,j} = \overline{\psi}_1(f_1, \ldots, f_{i-j}) \circ R_{rot} \circ \overline{\psi}_1(f_{i-j+2}, \ldots, f_i)$ are pseudorandom permutations, for $j = 1, 2, \ldots, k$ and $i - j \neq kl$, where $l = \lfloor \frac{i}{k} \rfloor$.

Proof Sketch: Note that type-1 transformation is a generalisation of DES-like permutations, and the effect of f_{i-j+1} in the inverse oracle gates and the effect of f_j in normal oracle gates can be removed individually by applying normal and oracle gates consequently, for $j = 1, 2, \ldots, k$ and $i - j \neq kl$, where $l = \lfloor \frac{i}{k} \rfloor$.

For justifying the theorem, two following claims should be proven to be true for each j:

1. If G_1 is super pseudorandom, then $G_{2,j}$ and $G_{3,j}$ are pseudorandom.

2. If $G_{2,j}$ and $G_{3,j}$ are pseudorandom, then G_1 is super pseudorandom.

The validity of the above claims can be checked for each j easily. For instance, consider $j = 1$, then $G_{2,1} = \psi_1(f_i, \ldots, f_2) \circ L_{rot}$ and $G_{3,1} = \overline{\psi}_1(f_1, \ldots, f_{i-1}) \circ R_{rot}$. Similar to the proof of Lemma 2, it can be shown that $G_{2,1}$ and $G_{3,1}$ are independent if and only if they are pseudorandom. In addition, similar to the proof of Theorem 1, it can be shown that if $G_{2,1}$ and $G_{3,1}$ are independent, G_1 is super pseudorandom. For $j = 2, \ldots, k$ the proofs can also be obtained by repeating a similar method and procedure. Note that when $i - j = \lfloor \frac{i}{k} \rfloor k$, the effect of f_{i-j+1} would not be removed with inverse and normal oracle gates, due to the structure of type-1 transformations, so it is not necessary to prove the above claims for it. Since all possible reductions of ψ_1 and $\overline{\psi}_1$ would remain pseudorandom having even super distinguishing circuits, then ψ_1 would be a super pseudorandom permutation. □

Although it was already stated in Lemma 4 that $2k - 1$ rounds of type-1 permutation ψ_1 give a pseudorandom permutation, it would not follow immediately that $3k - 2$ rounds of this transformation would yield a super pseudorandom permutation since $2k - 1$ rounds, or even $3k - 2$ rounds, of its inverse transformation $\overline{g}$ is not pseudorandom. It can be shown that $k(k - 1) + 1$ rounds of $\overline{g}$ is pseudorandom, which has been stated in the following lemma formally.

Lemma 6 Let $\overline{\psi}_1$ be a permutation defined by

$$\overline{\psi}_1(f_{k^2-k+1}, \ldots, f_2, f_1) - \overline{g}_{1,k^2-k+1} \circ \cdots \circ \overline{g}_{1,2} \circ \overline{g}_{1,1}$$

Let Q be a polynomial in n and C_{kn} be an oracle circuit with $Q(n) < 2^n$ oracle gates, then

$$| \operatorname{Prob}\{C_{kn}[P_{kn}] = 1\} - \operatorname{Prob}\{C_{kn}[\overline{\psi}_1(f_{k^2-k+1}, \ldots, f_2, f_1)] = 1\} | \leq \frac{(k^2 - 1)Q^2(n)}{2^{n+1}}$$

where $f_1, f_2, \ldots, f_{k^2-k+1} \in_r F_n$.

Proof : The proof of this lemma is very similar to the proof of Lemma 4, presented in [7], and is omitted here. □

Since, according to Theorem 3, by application of normal and reverse oracle gates, the effect of only $k-1$ rounds of ψ and only $k-1$ rounds of $\overline{\psi}_1$ can be removed, $k^2 = k^2 - k + 1 + (k-1) > 2k - 1 + (k-1)$ rounds of type-1 transformation can resist against super distinguishing circuits. This is stated in the following theorem formally.

Theorem 4 *Let Q be a polynomial in n and SC_{kn} be a super distinguishing circuit with $Q(n) < 2^n$ normal and inverse oracle gates, then*

$$| \mathrm{Prob}\{SC_{kn}[P_{kn}] = 1\} - \mathrm{Prob}\{SC_{kn}[\psi_1(f_{k^2}, \ldots, f_2, f_1)] = 1\} | \leq \frac{(k^2 - 1)Q^2(n)}{2^n}$$

where $f_1, f_2, \ldots, f_{k^2} \in_r F_n$.

Proof: To obtain a pseudorandom permutation, $G_{2,j}$ and $G_{3,j}$ should be pseudorandom for $j = 1, 2, \ldots, k-1$, where for $\psi_1(f_{k^2}, \ldots, f_2, f_1)$,

$$G_{2,j} = \psi_1(f_{k^2}, \ldots, f_{j+1}) \circ L_{rot} \circ \psi_1(f_{j-1}, \ldots, f_1)$$

and

$$G_{3,j} = \overline{\psi}_1(f_1, \ldots, f_{k^2-j}) \circ R_{rot} \circ \overline{\psi}_1(f_{k^2-j+2}, \ldots, f_{k^2})$$

$G_{2,j}$ is partitioned into two parts where a part always consists more than $2k-1$ rounds, so in the normal oracles even if the effect of the other $k-1$ rounds of $\psi 1$ could be removed, the remaining would maintain pseudorandomness. $G_{3,j}$ is also partitioned into two parts where a part always consists at least $k^2 - k + 1$ rounds, so in the inverse oracles even if the effect of the other $k-1$ rounds of $\overline{\psi}_1$ could be removed, the remaining would maintain pseudorandomness. Hence, $G_{2,j}$ and $G_{3,j}$ are pseudorandom for all $j = 1, 2, \ldots, k-1$, then G_1 is a super pseudorandom permutation. The probability that a super distinguisher circuit outputs 1, in the worst case, is equal to the probability that a distinguishing circuit for ψ_1 outputs 1, plus the probability that a distinguishing circuit for $\overline{\psi}_1$ outputs 1 and is equal to $\frac{(k^2-1)Q^2(n)}{2^n}$ □

7 Conclusions and Open Problems

If a block cryptosystem is super pseudorandom it is secure against chosen plaintext / ciphertext attack which is a much stronger attack than plaintext attack. We showed that a cryptosystem which consists of DES-like permutations and is secure against plaintext attack can be enhanced to a super pseudorandom cryptosystem by only applying one more random function with adding one more round of DES-like permutations. A problem is *how to construct a super pseudorandom permutation from a single pseudorandom function.*

It can be shown that $\psi_1(f_{k-1}, \ldots, f_1, f_k, f_{k-1}, \ldots, f_1)$ is not pseudorandom although it consists of $2k - 1$ rounds of DES-like permutation. On the other hand it can be conjectured that $\psi_1(f_2, \ldots, f_2, f_1, \ldots, f_1)$, where f_1 is used in k rounds and f_2 is used in $k - 1$, is pseudorandom. Another open problem is *what is the minimum number of random functions needed to achieve super pseudorandomness with k^2 rounds of type-1 transformations.*

ACKNOWLEDGMENT

We would like to thank Cathy Newberry and other members of CCSR for their help and assistance during the preparation of this work. The first author would also like to thank the organising committee of ASIACRYPT '91 for their generous support.

References

[1] O. Goldreich, S. Goldwasser, and S. Micali. How to construct random functions. *Journal of the ACM*, 33(4):792–807, 1986.

[2] Michael Luby and Charles Rackoff. How to Construct Pseudorandom Permutations from Pseudorandom Functions. *SIAM Journal on Computing*, 17(2):373–386, 1988.

[3] Y. Ohnishi. A study on data security. Master's thesis, Tohoku University, 1988. in Japanese.

[4] Josef Pieprzyk. How to Construct Pseudorandom Permutations from Single Pseudorandom Functions. In *Advances in Cryptology - EUROCRYPT '90*, volume 473 of *Lecture Notes in Computer Science*, pages 140–150. Springer-Verlag, 1991.

[5] A. C. Yao. Theory and applications of trapdoor functions. In *the 23rd IEEE Symposium on the Foundations of Computer Science*, pages 80–91, 1982.

[6] Yuliang Zheng, Tsumoto Matsumoto, and Hideki Imai. Impossibility and Optimality Results on Constructing Pseudorandom Permutations. In *Advances in Cryptology - EUROCRYPT '89*, volume 434 of *Lecture Notes in Computer Science*, pages 412–422. Springer-Verlag, 1990.

[7] Yuliang Zheng, Tsumoto Matsumoto, and Hideki Imai. On the Construction of Block Ciphers Provably Secure and Not Relying on any Unproved Hypotheses. In *Advances in Cryptology - CRYPTO '89*, volume 435 of *Lecture Notes in Computer Science*, pages 461–480. Springer-Verlag, 1990.

A Construction of a Cipher

From a Single Pseudorandom Permutation

Shimon Even[1] and Yishay Mansour[2]

February 7, 1992

Abstract

Shannon defined a random cipher as a collection of randomly chosen permutations, one for each value of the key.

We suggest a scheme for a block cipher which uses only one randomly chosen permutation, F. The key, consisting of two blocks, K_1 and K_2 is used in the following way: The message block is XORed with K_1 before applying F, and the outcome is XORed with K_2, to produce the cryptogram block. This removes the need to store, or generate a multitude of permutations.

Although the resulting cipher is not random, we claim that it is secure. First, it is shown that if F is chosen randomly then, with high probability the scheme is secure against any polynomial-time algorithmic attack. Next, it is shown that if F is chosen pseudorandomly, the system remains secure against oracle-type attacks.

The scheme may lead to a system more efficient than systems such as the DES and its siblings, since the designer has to worry about one thing only: How to implement one pseudorandomly chosen permutation. This may be easier than getting one for each key.

[1]Comp. Sci. Dept., Technion, Israel Institute of Technology, Haifa, Israel 32000. Supported by the Fund for the Promotion of Research at the Technion, and by Bellcore, Morristown, NJ. E-address: even@cs.technion.ac.il

[2]IBM T.J. Watson Research Center, Yorktown Heights, NY. E-address: mansour@watson.ibm.com.

1   Introduction

Following Shannon, [1], a *cipher*, C, consists of:

1. A finite set of *messages* (and cryptograms), **M**.

2. A finite set of *keys*, **K**.

3. Each key $\kappa \in K$ is assigned a permutation $\Pi_\kappa: M \mapsto M$.

If each of the permutations Π_κ is chosen randomly, with uniform probability, from the set of all $|M|!$ permutations, then C is called a *random cipher*. If $M = \{0, 1\}^n$ then C is called a *block-cipher*. The DES, [2], is an attempt to realize an approximation of a random block-cipher.

We propose a block-cipher scheme where only one permutation is randomly chosen and used. The key modifies the permutation in a simple, and fast to implement way, but the permutations on the messages for different keys are not independent. Thus, our system is not a random cipher, not even an attempt to approximate one. Yet we show that, from the point of view of an algorithmic adversary whose time is bounded polynomially in n, the situation is similar to the one when a random block-cipher is used; i.e. the probability of the adversary to crack the system is negligible.

Furthermore, even if instead of choosing a random permutation, we choose it pseudorandomly, every algorithmic polynomially bounded attack on the system still has a negligible probability to succeed.

It is easy to see that if one has a family of pseudorandom permutations then it can be used to build a secure cipher; simply choose for each key one of the permutations of the family. The novelty of our scheme is that we use only one permutation.

2   The Scheme

Let $\{0, 1\}^n$ denote the set of binary words of length n, let F be a common and publically known permutation on $\{0, 1\}^n$ and F^{-1} be its inverse. It is assumed that, for any given $x \in \{0,1\}^n$, it is easy to get $F(x)$ or $F^{-1}(x)$, either by a direct computation or by using an easily and commonly accessible black-box (oracle).

A *key* consists of two components, K_1 and K_2, each chosen at random from $\{0,1\}^n$. Initially, it is assumed that the key is known to the legitimate parties only; all other parties have no knowledge about it. Also, it is assumed that the key remains fixed and is used, by the legitimate parties, to encrypt messages and decrypt cryptograms, repeatedly, for a relatively long time.

The *encryption* $E(M)$ of a *message* $M \in \{0,1\}^n$, is performed by

$$E(M) = F(M \oplus K_1) \oplus K_2,$$

where $\oplus$ denotes the bit-by-bit exclusive-or operation. The corresponding *cryptogram* C is defined to be $E(M)$.

The *decryption* of a *cryptogram* $C \in \{0,1\}^n$, is performed by

$$D(C) = F^{-1}(C \oplus K_2) \oplus K_1.$$

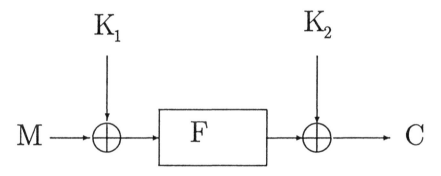

It is easy to verify, that for every $M \in \{0,1\}^n$, $D(E(M)) = M$.

3 Definition of Security

The two most important applications of conventional crypto-systems are concealment of messages from eavesdroppers and authentication of the identity of correspondents. We assume that F is a random (i.e. randomly chosen), or pseudorandom permutation, and investigate the security of the system from these two points of view.

Our aim is to model an adversary who, for a while, can get the system to encrypt messages and decrypt cryptograms for him, as well as use F, as a black-box, in both directions, but has no direct access to either K_1 or K_2. For this reason the access of the adversary to this information is modeled by oracles, which only answer queries of a particular nature.

Of the different notions of security we choose two. We show that for our system, these two notions, defined as problems, are hard to solve. We believe that this demonstrates the security of our system.

The *cracking problem*, CP, is an attempt (by an adversary) to decode a given encryption $C_0 = E(M_0)$, without any a priori knowledge of the key. The algorithm, employed by the adversary, has access to the following four oracles:

1. *F-oracle:* Presented with $x \in \{0,1\}^n$, the oracle supplies $F(x)$.

2. *F^{-1}-oracle:* Presented with $x \in \{0,1\}^n$, the oracle supplies $F^{-1}(x)$.

3. *E-oracle:* Presented with $M \in \{0,1\}^n$, the oracle supplies $E(M)$.

4. *(C_0-Restricted) D-oracle:* Presented with $C \in \{0,1\}^n$ (such that $C \neq C_0$) the oracle supplies $D(C)$.

The algorithm is successful if it outputs M_0. The *success probability* of the algorithm is the probability that on a randomly chosen encryption $C_0 = E(M_0)$ it outputs M_0, where all C_0 are equally likely.

The cracking problem is sometimes called *chosen plaintext/ciphertext attack* or *two-sided attack*.

In the *existential forgery problem*, EFP, the adversary has access to four oracles: F-oracle, F^{-1}-oracle, E-oracle and (unrestricted) D-oracle. The latter is defined as follows: Presented with (any) $C \in \{0,1\}^n$, the oracle supplies $D(C)$. The task is to find a *new* pair $< M, C >$, $C = E(M)$; i.e. it does not consist of a query and an answer, as previously supplied by either the E-oracle or the D-oracle.

Assuming F is a pseudo-random permutation – i.e. it is not distinguishable from a random permutation by any polynomially time-bounded randomized algorithm – our purpose is to prove that both CP and EFP are hard. To specify what we mean by "hard" we need a few more definitions.

Let $f(n)$ be a function defined from the positive integers to the interval $[0,1]$. We say that $f(n)$ is *polynomially negligible* if for every polynomial $p(n)$ there is an n_0, such that if $n > n_0$ then $f(n) < 1/p(n)$.

Assume that the adversary employs a randomized algorithm, whose time is polynomially bounded (in n), to solve one of these problems. We say that the problem is *hard* if, for every such algorithm, the success probability is polynomially negligible; the probability is taken over all choices made in the design on the system (i.e. the choice of F), the keys and coin-flips performed by the algorithm.

Instead of proving that CP and EFP are hard separately, we first reduce the EFP to CP; i.e. we show that the existence of a (successful) CP attack implies the existence of an EFP attack.

Theorem 3.1 *If there exists a CP attack which runs in time t and its success probability is ϵ, then there is an EFP attack which runs in time t and its success probability is ϵ/t.*

Proof: We show how to construct an attacking algorithm A, for the EFP, making use of a procedure P; P performs the CP attack in time t and its success probability is ϵ.

We use the fact that the range of the encryption is samplable; for our encryption scheme this holds trivially, since the range is $\{0,1\}^n$.

Also, we may assume, without loss of generality, that P queries the E-oracle on its output; i.e. the procedure P, while attempting to decrypt its input C_0, checks the correctness of its output M, by feeding M to the E-oracle, and comparing $E(M)$ with C_0. If P inverts a cryptogram C_0 successfully, there is a *critical* time $i \leq t$, such that at time i, P queries the E-oracle about M_0, and at no prior time M_0 has been queried.

We construct A as follows.

1. Randomly, choose a cryptogram C_0.

2. Feed it, as an input, to P.

3. Randomly, with uniform distribution, choose $1 \leq \tau \leq t$.

4. Let P run exactly $\tau - 1$ steps. If at the τ'th step, P queries the E-oracle about a value M', output $< M', C_0 >$, (without querying the E-oracle).

The probability that A generates a legitimate pair, i.e. $M' = D(C_0)$, is at least ϵ/t. This follows from the fact that with probability ϵ, P inverts C_0 successfully, and with probability $1/t$, P has been stopped at the critical time. □

Corollary 3.2 *If for every polynomial-time EFP attack the success probability is polynomially negligible, then for every polynomial-time CP attack the success probability is polynomially negligible.*

Proof: Let us prove the contrapositive. Assume there is a polynomial-time, $t(n)$, algorithm P which attacks CP, and its probability of success in not polynomially negligible. Thus, there is a polynomial $p(n)$ such that for every n_0, there exists an $n \geq n_0$ such that the probability of P to succeed in solving CP is greater than or equal to $1/p(n)$. By Theorem 3.1, there is a $t(n)$-time algorithm A, which solves the EFP problem with probability $1/(p(n) \cdot t(n))$. Thus, there is a polynomial $p'(n) = p(n) \cdot t(n)$ such that for every n_0, there exists an $n \geq n_0$ such that the probability of A to succeed in solving EFP is greater than or equal to $1/p'(n)$. This means that the probability of A to be successful in solving EFP is not polynomially negligible. □

4 The immunity of the system when F is truly random

In this section we assume that the permutation F is a truly random permutation, i.e. it has been chosen randomly, out of the set of all $2^n!$ permutations, where all permutations are equally likely to be picked. Under this assumption, we show that solving EFP for our system is *hard*; i.e. that for every polynomial-time EFP attack on our system, the success probability is polynomially negligible. By Corollary 3.2, solving CP is also hard. In fact we prove a stronger result, namely, any algorithm that makes only a polynomial number of queries has an exponentially small probability of success, regardless of its running time.

Recall that the adversary asks queries of two forms:

1. E/D queries. For a given message M_i, the E oracle returns $E(M_i)$, or for a given a cryptogram $E(M_i)$, the D-oracle returns M_i. We denote such a query by the pair $< M_i, E(M_i) >$.

2. F/F^{-1} queries. For a given value A_j, the F-oracle returns $F(A_j)$, or for a given $F(A_j)$, the F^{-1}-oracle returns A_j. We denote such a query by the pair $< A_j, F(A_j) >$.

An algorithm A for EFP asks various queries of the four types and then computes a pair $< M, E(M) >$ which it has not queried; i.e. a new E/D query. We show that for every A, if it asks only polynomially many queries, then its probability to succeed is exponentially small.

Consider two E/D queries,) $< M_1, C_1 >$ and $< M_2, C_2 >$. If $M_1 = M_2$ or $C_1 = C_2$, then we say that the two queries overlap. Two overlapping queries, $< M_1, C_1 >$ and $< M_2, C_2 >$ are identical if $M_1 = M_2$ and $C_1 = C_2$. Note that the replies of the (genuine) E-oracle and D-oracle are such that if the resulting query-pairs overlap, then they are identical. The situation with the F/F^{-1} queries is similar.

Note that the adversary's algorithm gets no new information from overlapping queries, and that its polynomial-time behavior remains polynomial if it is required to remember all previous query pairs and look them up, if necessary. Thus, without loss of generality, we may assume that all queries are non-overlapping.

Our proof methodology is to examine the amount of ignorance of the adversary, after asking the queries and getting the corresponding answers, and before producing the answer, $< M, E(M) >$. We will show that his information allows for too many answers, consistent with the information he has, and therefore, his answer is likely to be false.

First, we investigate the number of possible choices of the keys, K_1 and K_2, and the amount of freedom in the choice of F, such that these choices are consistent with a given set of queries.

We start by defining when a key K_1 is bad. This definition depends only on the set of queries. Consider all E/D queries, $< M_i, E(M_i) >$. Since they are non-overlapping, all M_i's are different, and therefore all the corresponding values $M_i \oplus K_1$ (potential inputs to F) are different. Also, since all F/F^{-1} queries, $< A_j, F(A_j) >$, are non-overlapping, all A_j's are different. We say that key K_1 is *bad*, if there are queries $< M_i, E(M_i) >$ and $< A_j, F(A_j) >$, such that $M_i \oplus K_1 = A_j$; otherwise K_1 is *good*. Similarly, we say that a specific key K_2 is bad, if there are queries $< M_i, E(M_i) >$ and $< A_j, F(A_j) >$, such that $F(A_j) \oplus K_2 = E(M_i)$, otherwise K_2 is good. Note that if all queries have been answered "honestly", with respect to a given triple $< K_1, F, K_2 >$, then K_1 is good iff K_2 is good.

Lemma 4.1 *Assume there are l E/D queries and m F/F^{-1} queries. The number of bad K_1 (K_2) keys is at most lm.*

Proof: A key K_1 is bad if there are queries i and j such that, $M_i \oplus K_1 = A_j$, or alternately, $K_1 = A_j \oplus M_i$. Therefore, at most lm keys K_1 are bad. A similar argument holds for K_2. □

Lemma 4.1 implies that for every set of l E/D queries and m F/F^{-1} queries, there are at least $2^n - lm$ good K_1 keys and as many good K_2 keys.

Lemma 4.2 *If, with respect to a given set of queries, K_1 and K_2 are good, then there is a permutation Π such that the triple $< K_1, \Pi, K_2 >$ is consistent with all queries.*

Proof: Consider the set of pairs $< x, y >$, such that $x, y \in \{0, 1\}^n$ and either $< x, y >$ is equal to one of the given F/F^{-1} query pairs, or $M_i \oplus K_1 = x$ and $y \oplus K_2 = E(M_i)$, for one of the given E/D query pairs $< M_i, E(M_i) >$. Since K_1 and K_2 are good, every two pairs in the set share neither their first component, nor their second component. Thus, the set of pairs can be complemented to be a permutation Π on $\{0, 1\}^n$. □

Theorem 4.3 *The probability of an algorithm A to solve the EFP problem, when F is randomly chosen, is bounded by*

$$\frac{2lm}{2^n - lm} + \frac{1}{2^n - (l + m)}, \tag{1}$$

where l is the number of E/D queries and m is the number of F/F^{-1} queries.

Proof: Let us describe a sequence of three scenarios. The first scenario is that depicted in the statement of the theorem. As we move from one scenario to the next, we shall argue that the success probability of the adversary does not decrease. Finally, we shall show that the success probability of the adversary in the last scenario is bounded as in 1.

Scenario 1: Consider the set of triples, $< K_1, F, K_2 >$, with a uniform distribution. Choose one triple, and answer all A's queries accordingly. A supplies $< M, C >$. Its answer is a success if $< M, C >$ has not been an E/D query and $C = F(M \oplus K_1) \oplus K_2$.

Clearly, this is just a rephrasing of the real scenario, the one to which the theorem relates.

Scenario 2: As in Scenario 1, choose one triple $\beta =< K_1', F', K_2' >$ randomly and uniformly. Answer all A's queries accordingly. Contrary to Scenario 1, consider now the subset of all "good" triples; i.e. those which are consistent with all the answers given to A's queries. Choose one good triple, $\alpha =< K_1, F, K_2 >$, randomly and uniformly. From there on, continue as in Scenario 1.

Let us show that from A's point of view, nothing has changed. The probability distribution of the possible conversations between A and the oracles is the same, and so is A's success probability.

First, let us assume that A is a deterministic algorithm. Denote by Γ the conversation between A and the oracles. Probabilities which refer to the Scenario 1 (2) will be indexed 1 (2). Since α and β are both in the set of triples consistent with Γ, and since the (new) triple, for A to beat, is chosed in Scenario 2 randomly and uniformly, it is clear that

$$P_2(\alpha|\Gamma) = P_2(\beta|\Gamma).$$

What we want to show is that the conditional probabilities of α and β are the same in Scenario 1, as well. Note that

$$P_1(\Gamma|\alpha) = P_1(\Gamma|\beta)(= 1).$$

Thus,

$$P_1(\alpha|\Gamma) = \frac{P_1(\alpha) \cdot P_1(\Gamma|\alpha)}{P_1(\Gamma)} = \frac{P_1(\beta) \cdot P_1(\Gamma|\beta)}{P_1(\Gamma)} = P_1(\beta|\Gamma).$$

Namely, given Γ, from the adversary's point of view α and β are equally likely to be the triple he has to beat, in both scenarios, and his chances to succeed do not change by the change of the scenario.

Next, let us show that the claim remains valid if A is randomized. Without loss of generality we may assume that A runs as follows: First it tosses a fair coin sufficiently many times, and records the sequence of outcomes. Next it uses a deterministic algorithm whose behavior is affected both by the recorded sequence and the answers of the oracles. Since the claim holds for every sequence of coin tosses, it follows that the distributions of the conversations are the same in the two scenarios, and A's success probability is the same.

This completes the proof that Scenario 1 and Scenario 2 are equivalent from A's point of view.

Scenario 3: As in Scenario 2, choose one triple $\beta = < K_1', F', K_2' >$ randomly and uniformly. Answer all A's queries accordingly. Next, consider the subset of all good triples. Choose one, $\alpha = < K_1, F, K_2 >$, randomly and uniformly. The remainder of the process is now changed. A supplies M, but not the "matching" C. The keys K_1 and K_2 are now revealed to the adversary, who can now abandon A and use any means of computation he wishes, but has no access to the oracles. The adversary attempts to supply the matching C.

It is obvious that divulging K_1 and K_2, before the adversary's attempt to guess the matching C, can only increase the adversary's success probability. Let us derive a bound on this probability.

To simplify the derivation of the upper bound we make a further allowance to the adversary. If at any time a query causes K_1' (and K_2') to become bad, the adversary is declared victorious. Let us bound the probability that a query turns K_1' from good to bad.

Assume that up to this query there have been a E/D queries and b F/F^{-1} queries, and the query in question is to the E oracle. There are at least $(2^n - ab)^2$ good $< K_1, K_2 >$ key pairs left. For each of these pairs, there are b pairs of F directly committed (and known to the adversary), due to the F/F^{-1} queries, and a pairs of F implicitly committed (but not known to the adversary), due to the E/D queries. The remaining $2^n - (a+b)$ pairs of F are not yet committed and can be filled-in in $(2^n - (a+b))!$ different ways. From the adversary's point of view, all these triples, $< K_1, F, K_2 >$ are equally likely to be the secret one. (This can be proved, using Bayes' theorem, as was done above.) Thus, each of the good K_1's is likely to be the secret K_1'. Every E-query M, (different from all M's of the previous a E/D queries) will add exactly b bad K_1 keys. Thus, no matter how the adversary chooses the query M, his chance of making K_1' bad is bounded by

$$\frac{b}{2^n - ab}.$$

The analysis for a query to the D oracle is almost identical, yielding exactly the same bound. In the case of an F query, or an F^{-1} query, a similar analysis yields the bound

$$\frac{a}{2^n - ab}.$$

It follows that the probability of A's query to make K_1' bad by a single E/D query is bounded by $\frac{m}{2^n - lm}$, and the probability of this happening during any of the l E/D queries is bounded by $\frac{lm}{2^n - lm}$. The same expression bounds the probability of making K_1' bad by any of the m F/F^{-1} queries. Thus, the probability of K_1' turning bad during the conversation is bounded by

$$\frac{2lm}{2^n - lm}.$$

This explains the first term of 1.

From now on we assume that K_1' (and K_2') never becomes bad.

After the adversary declares M, and the chosen and good $< K_1, K_2 >$ are revealed to him, all possible completions of the uncommitted $2^n - (l+m)$

pairs of F are equally likely, from the adversary's point of view. Thus, his chance of guessing the mate of $M \oplus K_1$ are not better than

$$\frac{1}{2^n - (l + m)}.$$

This explains the second term of 1. □

Corollary 4.4 *Consider our system, with a randomly chosen F. For every polynomially bounded algorithm to solve EFP, the probability of success is polynomially negligible.*

5 The immunity of the system when F is pseudorandom

In this section we show that our system remains secure even if F is just known to be chosen pseudorandomly.

As before, we assume that the adversary has access to the 4 oracles, but has no access to the innards of the box which implements F. This issue is meaningless when F is chosen randomly, but when it is chosen pseudorandomly, the difference may be crucial. Our claims are restricted to the oracle-type attacks.

Let S_{2^n} be the set of all permutations on $\{0, 1\}^n$. A probability distribution, D_n, assigning probabilities to the elements of S_{2^n}, is said to be *uniform* if for every $\Pi \in S_{2^n}$, $Prob(\Pi) = \frac{1}{2^n!}$.

A sequence of distributions, $\Lambda = \{D_n | n \geq 1\}$, where D_n is a distribution on S_{2^n}, is called an *ensemble*. An ensemble is said to be *uniform* if each of its elements is a uniform distribution.

A *test*, $T(n, \Lambda)$, has the following characteristics:

1. T is a randomized algorithm which runs in time polynomial in n,

2. T has access to a Π-oracle and a Π^{-1} oracle, where Π is chosen from S_{2^n}, according to the distribution $D_n \in \Lambda$,

3. T outputs "random" or "nonrandom".

Let $P(T, n, \Lambda)$ be the probability that T outputs "random", on input (n, Λ), where the probability is taken over the choices of Π and T's coin tosses.

Let Λ be a uniform ensemble. We say that an ensemble $\Lambda' = \{D'_n | n \geq 1\}$ is *pseudorandom* if for every test T and every polynomial $p(n)$, there is an n_0 such that if $n > n_0$ then

$$|P(T, n, \Lambda) - P(T, n, \Lambda')| < \frac{1}{p(n)}.$$

We say that F is *pseudorandomly* chosen, if there exists a pseudorandom ensemble $\Lambda = \{D_n | n \geq 1\}$, and for the chosen n, F is chosen from S_{2^n} according to D_n.

Theorem 5.1 *If F is pseudorandomly chosen, then for every polynomially bounded algorithm to solve EFP for our system, the probability of success is polynomially negligible.*

Proof: By contradiction. Assume there is a pseudorandom ensemble $\Lambda' = \{D'_n | n \geq 1\}$, which characterizes the choice of F. Also, let A be a polynomial time randomized algorithm which solves EFP when F is chosen as above, and its success probability is not polynomially negligible. Thus, there is a polynomial $p(n)$ such that for every n_0, there exists an $n \geq n_0$ such that the probability of A to succeed in solving EFP is greater than or equal to $1/p(n)$.

Consider now the following test, $T(n, \Lambda)$, where $\Lambda = \{D_n | n \geq 1\}$. Given n, choose K_1 and K_2 from $\{0,1\}^n$, randomly, and choose Π from S_{2^n} according to the distribution D_n. Now apply A. When A has an F/F^{-1} query, use the Π-oracle, or the Π^{-1}-oracle, as the case may be, to answer A's query. When A has an E-query M_i, present to the Π-oracle the query $M_i \oplus K_1$, take its answer $\Pi(M_i \oplus K_1)$, and answer A's query with $\Pi(M_i \oplus K_1) \oplus K_2$. The D-queries of A are handled similarly, using the Π^{-1}-oracle. Finally, when A produces the new E/D query, $< M_0, C_0 >$, T presents $M_0 \oplus K_1$ as a query to the Π-oracle. If $\Pi(M_0 \oplus K_1) \oplus K_2 = C_0$ then T's output is "nonrandom", otherwise its output is "random".

Consider $T(n, \Lambda')$. From the choice of A, for every n_0, there exists an $n \geq n_0$ such that the probability of A to "guess" $< M_0, C_0 >$ correctly is greater than or equal to $1/p(n)$. Thus, the probability of $T(n, \Lambda')$ to output "nonrandom", for such inputs, is also greater than or equal to $1/p(n)$.

By Corollary 4.4, if Λ is uniform, A's probability to "guess" $< M_0, C_0 >$ correctly, i.e. consistently with K_1, K_2 and Π, is polynomially negligible. In other words, for every polynomial $q(n)$, there is an n_0, such that if $n > n_0$ then A's probability of success is less than $1/q(n)$. Therefore, T's probability to output "nonrandom", on such inputs, is less than $1/q(n)$. By considering $q(n) = 2 \cdot p(n)$ we reach a contradiction, and the theorem follows. $\square$

Acknowledgements

The authors would like to thank Oded Goldreich and Silvio Micali for their encouragements, Shai Ben-David for his interest, Reuven Bar-Yehuda and Guy Even for their help with the proof of Theorem 4.3, Benny Chor and Refael Heiman for their comments.

References

[1] C.E. Shannon, "Communication Theory of Secrecy Systems", *Bell System Tech. J.*, Vol. 28, 1949, pp. 656-715.

[2] National Bureau of Standards, "Data Encryption Standard", *Federal Information Processing Standard*, U.S. Department of Commerce, FIPS PUB 46, Washington, DC, 1977.

[3] M. Luby and C. Rackoff, "How to Construct Pseudorandom Permutations from Pseudorandom Functions", *SIAM J. on Computing*, Vol. 17, No. 2, 1988, pp. 373-386.

Postscript

In a recent note, "Limitations of the Even-Mansour Construction", by J. Daemen (see the Rump-Session section of these proceedings), the author argues that our construction is wasteful of key bits. He conclude: "In our opinion the complexity theoretical way of thinking encourages poor design. The Even-Mansour construction is an example."

The scientific part of the Daemen's note is interesting. Ignoring his technical mistakes, which can be corrected, he shows that our scheme, which uses $2n$ key-bits, can be cracked in time $O(2^n)$, using Known Plaintext Attack, and in space and time $O(2^{n/2})$, using Chosen Plaintext Attack. Our Theorem 4.3 implies a lower bound of $2^{n/2}$ on the cracking time. (We have no hidden constant factors, as Daemen seems to imply. The constant depends in a straight forward way on the probability of success one wants to achieve. In fact, the approach taken by Daemen does have some hidden constant factors, which he thinks will simply "go away" if he avoid the big-O notation...) In short, we now know that for the scheme to be secure, n must be chosen to be high enough to make $2^{n/2}$ infeasible.

Daemen concludes, that since our system uses $2n$ key-bits but its cracking time is only $O(2^{n/2})$, it is wasteful of key-bits and is therefore useless. We do not consider random bits a precious commodity. One can decide to use 400 key bits ($n = 200$), and be sure that it will take about 2^{100} computational steps to crack the system. This is secure enough for any purpose. Using 400 key-bit is not considered excessive. For many practical systems, such as RSA, more bits are recommended and used.

Optimal Perfect Randomizers

Josef Pieprzyk *
Babak Sadeghiyan

Department of Computer Science
University College
University of New South Wales
Australian Defence Force Academy
Canberra, ACT 2600, AUSTRALIA

Abstract

The work examines a class of randomizers built using concatenation of several layers of Luby-Rackoff elementary randomizers. First we examine some properties of Luby and Rackoff randomizers. Next we discuss the quality of randomizers with the concatenation of several Luby-Rackoff randomizers. Finally, the main result of the work is presented which proves that the concatenation of two layers of modified L-R randomizers is a perfect randomizer.

1 Introduction

Yao [3] formulated a basic concept of pseudorandomness from a complexity theory point of view. He also introduced the concept of indistinguishability of pseudorandom generators from truly random ones. There are many different constructions of pseudorandom bit generators using one-way functions. Goldreich, Goldwasser and Micali [1] showed that having a pseudorandom bit generator, one can design a pseudorandom function generator. Luby and Rackoff [2] used three different pseudorandom functions and three rounds of the Data Encryption Standard (DES) to build a pseudorandom permutation generator. In particular, they proved that the quality of pseudorandom permutation generators with three rounds of DES and three different pseudorandom functions, depends upon the quality of the pseudorandom functions and the structure itself. They also showed that structures with two DES rounds are "bad" as these

---
*Support for this project was provided in part by TELECOM Australia under the contract number 7027 and by the Australian Research Council under the reference number A48830241.

permutations can always be distinguished from the truly random function in polynomial time. The structure with three DES rounds is sound as the resulting permutation can be distinguished no better than with probability $\frac{m^2}{2^n}$ (where m is the number of input/output samples of the permutation or the number of oracle gates).

In this paper, we consider the concatenation of Luby-Rackoff structure and we are going to prove that two concatenations of L-R randomizers (six DES rounds with four random functions and two random permutations) guarantees that the structure is not transparent for input.

2 Preliminaries

First we introduce necessary notations and definitions. Let $I_n = \{0,1\}^n$ be the set of all 2^n binary strings of length n. For $a, b \in I_n$, $a \oplus b$ stands for bit-by-bit exclusive-or of a and b. The set of all functions from I_n to I_n is F_n, i.e.,

$$F_n = \{f \mid f : I_n \to I_n\}$$

It consists of 2^{n2^n} elements. If we have two functions $f, g \in F_n$, their composition $f \circ g$ is denoted as

$$f \circ g(x) = f(g(x))$$

for all $x \in I_n$. The set of all permutations from I_n to I_n is P_n, i.e.,

$$P_n \subset F_n$$

Definition 2.1 *For a function $f \in F_n$, we define the DES-type (Feistel type) permutation associated with f as*
$$D_{2n,f}(L, R) = (R, L \oplus f(R))$$
where L and R are n-bit strings ($L, R \in I_n$) and $D_{2n,f} \in P_{2n}$.

Given a sequence of functions $f_1, f_2, \cdots, f_i \in F_n$, we can determine the concatenation of their DES-like permutations ψ and

$$\psi(f_1, f_2, \cdots, f_i) = D_{2n,f_i} \circ D_{2n,f_{i-1}} \circ \cdots \circ D_{2n,f_1}$$

Now, $\psi(f_1, f_2, \cdots, f_i) \in P_{2n}$.

Properties of the permutation $\psi(f_1, f_2, \cdots, f_i)$ depend upon the selection of functions f_j for $j = 1, \cdots, i$. There are two possible cases:

- all the functions are different pseudorandom ones, i.e., $f_i \in_{PS} F_n$. The resulting permutation ψ is pseudorandom for $i = 3, 4, \cdots$ (see Luby and Rackoff [2]);

- all the functions are random ones, i.e. $f_i \in_R F_n$. The resulting permutation is called a *randomizer* (it will be used later to assess the quality of the structure).

To draw some conclusions about the quality of the structured permutations (based on either pseudorandom or truly random functions), it is necessary to introduce the notion of indistinguishability (see Yao [3]). Intuitively, a given structured permutation is indistinguishable from a truly random one if there is no statistical test which can be used to identify which one is which assuming that a polynomial-size sample of input-output pairs is known. Such a statistical test is also called a distinguishing circuit or simply a distinguisher.

Definition 2.2 *A distinguishing circuit (or distinguisher) C_n is an acyclic circuit which consists of Boolean gates (AND, OR, NOT), constant gates ("0" and "1") and oracle gates. The circuit has one bit output only. Oracle gates accept binary inputs of length n and generate outputs of the same length. Each oracle gate is evaluated using some permutation from P_n.*

Definition 2.3 *A family of distinguishing circuits for a permutation generator $p= \{p_{n_i}; i = 1, 2, \cdots\}$ is an infinite sequence of circuits $C_{n_1}, C_{n_2}, \cdots (n_1 < n_2 < \cdots)$ such that for two constants c_1 and c_2 and for each parameter n, there exists a circuit C_n which has the following properties:*

- *the size of C_n is smaller than n^{c_1} (the size is defined as the number of all connections between gates);*

- *let $Pr[C_n(P_n)]$ be the probability that the output bit of C_n is one when a permutation is randomly selected from P_n and used to evaluate the oracle gates. Let $Pr[C_n(p_n)]$ be the probability that the output bit of C_n is one when a permutation p_n is used to evaluate the oracle gates. The probability of distinguishing the truly random permutation from the permutation generated by p_n, is*

$$| Pr[C_n(P_n)] - Pr[C_n(p_n)] | \geq \frac{1}{n^{c_2}}$$

Definition 2.4 *A function f is said to be random (denoted as $f \in_R F_n$) if for any fixed argument $x \in \{0, \cdots, 2^n - 1\}$, $f(x)$ is an independent and uniformly distributed random variable.*

Goldreich et al. [1] defined pseudorandom functions (or permutations) as poly-random collections that satisfy the three following properties: indexing, polynomial time evaluation, and indistinguishability from truly random functions.

Luby and Rackoff [2] proved that $\psi(f, g, h)$ is a pseudorandom permutation provided $f, g, h \in F_n$ are three different pseudorandom functions. It also means that the implementation of such permutations for a given parameter n takes polynomial-size computer resources (i.e. they are easy to implement).

If a randomizer ψ does not have a distinguisher, then the difference between probabilities

$$| Pr[C_n(F_n)] - Pr[C_n(\psi)] | \leq \frac{p(m)}{2^n} \tag{1}$$

can be made as small as requested by selecting a large enough parameter n, where m is the number of oracle gates in C_n. The formula (1) gives the upper bound on the successful distinguishing of ψ from the random function F_n.

3 Luby-Rackoff randomizers

Luby and Rackoff [2] considered a permutation generator $\psi(f, g, h)$ with three rounds of DES and three random functions f, g, h. They proved that

$$| Pr[C_{2n}(F_{2n})] - Pr[C_{2n}(\psi)] | \le \frac{m^2}{2^n} \qquad (2)$$

where m is the number of oracle gates and $m \le 2^n$. They showed that the structure $\psi(f, g, h)$ can be "transparent" to the input and proved the necessary conditions for "a leakage" of input information to the output. If there is no leakage of the input to the output (they called this ω is preserving), the distinguisher cannot make any sensible decision about the generator used to evaluate the oracle gates (it may be either F_{2n} or ψ). $\underline{\omega \text{ is preserving}}$ means that all the outputs of oracle gates are independent from the input (and from each other).

$\underline{\omega \text{ is NOT preserving}}$ means that there is at least one pair of oracle gates such that its outputs are related to its inputs and this relation can be used to distinguish ψ from F_{2n} (with a small probability). Note that if $\underline{\omega \text{ is preserving}}$ in a distinguishing circuit, then the distinguisher cannot find any pair of oracle gates with outputs related to its inputs.

Luby and Rackoff showed that the leakage of the input happens only in two cases when $\underline{Y \text{ is bad}}$ or $\underline{X \text{ is bad}}$.

The first case occurs if $\underline{Y \text{ is bad}}$, i.e. there is a pair of oracle gates (O_i, O_j) such that the random function g collides. Figure 1 shows it. It happens when the inputs random variables $R = R_i = R_j$ (but $L_i \ne L_j$). It is obvious that

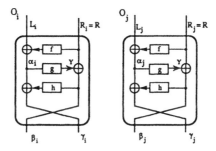

Figure 1: Y is bad in two oracle gates (O_i, O_j)

$$\alpha_i = L_i \oplus f(R) \quad \text{and} \quad \alpha_j = L_j \oplus f(R)$$

and as $L_i \ne L_j$, $\alpha_i \ne \alpha_j$. It means that the random function g assigns two independent random variables $g(\alpha_i)$ and $g(\alpha_j)$, and the outputs β_i, β_j are independent from the input. The outputs γ_i, γ_j are also independent only if the random variables $g(\alpha_i)$, $g(\alpha_j)$ take on different values. Otherwise, if $\beta_i = \beta_j$ (this may happen with the probability $\frac{1}{2^n}$ for a single pair of oracle gates), γ_i, γ_j are related. This may happen only if the function g collides, i.e.

$$g(\alpha_i) = g(\alpha_j) = Y$$

and then

$$\gamma_i = \alpha_i \oplus h(Y)$$
$$\gamma_j = \alpha_j \oplus h(Y)$$

The second possibility of the input information leakage to the output happens when X is bad (see [2] page 383) - Figure 2. It can happen only if $R_i \neq R_j$. The random function f assigns two independent random variables $f(R_i)$, $f(R_j)$ and as the result the output variables γ_i, γ_j are independent from the input. The input information can pass only through β outputs if $\alpha = \alpha_i = \alpha_j$ (this happens with the probability $\frac{1}{2^n}$ for a single pair of gates). Then

$$\beta_i = R_i \oplus g(\alpha)$$
$$\beta_j = R_j \oplus g(\alpha)$$

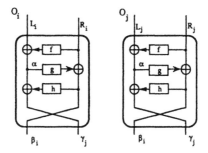

Figure 2: X is bad in two oracle gates (O_i, O_j)

The distinguisher can apply two strategies: the first one *"hunting for (Y is bad)"* or the second one *"hunting for (X is bad)"*. Luby and Rackoff [2] calculated that the probability P_Y that Y *is bad* in at least one pair of gates is

$$P_Y \leq \frac{m(m-1)}{2} \frac{1}{2^n}$$

where $\frac{m(m-1)}{2}$ is the number of different pairs of gates if the distinguisher has m oracle gates.

In the second strategy, the distinguisher selects different R_i for all oracle gates and the probability P_X that X *is bad* in at least one pair of oracle gates is

$$P_X \leq \frac{m(m-1)}{2} \frac{1}{2^n}$$

Obviously if a distinguisher applies some mixed strategy, then

$$Pr[\omega \text{ is NOT preserving}] < P_Y + P_X \leq \frac{m^2}{2^n}$$

Now consider a randomizer $\Psi_2 = \psi_1(f, g, h) \circ \psi_2(f, g, h)$ which is constructed from two L-R randomizers. ω is NOT preserving in Ψ_2 if there is at least one pair of oracle gates O_i, O_j for which Y is bad or X is bad in the first randomizer ψ_1. Figure 3 shows the pair with Y is bad

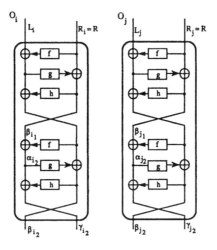

Figure 3: Y is bad in two oracle gates (O_i, O_j) for Ψ_2

(note that $R_i = R_j = R$). Clearly the outputs γ_{i_2}, γ_{j_2} are independent from the input. β_{i_2} and β_{j_2} , however, may be related if $\beta_{i_1} = \beta_{j_1}$ (with the probability $\frac{1}{2^n}$ for a single pair of oracles (O_i, O_j)) and $\alpha_{i_2} = \alpha_{j_2}$ (this happens with the probability $\frac{1}{2^n}$ in a single pair of oracle gates). Therefore the probability of Y being bad in a single pair of oracle gates is $\frac{1}{2^{2n}}$. Considering the all possible pairs of gates, we can conclude that the probability P_Y of $\underline{Y \text{ is bad}}$ in Ψ_2 is

$$P_Y = \frac{m(m-1)}{2} \frac{1}{2^{2n}}$$

The second case when $\underline{X \text{ is bad}}$ in ψ_1 is presented in Figure 4 ($R_i \neq R_j$). Clearly, the outputs β_{i_2}, β_{j_2} are independent from the input. If $\underline{X \text{ is bad}}$ in (O_i, O_j), then $\alpha_{i_1} = \alpha_{j_1} = \alpha$ and it results that $\beta_{i_1} = R_i \oplus g(\alpha)$, $\beta_{j_1} = R_j \oplus g(\alpha)$. The function h in ψ_1 generates two independent random variables. Thus $\gamma = \gamma_{i_1} = \gamma_{j_1}$ with the probability $\frac{1}{2^n}$ and the relation to the input continues. The random function g in ψ_2 assigns two independent random variables and the outputs γ_{i_2}, γ_{j_2} are related only if $\beta = \beta_{i_2} = \beta_{j_2}$ (with the probability $\frac{1}{2^n}$). Therefore $\underline{X \text{ is bad}}$ in Ψ_2 for a single pair of oracle gates with probability $\frac{1}{2^{3n}}$.

Now if a distinguisher uses some strategy to tell apart the tested permutation generator, it succeeds with some probability and

$$Pr[\omega \text{ is NOT preserving in } \Psi_2] \leq \frac{m(m-1)}{2} \left(\frac{1}{2^{2n}} + \frac{1}{2^{3n}} \right)$$

Hence, we have proven the following theorem.

Theorem 3.1 *The randomizer $\Psi_2 = \psi_1(f, g, h) \circ \psi_2(f, g, h)$, where $f, g, h \in_R F_n$, does not have a distinguisher and*

$$| Pr[C_{2n}(P_{2n})] - Pr[C_{2n}(\Psi_2)] | \leq \frac{m^2}{2} \left(\frac{1}{2^{2n}} + \frac{1}{2^{3n}} \right) \tag{3}$$

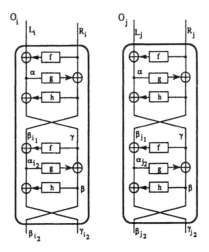

Figure 4: X is bad in two oracle gates (O_i, O_j) for Ψ_2

where $m \leq 2^n$ is the number of oracle gates in the distinguisher.

It is easy to generalize the previous theorem for the concatenation of $k = 2, 3, 4, \cdots$ L-R randomizers. Such a generator $\Psi_k = \underbrace{\psi_1(f_1, g_1, h_1) \circ \ldots \circ \psi_k(f_k, g_k, h_k)}_{k}$ has better quality as the parameter k grows.

Theorem 3.2 *The randomizer Ψ_k, where $f_i, g_i, h_i \in_R F_n$ $(i = 1, \ldots, k)$, does not have a distinguisher and*

$$| Pr[C_{2n}(P_{2n})] - Pr[C_{2n}(\Psi_k)] | \leq \frac{m^2}{2} \left(\frac{1}{2^{kn}} + \frac{1}{2^{(2k-1)n}} \right) \tag{4}$$

where $m \leq 2^n$ is the number of oracle gates in the distinguisher.

4   Perfect randomizers

From the discussion in the previous section, we can see that the concatenation of L-R randomizers does not have a distinguisher with m oracle gates ($m \leq 2^n$). There is always a small probability of success no matter how many elementary randomizers are used (having $m \leq 2^n$ oracle gates). This leads us to a contradiction since for any parameter n, there is a finite number (it may be exponential in n) of concatenations after which it generates the alternating group A_{2n} $(A_{2n} \subset P_{2n})$. Thus there must be a better way to design a permutation generator.

In this section, we show how to improve the L-R randomizer to obtain a so called perfect randomizer. First we define the notion of perfectness.

Definition 4.1 *A randomizer is perfect if for all oracle gates used by the distinguisher, their outputs are independent from their inputs and independent from each other.*

The next lemma shows that a change in the L-R randomizer structure does not deteriorate its quality. At the same time the modified randomizer behave more regularly - the right hand outputs (γ-outputs) are always independent from the input.

Theorem 4.1 *Given a randomizer $\psi(f, g^*, h)$, where f, h are random functions ($f, h \in_R F_n$) and g^* is a random permutation ($g^* \in_R P_n$), then $\psi(f, g^*, h)$ does not have a distinguisher and*

$$| Pr[C_{2n}(P_{2n})] - Pr[C_{2n}(\psi)] | \leq \frac{m^2}{2^n} \tag{5}$$

where $m \leq 2^n$ is the number of oracle gates in the distinguisher.

Proof: We are going to follow the main idea of the proof given by Luby and Rackoff [2]. Using their notations we can say that ω is NOT preserving in ψ if Y is bad or X is bad.

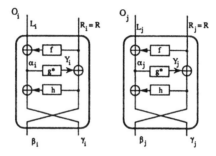

Figure 5: Y is bad in two oracle gates (O_i, O_j) for $\psi(f, g^*, h)$

If Y is bad (see Figure 5) in a pair of oracle gates (O_i, O_j), then $R = R_i = R_j$ and $\alpha_i \neq \alpha_j$. Therefore the random permutation g^* assigns two different random variables Y_i, Y_j (which never collide) and the random function h generates two independent random variables, i.e. the outputs γ_i, γ_j are independent from the input. The distinguisher, however, can work on β_i, β_j as they are generated according to different probability distribution (they are random permutations). Clearly, β_i, β_j are always different if oracle gates are evaluated by ψ. If all the oracle gates are evaluated by F_{2n}, then β_i may collide ($i = 1, 2, \ldots, m$). The probability that β_i do not collide (when oracle gates are evaluated by F_{2n}) is

$$\frac{2^n!}{2^{nm}(2^n - m)!}$$

Thus the probability P_Y that the distinguisher may succeed (if Y is bad) when it finds a collision (oracle gates are evaluated by F_{2n}) is

$$P_Y = 1 - \frac{2^n!}{2^{nm}(2^n - m)!} \leq \frac{m(m+1)}{2^{n+1}}$$

Consider the second case when $\underline{X\text{ is bad}}$ $(R_i \neq R_j)$. This case is identical to that in Figure 2. The random function f assigns two independent random variables $f(R_i)$ and $f(R_j)$. The outputs γ_i, γ_j are independent from the input. The probability P_X is precisely the same as for the original L-R randomizer. Therefore

$$Pr[\omega \text{ is NOT preserving in } \Psi_2] \leq P_X + P_Y \leq \frac{m(m-1)}{2}\frac{1}{2^n} + \frac{m(m+1)}{2^{n+1}}$$

and the final result follows.

□

Now we are ready for the main theorem of the paper.

Theorem 4.2 *The randomizer* $\Psi_2^* = \psi_1(f_1, g_1^*, h_1) \circ \psi_2(f_2, g_2^*, h_2)$ *is perfect if the number of oracle gates* $m \leq 2^n$ $(f_1, f_2, h_1, h_2 \in_R F_n$ *and* $g_1^*, g_2^* \in_R P_n)$.

Proof: Following the Luby and Rackoff proof [2], we consider all $\frac{m(m-1)}{2}$ possible pairs of oracle gates. If none of the pairs is transparent to the input, it means that Ψ_2 is perfect (or in Luby and Rackoff terms $\underline{\omega \text{ is preserving}}$).

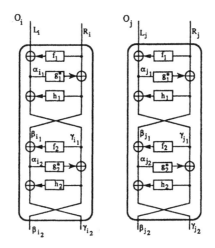

Figure 6: Two oracle gates (O_i, O_j) evaluated by Ψ_2^*

Consider a single pair of oracle gates (O_i, O_j). We are going to show that its outputs $(\beta_{i_2}, \beta_{j_2})$ and $(\gamma_{i_2}, \gamma_{j_2})$ are independent from the input variables (see Figure 6). According to the previous lemma, the outputs γ_{i_1}, γ_{j_1} are always independent from the input so are the outputs $(\beta_{i_2}, \beta_{j_2})$.

Now take the randomizer ψ_2 which is fed by two pairs $\beta_{i_1}, \gamma_{i_1}$ and $\beta_{j_1}, \gamma_{j_1}$. There are two possible cases:

1. γ_{i_1} and γ_{j_1} take on different values (this happens with the probability $1 - \frac{1}{2^n}$). It turns out that f_2 assigns two independent random variables and γ_{i_2}, γ_{j_2} are independent from the input.

2. γ_{i_1} and γ_{j_1} have the same value (this happens with the probability $\frac{1}{2^n}$). Thus $\alpha_{i_2} \neq \alpha_{j_2}$ and the permutation g^* generates two different values. Finally, the random function h_2 makes γ_{i_2}, γ_{j_2} independent from the input.

Random permutations g_1^*, g_2^* play an important role as far as single randomizers are concerned. However, Ψ_2 randomizers can be substituted by a fixed permutation for instance the identity permutation $g_1^* = g_2^* = 1$. It is easy to prove that the following corollary is true.

Corollary 4.1 *The randomizer* $\Psi_2^{**} = \psi_1(f_1, 1, h_1) \circ \psi_2(f_2, 1, h_2)$ *is perfect if the number of oracle gates* $m \leq 2^n$ *and* f_1, f_2, h_1, h_2 *are four different random functions.*

The structure Ψ_2^{**} is optimal as it uses six DES rounds and four different random functions. The constant permutation $g^* = 1$ switches the random functions. If the right hand halves of the inputs for a pair of oracle gates are different, it selects the upper random function. Otherwise, it uses the lower random function.

Perfectness implies that the randomizer $\Psi_2^* = \psi_1(f_1, g_1^*, h_1) \circ \psi_2(f_2, g_2^*, h_2)$ does not leak any information about the input to the output. Luby and Rackoff called this case $\underline{\omega\ is\ preserving}$ which implies

$$| Pr[C_{2n}(F_{2n})] - Pr[C_{2n}(\psi)] | = 0$$

This conclusion is obviously wrong. However, it is true that if $\underline{\omega\ is\ preserving}$, the distinguisher works (without any loss of its efficiency) on the knowledge of the output only (it obviously selects different input values but their values are not important).

Such a case happens if one wants to distinguish truly random permutation from a truly random function. A distinguisher can look for collisions for the random function (which never happen for the random permutation). The input values put to the oracle gates do not matter as long as they are different.

5   Conclusions

Considering Theorem 4.2 and the randomizer Ψ_2^* given there, we may draw the following conclusions:

1. As outputs of all oracle gates (evaluated by Ψ_2^*) are independent random variables, so the knowledge of their inputs does not provide any useful information to the distinguisher (it is enough to the distinguisher to know that inputs are different for all oracle gates). It means that the distinguisher works only if it can extract some information about the structure from the outputs of oracle gates.

2. The requirement about the random permutations g_1^*, g_2^* used in ψ_1 and ψ_2 can be relaxed as their randomness is not exploited. It is applied to exclude (Y *is bad in* ψ) from happening so the random function h is used for different values and generates two independent random variables. If we accept the identity permutation ($g^* = 1$), as the permutation g^*, the modified Ψ_2^{**} structure is $\Psi_2^{**} = \psi_1(f_1, 1, h_1) \circ \psi_2(f_2, 1, h_2)$ and

$$\mid Pr[C_{2n}(\Psi_2^*)] - Pr[C_{2n}(\Psi_2^{**})] \mid = 0$$

3. If we use pseudorandom, instead of random functions ($f_1, h_1, f_2, h_2 \in_{PS} F_{2n}$) and a pseudorandom permutation $g_1^*, g_2^* \in_{PS} P_n$, then the generator Ψ_2^* is also pseudorandom. Moreover, its quality exclusively depends upon the quality of the pseudorandom functions used. The structure of Ψ_2^* is sound.

4. Most of the DES-type cryptosystems use the structure $\psi(f_1, f_2, \cdots, f_k)$, where f_i ($i = 1, \cdots k$) are functions which are generated by a short cryptographic key (they are neither random or pseudorandom). The structure $\psi(f_1, 1, f_2, f_3, 1, \cdots f_{k-1}, 1, f_k)$ is better as far as random functions are applied.

5. Let $\Psi_k = \underbrace{\psi(f_1, g_1, h_1) \circ \ldots \circ \psi(f_k, g_k, h_k)}_{k}$ and $\Psi_2^* \circ \Psi_k$ (for some $k = 1, 2, \ldots$). The two permutation generators Ψ_k^* and $\Psi_2^* \circ \Psi_k$ are not distinguishable and

$$\mid Pr[C_{2n}(\Psi_2^*)] - Pr[C_{2n}(\Psi_2^* \circ \Psi_k)] \mid = 0$$

(a distinguishing circuit has $m \leq 2^n$ oracle gates).

6. The alternate group A_{2n} of the group of all permutations P_{2n} can be generated using a finite number of concatenations of ψ. Thus

$$\mid Pr[C_{2n}(P_{2n})] - Pr[C_{2n}(\Psi_2^*)] \mid = 0$$

Note that $\mid Pr[C_{2n}(F_{2n})] - Pr[C_{2n}(\Psi_2^*)] \mid > 0$ as it is possible to design a distinguisher which can tell apart F_{2n} from Ψ_2^* with a small probability. The distinguisher tries to get the same output in two different oracle gates for two different messages. It may succeed only if the oracle gates are evaluated by F_{2n} (for oracle gates evaluated by Ψ_2^*, any output is different for different input).

ACKNOWLEDGMENT

We would like to thank Yuliang Zheng for his critical comments and help. We also thank Professor Jennifer Seberry for her continuous support.

References

[1] O. Goldreich, S. Goldwasser, and S. Micali. How to construct random functions. *Journal of the ACM*, 33(4):792–807, October 1986.

[2] M. Luby and Ch. Rackoff. How to construct pseudorandom permutations from pseudorandom functions. *SIAM Journal on Computing*, 17(2):373–386, April 1988.

[3] Andrew C. Yao. Theory and application of trapdoor functions. In *Proceedings of the 23rd IEEE Symposium on Fundation of Computer Science*, pages 80–91, New York, 1982. IEEE.

A General Purpose Technique for Locating Key Scheduling Weaknesses in DES-like Cryptosystems

(Extended Abstract)

Matthew Kwan - mkwan@cs.adfa.oz.au
Josef Pieprzyk - josef@cs.adfa.oz.au

Centre for Computer Security Research
Department of Computer Science
University of New South Wales
Australian Defence Force Academy
Canberra ACT 2600
AUSTRALIA

ABSTRACT

The security of DES-style block ciphers rests largely upon their non-linear S-boxes. If different pairs of input data and key can produce identical inputs to all of a cipher's S-boxes, then for those pairs the system is weakened. A technique is described here which enables a cryptanalyst to find how many of these pairs, if any, exist for a given cryptosystem, and how to exploit those pairs under a chosen plaintext attack.

Introduction

Cryptosystems, in this case block ciphers, are used to encrypt and decrypt blocks of data, under the control of some key. The *chosen plaintext* strength of a cryptosystem is determined by the number of encryptions which are required to find the key, or a functionally equivalent key, given that the algorithm is known and the attacker can find the ciphertext value for any chosen input plaintext.

Ideally, the strength of a cryptosystem should be two to the power of the number of bits in the key - i.e. the only way to find the correct key is to do an exhaustive search of all possible keys.

For convenience, the strength of a cryptosystem will be defined by the number of effective bits in the key - in other words $\log_2$ of the number of operations required to find the key. As an example, the NBS Data Encryption Standard (DES) cryptosystem has a 56-bit key [1], but because of the weakness

$$DES(\overline{K},\overline{P}) = \overline{DES(K,P)}$$

we can reduce the key search space by half under a chosen plaintext attack [2]. Thus DES is said to have 55 bits of strength. Conversely, it is also said to have one bit of weakness, since it is one bit weaker than the size of its key.

The technique described in this paper is used to find the number of bits of that kind of weakness in DES or Feistel-like cryptosystems, and other cryptosystems which use the exclusive-OR operation to mix key and plaintext values together. The weaknesses are caused by linear (under XOR) dependencies between key and plaintext bits, and can be used to greatly reduce the search required to find the key [3].

Cryptosystem Security

Definition : In this paper a linear function f is one for which

$$f(a \oplus b) = f(a) \oplus f(b)$$

Examples are bit permutation functions (P-boxes) and expansion functions (E-boxes). By themselves, these functions are cryptographically insecure, since they are easily inverted.

The strength of a cryptosystem lies almost entirely in its non-linear functions, called S-boxes. Examples of these functions are arithmetic addition and Galois field exponentiation, although in many cases a pre-calculated lookup table is used.

Typically, cryptosystems use a combination of linear and non-linear functions to create what is hopefully a very large non-linear function. But if the effects of the S-boxes can be circumvented, then the cryptosystem becomes linear, and thus insecure.

S-boxes can be circumvented if they can be made to act in a linear fashion. This simplest way to do this is to find key and plaintext pairs (call them KP pairs) which, although different, provide identical input to the S-boxes. Then, for these KP pairs, the S-boxes will all produce identical outputs, and the cryptosystem outputs will only differ by some linear function of the differences between these pairs. A well known example of this is the DES weakness, described above. When the key and plaintext are inverted (i.e. they differ by $ffffffffffffff_{16}$ and $ffffffffffffffff_{16}$ respectively), the ciphertext output always differs by the fixed value $ffffffffffffffff_{16}$.

If, for a given KP pair, there are $n-1$ other KP pairs which produce identical inputs to the S-boxes, then the cryptosystem's security has been weakened by $\log_2 n$ bits. The technique described here will find the number of those bits of weakness in a cryptosystem, and will provide information on how to exploit those weaknesses. Naturally, the information can also be used to remove the weaknesses by pinpointing flaws in the cryptosystem's design.

S-box Inputs

For the technique to be effective against a particular cipher, each input bit of the S-boxes must be expressible as an XOR sum of key bits, plaintext bits, and S-box output bits. Now, since the aim of the technique is to provide identical inputs to the S-boxes, we know that the S-box outputs will also be identical. Thus we can disregard them as inputs since they will remain unchanged from one KP pair to the next.

For convenience, the bits of the KP pair will be referred to by the variables $x_1 .. x_n$, where x_i take the value 0 or 1, and n is the combined number of bits of the key and plaintext. For example, with DES, $n = 56+64 = 120$, since DES has a 56-bit key and a 64-bit plaintext.

So we end up with a series of equations for the input bits to the S-boxes of the form

$$s = x_i \oplus .. \oplus x_j$$

What we wish to do now is change these x values, yet have the values of the equations remain constant. If the changes in the x values are represented by Δx, then we desire that

$$(x_i \oplus \Delta x_i) \oplus .. \oplus (x_j \oplus \Delta x_j) = x_i \oplus .. \oplus x_j$$

for each S-box input bit.

Simplifying the above leaves us with the set of equations

$$\Delta x_i \oplus .. \oplus \Delta x_j = 0$$

Now it is simply a matter of solving for these Δx values. This is done most conveniently by expressing the equations as a matrix of zeroes and ones, and transforming the matrix to Reduced Row Echelon (RRE) form. Using DES as an example, we create a matrix with 120 columns (corresponding to the 56 key and 64 plaintext input bits) and 768 rows (corresponding to 16 rounds each with 48 S-box input bits). In effect, there are 768 equations to solve, with 120 variables.

Since we are dealing with exclusive-OR rather than arithmetic addition, converting the matrix to RRE is relatively simple. All the values are either zero or one, so there is no need for division or checking for underflow errors. If two rows have their leftmost non-zero values in the same column, then the rows are simply XORed together, and the result stored in the lower row. In other words, it is identical to arithmetic RRE, except that the addition and subtraction operations have been replaced with XOR.

After solving the matrix we now have at most n non-zero rows (at most 120 for DES). To find the number of independent bits (and thus the number of bits of weakness), we strip away the leftmost set bit from each of the remaining equations and inclusive-OR the remaining bits together into a single row. The bits in this row are the independent bits, and the number of them gives the number of bits of weakness.

Let's say there are m independent bits. Then for the 2^m values that these bits can take, the remaining dependent bits Δx_i can be set so that the equations hold. Thus there are 2^m sets of Δx_i such that the inputs to the S-boxes, given by

$$(x_i \oplus \Delta x_i) \oplus .. \oplus (x_j \oplus \Delta x_j)$$

will remain unchanged. These sets can be used to reduce an exhaustive search by a factor of 2^m, as will be shown in the DES example.

An Example - DES

The technique can be applied to the NBS Data Encryption Algorithm, DES, as shown in figure 1. DES is a 16 round Feistel block cipher, with a 64-bit plaintext and a 56-bit key. In each round 48 bits of the key are exclusive-ORed with a 48 bit expansion of 32 bits of the plaintext, and the result fed into the S-boxes. When each of these XORs are represented as individual equations, we obtain 768 equations with 120 variables.

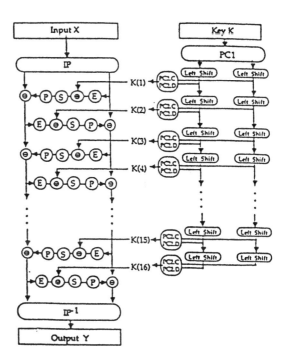

Figure 1. DES encryption algorithm

Using the variables $x_0..x_{63}$ to represent the plaintext bits (from MSB to LSB), and the variables $x_{64}..x_{119}$ for the key bits, we get 768 equations as follows

$$\Delta x_6 \oplus \Delta x_{72} = 0$$

$$\Delta x_{56} \oplus \Delta x_{108} = 0$$

$$\Delta x_{48} \oplus \Delta x_{93} = 0$$

etc.

When these equations are solved as described earlier, we end up with the following 119 equations

$$\Delta x_0 \oplus \Delta x_{119} = 0 \tag{1.a}$$

$$\Delta x_1 \oplus \Delta x_{119} = 0 \tag{1.b}$$

$$\Delta x_2 \oplus \Delta x_{119} = 0 \tag{1.c}$$

etc.

Stripping off the leftmost variable of these equations, then merging the resulting matrix rows together, we are left with a single bit in column 119. This bit corresponds to the one bit weakness in DES.

The variable Δx_{119} is the independent variable, with all the other variables dependent upon it, as can be seen in equations (1.a-c). With the two values it can take, 0 and 1, and with the other variables set accordingly, these two sets of variables can be XORed with any input KP pair and produce identical inputs to all the S-boxes.

The case where $\Delta x_{119} = 0$ is the trivial case where there is no change to the KP pair. The other case, where $\Delta x_{119} = 1$ and thus all the other variables $\Delta x_i = 1$, causes both the key and plaintext in the KP pair to be inverted, yet still provide identical inputs to the S-boxes.

Looking at figure 1, it can be seen that the ciphertext output can be expressed as the XOR sum of the plaintext and all the S-box outputs. Since the plaintext has been XORed with $ffffffffffffffff_{16}$, and the S-box outputs haven't changed, the ciphertext will also change by $ffffffffffffffff_{16}$.

Hence the inversion weakness where

$$C = DES(K,P)$$

implies that

$$\overline{C} = DES(\overline{K},\overline{P})$$

or alternatively

$$C = \overline{DES(\overline{K},\overline{P})} \tag{3}$$

This weakness can be exploited under a chosen plaintext attack to reduce the key search by half. Assume that we are trying to find the secret key K_s. This key can only be accessed via the function $DES(K_s,P)$, where P is an arbitrary plaintext value of our choice.

First we calculate two ciphertexts C_0 and C_1, such that

$$C_0 = DES(K_s,P)$$

and

$$C_1 = \overline{DES(K_s,\overline{P})} \tag{4}$$

Then we begin a theoretical brute force search for the key, trying all key values K_b. For each K_b, we calculate the ciphertext C_b, where

$$C_b = DES(K_b,P)$$

We then compare C_b with both C_0 and C_1. If it equals C_0, then clearly $K_s = K_b$, and we have the required result. If $C_b = C_1$, then referring to equations (3) and (4), we know that $K_s = \overline{K_b}$, again producing the required result. If neither value matches, then we have eliminated two possible keys at the expense of only one encryption. Thus it is possible to search the entire key space in half the normal time, so long as we ensure than no two keys K_b are the inverse of each other (most easily done by ensuring that the independent bits remain a constant value for all K_b).

Another Example - LOKI

LOKI [4] is a DES-style block cipher developed in recent years by the Centre for Computer Security Research at the University of New South Wales. Its aim was to provide similar security to DES, but with a simpler structure to facilitate analysis and implementation, as can be seen in figure 2.

Features of the LOKI algorithm of interest to this paper are the XORing of the key with the plaintext at the beginning and end of the algorithm, and the key schedule, which involves 12 bit rotations of the left and right halves of the key. Another thing to note is that because the key halves are swapped at the same rate as the plaintext halves, the left half of the key is only mixed with the right half of the plaintext, and the left half of the plaintext is only mixed with the right half of the key. A combination of these features results in 8 bits of weakness [56].

The input to the S-boxes in each of LOKI's rounds can be expressed as the XOR sum of a plaintext bit and two key bits (one from the XOR at the beginning, the other from the key schedule). There are 16 rounds, each with 32 bits entering the S-boxes, so there are 512 equations to be solved in order to find the number of independent bits. Since LOKI's plaintext and key are both 64-bit, there are 128 variables. Using variables $x_0 .. x_{63}$ to represent the plaintext bits, from MSB to LSB, and $x_{64}..x_{127}$ to represent the key, we get the following equations

$$x_{32} \oplus x_{64} \oplus x_{96} = 0$$

$$x_{33} \oplus x_{65} \oplus x_{97} = 0$$

$$x_{34} \oplus x_{66} \oplus x_{98} = 0$$

etc.

After solving the matrix, we are left with 120 equations, namely

$$x_0 \oplus x_{92} \oplus x_{124} = 0$$

$$x_1 \oplus x_{93} \oplus x_{125} = 0$$

$$x_2 \oplus x_{94} \oplus x_{126} = 0$$

etc.

When these rows are stripped and merged we are left with 8 bits, corresponding to the variables $x_{92}-x_{95}, x_{124}-x_{127}$. These variables can be exploited to produce 8 bits of weakness, since there are $2^8 = 256$ values they can take which satisfy the equations.

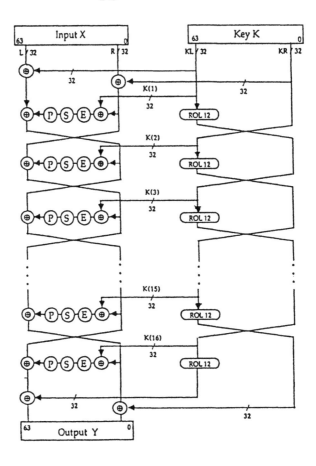

Figure 2. LOKI encryption algorithm

Analysis of the key and plaintext bit dependencies reveals the weakness

$$LOKI\,(K \oplus mmmmmmmmmmnnnnnnnn, P \oplus pppppppppppppppp) = LOKI\,(K,P) \oplus ppp \cdots ppp$$

where m and n are 4 bit values, and $p = m \oplus n$. Note that for any p there are 16 pairs of m and n for which this holds, which implies that there are 16 different key values which are functionally equivalent. This leads to 4 bits of weakness under a *known* plaintext attack, since 4 bits of the key are redundant.

As an aside, known-plaintext weaknesses can be detected using the technique described in this paper by solving for key bits only (i.e. forcing the plaintext bits to remain constant). In other words, set up equations corresponding to the XOR sum of key bits at the inputs to S-boxes, and solve accordingly. When this is done for DES there is only the trivial solution where all values are 0, as would be expected, since the DES weakness is a chosen-plaintext weakness.

Continuing with the analysis of LOKI, we perform a chosen plaintext attack on the key K_s by taking an arbitrary plaintext P, and generating the ciphertext values C_p for $p = 0..15$ such that

$$C_p = LOKI(K_s, P \oplus pppppppppppppppp_{16}) \oplus pppppppppppppppp_{16}$$

We then carry out a theoretical brute force search, similar to that used in the DES example. The 16 C values, combined with the 4 bit redundancy described above, contribute to reducing the search space by a factor of 256. Thus LOKI only has 56 bits of strength under a chosen plaintext attack.

In response to these weaknesses the LOKI algorithm has undergone redesign, with the new version removing the initial and final XOR of the key, using 12 and 13 bit rotations in the key schedule, and swapping the key halves after two rounds, rather than every round. Analysis of these modifications reveals only one bit of weakness, which results in an inversion weakness the same as for DES, and leaves an effective key strength of 63 bits.

Software Implementation

Being a very systematic technique, it was implemented in software at an early stage. This was particularly useful in the redesign of the LOKI key schedule, which proceeded largely by trial and error. Using automated analysis, with a convenient user interface, greatly sped up the evaluation of new designs.

Since the software already had access to all the information about a cipher's key schedule, it required only minimal effort to extend the software to also search for decrypting key pairs. In other words, a pairs of keys K_1 and K_2 such that encrypting a plaintext P with key K_1 then encrypting the result with K_2 yields the initial value P. In situations where $K_1 = K_2$, the key is called self-decrypting. Self-decrypting keys can be exploited to break some message hashing schemes, and lead to very short cycles in the Output Feedback mode of most ciphers, so their detection is very important.

The decrypting pairs are brought about by keys which generate identical subkeys, but in reverse order. For example, in a 16-round block cipher such as DES, the round 1 subkey of K_1 equals the round 16 subkey of K_2, the round 2 subkey of K_1 equals the round 15 subkey of K_2, and so on. Describing these conditions as a series of equations and solving them is a relatively trivial matter, especially since the matrix solving routines can be reused.

The program was tested on a description of KSX, provided by its designer Richard Outerbridge. KSX is essentially DES with a 64-bit key and a redesigned key schedule [7]. The (slightly edited) output of the analysis is as follows.

```
Plaintext/key XOR redundancies
------------------------------
0000000000000000 0000000000000000
ff00ff00ff00ff00 aaaaaaaaaaaaaaaa
00ff00ff00ff00ff 5555555555555555
ffffffffffffffff ffffffffffffffff

Decrypting key pairs (* denotes self-decrypting)
------------------------------------------------
0000000000000000 0000000000000000  *
0808080808080808 8080808080808080
0404040404040404 4040404040404040

[ 230 pairs deleted ]

7979797979797979 9797979797979797
7575757575757575 5757575757575757
7d7d7d7d7d7d7d7d d7d7d7d7d7d7d7d7
7373737373737373 3737373737373737
7b7b7b7b7b7b7b7b b7b7b7b7b7b7b7b7
7777777777777777 7777777777777777  *
7f7f7f7f7f7f7f7f f7f7f7f7f7f7f7f7
f0f0f0f0f0f0f0f0 0f0f0f0f0f0f0f0f
f8f8f8f8f8f8f8f8 8f8f8f8f8f8f8f8f
f4f4f4f4f4f4f4f4 4f4f4f4f4f4f4f4f
fcfcfcfcfcfcfcfc cfcfcfcfcfcfcfcf
f2f2f2f2f2f2f2f2 2f2f2f2f2f2f2f2f
fafafafafafafafa afafafafafafafaf
f6f6f6f6f6f6f6f6 6f6f6f6f6f6f6f6f
fefefefefefefefe efefefefefefefef
f1f1f1f1f1f1f1f1 1f1f1f1f1f1f1f1f
f9f9f9f9f9f9f9f9 9f9f9f9f9f9f9f9f
f5f5f5f5f5f5f5f5 5f5f5f5f5f5f5f5f
fdfdfdfdfdfdfdfd dfdfdfdfdfdfdfdf
f3f3f3f3f3f3f3f3 3f3f3f3f3f3f3f3f
fbfbfbfbfbfbfbfb bfbfbfbfbfbfbfbf
f7f7f7f7f7f7f7f7 7f7f7f7f7f7f7f7f
ffffffffffffffff ffffffffffffffff  *
```

The analysis reveals two bits of weakness in the key schedule, or four KP pairs. This means that the effective key strength of KSX under a chosen-plaintext attack is 62 bits, not 63 bits as was previously thought. In addition, KSX has 256 decrypting pairs, 16 of which are self decrypting. This does not compare favourably with the original DES key schedule, which had only 16 pairs, 4 self-decrypting.

Conclusion

The technique of solving equations at the inputs to S-boxes is very useful for highlighting systematic weaknesses in a cryptosystem's key and plaintext schedule. The mathematical nature of the equation solving gives reliable evidence that the bits of weakness found are the *only* bits of weakness under this kind of attack, and thus give a good indication of the strength of this aspect of security.

While this method of attack should not be used as the sole test of a cryptosystem's security, it does highlight one of the more easily exploited weaknesses, and as such should be a useful tool in any cryptanalyst's arsenal.

Acknowledgements

Thanks to the crypt group for providing equipment, support, and advice. And to Lawrie Brown for letting me read his references. This work has been supported by ARC grant A48830241, ATERB, and Telecom Australia research contract 7027.

References

1. NBS, "Data Encryption Standard (DES)," FIPS PUB 46, US National Bureau of Standards, Washington, DC, JAN 1977.

2. M. Hellman, R. Merkle, R. Schroppel, L. Washington, W. Diffie, S. Pohlig, and P. Schweitzer, *Results of an Initial Attempt to Cryptanalyze the NBS Data Encryption Standard,* Stanford University, September 1976.

3. David Chaum and Jan-Hendrik Evertse, "Cryptanalysis of DES with a Reduced Number of Rounds Sequences of Linear Factors in Block Ciphers," in *Advances in Cryptology-Crypto 85,* Lecture Notes in Computer Science, vol. 218, pp. 192-211, Springer-Verlag.

4. Lawrence Brown, Josef Pieprzyk, and Jennifer Seberry, "LOKI - A Cryptographic Primitive for Authentication and Secrecy Applications," in *Advances in Cryptology: Auscrypt'90,* Lecture Notes in Computer Science, vol. 453, pp. 229-236, Springer-Verlag, 1990.

5. Lars Ramkilde Knudsen, "Cryptanalysis of LOKI," in *Advances in Cryptology - proceedings of ASIACRYPT'91,* 1991.

6. E. Biham and A. Shamir, "Differential Cryptanalysis of Snefru, Khafre, REDOC-II, LOKI and Lucifer," in *Advances in Cryptology - proceedings of CRYPTO'91,* 1991.

7. Richard Outerbridge, "Some Design Criteria for Feistel-Cipher Key Schedules," *Cryptologia,* vol. 10, no. 3, pp. 142-156, JUL 1986.

Results of Switching-Closure-Test on FEAL

(Extended abstract)

Hikaru Morita   **Kazuo Ohta**   **Shoji Miyaguchi**

NTT Laboratories

NTT R&D Center (Room 309A), 1-2356 Take Yokosuka Kanagawa 238-03 Japan

Phone: +81-468-59-2514   FAX: +81-468-59-3858   E-mail: morita@sucaba.ntt.jp

Abstract

The closure tests, CCT and MCT, were introduced to analyze the algebraic properties of cryptosystems by Kaliski et al. [KaRiSh]. If a cryptosystem is closed, the tests give the same results "Fail" and the cryptosystem might be breakable. Though CCT requires much less memory and time than MCT, we cannot apply CCT to check cryptosystems having the same data and key block lengths such as FEAL with non-parity mode. Because CCT utilizes the differences in data and key block lengths.

Though CCT experiments performed by Kaliski et al. detected that DES is not closed, how should FEAL be checked? Does FEAL pass in MCT? Since MCT needs a lot of memory and time, to check FEAL, we developed a switching closure test SCT [MoOhMi], which is practical version of MCT. In this paper, by using SCT, it is confirmed that FEAL is not closed with high probability.

1. Introduction

To find the algebraic structure of cryptosystems in general, Kaliski et al. [KaRiSh] proposed two closure tests: CCT (cycling closure test) and MCT (meet-in-the-middle closure test). These tests can detect features such as algebraic closure. Moreover, they also proposed two cryptattack methods based on the algebraic features.

Generally, both CCT and MCT can determine if a cryptosystem is closed or not. If a cryptosystem is closed, they give the same results "Fail", which means the cryptosystem might be breakable. However, if a cryptosystem is not closed, you cannot be sure that they will give the same results because it isn't known whether they can detect the same algebraic structure or not. When each closure test detects that a cryptosystem is not closed, we say the cryptosystem "Passes." A "Passed" cryptosystem is secure against cryptattack methods based on the closure property. CCT experiments performed by Kaliski et al. detected that DES is not closed.

Our interest was that MCT might prove to be a fertile avenue for cryptographic research. MCT offers the possibility of extracting information from a not-closed cryptosystem that would allow the cryptosystem to be broken. However, MCT needs an excessive amount of memory. On the other hand, though CCT requires much less memory, we cannot apply CCT to a cryptosystem having the same data and key block lengths such as FEAL with non-parity mode. Because CCT utilizes the property that if a cryptosystem is closed, the cycling period approaches $O\left(\sqrt{|K|}\right)$, otherwise it approaches $O\left(\sqrt{|M|}\right)$ where K and M are key and message space, respectively, and $|S|$ indicates the number of elements of a set S. Consequently, we presented a switching closure test (SCT) [MoOhMi] so that MCT's memory requirements are dropped from $O\left(\sqrt{|K|}\right)$ to a constant value. While up to now the memoryless method has been applied to collision search [QuDe], we applied it to a closure test for cryptosystems.

We wanted to know if all FEAL-N including FEAL-8 fails in MCT or not. If FEAL fails in MCT, FEAL may be broken by the known-plaintext attack with high probability because MCT can be easily applied to the known-plaintext attack. We tested FEAL by using SCT.

Section 2 shows SCT procedures. Section 3 presents experimental results of FEAL-8. This paper is concluded in Section 4.

2. Switching Closure Test [MoOhMi]
2.1 Background

When a cryptosystem is defined by $\Pi = (K, M, C, T)$, where K, M, and C are key space, message space, and ciphertext space, respectively and T is a set of all encryption transformations defined by $T \equiv \{E_k | k \in K\}$ where the ciphertext $c = E_k(p), (p \in M, c \in C)$, its algebraic system is $(T, \cdot)$ where $E_{k_2} \cdot E_{k_1}(p) \equiv E_{k_2}\left(E_{k_1}(p)\right)$ for $\forall p \in M$.

If the set T is closed under the product operation "$\cdot$", the double-encryption transformation $E_{k_2} \cdot E_{k_1}$ equivalent to a given encryption transformation E_k for any plaintext can be found. Thus, you can decrypt ciphertext c by using double-encryption key pair (k_1, k_2) instead of the real key k, and it is known that the algebraic system $(T, \cdot)$ constitutes a group.

If the system $(T, \cdot)$ is a group, it is not difficult to find pairs of double-encryption transformations. This is because, for arbitrary keys k and k_1, there exists a key k_2 such that $E_k = E_{k_2} \cdot E_{k_1}$. Though the number of all double-encryption pairs is the square of key size $|K|$, the number of double-encryption key pairs (k_1, k_2) equivalent to the encryption key k is the key size $|K|$. Consequently, if you search $|K|$ times, you should find one equivalent double-encryption key pair. Therefore, we do not need a large number of search operations.

In the meet-in-the-middle strategy, r_1 values of $E_{k_1}(p)$ and r_2 values of $E_{k_2}^{-1}(c)$ (inverse transformation of E_{k_2}) are generated like this. Then, values of $\{E_{k_1}(p)\}$ and $\{E_{k_2}^{-1}(c)\}$ are matched against each other. If you select $r_1 = r_2 = \sqrt{|K|}$, you can check $|K|$ pairs. The meet-in-the-middle strategy can reduce the order of $|K|$ to $\sqrt{|K|}$. So, if the cryptosystem has an algebraic structure such that $(T, \cdot)$ forms a group, a key pair found by this strategy becomes equivalent to the real key for an arbitrary plaintext with high probability.

2.2 SCT Procedure

SCT is constructed around two important techniques.

One is the switching function $h(x)$ defined by

$$h(x) = \begin{cases} f(x) & \text{if } f_c(x) \text{ is true,} \\ g(x) & \text{if } f_c(x) \text{ is false,} \end{cases}$$

where $f_c(x)$ is a conditional function which generates true or false with 50-% probability for each value of x. Functions f and g are related to the encryption function E as shown in Fig. 1. Functions F_E and F_D are arbitrary functions having one-to-one relationship between input and output.

The other technique is the intersection search to find $x_i = x_j$ and $x_{i-1} \neq x_{j-1}$ for sequences $\{x_0, x_1, \cdots\}$ generated by $x_{i+1} = h(x_i)$. It can be achieved with a constant memory.

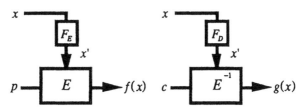

Fig. 1 Definitions of f and g

3. Experimental results on FEAL-8
3.1 Experimental condition

SCT's efficiency depends mainly on the delay of the switching function $h(x)$. When functions, $f_c(x)$, $F_E(x)$, and $F_D(x)$ are determined as

$$f_c(x) = (\text{true if } x \text{ is "even," false if } x \text{ is "odd"}),$$

$$F_E(x) = x, F_D(x) = x,$$

experimentally, generating frequencies of $f(x)$ and $g(x)$ using FEAL-8's cipher functions are less than 0.1 % different. Thus, $h(x)$ is switched to $f(x)$ or $g(x)$ with 50%. The delay of $h(x)$ is about 31 μsec on a 27 MIPS machine (e.g., SparcStation 2) using a C language program. The most important factor is computing power, since SCT needs only a small amount of memory and can be parallelized because sequences starting from various points can be independently generated.

3.2 Parameters and results

The known-plaintext pair shown in Note in Table 1 was tested where p, k, and c are plaintext, hidden real key, and ciphertext, respectively. We selected Strategy #1 in [MoOhMi] in a software environment using WSs which means that many sequences are generated by using $h(x)$ for the same fixed p, k, and c.

For example, the directed graph for sequences in Fig.2 and values of the intersection points between sequences in Table 1 are shown. In Fig.2, the sequence started from 0 meets directly the loop L0. In the Table 1, the sequence is indicated by 0-L0, its length is 66×2^{24}, and the intersection point between the sequence 0-L0 and the loop L0 has 3f73083c 3bc3a2ce as $h(x) = h(y)$ for $x=$ f39350b8 06424393 and $y=$ 128e714c a22415f2 in hexadecimal representation. Though this is a meeting point of $g(x) = f(y)$, a key pair (x, y) is not equivalent to the hidden key k. Because $g(x) \neq f(y)$ for any plaintext pair.

In the example shown in Fig.2 and Table 1, we generated 1126×2^{24} f or g functions. Though we find 11 intersection point pairs (x, y) between $f(x) = g(y)$ or $g(x) = f(y)$, they are not equivalent key pairs to the hidden key k. Therefore, if you select a null hypothesis that FEAL-8 is algebraically closed, the hypothesis is rejected with a level of 0.8-% error probability. The value of 0.8 % is calculated by using $\exp\left(-\dfrac{r_1 r_2}{m}\right)$;

$$\exp\left(-\frac{\left(563 \times 2^{24}\right)^2}{2^{64}}\right) \approx 0.8\%,$$

where $r_1 = r_2 = 563 \times 2^{24}$ and $m = 2^{64}$. (The probability for the meet-in-the-middle strategy is precisely studied in [Ni].)

Sequence	Length ($\times 2^{24}$)	x_{i-1}	y_{j-1}	$x_i = y_j$
0-L0	66	13 93 50 b8 06 42 43 93	12 8e 71 4c a2 24 15 f2	3f 73 08 3c 3b c3 a2 ce
1-L0	102	fd a2 83 11 3c 9d 04 80	71 c7 6a 2b 08 d6 61 11	d8 1e 88 5d 35 71 f2 d4
2-3	24	91 c1 d3 d5 22 83 e4 8b	a9 c2 b3 2d 61 bf 4e 69	a8 1c ef 1d dd ca 28 29
3-L0	174	2a ab a1 63 eb 58 68 4b	19 dc 55 3f de 00 d1 91	55 af 6c 8d c4 cf 5d c8
4-L0	35	d7 79 53 0b 66 c7 81 8f	37 27 64 ec 29 f2 ba c5	11 d3 04 61 d6 67 f0 3d
5-c	9	c1 38 1c 32 f5 5b 8d 16	4d 25 d1 09 e4 4e 8d 7b	d4 8c 59 9e 14 b3 86 ca
6-3	33	18 49 4d c2 00 d5 d5 c9	fd c8 c0 ff d9 9b 03 26	6c a3 21 af 3d 51 93 d6
7-f	2	58 cf da e5 ad 5f f7 af	1a 63 45 56 03 59 6b e6	cc ba 66 69 f1 85 b3 d0
8-3	119	c4 b7 bb 71 cc 7d 37 39	cd 89 80 df a5 cb ac bf	6b 2a b3 a1 ee b4 44 f5
9-0	24	33 62 3b d5 15 6b 8d a8	60 21 81 f8 b8 15 6e 6f	ee 5e 9c f9 c6 2c d8 7b
a-8	6	c8 18 ce a0 39 b3 2b e9	6f bf a8 52 46 33 35 f1	81 68 cc 5c 5f 1e df ad
b-d	18	b3 4c d4 ce 4d 80 90 09	27 ee a2 9e 76 a0 3f 2a	bd 7a 93 8d 68 a5 bf 42
c-L0	40	19 c1 28 34 a4 34 e3 41	c3 08 db 47 21 33 7b 90	57 23 e9 ae e9 d9 e7 d7
d-L0	91	41 1f e8 19 b1 c4 2d ee	0e 16 98 75 5a 43 ae ed	cb 4e 1a b7 64 43 bb d7
e-L0	107	68 e6 ea e6 14 97 76 57	98 ac cf d9 92 1a af ff	65 97 e2 d8 6c bf d5 1e
f-L1	57	eb a5 1c e9 e3 27 48 75	9c 12 76 c8 dc 66 df a8	16 b2 15 8f 05 91 2f 3a
10-f	4	14 c9 98 7d 6b 1f fb 89	1c 81 17 02 5b ad bd d9	b7 13 1b cd 86 b1 eb 41
11-f	52	b8 7f cb 6a 66 4a 2c 7b	77 2c cc a4 3b 26 2a 52	98 db 1b b0 8f b9 18 7a
L0	147			
L1	16			

Note: Length is omitted below 2^{24}.

p=00 00 00 00 00 00 00 00, k=01 23 45 67 89 ab cd ee, c=84 5b 58 a5 41 69 89 f4

Table 1 The values of intersection points

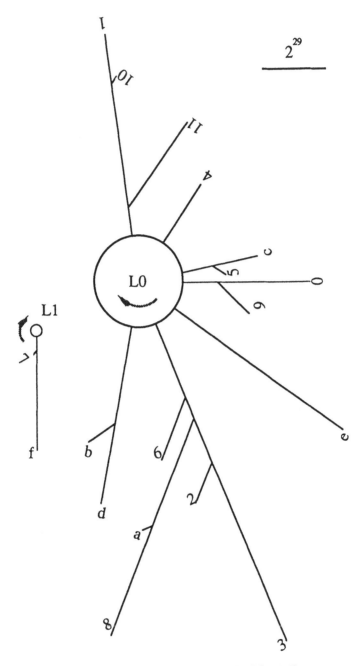

$$\frac{2^{29}}{}$$

Fig. 2 Directed graph of FEAL-8

4. Conclusion

We applied SCT to FEAL-8. FEAL-8 passes SCT with an extremely small probability of error.

Acknowledgment

The authors would like to thank Mr. Sadami Kurihara for his helpful suggestions and Mr. Kazuyoshi Takagi for his improving our software in SCT experiments.

References

[KaRiSh] B. S. Kaliski Jr., R. L. Rivest, and A. T. Sherman: "Is the Data Encryption Standard a Group? (Results of Cycling Experiments on DES)," J. Cryptology, 1, 1, pp.3-36, 1988.

[MoOhMi] H. Morita, K. Ohta, and S. Miyaguchi: "A Switching Closure Test to Analyze Cryptosystems," Abstracts of CRYPTO'91, pp.4.14-4.18, Aug. 1991.

[Ni] K. Nishimura: "Probability to Meet in the Middle," J. Cryptology, 2, 1, pp.13-22, 1990.

[QuDe] J.-J. Quisquater and J.-P. Delescaille: "How Easy is Collision Search. New Results and Applications to DES," Advances in Cryptology-CRYPTO'89, Proceedings, pp.408-413, Springer-Verlag, 1990.

[ShMi] A. Shimizu and S. Miyaguchi: "Fast Data Encipherment Algorithm FEAL," Advances in Cryptology-EUROCRYPT'87, Proceedings, pp.267-278, Springer-Verlag, 1988.

See also S. Miyaguchi, et al.:"Expansion of FEAL Cipher," NTT Review, 2, 6, pp.117-127, Nov. 1990.

IC-CARDS AND TELECOMMUNICATION SERVICES

Jun-ichi Mizusawa
Faculty of Engineering, The University of Tokyo
7-3-1 Hongo, Bunkyo-ku, Tokyo 113· Japan

Telecommunication service is the major concern among public network service providers. New technologies are required realizing new type of network services. One of the prospective markets is the personal communication. UPT (Universal Personal Telecommunication) service is now under study in the international standard organization CCITT. This paper introduces one view of the personalized telecommunication services. At the beginning the definition is given. Then a lot of service ideas are listed. The result of questionnaire on personalized services are explained. Finally principal study items are discussed.

1. PREFACE

The contents of this paper is the summary of the report of "Study Committee on Personalized Telecommunication Services".[1] The workshop started last year, and the report was published last September. The activity was organized by MPT (Ministry of Post and Telecommunication). The author joined the committee activities as a committee chairman.

Unlike general technical paper, I prepared this paper mainly with figures. I excuse at first the explanatory sentences are short. This is partly because I could not have enough time to prepare elaborated sentences, and partly because I have no confidence in writing in English.

2. THE CONCEPT OF PERSONALIZED TELECOMMUNICATION

The personalization of telecommunication services is defined as introduction of person indicating numbers for network services explained in Figure 1. The personal communication number has both advantages and disadvantages when we consider telecommunication circumstances. (Figure 2) It is pointed out that personal aspect of individual life has plural categories. (Figure 3) The study committee derived three key words for accelerating personalization of telecommunication services, i.e., Reliable, Gentle and Flexible as it is shown in Figure 4.

3. PERSONALIZED TELECOMMUNICATION SERVICES

In order to categorize the personal telecommunication services, six types are listed as it is shown in Figure 5. IC-card is a quite useful tool for implementing a personal communication number for telecommunication services. Service ideas are listed as distinguishing six types in Table 1 and 2. The personalized communication has four major aspect which should be carefully studied, including privacy and security issues.

(Figure 6) Three service examples are explained with figures which got the best reputation among personalized service examples listed on the questionnaire. (Figure 7-9)

4. THE RESULTS OF QUESTIONNAIRE ON PERSONALIZED SERVICES

The questionnaire on the personalized telecommunication services was planned. The results obtained from several hundreds of answers shows that the personalized services are attractive for users. Figure 10 shows users preferences for existing telecommunication services. Figure 11 shows users preferences for the personalized services. Figure 12 clearly indicates that the personalized service preferences changes according to the situation such as on business, private use and student. The five typical service supported ratio exceeds 50% as shown in Figure 13.

5. STUDY ITEMS FOR FURTHER PROGRESS

The telecommunication service providers are very encouraged by the results of the questionnaire. But it is also true that there are so many issues to be overcomed. Table 3 lists principal technical study items. Table 4 shows privacy and security issues which requires further discussions. On the other hand, a kind of personalized telecommunication services has been expanding so rapidly such as mobile telephone, portable telephone, cordless telephone, pagers, etc.. Figure 14 illustrates two way of network evolution for constructing fully equipped personalized telecommunication network.

REFERENCES

(1) "The Issues of Telecommunication Personalization": The Report of Study Committee on Personalized Telecommunication Services, MPT July 1991

Note 1: (1) is available in Japanese only.
Note 2: Table 1, 2, Figure 1,7-12 are directly translated from the report[1]. Other tables and figures are original of the author.

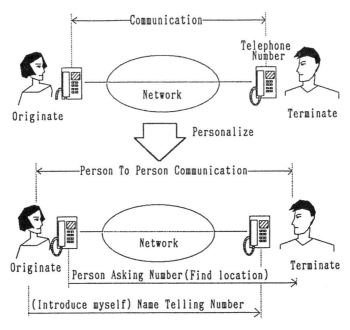

Figure 1 Personalization of Telecommunication

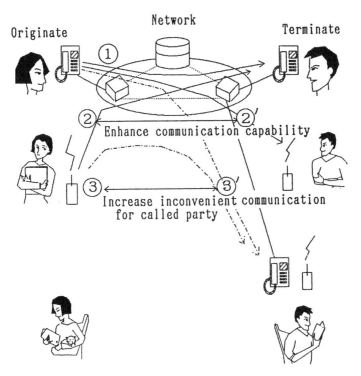

Figure 2 Effects of Personalization

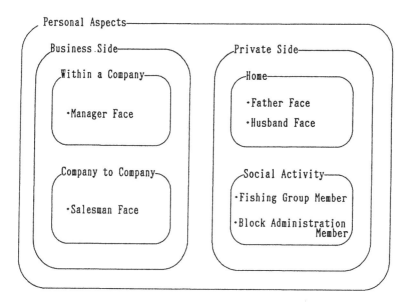

Figure 3   Personal Aspects

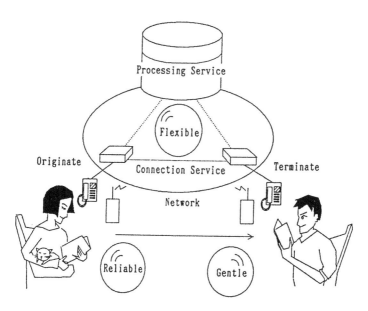

Figure 4   Communication Services Intelligece

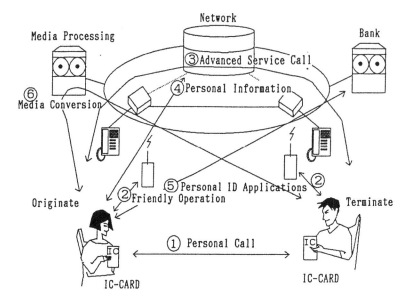

Figure 5　Types of Personalized Telecommunication Services

Type	Service Examples
① Personal Call	Caller Name Display
	Chase Call
	Callee Specified Call
	Business Position Specified Call
	International Personal Communication
	Personal Portable Telephone
② Friendly Operation	Voice Dialling
	AI Dialling
	Pen-name Calling
	Private Call
	Background Voice & Video
③ Advanced Service Call	Terminate Call Screening
	Group Communication
	Virtual My-telephone
	Camp-on
	A Variety of Charging Menu
	Democratic Call
	Call Reservation
	Call Management

Table 1   Connection Service

Type	Service Examples
④ Personal Information	Electronic Secretary
	Personal Mail-box
	Message Receipt Notification
	Location Display
	Personal Database Access
⑤ Personal ID Applications	Tele-control
	Electronic Settlement
⑥ Media Conversion	Electronic Translation
	Media Conversion

Table 2   Processing Service

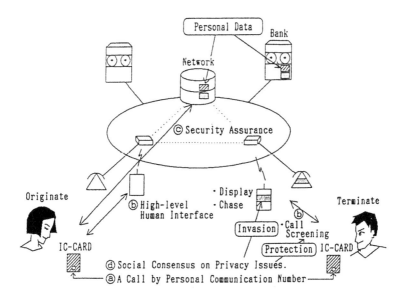

Figure 6   Characteristics of Personalized Communication

Figure 7   Caller Name Display

260

Personal Communication Number dialling establishes
direct communication line wherever a callee is.

Figure 8   Callee Specified Call

The selection of caller name, time & day, is possible.
Screened calles are guided to message services,
call trnsfer, etc.

Figure 9   Terminate Call Screening

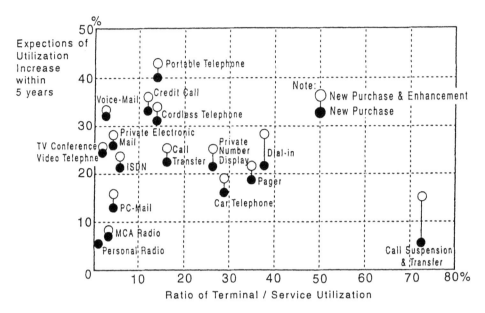

Figure 10 Users Status and Expectations of Present Communication Services.

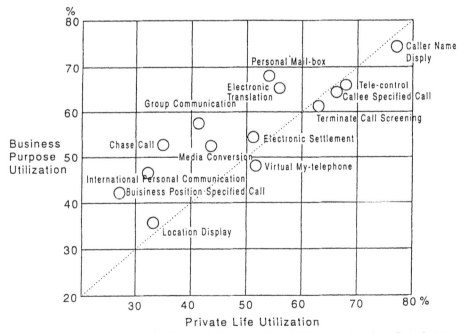

Figure 11 Users Expectations of Personalized Communication Services

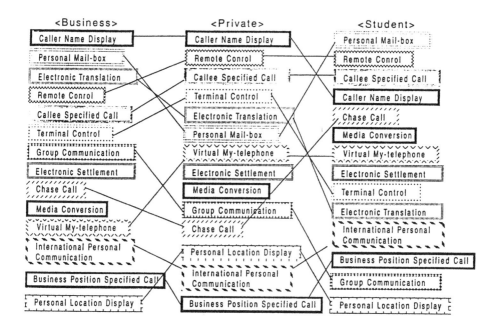

Figure 12 Services Ranking of Users Expectations

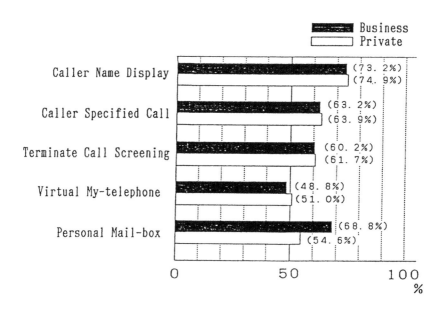

Figure 13 Demand of Typical Services

Study Items
Network Technology
Introduction of Common Channel Signalling to all switching nodes
Popularization of ISDN
Digitalization of the network
The introduction of Intelligent Network technologies
Radio-wave Technology
Efficient wave utilization, Error correcting coding
Zone pursuit algorithm of switching control
Terminal Technology
Light, small volume, low power consumption terminal
Security Technology
Personal indentification by digital signature
Keep information security by cryptograph technology

Table 3 Technical Study Items

Study Items
Formation of Social Consensus
Listing of Secured Object
Improve and Populalization of Security System
Risk Assessment of Personalized Communication
Validity Evaluation of Security Mechanism

Table 4 Privacy & Security Issues

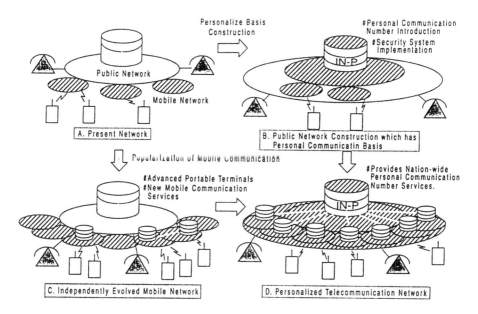

Figure 14 Network Evolution

Cryptanalysis of several conference key distribution schemes

Atsushi SHIMBO *   Shin-ichi KAWAMURA †

Toshiba Corporation, Research and Development Center
1, Komukai Toshiba-cho, Saiwai-ku,
Kawasaki, 210. Japan

Abstract

At the Eurocrypt'88 meeting, Koyama and Ohta proposed a conference key distribution scheme, which was an improved protocol of their earlier version. The authors show that their schemes, constructed for star and complete graph networks, are not secure. Another key distribution scheme, which can be used for conference key distribution, proposed at the Globecom'90 meeting by Chikazawa and Inoue, is not secure either.

1 Introduction

From the Diffie and Hellman's foresighted paper [6], several key distribution schemes have been proposed in order to solve the key sharing problem arising in the secret key cryptosystem. Among various cryptologic protocols, the key distribution protocol is one of the most important, because it can be used not only for message confidentiality but also for message authentication and peer entity authentication in a secret key cryptosystem. We discuss, in this paper, the group key distribution category in which plural users share a common secret key.

A typical group key distribution scheme has been proposed by Koyama and Ohta. The first version of their scheme [2] was proposed at the Crypto'87 meeting. They proposed three different schemes, one for a ring network, one for a complete graph network, and the other for a star network. However, soon after their proposal, the scheme for a complete graph network and the one for a star network were attacked by Yacobi [3]. In order to counter Yacobi's attack, Koyama and Ohta improved the authentication part of those schemes and proposed their improved versions [1] at the Eurocrypt'88 meeting. For these improved versions, no attacking method has been reported so far.

However, we have found that neither the improved scheme for a star network nor the one for a complete graph network is secure yet. Concrete attacking methods against them

---

*E-mail: shimbo@csl.rdc.toshiba.co.jp
†E-mail: kawamura@csl.rdc.toshiba.co.jp

will be shown in this paper. The proposed attack requires only two members' conspiracy for disclosing a sender's secret information. The idea of this attack is also applicable to the scheme proposed by Chikazawa and Inoue [4] at the Globecom'90 meeting, as long as their scheme is used for group key distribution. The attacking method has been discovered during the examination of our proposed scheme [14] 's security.

The organization of this paper is as follows: Section 2 presents a brief description of the Koyama and Ohta conference key distribution scheme. Section 3 shows concrete attacks against the Koyama and Ohta scheme. Section 4 gives a summary of Chikazawa and Inoue group key distribution scheme. An attack against their scheme is described in Section 5. In Section 6, a countermeasure for this attack is considered. Finally, Section 7 gives the conclusion.

2 Koyama and Ohta scheme

Koyama and Ohta's conference key distribution scheme for a star network is briefly reviewed in this section.

A trusted center generates the following information:

- Three large primes p, q, and r. The partial product $N = pq$, and a system's public modulus Nr.

- Integers (e, d) which satisfy the following congruence:

$$ed \equiv 1 \bmod L, \text{ where } L = lcm((p-1), (q-1), (r-1)).$$

- An integer g which is a primitive element over $GF(p)$, $GF(q)$, and $GF(r)$.

- Secret information S_i of user i, whose identification information is I_i, which is given by

$$S_i = I_i^d \bmod Nr.$$

The above information is classified into three categories: a secret system key (p, q, d), a public system key (N, r, g, e), and a secret user key S_i for user i.

A conference key is generated by the following procedures. Let one user be "a pivot user", who communicates with all the other users belonging to the group. The procedures between the pivot user, user 1, and one of the other users, user j $(2 \leq j \leq m)$, are summarized as follows.

Step 1: user j's procedure

- Choose a random number P_j and its reciprocal $\overline{P_j}$, which satisfy the following congruence:

$$P_j \overline{P_j} \equiv 1 \bmod (r-1).$$

- Compute the following (X_j, Y_j):

$$X_j = g^{eP_j} \bmod Nr,$$

$$Y_j = S_j g^{X_j P_j} \bmod Nr.$$

- Send (I_j, X_j, Y_j) to user 1.

Step 2: user 1's procedure

- Check whether the following congruence holds:

$$Y_j^e / X_j^{X_j} \equiv I_j \bmod Nr.$$

If the congruence holds, the following procedures are executed.

- Choose a secret random number R_1, which is commonly applied to all j's.

- Compute the following (A_{1j}, B_{1j}):

$$A_{1j} = X_j^{eR_1} \bmod Nr,$$

$$B_{1j} = S_1 X_j^{A_{1j} R_1} \bmod Nr.$$

- Compute the common key K by

$$K = g^{e^2 R_1} \bmod r.$$

- Send (I_1, A_{1j}, B_{1j}) to user j.

Step 3: user j's procedure

- Check whether the following congruence holds:

$$B_{1j}^e / A_{1j}^{A_{1j}} \equiv I_1 \bmod Nr.$$

If the congruence holds, user j verifies that the message is from user 1.

- Compute the common key K by:

$$
\begin{aligned}
K &= A_{1j}^{\overline{P_j}} \bmod r \\
&= g^{e^2 R_1} \bmod r.
\end{aligned}
$$

3 Attack against the Koyama and Ohta scheme

3.1 Attacking method

The attack requires a conspiracy between two users, other than the pivot user, belonging to the group. The attackers' aim is to disclose the pivot user's secret information S_1.

A concrete attacking method is as follows. Without any loss of generality, user 2 and user 3 are assumed to be the conspirators.

Step 1: user 2's procedure

- Get P_3 and A_{13} from user 3.

- Compute the following information T_2:

$$T_2 = B_{12}^{P_3 A_{13}} \bmod Nr$$
$$= S_1^{P_3 A_{13}} g^{e P_2 P_3 A_{12} A_{13} R_1} \bmod Nr.$$

Step 2: user 3's procedure

- Get P_2 and A_{12} from user 2.

- Compute the following information T_3:

$$T_3 = B_{13}^{P_2 A_{12}} \bmod Nr$$
$$= S_1^{P_2 A_{12}} g^{e P_2 P_3 A_{12} A_{13} R_1} \bmod Nr.$$

Step 3: procedure for deriving S_1

- Compute the following information T, which is the secret information S_1 raised to a known exponent:

$$T = T_2/T_3 \bmod Nr$$
$$= S_1^{P_3 A_{13} - P_2 A_{12}} \bmod Nr.$$

- Derive S_1, noting that the attackers know the following relation:

$$S_1^e = I_1 \bmod Nr.$$

If the number $(P_3 A_{13} - P_2 A_{12})$ and e are relatively prime, it is possible to apply the attacking method based on the Euclidean algorithm to derive the secret information S_1. The Euclidean algorithm states: if c is relatively prime to e, there exist integers a and b satisfying

$$ae + bc = 1.$$

Letting c denote $(P_3 A_{13} - P_2 A_{12})$, the integers (a, b) are obtained by the Euclidean algorithm. Then, S_1 is derived by

$$I_1^a T^b = S_1^{ae+bc} \bmod Nr$$
$$= S_1 \bmod Nr.$$

For the above attacking method to be effective, the attackers should so select those P_2 and P_3 that they are relatively prime. Otherwise, the probability that $c = (P_3 A_{13} - P_2 A_{12})$ has a small prime factor will be fairly high. Then, if e is selected randomly, the probability that c is not relatively prime to e will also be fairly high.

3.2 Probability of successful attack

The probability of successful attack can be roughly estimated as follows.

Let us make the following assumption on the distribution of $A_{1j} = g^{e^2 P_j R_1} \bmod Nr$ $(j = 2, 3, \cdots, m)$: if P_j $(j = 2, 3, \cdots, m)$ are randomly selected on condition that P_j's are relatively prime each other, and R_1 is randomly selected, then each value $A_{1j} \bmod p_i$ is uniformly distributed between 0 and $p_i - 1$, where p_i is any prime number smaller than Nr. Furthermore, the values $A_{1j} \bmod p_i$'s $(j = 2, 3, \cdots, m)$ are independent each other.

The above assumption is almost equivalent to saying that the A_{1j}'s are random numbers between 0 and $Nr - 1$. For simplicity, we will investigate the case where $p = 2p' + 1$, $q = 2q' + 1$, and $r = 2r' + 1$, p', q', and r' being prime numbers. Then $L = 2p'q'r'$ holds. From the assumption, the probability that $P_3 A_{13} - P_2 A_{12} = 0 \bmod p_i$ is given by $1/p_i$.

First, let us investigate a case where the public exponent e is selected randomly on condition that e and $L = 2p'q'r'$ are relatively prime. The probability that $e = 0 \bmod p_i$ is $1/p_i$ except for the case where $p_i = 2, p', q'$, or r'. From the above preparation, the success probability is approximately given by

$$\prod_{\substack{p_i \in \{prime\} \\ p_i \neq 2, p', q', r'}} 1 - \frac{1}{p_i^2}.$$

Now Euler's infinite product representation:

$$\prod_{p_i \in \{prime\}} 1 - \frac{1}{p_i^k} = \frac{1}{\zeta(k)},$$

where $\zeta()$ denotes the Riemann's ζ function, can be applied. It is known that $\zeta(2) = \pi^2/6$. Then the probability of successful attack is

$$\frac{1}{\zeta(2)} \cdot \frac{1}{1 - 2^{-2}} \cdot \frac{1}{1 - p'^{-2}} \cdot \frac{1}{1 - q'^{-2}} \cdot \frac{1}{1 - r'^{-2}} \approx \frac{1}{(1 - 2^{-2}) \cdot \zeta(2)}$$
$$\approx 0.8.$$

Next, let us consider a highly probable case; a case where the public exponent e is a small prime number. In such a case, the success probability is approximated by $1 - 1/e$. For example, if $e = 3$, the probability is $2/3$. In the Reference [1], Koyama and Ohta recommended to set e to be a prime such that $Nr/2 < e < Nr$. In this case, the probability is greater than $1 - 2/Nr$; i.e., it is very close to 1.

3.3 Extension of the attack

The success probability can be further improved by either one of the following two ways.

(1) Increasing the number of conspirators

If three members, for example user 2, user 3, and user 4, conspire, they can get the following three pieces of information:

- $S_1^e \bmod Nr$,

- $S_1^{P_3 A_{13} - P_2 A_{12}} \bmod Nr$,

- $S_1^{P_4 A_{14} - P_2 A_{12}} \bmod Nr$.

The first one is public. The second is derived through the conspiracy by user 2 and user 3, and the third through the conspiracy by user 2 and user 4. Note that the above three are independent each other. The conspirators can get no more information which is independent of the above three.

The condition for a successful attack is that the three exponents e, $(P_3 A_{13} - P_2 A_{12})$, and $(P_4 A_{14} - P_2 A_{12})$ are relatively prime. Let us assume that the assumption on A_{1j}'s distribution mentioned in Section 3.2 holds. For simplicity, we will analyze the case where $p = 2p' + 1$, $q = 2q' + 1$, and $r = 2r' + 1$; p', q', and r' being prime. When e is selected randomly on condition that e is relatively prime to $L = 2p'q'r'$, the success probability is approximately given by

$$\frac{1}{(1 - 2^{-3}) \cdot \zeta(3)} \approx 0.95.$$

In general, if k members conspire and the parameter e is selected randomly, then the success probability is approximated by

$$\frac{1}{(1 - 2^{-k}) \cdot \zeta(k)}.$$

When e is selected to be a prime, the success probability is approximately given by

$$1 - 1/e^{k-1}.$$

(2) Iterating the attack by two conspirators

If the same user, user 1, acts twice as a pivot in the group key distribution, the two conspirators, user 2 and user 3, can improve their success probability by iterating the attack. In the first session the attackers choose an integer pair (P_2, P_3), and in the second session they choose a different integer pair (P_2', P_3'). Then, the following three pieces of information are obtained:

- $S_1^e \bmod Nr$,

- $S_1^{P_3 A_{13} - P_2 A_{12}} \bmod Nr$,

- $S_1^{P_3' A_{13}' - P_2' A_{12}'} \bmod Nr$.

The first one is public. The second is derived from the first session and the third from the second session.

The attackers can derive S_1, if the exponents e, $(P_3 A_{13} - P_2 A_{12})$, and $(P_3' A_{13}' - P_2' A_{12}')$ are relatively prime. Note that the number of independent pieces of information is $(k+1)$, if the attack is iterated k times.

The success probability can be analyzed under the same assumption as that mentioned in Section 3.2. If the attack is iterated k times by the two conspirators and the parameter e is selected randomly, then the success probability is approximately given by

$$\frac{1}{(1 - 2^{-(k+1)}) \cdot \zeta(k + 1)}.$$

When e is selected as a prime, the success probability is approximately given by

$$1 - 1/e^k.$$

Koyama and Ohta modified the conference key distribution scheme described earlier into another scheme for a complete graph network by making all the members play the pivot user's role in turn. More precisely, the protocol described in Section 2 is iterated between all the combinations of two users, user i and user j $(1 \leq i, j \leq m, i \neq j)$, user i being the pivot. Therefore, their scheme for a complete graph, too, can be effectively attacked by either one of the above mentioned methods.

3.4 Small example of the attack

Assume that $p = 7, q = 11$, and $r = 89$. Then $N = 77$ and $Nr = 6853$.
Let $(e, d) = (41, 161)$, and $g = 4275$.
Assume that the users' identification information is as $I_1 = 10, I_2 = 15$, and $I_3 = 18$.
Then, the users' secret information comes out to be as $S_1 = 1286, S_2 = 5944$, and $S_3 = 3670$.

User 2 chooses his secret number $P_2 = 5$, then he gets $(X_2, Y_2) = (2980, 5394)$.
User 3 chooses his secret number $P_3 = 13$, then he gets $(X_3, Y_3) = (332, 2591)$.

Let user 1 select his secret number $R_1 = 27$. The information transmitted to user 2 comes out to be as $(A_{12}, B_{12}) = (4465, 6491)$, and that to user 3 as $(A_{13}, B_{13}) = (2295, 4412)$.

User 2 and user 3 conspire and calculate

$$B_{12}^{P_3 A_{13}} / B_{13}^{P_2 A_{12}} = 5413 \bmod 6853,$$

which is congruent with

$$S_1^{P_3 A_{13} - P_2 A_{12}} = S_1^{7510}.$$

On the other hand, the public information corresponding to S_1 is calculated as

$$S_1^{41} = 10 \bmod 6853.$$

Since the exponents 7510 and 41 are relatively prime, the attackers can derive S_1 by using the Euclidean algorithm as

$$
\begin{aligned}
S_1 &= 5413^6 \cdot 10^{-1099} \bmod 6853 \\
&= 1286.
\end{aligned}
$$

4 Chikazawa and Inoue scheme

This section reviews the Chikazawa and Inoue key distribution scheme, which is the next target for a similar attack.

A trusted center generates the following information and the public functions:

- Large primes p and q, and the product, which is a public modulus, $N = pq$.

- The Carmichael function of N, that is $L = lcm((p-1), (q.-1))$.

- Integers (e, d) which satisfy the following congruence:

$$ed \equiv 1 \bmod L.$$

- An integer g which is a primitive element over $GF(p)$ and $GF(q)$.

- A one-way function h which is used for converting the user i's identification information ID_i into an n-bit binary vector $\mathbf{I}_i$:

$$\mathbf{I}_i = h(ID_i) = (y_{i1}, y_{i2}, ..., y_{in}), \text{ where } y_{il} \in \{0, 1\}.$$

- A one-way function f which is used for generating time-variable information from the clock data. This clock data is used for detecting the replay attack.

- A secret vector $\mathbf{a}$: $\mathbf{a} = (a_1, a_2, \cdots, a_n)$, where each element $a_l \in Z_L$.

- A public vector $\mathbf{G}$:

$$\mathbf{G} = (g^{a_1} \bmod N, g^{a_2} \bmod N, \cdots, g^{a_n} \bmod N).$$

- Secret information v_i for user i generated by the following procedures:
  First, determine u_i by

$$
\begin{aligned}
u_i &= \mathbf{a} \cdot \mathbf{I}_i \bmod L \\
&= \sum_{l=1}^{n} a_l \cdot y_{il} \bmod L.
\end{aligned}
$$

Then, information v_i is determined by solving the following congruence:

$$u_i \cdot v_i \equiv 1 \bmod L.$$

- The other secret information S_i for user i is

$$S_i = ID_i^d \bmod N.$$

As a result, a secret system key $(p, q, L, d, \mathbf{a}, u_i(i = 1, 2, .., m))$, a public system key $(N, g, e, \mathbf{G})$, and a secret user key (v_i, S_i) for user i are generated.

A group key is distributed by the following procedures. In this scheme, too, an arbitrary user becomes a pivot. Letting user 1 be the pivot, we summarize the procedures between user 1 and user j.

Step 1: user 1's procedure

- Calculate $\mathbf{I}_j$ from user j's identification information ID_j by

$$\begin{aligned} \mathbf{I}_j &= h(ID_j) \\ &= (y_{j1}, \cdots, y_{jn}). \end{aligned}$$

- Generate user j's public information G_j by

$$\begin{aligned} G_j &= \prod_{l=1}^{n}(g^{a_l} \bmod N)^{y_{jl}} \bmod N \\ &= g^{u_j} \bmod N. \end{aligned}$$

- Generate a random number r, and compute the following (C_{j1}, C_{j2}) with a clock value t:

$$C_{j1} = S_1 G_j^{rf(t)} \bmod N,$$
$$C_{j2} = G_j^{re} \bmod N.$$

- Compute the common key K by

$$K = g^{re} \bmod N.$$

- Send $(ID_1, C_{j1}, C_{j2}, t)$ to user j.

Step 2: user j's procedure

- Check the clock value t sent by user 1.

- Determine whether the information (C_{j1}, C_{j2}) satisfies the following relation:

$$ID_1 \equiv C_{j1}^e / C_{j2}^{f(t)} \bmod N.$$

- Compute the common key K, using user j's secret information v_j, by

$$\begin{aligned} K &= C_{j2}^{v_j} \bmod N \\ &= g^{re} \bmod N. \end{aligned}$$

5   Attack against the Chikazawa and Inoue scheme

The Chikazawa and Inoue scheme is also breakable in a similar way. Assume that user 2 and user 3 conspire.

- User 2 computes the following information T_2:

$$\begin{aligned} T_2 &= C_{21}^{v_2} \bmod N \\ &= S_1^{v_2} g^{rf(t)} \bmod N. \end{aligned}$$

- User 3 computes T_3 by

$$\begin{aligned} T_3 &= C_{31}^{v_3} \bmod N \\ &= S_1^{v_3} g^{r f(t)} \bmod N. \end{aligned}$$

- User 2 and user 3 conspire to compute

$$T_2/T_3 = S_1^{v_2 - v_3} \bmod N.$$

- Derive S_1 using the Euclidean algorithm. Note that the attackers know the following relation.

$$S_1^e = ID_1 \bmod N.$$

Now, if the two exponents e and $(v_2 - v_3)$ are relatively prime, the secret information S_1 is revealed. Note that the exponent $(v_2 - v_3)$ is precomputable if the attackers exchange their secret information v_2 and v_3. Therefore, an attacker is able to investigate whom to choose as his partner.

When an attacker selects his partner randomly, the success probability can be analyzed similarly to the Koyama-Ohta scheme. Assume that $p = 2p' + 1$ and $q = 2q' + 1$, where p' and q' are prime numbers. If e, v_2 and v_3 are selected randomly, the probability of successful attack is approximated by $8/\pi^2 \approx 0.8$. If e is set to be a prime, the probability is $1 - 1/e$.

When k members, who are randomly selected, conspire and the parameter e is randomly selected, the success probability is approximately

$$\frac{1}{(1 - 2^{-k}) \cdot \zeta(k)}.$$

When e is selected as a prime, the success probability is approximately

$$1 - 1/e^{k-1}.$$

6 Considerations and Countermeasures

Let us consider what causes the disclosure of pivot user's secret information in the above group key distribution schemes. The schemes have the following common features:

- One user, being a pivot, distributes a common key to the other members by a star type communication.

- The authentication scheme based on the extended Fiat-Shamir protocol [9] [10] [11] is used. The pivot user plays a prover, while other users play verifiers. The authentication scheme is iterated between the pivot user and each member.

- The randomized information in the extended Fiat-Shamir scheme is utilized to send a common key secretly. Concretely, the randomized information corresponds to the A_{1j} in the Koyama-Ohta scheme, and to the C_{j2} in the Chikazawa-Inoue scheme. A common key is encrypted into a ciphertext that can be deciphered only by a legitimate verifier, and the ciphertext is used as the randomized information. (This idea was later formalized into a methodology of utilizing the randomness of the zero-knowledge proof by Okamoto and Ohta [5].)

Here let us remind ourselves of the feature of the Fiat-Shamir type zero-knowledge proof. In such a scheme, the prover's random number, which determines the randomized information, conceals the secret information of the prover. The random number must not be a fixed value; otherwise the secret may be disclosed. This property is utilized for proving "the soundness" [8] and for ensuring the disposability of electronic cash [13].

In the target schemes, each of the prover's random numbers determines the encrypted result of a common key for each verifier. Since the common key is fixed during one group key distribution session, the prover's "random numbers" are not independent even though they seem to be so at first sight. The seemingly random numbers can be transformed into a fixed value by the conspiracy of plural members in the group who have secret transformation keys. Then, they can derive new knowledge on the pivot user's secret by eliminating the fixed value term from plural pieces of authentication information. Finally, the new knowledge results in the derivation of the secret. Note that the prover's "random numbers" seem to be random and independent each other for a member who doesn't belong to the group (this is assumed in analyzing the probability of successful attack in Section 3.2), but they are not independent for the members in the group.

An easy way to counter an attack of this kind is to separate the ciphertext of the common key from the authentication information in the group key distribution scheme. Concretely, a pivot user encrypts a common key into the ciphered key data and generates the extended Fiat-Shamir signature on the ciphered key data using a random number for each proof. This countermeasure, however, requires that the pivot user send enciphered information and authentication information separately. This brings about some degradation in the transmission efficiency.

Since the proposed attacks are based on the conspiracy of plural verifiers, the target schemes seem to be secure as long as they are used for the two-party key distribution.

7   Conclusion

This paper have investigated attacks against the Koyama-Ohta conference key distribution scheme for either a star or a complete graph network, and those against the Chikazawa-Inoue group key distribution scheme. As a result, it has been found that these schemes are not secure for group key distribution, and they are likely to be usable only for two-party key distribution. One countermeasure is to separate the encryption part from the sender authentication part. This countermeasure, however, slightly degrades the transmission efficiency compared with the two-party case. It is an open problem to find a group key distribution scheme with higher transmission efficiency.

References

[1] K.Koyama and K.Ohta, "Security of Improved Identity based Conference Key Distribution Systems", Lecture Notes in Computer Science (LNCS), Advances in Cryptology - EUROCRYPT'88, Springer-Verlag, pp.11-19 (1988).

[2] K.Koyama and K.Ohta, "Identity-based conference key distribution systems", LNCS, Advances in Cryptology - CRYPTO'87, Springer-Verlag, pp.175-184 (1987).

[3] Y.Yacobi, "Attack on the Koyama-Ohta identity based key distribution scheme", LNCS, Advances in Cryptology - CRYPTO'87, Springer-Verlag, pp.429-433 (1987).

[4] T.Chikazawa and T.Inoue, "A new key sharing system for global telecommunications", Proceedings of GLOBECOM'90, pp.1069-1072 (1990).

[5] T.Okamoto and K.Ohta, "How to utilize the randomness of zero-knowledge proofs", Proc. of CRYPTO'90 (1990).

[6] W.Diffie and M.Hellman, "New directions in cryptography", IEEE Trans. of Inf. Theory, IT-22, 6, pp.644-654 (1976).

[7] A.Fiat and A.Shamir, "How to prove yourself: Practical solutions to identification and signature problems", LNCS, Advances in Cryptology - CRYPTO'86, Springer-Verlag, pp.186-194 (1986).

[8] U.Feige, A.Fiat and A.Shamir, "Zero knowledge proofs if identity", STOC, pp.210-217 (1987).

[9] K.Ohta and T.Okamoto, "A modification of the Fiat-Shamir scheme", LNCS, Advances in Cryptology - CRYPTO'88, Springer-Verlag, pp.232-243 (1988).

[10] L.C.Guillou and J.J.Quisquater, "A practical zero-knowledge protocol fitted to security microprocessor minimizing both transmission and memory", LNCS, Advances in Cryptology - EUROCRYPT'88, Springer-Verlag, pp.123-128 (1988).

[11] K.Ohta, "Efficient identification and signature scheme", Electronics Letters, 24, 2, pp.115-116 (1988).

[12] G.J.Simmons, "A 'weak' privacy protocol using the RSA crypto algorithm", Cryptologia, 7, 2, pp.180-182 (1983).

[13] T.Okamoto and K.Ohta, "Disposable zero-knowledge authentications and their applications to untraceable electronic cash", LNCS, Advances in Cryptology - CRYPTO'89, Springer-Verlag, pp.481-496 (1989).

[14] S.Kawamura and A.Shimbo, "A one-way key distribution scheme based on a Fiat-Shamir digital signature", (written in Japanese) IEICE, 1991 Spring National Convention Record, A-292 (1991).

Revealing Information with Partial Period Correlations (extended abstract)

Andrew Klapper   Mark Goresky

Northeastern University, College of Computer Science, Boston, MA 02115, U.S.A.

1   Introduction

In several applications in modern communication systems periodic binary sequences are employed that must be difficult for an adversary to determine when a short subsequence is known, and must be easy to generate given a secret key. This is true both in stream cypher systems, in which the binary sequence is used as a pseudo-one-time-pad [11], and in secure spread spectrum systems, in which the sequence is used to spread a signal over a large range of frequencies [10]. While theoreticians have long argued that such security can only be achieved by sequences satisfying very general statistical test such as Yao's and Blum and Micali's next bit test [12,2], practitioners are often satisfied to find sequences that have large linear complexities, thus ensuring resistence to the Berlekamp-Massey algorithm [8]. Linear feedback shift registers are devices that can easily generate sequences with exponentially larger period than the size of their seeds [6], though with small linear complexity. Thus much effort has gone into finding ways of modifying linear feedback shift registers so that the sequences they generate have large linear complexities, typically by adding some nonlinearity.

Chan and Games [4] have suggested using a class of sequences called geometric sequences for these purposes. Geometric sequences are derived from m-sequences over a finite field $GF(q)$ by applying a (nonlinear) map from $GF(q)$ to $GF(2)$. Chan and Games showed that for q odd, geometric sequences have high linear complexities. For this reason, these sequences have been employed in commercial applications, typically with enormous periods. More recently Chan, Goresky, and Klapper [5] derived formulas for the periodic autocorrelation function of a geometric sequence and, in some cases, (with q even) for the periodic cross-correlation function of a pair of geometric sequences with the same period. Knowledge of these correlation function values is essential for applications involving spread spectrum systems. Furthermore, Brynielsson [3] derived a formula for the linear complexity of a geometric sequence when q is even, showing that such sequences can be constructed with moderately large linear complexities.

The purpose of this paper is to show that a certain statistical attack, partial period autocorrelation attack, can be used to obtain critical information about geometric sequences from knowledge of a relatively small subsequence when q is odd. Specifically, for a sequence of period $q^n - 1$, with $q \geq 5$, q odd, and $n \geq 10$, if q^8 bits of the sequence are known, then q can be determined with high probability (n must be at least 11 if $q = 3$). While this does not as yet allow us to determine the sequence, it does render their security questionable.

Suppose a small subsequence is known. We show that if a small shift of this subsequence is taken, then the correlation between the original subsequence and its shift is close to q^2 when q is odd. More specifically, we show that the expected correlation (letting the starting point of the shift vary and keeping the size of the subsequence fixed) is approximately q^2, and that under the hypotheses on q and n given above, the variance is sufficiently small that the correlation is close to its expectation with high probability (this is a consequence of Chebyshev's inequality [1]). One way to view our resluts is that we have introduced a new statistical test that a sequence must satisfy in order to be secure – the variance of the partial period autocorrelation must be high for small subsequences.

Most of the difficulty in acquiring these results lies in calculating the variance. A critical part of this calculation involves understanding how uniformly each orbit of the action of $GL_2(GF(q))$ on $GF(q^n)$ by fractional linear transformations is distributed in $GF(q^n)$. In some cases we make estimates that we expect can be improved, thus decreasing the number of bits required for success. It is unlikely, however that we can get by with fewer than q^6 bits.

A second application of our results on the partial period autocorrelation of geometric sequences is to spread spectrum CDMA systems [10]. In such systems a signal is distributed over a large number of channels using a pseudo-random binary sequence to determine parity. The signal is then recovered by computing an (unshifted) autocorrelation. To avoid phase shift interference it is desirable that the shifted partial period autocorrelation values be small. Partial period correlations are generally quite difficult to compute, so in practice families of sequences are found that have low full period autocorrelations. These families are then searched for sequences with low partial period autocorrelation. Such searches are slow and may be unsuccessful. The expected partial period autocorrelation always differs from the full period autocorrelation by a factor depending only on the size of the subsequence used, so is low if the full period autocorrelation is low. This is only useful, however, if the variance is low enough to ensure the partial period autocorrelation is low with high probability. Our results show that this is the case for geometric sequences, making them strong candidates for use in spread spectrum systems.

The proofs of all lemmas are omitted and will later appear in a full version of the paper.

2   Geometric Sequences and Correlations

In this section we recall the definition of the geometric sequences and some of their basic properties, and the definition of full and partial period correlation functions of periodic sequences. Throughout this paper, q will denote a fixed power of a fixed prime p, and $GF(q)$ will denote the Galois field of q elements. See Lidl and Niederreiter's or McEliece's book [7,9] for background on finite fields.

Definition 1 (Chan and Games [4]) *Let n be a positive integer and let α be a primitive element of $GF(q^n)$. The sequence $U_i = Tr_q^{q^n}(\alpha^i)$ is a q-ary m-sequence. Let f be a (possibly nonlinear) function from $GF(q)$ to $GF(2)$. The binary sequence S whose ith term is*

$$S_i = f(Tr_q^{q^n}(\alpha^i)).$$

is called a geometric sequence.

Geometric sequences with q odd have been used in commercial applications where easily generated sequences with large linear complexities are needed. Recall that the m-sequence U can be generated by a linear feedback shift register over $GF(q)$ of length n, so the geometric sequence S is easy to generate if the feedforward function f is easy to compute. Such a geometric sequence is a $(q^n - 1)$-periodic binary sequence. The periodic autocorrelation function $\mathcal{A}_S(\tau)$ of S is the function whose value at τ is the correlation of the τ-shift of S with itself.

$$\mathcal{A}_S(\tau) = \sum_{i=1}^{q^n-1} (-1)^{S_{i+\tau}}(-1)^{S_i}$$

The partial autocorrelation of a sequence is defined by limiting the range of values in the sum to a fixed window. It is parametrized by the start position k and length D of the window, as well as the shift τ. Precisely

$$\mathcal{A}_{\mathbf{S}}(\tau, k, D) = \sum_{i=k}^{D+k-1} (-1)^{S_{i+\tau}}(-1)^{S_i} \quad .$$

We next recall a result due to Chan, Goresky, and Klapper [5] regarding the autocorrelation of a geometric sequence. We use the notation $F(x) = (-1)^{f(x)}$, $I(f) = \sum_{x \in GF(q)} F(x)$, the imbalance[1] of f, and $\Delta_a(f) = \sum_{x \in GF(q)} F(ax)F(x)$, the short autocorrelation function[2] of f.

Theorem 1 *Then the values for the periodic autocorrelation of a geometric sequence* $\mathbf{S}$ *are:*

1. $\mathcal{A}_{\mathbf{S}}(\tau) = q^{n-2}I(f)^2 - 1$, *if* $\alpha^\tau \notin GF(q)$.

2. $\mathcal{A}_{\mathbf{S}}(\tau) = q^{n-1}\Delta_{\alpha^\tau}(f) - 1$, *if* $\alpha^\tau \in GF(q)$.

Note that, letting $\nu = (q^n - 1)/(q - 1)$, the second case of the theorem, $\alpha^\tau \in GF(q)$, occurs exactly when ν divides τ.

If p equals two, f can be chosen to be balanced ($I(f) = 0$) and so that $\Delta_a(f) = 0$ for $a \neq 1$, $\Delta_1(f) = q$. This implies the shifted autocorrelation of $\mathbf{S}$ is -1, the minimum possible. Unfortunately, in this case the linear complexity is much smaller. If p is odd, then the imbalance is at least 1, so the autocorrelation is always large. Such sequences are used in commercial systems due to their enormous linear complexities. We will show, however, that their poor autocorrelations render such systems vulnerable. The idea is that if we know the autocorrelation, then we know q^n. Of course we will never be able to compute the full autocorrelation, since we will never see the full period of the sequence. However, if the partial period correlation is sufficiently well behaved for small enough windows, then we can get similar information by seeing only a small part of the sequence – seeing $D + \tau$ bits of the sequence allows us to compute a partial period autocorrelation with window D and shift τ. Such a short subsequence may be discovered, for example, by a known plaintext attack on a pseudo-one-time-pad system. Unfortunately, the partial period autocorrelation may vary considerably as the start position varies. We will show that the expected partial period autocorrelation (averaged over the starting position of the window) is closely related to the full period autocorrelation. We will also show that for certain window sizes the variance of the partial period autocorrelation (with fixed shift τ and window size D) is low enough that an adversary has high probability of discovering q. This is a consequence of Chebyshev's inequality [1] which says that a bound on the variance implies a bound on the probability that a particular partial period correlation is far from the expected partial period correlation.

We begin by showing that the expected partial period autocorrelation of any sequence can be determined from its full period autocorrelation. We denote the expectation of a random variable X by $\langle X \rangle$. All expectations are taken for fixed window size D and shift τ, assuming a uniform distribution on all start positions k.

---

[1] The imbalance of f is equal to the number of x for which f is zero minus the number of x for which f is one.

[2] If γ is a primitive element of $GF(q)$, and $a = \gamma^\sigma$, then $\Delta_a(f) - 1$ is the autocorrelation with shift σ of the sequence whose ith term is $f(\gamma^i)$.

Theorem 2 *Let* S *be a periodic binary sequences with period* N. *Then the expectation of the partial period autocorrelation of* S *is given by*

$$\langle \mathcal{A}_S(\tau, k, D) \rangle = \frac{D}{N} \mathcal{A}_S(\tau).$$

Proof: Straightforward.

Suppose S is a geometric sequence of period $q^n - 1$ with feedforward function $f : GF(q) \to GF(2)$, and that S is as balanced as possible, i.e., $I(f) = 1$. Then for shifts $0 < \tau < \nu$, the expected partial period autocorrelation of S is $\langle \mathcal{A}_S(\tau, k, D) \rangle = D(q^{n-2} - 1)/(q^n - 1)$.

We next consider the variance of the partial period autocorrelation. Recall that the variance of a random variable X is defined to be $\langle (X - \langle X \rangle)^2 \rangle = \langle X^2 \rangle - \langle X \rangle^2$, so we must determine the second moment $\langle \mathcal{A}_S(\tau, k, D)^2 \rangle$ of the partial period autocorrelation. We can reduce this determination to the determination of the cardinalities of certain fourfold intersections of hyperplanes (identifying $GF(q^n)$ with n-dimensional affine space over $GF(q)$) asstated in the following lemma. If $s \in GF(q)$, and $A \in GF(q^n)$, then we denote by H_A^s the hyperplane $\{x : Tr_q^{q^n}(Ax) = s\}$.

Lemma 1 *If* S *is a geometric sequence, then*

$$\langle \mathcal{A}_S(\tau, k, D)^2 \rangle = \frac{1}{q^n - 1} \sum_{i,j=0}^{D-1} \left(\sum_{s,t,u,v \in GF(q)} N_{i,j,\tau}(s,t,u,v) F(s) F(t) F(u) F(v) - 1 \right)$$

where

$$N_{i,j,\tau}(s,t,u,v) = |H_{\alpha^i+\tau}^s \cap H_{\alpha^i}^t \cap H_{\alpha^j+\tau}^u \cap H_{\alpha^j}^v|.$$

Thus we must determine the values of $N_{i,j,\tau}(s,t,u,v)$.

3 Intersections of Hyperplanes

There are five possible values for $N_{i,j,\tau}(s,t,u,v)$: q^{n-4}, q^{n-3}, q^{n-2}, q^{n-1}, and 0. The following lemma will be useful in determining which case occurs. If $\{A_m\}$ are elements of a vector space over $GF(q)$, by abuse of notation we denote by $\dim\{A_m\}$ the dimension over $GF(q)$ of the span of $\{A_m\}$.

Lemma 2 *Let* $A_1, A_2, A_3, A_4 \in GF(q^n)$ *and* $s_1, s_2, s_3, s_4 \in GF(q)$. *Then* $|\cap_{m=1}^4 H_{A_m}^{s_m}| = q^{n-r}$ *if and only if* $\dim\{A_m\} = r$, *and whenever* $\{A_m\}$ *satisfy a linear equation* $\sum_{m=1}^4 a_m A_m = 0$, $\{a_m\} \in GF(q)$, $\{s_m\}$ *satisfy the same relation, i.e.,* $\sum_{m=1}^4 a_m s_m = 0$. *Otherwise the intersection is empty. For a given set* $\{A_m\}$ *whose span has dimension* r, *there are* q^r *sets* $\{s_m\}$ *for which* $\cap_{m=1}^4 H_{A_m}^{s_m} \neq \emptyset$.

Let $A = \alpha^i$, $B = \alpha^j$, and $C = \alpha^\tau$ in the sum we derived for the second moment of the partial period correlation. We need to determine $\dim\{A, AC, B, BC\}$, which can be 1, 2, 3, or 4. We consider 4 to be the generic case and consider when the other four cases occur.

3.1 Dimension 1:

This occurs when every element is a $GF(q)$ multiple of every other element. Thus $\alpha^\tau, \alpha^{j-i} \in GF(q)$. If the window size D is in the range $0 < D < \nu$, then this case cannot occur.

3.2 Dimension 2:

This occurs if AC, A, BC, B satisfy two linearly independent equations, say

$$
\begin{aligned}
aAC + bA + cBC + dB &= 0 \qquad\qquad (1) \\
eAC + fA + gBC + hB &= 0,
\end{aligned}
$$

where $a, b, c, d, e, f, g, h \in GF(q)$ and (a, b, c, d) and (e, f, g, h) are independent vectors.

If $C \in GF(q)$, then the span of $\{AC, A, BC, B\}$ equals the span of $\{A, B\}$. $\mathrm{Dim}\{A, AC, B, BC\}$ is two if A/B is not in $GF(q)$, and is one otherwise.

If C is not in $GF(q)$, then we can use each of these equations to write B/A as the result of applying to C a fractional linear transformation with coefficients in $GF(q)$:

$$
\frac{B}{A} = -\frac{aC + b}{cC + d} = -\frac{eC + f}{gC + h}.
$$

We can use the second equation to find a quadratic equation over $GF(q)$ satisfied by C. This equation is degenerate if and only if $B/A \in GF(q)$. Thus if B/A is not in $GF(q)$, then C is in $GF(q^2) - GF(q)$. Conversely, suppose C is in $GF(q^2) - GF(q)$. If $\dim\{A, AC, B, BC\}$ is less than four, then $\{AC, A, BC, B\}$ satisfy a linear equation. The quadratic equation satisfied by C can then be used to produce a second, independent linear equation. Hence $\dim\{A, AC, B, BC\}$ is two (it must be at least two since $C \notin GF(q)$).

3.3 Dimension Three:

As a consequence of the preceding subsection, $\dim\{A, AC, B, BC\}$ can only be three if C is not a root of a quadratic equation over $GF(q)$. Moreover, we must have a single equation

$$
aAC + bA + cBC + dB = 0,
$$

or, equivalently,

$$
B = \frac{aC + b}{cC + d} A.
$$

As before, $(aC + b)/(cC + d)$ is in $GF(q)$ (and hence $\dim\{A, AC, B, BC\}$ is two) if and only if $ad - bc = 0$.

There is an action of the general linear group over $GF(q)$ of rank two, $G = GL_2(GF(q))$, on $GF(q^n)$ which we shall make use of. Recall that this group is the multiplicative group of two by two matrices over $GF(q)$ with nonzero determinate. The group acts by fractional linear transformations. That is, the matrix

$$
M = \begin{pmatrix} a & b \\ c & d \end{pmatrix}
$$

acts on the element $C \in GF(q)$ by

$$C \mapsto \frac{aC + b}{cC + d} = M(C).$$

It is straightforward to check that if $M, N \in GL_2(GF(q))$, then $(MN)(C) = M(N(C))$. Recall that when a group G acts on a set W, the _G-orbit_ of an element $x \in W$ is the set $\{M(x) : M \in G\} = $ orbit(x). Our analysis of the conditions under which the various dimensions occur is summarized in the following table:

	Dim = 1	Dim = 2	Dim = 3
$C \in GF(q)$	$B/A \in GF(q)$	$B/A \notin GF(q)$	–
$C \in GF(q^2) - GF(q)$	–	$B/A \in GF(q) \cup \text{orbit}(C)$	–
$C \in GF(q^n) - GF(q^2)$	–	$B/A \in GF(q)$	$B/A \in \text{orbit}(C)$

In all other cases $\dim\{A, AC, B, BC\}$ is four.

4 Variance of Partial Period Correlations

Returning to our computation of the variance of the partial period correlation of geometric sequences, we need to know, for given $0 \leq i, j < D$, whether $\alpha^{i-j} \in \text{orbit}(\alpha^{\tau})$. We break down our analysis depending upon whether α^{τ} is in $GF(q)$, $GF(q^2) - GF(q)$, or $GF(q^n) - GF(q^2)$.

4.1 $\alpha^{\tau} \in GF(q)$

In this case $N_{i,j,\tau}(0,0,0,0) = q^{n-2}$ if $\alpha^{i-j} \notin GF(q)$, $N_{i,j,\tau}(0,0,0,0) = q^{n-1}$ if $\alpha^{i-j} \in GF(q)$. We have $\alpha^{i-j} \in GF(q)$ if and only if ν divides $i - j$. Thus for a given i, $0 \leq i < D$, the number of j, $0 \leq j < D$, such that the i, j term contributes to the sum is the number of j in this range such that ν divides $i - j$. This number is

$$\left\lfloor \frac{D - i - 1}{\nu} \right\rfloor + \left\lfloor \frac{i}{\nu} \right\rfloor + 1 \leq \frac{D-1}{\nu} + 1.$$

We can bound the second moment as follows

$$
\begin{aligned}
\langle A_{\mathsf{S}}(\tau, k, D)^2 \rangle &= \frac{1}{q^n - 1} \sum_{i,j=0}^{D-1} \Big(\sum_{t, v \in GF(q)} N_{i,j,\tau}(\alpha^{\tau} t, t, \alpha^{\tau} v, v) F(\alpha^{\tau} t) F(t) F(\alpha^{\tau} v) F(v) - 1 \Big) \\
&= \frac{1}{q^n - 1} \Big(\sum_{i,j=0}^{D-1} \Big(\sum_{t, v \in GF(q)} q^{n-2} F(\alpha^{\tau} t) F(t) F(\alpha^{\tau} v) F(v) - 1 \Big) \\
&\quad + \sum_{\substack{0 \leq i,j < D \\ \nu|(i-j)}} (q^n - \sum_{t, v \in GF(q)} q^{n-2} F(\alpha^{\tau} t) F(t) F(\alpha^{\tau} v) F(v)) \Big) \\
&\leq \frac{D^2}{q^n - 1} (q^{n-2} \Delta_{\alpha^{\tau}}(f)^2 - 1) + \frac{D}{q^n - 1} \Big(\frac{D-1}{\nu} + 1 \Big) (q^n - q^{n-2} \Delta_{\alpha^{\tau}}(f)^2).
\end{aligned}
$$

The expectation in this case is $\frac{D}{q^n-1}(q^{n-1}\Delta_{\alpha^\tau}(f)-1)$. Therefore the variance is

$$
\begin{aligned}
V(\mathcal{A}_S(\tau,k,D)) &= \langle \mathcal{A}_S(\tau,k,D)^2\rangle - \langle \mathcal{A}_S(\tau,k,D)\rangle^2 \\
&\le \frac{D^2}{q^n-1}(q^{n-2}\Delta_{\alpha^\tau}(f)^2-1) + \frac{D}{q^n-1}(\frac{D-1}{\nu}+1)(q^n-q^{n-2}\Delta_{\alpha^\tau}(f)^2) \\
&\quad - \frac{D^2}{(q^n-1)^2}(q^{n-1}\Delta_{\alpha^\tau}(f)-1)^2 \\
&= \frac{q^{n-2}D}{q^n-1}((\frac{D-1}{\nu}+1)(q^2-\Delta_{\alpha^\tau}(f)^2) - \frac{D}{q^n-1}(q-\Delta_{\alpha^\tau}(f))^2) \\
&\le \frac{q^n D}{q^n-1}(\frac{D-1}{\nu}+1).
\end{aligned}
$$

In particular, if $D \le \nu$, then the variance is bounded above by $2q^n D/(q^n-1)$.

4.2 $\alpha^\tau \in GF(q^2) - GF(q)$

If $x \in GF(q^2)$, and $M \in G$, then $M(x) \in GF(q^2)$. If, moreover, $x \notin GF(q)$, then x is a generator for $GF(q^2)$ over $GF(q)$, that is, every element of $GF(q^2)$ can be written in the form $(ax+b)/(cx+d)$ for some $a,b,c,d \in GF(q)$. It follows that $N_{i,j,\tau}(0,0,0,0)$ is q^{n-2} if $\alpha^{i-j} \in GF(q^2)$, i.e., if $\nu_2 = (q^n-1)/(q^2-1)$ divides $i-j$, and is q^{n-4} otherwise. As above, we can bound the second moment:

$$
\langle \mathcal{A}_S(\tau,k,D)^2\rangle \le \frac{D^2}{q^n-1}(q^{n-4}I(f)^4-1) + \frac{D}{q^n-1}(\frac{D-1}{\nu_2}+1)(q^n-q^{n-4}I(f)^4).
$$

The expectation in this case is $\frac{D}{q^n-1}(q^{n-2}I(f)^2-1)$. Therefore the variance is bounded by:

$$
V(\mathcal{A}_S(\tau,k,D)) \le \frac{q^n D}{q^n-1}(\frac{D-1}{\nu_2}+1).
$$

In particular, if $D \le \nu$, then the variance is bounded above by $(q+1)q^n D/(q^n-1)$.

4.3 $\alpha^\tau \in GF(q^n) - GF(q^2)$

Unfortunately, in this case the situation more complex. In general, the G-orbit is not uniformly distributed in $GF(q^n)$, so for a fixed i, the number of j in a window with $\dim\{\alpha^{i+\tau},\alpha^i,\alpha^{j+\tau},\alpha^j\}=3$ is not proportional to the size of the window. We settle here for a cruder estimate, based on the structure of the group G. We first determine the size of the G-orbit of α^τ.

Lemma 3 *If $x \in GF(q^n) - GF(q)$, then the G-orbit of x has cardinality $q^3 - q$.*

We will next decompose the elements of G into the composition of certain simple types of matrices with scalar multiplication. Since scalar multiplication by elements of $GF(q)$ moves elements large distances, this will allow us to bound the number of elements in an orbit that are in a given small window.

For matrices M and N, we write $M \sim N$ if M and N define the same transformation (i.e., the matrices differ by a scalar multiple). Let

$$M = \begin{pmatrix} a & b \\ c & d \end{pmatrix}$$

be an element of G, so $\delta = ad - bc \neq 0$. First suppose $a \neq 0$. Then

$$M \sim \begin{pmatrix} a^2/\delta & ab/\delta \\ ac/\delta & ad/\delta \end{pmatrix} = \begin{pmatrix} a^2/\delta & 0 \\ 0 & 1 \end{pmatrix} \begin{pmatrix} 1 & b/a \\ ac/\delta & bc/\delta + 1 \end{pmatrix}.$$

Letting $S_x = \{(x + b)/(cx + bc + 1)\}$, we have shown that $M(x)$ is a scalar multiple of an element of S_x.

On the other hand, suppose $a = 0$. Then $b \neq 0$ and $c \neq 0$, so

$$M \sim \begin{pmatrix} 0 & b/c \\ 1 & d/c \end{pmatrix} = \begin{pmatrix} b/c & 0 \\ 0 & 1 \end{pmatrix} \begin{pmatrix} 0 & 1 \\ 1 & d/c \end{pmatrix}.$$

Let $T_x = \{1/(x + d)\}$. Then in this case $M(x)$ is a scalar multiple of some element of T_x. We have shown that an arbitrary element of the orbit of x is a scalar multiple of some element of $S_x \cup T_x$.

Consider a window of size ν. If y is any element of $GF(q^n)$, then there is a unique $a \in GF(q)$ such that ay is in the given window. Therefore, for each element y of $S_x \cup T_x$, there is a unique scalar multiple of y, i.e., a unique element of the orbit of x, in the given window. Thus we have

Proposition 1 *If $x \in GF(q^n) - GF(q^2)$, then the intersection of the orbit of x with a window of size at most ν has cardinality at most $|S_x \cup T_x| = q^2 + q$.*

It follows that if $D \leq \nu$, then the number of i, j such that $0 \leq i, j < D$ and $N_{i,j,\tau}(s, t, u, v) = q^{n-3}$ or 0 is at most $D(q^2 + q)$ (or D^2 if $D < (q^2 + q)$). The number of s, t, u, v for which we get q^{n-3} is q^3. Moreover, $N_{i,i,\tau}(s, s, u, u) = q^{n-2}$, and $N_{i,i,\tau}(s, t, u, v) = 0$ if $s \neq t$ or $u \neq v$. In all other cases $N_{i,j,\tau}(s, t, u, v) = q^{n-4}$. Thus if $D \leq \nu$, as above we can bound the second moment:

$$\langle \mathcal{A}_S(\tau, k, D)^2 \rangle \leq \frac{D^2}{q^n - 1}(q^{n-4}I(f)^4 - 1) + \frac{D(q^2 + q + 1)}{q^n - 1}(q^n - q^{n-4}I(f)^4).$$

The expectation in this case is $\frac{D}{q^n-1}(q^{n-2}I(f)^2 - 1)$. Therefore the variance is bounded by:

$$V(\mathcal{A}_S(\tau, k, D)) \leq \frac{q^n(q^2 + q + 1)D}{q^n - 1}.$$

Theorem 3 *For any τ, if $D \leq \nu$, then the variance of the partial period autocorrelation of a geometric sequence with shift τ and window D is bounded above by $(q^2 + q + 1)q^n D/(q^n - 1)$.*

5 Cracking Cryptosystems

In this section we describe how the results of the previous sections can be used to obtain critical information on pseudorandom sequences used in stream cyphers. Specifically, using a known plaintext attack, we are able determine q with high probability. We will make use of Chebyshev's inequality [1].

Proposition 2 (Chebyshev's Inequality) *If X is a random variable with expectation ϵ and variance σ^2, then for any k,*

$$Prob\{|X - \epsilon| > k\} < \sigma^2/k^2.$$

First suppose q is odd. Suppose f is a balanced feedforward function (that is, $I(f) = 1$). Geometric sequences are generally taken with $I(f) = 1$ so they are statistically random. For small enough windows ($D \leq \nu$) the expected partial period autocorrelation is $D(q^{n-2} - 1)/(q^n - 1)$, or approximately D/q^2. Thus we can hope to determine q. As we determine bits of the sequence, we can take a small shift and compute a partial period autocorrelation. We must have the value of the partial period autocorrelation close enough to its expectation to unambiguously recover q.

A difficulty is that we must recover q without knowing n. For each q there will be an interval $I_q = (a_q, a_{q-1})$ such that if the computed value of the partial period correlation is in I_q, then we assume that q is being used by the generator of the sequence. In order to have a high probability of success, we must chose I_q so that there is a large k such that for each n, the interval of radius k around the expected partial period autocorrelation lies entirely within I_q. To simplify things, we assume $n \geq 4$. As it turns out, the statistics prevent us from successfully determining q if $n \leq 3$, so this restriction is of no importance.

For a given q, the sequence $d_{q,n} = D(q^{n-2} - 1)/(q^n - 1)$ is an increasing sequence with limit D/q^2. Let c_q be the center of the smallest interval containing all these points ($n \geq 4$). That is

$$c_q = \frac{1}{2}\left(d_{q,n} + \frac{D}{q^2}\right).$$

We then let a_q be the point midway between c_{q+1} and c_q, that is $a_q = (c_{q+1} + c_q)/2$, and $I_q = (a_q, a_{q-1})$ (note that $c_{q+1} < c_q$).

Lemma 4 *Let*

$$k = \frac{D(q-1)}{q^2(q+1)^2}.$$

If $n \geq 4$, then $d_{q,n} \in (a_q + k, a_{q-1} - k)$. Hence $(d_{q,n} - k, d_{q,n} + k) \subseteq I_q$.

The algrithm for determining q is then: compute the partial period correlation. If the result is in I_r, then assume that $q = r$. Applying Chebyshev's inequality, we have

Theorem 4 *If $n \geq 3$, and S is a geometric sequence of period $q^n - 1$, then a partial autocorrelation attack with a window $D \leq \nu = (q^n - 1)/(q - 1)$ will fail to determine q with probability at most*

$$\frac{q^{n+4}(q^2 + q + 1)(q + 1)^4}{D(q^n - 1)(q - 1)^2}.$$

If n is small, this probability will be larger than one, so Chebyshev's inequality does not tell us whether the atack has a positive probability of determining q. However, we have

Corollary 1 *If n is large enough that*

$$\frac{q^{n+4}(q^2 + q + 1)(q + 1)^4}{(q^n - 1)(q - 1)^2} < \frac{q^n - 1}{q - 1}$$

then using a window of size

$$\frac{q^{n+4}(q^2 + q + 1)(q + 1)^4}{(q^n - 1)(q - 1)^4} \sim q^8$$

gives a positive probability of determining q.

The conditions of the corollary are satisfied if $n \geq 10$, and $q \geq 5$; or $n \geq 11$, and $q \geq 3$; or $n \geq 14$, and $q \geq 2$.

6 Conclusions

We have shown that geometric sequences based on m-sequences over a finite field $GF(q)$ of odd characteristic exhibit vulnerability to a partial period correlation attack when enough bits are known. Specifically, for sequences of period $q^n - 1$, if between about q^8 and $(q^n - 1)/(q - 1)$ bits are known, then q can be determined with high probability. If $n \leq 9$, then this condition is vacuous. However for q small the sequence would have small period. We conclude that q must be large. Even for large q, our results cut down on the number of usable bits of a geometric sequence. For example, if we chose $q = 17$ and $n = 16$, so that the sequence has period greater than 10^{18}, then we must see approximately 7 times 10^9 bits. If we received bits at 9600 baud, then it takes about 8 days to have positive probability of determining q.

There are two points where we have used estimates that might be improved. First, in evaluating the summation expression for the second moment, we make a worst case estimate of the smaller sum that each term contributes a plus one. It is possible that this can be improved by recognizing this sum as a higher order autocorrelation of a shorter sequence (of period $q - 1$).

Second, based on computer experiments, we belive that smaller estimates can be made of the number of points in the intersection of an orbit of the action of $GL_2(GF(q))$ with a window of size $D < (q^n - 1)/(q - 1)$. In particular, we conjecture that there is a small constant c (e.g. two or three) such that in a window of size less than q^{n-3}, there are at most cq elements of an orbit. Such a result would imply that only cq^6 bits are required to have positive probability of determining q.

Of course our results only allow the determination of q, not the entire sequence, or even q^n. It remains to be seen whether this information is enough to compromise sytems using geometric sequences, but it should make users wary.

Finally, the moral of this paper is that it is dangerous to rely on linear complexity as a measure of cryptographic security. There are many other statistical tests a sequence must pass – in this case, we have shown that the variance of the partial period autocorrelation must be high.

References

[1] H. BAUER, *Probability Theory and Elements of Measure Theory*, Holt, Rinehart and Winston, New York, 1972.

[2] M. BLUM AND S. MICALI, How to generate cryptographically strong sequences of pseudo-random bits, *SIAM Jo. Comput.* 13 (1984), 850-864.

[3] L. BRYNIELSSON, On the Linear Complexity of Combined Shift Registers, em in "Proceedings of Eurocrypt 1984," pp. 156-160.

[4] A. H. CHAN AND R. GAMES, On the linear span of binary sequences from finite geometries, q odd, *in* "Proceedings of Crypto 1986," pp. 405-417, Santa Barbara.

[5] A. H. CHAN, M. GORESKY, AND A. KLAPPER, Cross-correlations of geometric sequences and GMW sequences, *in* Proceedings of Marshall Hall Memorial Conference, Burlington, VT, 1990 and Northeastern University Technical Report NU-CCS-90-12.

[6] S. GOLOMB, *Shift Register Sequences*, Aegean Park Press, Laguna Hills, CA, 1982.

[7] R. LIDL AND H. NIEDERREITER, *Finite Fields, Encyclopedia of Mathematics vol. 20*, Cambridge University Press, Cambridge, 1983.

[8] J.L. MASSEY, Shift register sequences and BCH decoding, *IEEE Trans. Info. Thy.*, IT-15 (1969), pp. 122-127.

[9] R. MCELIECE, *Finite Fields for Computer Scientists and Engineers*, Kluwer Academic Publishers, Boston, 1987.

[10] M. SIMON, J. OMURA, R. SCHOLTZ, AND B. LEVITT, *Spread-Spectrum Communications, Vol. 1*, Computer Science Press, 1985.

[11] D. WELSH, *Codes and Cryptography*, Clarendon Press, Oxford, 1988.

[12] A. YAO, Theory and applications of trapdoor functions, *in* "Proc. 23rd IEEE Symp. on Foundations of Comp. Sci.," 1982.

Extended Majority Voting and Private-Key Algebraic-Code Encryptions

Joost Meijers*   Johan van Tilburg[†]

Abstract

In this paper, a private-key cryptosystem equivalent to the private-key cryptosystem proposed by Rao and Nam is analyzed. The main result is that the private-key cryptosystem is vulnerable to a so-called extended majority vote attack. This attack can be averted if one selects the predefined set of error vectors at random. As a consequence, the error vector generating process will become much easier.

1   Introduction

In [4], Rao and Nam describe a private-key cryptosystem based on error-correcting codes of length $n = 250$ bits or less with small minimum distance ($d \leq 6$). Their *private-key* cryptosystem is a modification of the well-known *public-key* cryptosystem proposed by McEliece [2] and will be referred to as RN-system. According to [4], the McEliece system and the RN-system belong to the class of algebraic-code encryptions.

Whereas the McEliece cryptosystem makes use of randomly chosen error vectors of low Hamming weight, the RN-system selects the error vectors at random from a predefined set $\mathcal{Z}$. According to [4, p. 830], the error vectors in $\mathcal{Z}$ should have Hamming weight approximately $n/2$ in order to avert an attack based on majority voting (MV).

In this paper, a private-key cryptosystem equivalent to the RN-system is analyzed. Further, the weight constraint on $\mathcal{Z}$ is modelled more rigidly. As a result, it appears that several weight constraints exist that avert an MV-attack.

An attack based on extended majority voting (EMV) is introduced. An EMV can be regarded as a generalized MV-attack, i.e., it considers more than one position for voting at the same time. It will be shown that the RN-system is vulnerable to an EMV-attack. It appears that this attack is essentially optimal if the error vectors in $\mathcal{Z}$ have Hamming weight $n/2$. As a consequence, the choice in [4] is actually the worst one. In order to prevent for the attack to succeed, the predefined set of error vectors has to be chosen at random, i.e., without any weight constraint.

This paper is organized as follows. In Section 2, we introduce the RN-system. In Section 3 a private-key cryptosystem equivalent to the RN-system is described. In Section 4 some

---

*Eindhoven University of Technology, Dept. of Mathematics and Computing Science, Eindhoven, the Netherlands.

[†]PTT Research, Dept. of Applied Mathematics and Signal Processing, P.O. Box 421, 2260 AK Leidschendam, the Netherlands.

basic definitions are given and the MV-attack is presented. Sections 5 and 6 focus on the ideas behind the proposed attack, while Sections 7, 8 and 9 deal with the actual EMV-attack. Finally, in Section 10 the results obtained are discussed and conclusions are drawn.

2 The RN-system

The RN-system as described in [4] can be summarized as follows. Let C be a linear $[n, k]$ code over $\mathbb{F}_2$ with code length n and dimension k. Let G be a $k \times n$ generator matrix of the code C and let H be an $(n-k) \times n$ parity check matrix such that $GH^T = 0$. The code C has exactly 2^k code words, all are linear combinations of the rows of G. Let $\underline{x} \in \mathbb{F}_2^n$, the set $\underline{x} + C = \{\underline{x} + \underline{c} | \underline{c} \in C\}$ is called a coset. Each coset contains 2^k vectors, and there are exactly 2^{n-k} different cosets. The vector $(\underline{x} + C)H^T = \underline{x}H^T$ is called the syndrome of $\underline{x}$ with respect to the code C. The Hamming weight $w_H(\underline{z})$ gives the number of nonzero coordinates of $\underline{z}$. Define a subset Z of $\mathbb{F}_2^n$ with cardinality $N \leq 2^{n-k}$, satisfying the following two properties.

1. *Weight property* - all vectors $\underline{z} \in Z$ have Hamming weight $w_H(\underline{z}) \approx \frac{n}{2}$.

2. *Syndrome property* - no distinct vectors lie in the same coset of C.

By Property 2, each vector $\underline{z} \in Z$ has a unique syndrome. Z will be referred to as the set of predefined *error* vectors. Let T be the syndrome-error table: $T = \{(\underline{z}H^T, \underline{z}) | \underline{z} \in Z\}$. Finally, let S be a $k \times k$ random non-singular matrix over $\mathbb{F}_2$ and let P be a random binary $n \times n$ permutation matrix.

A message $\underline{m} \in \mathbb{F}_2^k$ is *encrypted* into a ciphertext $\underline{c} \in \mathbb{F}_2^n$ by

$$\underline{c} = (\underline{m}SG + \underline{z})P, \qquad \underline{z} \in_R Z. \tag{1}$$

By convention, $\underline{z} \in_R Z$ means that $\underline{z}$ is chosen at random (according to an uniform distribution) from Z.

A ciphertext $\underline{c} \in \mathbb{F}_2^n$ is *decrypted* into a message $\underline{m} \in \mathbb{F}_2^k$ as follows.

- Calculate $\underline{c}' = \underline{c}P^T = \underline{m}'G + \underline{z}$, where $\underline{m}' = \underline{m}S$.

- Compute $\underline{c}'H^T$ to obtain the syndrome $\underline{z}H^T$. From the syndrome-error table T one finds the corresponding error vector $\underline{z}$. Let $\underline{m}'G = \underline{c}' + \underline{z}$ and recover $\underline{m}'$.

- Multiply $\underline{m}'$ by S^{-1} to obtain the message $\underline{m}$.

The matrices S, P, G (H) and the syndrome-error table T (Z) form the *secret* or *private* key.

3 An Equivalent Private-Key Cryptosystem

The RN-system is a private-key cryptosystem. As a consequence, the S and P matrices do not conceal a code or a decoding algorithm as they do in the McEliece public-key cryptosystem. As has been shown by Heiman [1], the two matrices are redundant. Consequently, the following private-key cryptosystem is *equivalent* to the RN-system and will be used in the cryptanalysis of the RN-system.

Let E be a $k \times n$ encryption matrix over $\mathbb{F}_2$ of rank k, with a right inverse E^{-R}. Let D be a corresponding $(n-k) \times n$ parity check matrix over $\mathbb{F}_2$ of rank $(n-k)$ such that $ED^T = 0$, i.e., the matrix D yields a mapping from $\mathbb{F}_2^n$ to $\mathbb{F}_2^{n-k}$. Select *at random* a set of error vectors $\mathcal{Z}$ from $\mathbb{F}_2^n$ satisfying the weight property as well as the syndrome property. Finally, compute the syndrome-error table $T = \{(\underline{z}D^T, \underline{z}) | \underline{z} \in \mathcal{Z}\}$.

A message $\underline{m} \in \mathbb{F}_2^k$ is *encrypted* into a ciphertext $\underline{c} \in \mathbb{F}_2^n$ by

$$\underline{c} = \underline{m}E + \underline{z}, \qquad \underline{z} \in_R \mathcal{Z}. \tag{2}$$

A ciphertext $\underline{c} \in \mathbb{F}_2^n$ is *decrypted* by applying the following three steps.

- Compute $\underline{c}D^T$ to obtain $\underline{z}D^T$.

- Recover $\underline{z}$ by means of the syndrome-error table T.

- Obtain the message $\underline{m}$ by letting $\underline{m} = (\underline{c} + \underline{z})E^{-R}$.

The matrix E (D) and the syndrome-error table T $(\mathcal{Z})$ form the secret or private key.

4   Majority Voting

According to [4, p. 830], an n bit error vector $\underline{z} \in \mathcal{Z}$ has to satisfy the weight property: $w_H(\underline{z}) \approx n/2$. Essentially without loss of generality, we assume n to be even throughout this paper. The weight property can be restated more rigidly as follows.

Definition 4.1 (Weight property) *Let $\underline{z} \in \mathcal{Z}$ be an n bit error vector, then the weight property* WP($\underline{z}$) *is given by*

$$\mathrm{WP}(\underline{z}) = |w_H(\underline{z}) - \frac{n}{2}| \le a, \tag{3}$$

where a is a non-negative constant. $\square$

Next, let $\mathcal{Z}^a$ be defined for a non-negative integer a $(0 \le a \le n/2)$ as

$$\mathcal{Z}^a = \{\underline{z} \in \mathbb{F}_2^n | \mathrm{WP}(\underline{z}) \le a\}. \tag{4}$$

Number the N different cosets Γ of the code generated by E from 1 to N $(= 2^{n-k})$, i.e, $\Gamma_1, \ldots, \Gamma_N$. Define $\Upsilon_j^a = \Gamma_j \cap \mathcal{Z}^a$ for all j.

Definition 4.2 (Set of error vectors) *The set of error vectors $\mathcal{Z}$ is a random subset of $\mathcal{Z}^a$ and satisfies the syndrome property, i.e.,*

$$\mathcal{Z} = \{\underline{z} \in_R \Upsilon_j^a | 1 \le j \le N\} \quad \text{such that} \quad \forall_{j,1 \le j \le N} \;\; |\Upsilon_j^a \cap \mathcal{Z}| = 1. \tag{5}$$

$\square$

Definition 4.3 (Symbol counter) *Let z_i be the i-th coordinate of error vector $\underline{z}$. For $1 \leq i \leq n$, and $x \in \mathbb{F}_2$, the symbol counter $SC_{i,x}(\underline{z})$ is defined as*

$$SC_{i,x}(\underline{z}) = \begin{cases} 1 - z_i, & \text{if } x = 0; \\ z_i, & \text{if } x = 1. \end{cases} \tag{6}$$

□

With a little ambiguity in notation, the symbol counter for the set $\mathcal{Z}$ is defined by

$$SC_{i,x}(\mathcal{Z}) = \sum_{\underline{z} \in \mathcal{Z}} SC_{i,x}(\underline{z}). \tag{7}$$

For $\mathcal{Z}^a$ we have $SC_{i,0}(\mathcal{Z}^a) = SC_{i,1}(\mathcal{Z}^a)$ for all a and i.

It is of interest to examine the case $SC_{i,0}(\mathcal{Z}) \neq SC_{i,1}(\mathcal{Z})$, $(1 \leq i \leq n)$, with respect to the following MV-attack. Consider L realizations of the discrete random variable $\underline{C} = \underline{m}E + \underline{Z}$. For $\lambda_i = SC_{i,0}(\mathcal{Z})/|\mathcal{Z}| < 1/2$, the (error) probability $P_e(\lambda_i, L)$ that among the L realizations the number of 1's on the i-th position does not exceed the number of 0's is given by

$$P_e(\lambda_i, L) = \sum_{j=\lceil \frac{L}{2} \rceil}^{L} \binom{L}{j} \lambda_i^j (1 - \lambda_i)^{L-j} \leq \left[\sqrt{4\lambda_i(1 - \lambda_i)} \right]^L. \tag{8}$$

Minority voting yields an estimate for the i-th position in the vector $\underline{m}E$. The probability $P_c(\lambda_i, L)$ of a correct estimate is given by

$$P_c(\lambda_i, L) = 1 - P_e(\lambda_i, L) \geq 1 - \left[\sqrt{4\lambda_i(1 - \lambda_i)} \right]^L. \tag{9}$$

Since $\lambda_i < 1/2$, it follows that $P_c(\lambda_i, L) \to 1$ for $L \to \infty$.

Repeat the above for all positions $1 \leq i \leq n$ and obtain an estimate of $\underline{m}E$. Repeat this procedure for k independent messages $\underline{m}$. Let the rows of a matrix M consist of the k messages and the rows of $\widehat{(ME)}$ consist of the k estimates. Then E follows from

$$ME = \widehat{(ME)} \quad \text{or} \quad E = M^{-1}\widehat{(ME)}. \tag{10}$$

With probability $P_c(L) = \prod_{i=1}^{n} P_c(\lambda_i, L)^k$ the matrix E is estimated correctly.

Theorem 4.4 *Let $\Pr\{Z_i^{(j)} = x\}$ denote the probability that the i-th coordinate of $\underline{z} \in_R \Upsilon_j^a$ equals the value $x \in \mathbb{F}_2$. Define $Z_i^N = \sum_{j=1}^{N} Z_i^{(j)}$. For $\epsilon > 0$ it holds that*

$$\forall_{n \geq 0} \forall_{i, 1 \leq i \leq n} \quad \Pr\{|\frac{Z_i^N}{N} - \frac{1}{2}| \geq \epsilon\} \leq \frac{1}{4N\epsilon^a}. \tag{11}$$

Proof. For each j, there exists exactly one (not necessarily different) k, $(1 \leq k \leq N)$, such that $\Upsilon_j^a = \Upsilon_k^a + 1$. As a consequence,

$$\forall_{i,1 \leq i \leq n} \quad E(\frac{Z_i^N}{N}) = \frac{1}{N} \sum_{j=1}^{N} E(Z_i^{(j)}) = \frac{1}{2}, \tag{12}$$

where E denotes the expectation operator.

For all i and j, the variance of $Z_i^{(j)}$, $\mathrm{Var}(Z_i^{(j)})$, can be upper bounded by $\sigma^2 = 0.25$. Hence,

$$\mathrm{Var}(\frac{Z_i^N}{N}) = \sum_{j=1}^{N} \frac{\mathrm{Var}(Z_i^{(j)})}{N^2} \leq \frac{\sigma^2}{N} = \frac{1}{4N}. \tag{13}$$

Applying Chebyshev's inequality proves the Theorem. □

Remark: If $\underline{1} \in C$, then all sets Υ_j^a are *balanced*, i.e., for all j: $\Upsilon_j^a = \Upsilon_j^a + 1$. In this case, for each set Υ_j^a and for all i it holds that $E(Z_i^{(j)}) = 0.5$ and $\mathrm{Var}(Z_i^{(j)}) = 0.25$.

From Theorem 4.4 it follows that an MV on an arbitrary position i $(1 \leq i \leq n)$ will yield a correct estimate with probability $\approx 1/2$ for all a. For this case it may be concluded that an MV-attack will not succeed.

5 Weight Distribution

Let $A = \{a_1, \ldots, a_s\}$ be a subset of the set $\{1, \ldots, n\}$, where $1 \leq s \leq n$. For all $\underline{z} \in \mathbb{F}_2^n$, let the subvector $\underline{z}_A \in \mathbb{F}_2^s$ of a vector $\underline{z}$ be defined by

$$\underline{z}_A = (z_{a_1}, \ldots, z_{a_s}) \quad \text{and} \quad (\underline{z}_A)_i = z_{a_i}, \quad i = 1, \ldots, s. \tag{14}$$

In the next definition, the symbol counter will be generalized.

Definition 5.1 (Vector counter) *Let $C \subseteq \mathbb{F}_2^n$, $\underline{v} \in \mathbb{F}_2^s$ and $1 \leq s \leq n$. For all sets $A \subseteq \{1, \ldots, n\}$ with $|A| = s$, let the vector counter $\mathrm{VC}_{A,\underline{v}}(C)$ denote the number of times vector $\underline{v}$ equals subvector $\underline{c}_A$, $\underline{c} \in C$. That is,*

$$\mathrm{VC}_{A,\underline{v}}(C) = |\{\underline{c} \in C | \underline{c}_A = \underline{v}\}|. \tag{15}$$

□

The set A extracts $|A| = s$ positions of $\underline{c} \in C$ on which $\mathrm{VC}_{A,\underline{v}}(C)$ looks for a match with subvector $\underline{v}$. For $s = 1$, the symbol counter is obtained.

Lemma 5.2 *Let $P_{\underline{Z}_A^a}(\underline{v}) = \Pr\{\underline{Z}_A^a = \underline{v}\}$ denote the probability that a realization of the discrete random variable $\underline{Z}_A^a$ equals $\underline{v} \in \mathbb{F}_2^s$. For Z^0, we have*

$$\forall_{\underline{v} \in \mathbb{F}_2^s} \quad P_{\underline{Z}_A^0}(\underline{v}) = \frac{\mathrm{VC}_{A,\underline{v}}(Z^0)}{|Z^0|} = \binom{n - |A|}{n/2 - w_H(\underline{v})} / \binom{n}{n/2}. \tag{16}$$

Proof. The vector $\underline{v}$ can be extended to a vector of length n and weight $n/2$ by inserting a vector of length $n - s$ and weight $n/2 - w_H(\underline{v})$ at the appropriate coordinates. There are exactly $\binom{n}{n/2}$ vectors $\underline{z} \in Z^0$ of weight $n/2$. $\qquad\square$

For fixed size of $\underline{z}$ (i.e., for fixed n) and for fixed $|A| = s$, the probability function $P_{\underline{Z}_A^0}(\underline{v})$ depends on the weight of $\underline{v}$ only. For $\underline{x}, \underline{y} \in F_2^s$ and $w_H(\underline{x}) = s - w_H(\underline{y})$, it follows that $P_{\underline{Z}_A^0}(\underline{x}) = P_{\underline{Z}_A^0}(\underline{y})$. This will be referred to as the *symmetry property*. Furthermore, it holds that

$$P_{\underline{Z}_A^0}(\underline{x}) < P_{\underline{Z}_A^0}(\underline{y}) \quad \begin{cases} \text{if } 0 \le w_H(\underline{x}) < w_H(\underline{y}) \le \lfloor s/2 \rfloor, \\ \text{or if } s \ge w_H(\underline{x}) > w_H(\underline{y}) \ge \lceil s/2 \rceil. \end{cases} \tag{17}$$

The second property will be called the *unequal weight distribution* of the subvectors. This in contrast to the case $\underline{z} \in Z^{n/2}$, where all $P_{\underline{Z}_A^{n/2}}(\underline{x})$ are equal. The unequal weight distribution will be the basis for attack.

Theorem 5.3 Let $\Pr\{\underline{Z}_A^{(j)} = \underline{v}\}$ denote the probability that the subvector $\underline{z}_A$ of $\underline{z} \in_R \Upsilon_j^a$ equals the value $\underline{v} \in \mathbb{F}_2^s$. Let $X_A^{(j)}(\underline{v}) = 1$, if $\underline{Z}_A^{(j)} = \underline{v}$, and 0 otherwise. Define $X_A^N(\underline{v}) = \sum_{j=1}^N X_A^{(j)}(\underline{v})$ for all $\underline{v}$. For $\epsilon > 0$ it holds that

$$\forall_{a \ge 0} \forall_{s, 1 \le s \le n} \forall_{|A|=s} \forall_{\underline{v} \in \mathbb{F}_2^s} \quad \Pr\{|\frac{X_A^N(\underline{v})}{N} - \sum_{j=1}^N \frac{VC_{A,\underline{v}}(\Upsilon_j^a)}{N|\Upsilon_j^a|}| \ge \epsilon\} \le \frac{1}{4N\epsilon^2}. \tag{18}$$

$\qquad\square$

The proof is similar to the proof of Theorem 4.4 and is therefore omitted.

The probability that $X_A^N(\underline{v})/N$ differs from its expectation by more than ϵ is bounded from above. However, this expectation depends on the structure of Υ_j^a. Without loss of generality, let the N different sets Υ_i^a be ordered such that

$$|\Upsilon_1^a| \le |\Upsilon_2^a| \le \ldots |\Upsilon_N^a|. \tag{19}$$

Suppose among the N sets Υ_i^a, there are N_1 with cardinality $|\Upsilon_1^a|$, N_2 with cardinality $|\Upsilon_{N_1+1}^a|$, etc.. Define $m_j = \sum_{i=1}^j N_i$ with $m_k = N$ and $m_0 = 0$. For sufficiently large N, it follows that

$$\sum_{j=1}^N \frac{VC_{A,\underline{v}}(\Upsilon_j^a)}{N|\Upsilon_j^a|} = \frac{1}{N} \sum_{j=1}^k |\Upsilon_{N_{m_j}}^a|^{-1} \sum_{i=m_{j-1}+1}^{m_j} VC_{A,\underline{v}}(\Upsilon_i^a) \simeq$$

$$\frac{1}{N} \sum_{j=1}^k |\Upsilon_{m_j}^a|^{-1} \left[N_j \frac{VC_{A,\underline{v}}(Z^a)}{|Z^a|} |\Upsilon_{m_j}^a| \right] = \frac{VC_{A,\underline{v}}(Z^a)}{|Z^a|}. \tag{20}$$

Equality holds if $|\Upsilon_1^a| = \ldots = |\Upsilon_N^a|$. The interpretation of this result is that for large N, $X_A^N(\underline{v})/N$ is close to $VC_{A,\underline{v}}(Z^a)/|Z^a|$, which depends solely on s (n and a are fixed).

Corollary 5.4 For $Z^{n/2}$, we have $|\Upsilon_1^{n/2}| = \ldots = |\Upsilon_N^{n/2}|$ and Equation (18) reduces to

$$\forall_{s, 1 \le s \le n} \forall_{|A|=s} \forall_{\underline{v} \in \mathbb{F}_2^s} \quad \Pr\{|\frac{X_A^N(\underline{v})}{N} - \frac{1}{2^s}| \ge \epsilon\} \le \frac{1}{4N\epsilon^2}. \tag{21}$$

$\qquad\square$

From this corollary it follows that all subvectors $\underline{v}$ are expected equally likely to appear in N realizations of Z. For this case it may be concluded that an LMV-attack (see Section 7) will not succeed.

6 Error Probability

Due to the unequal weight distribution (17), the zero and all-one subvector are expected to occur less frequently than any other subvector. Let P_{e_A} denote the (error) probability that a minority vote on the A positions of N realizations $\underline{C}_A = (\underline{m}E + \underline{Z})_A$ yields a vector other than the zero or all-one vector. Obviously, it holds that

$$P_{e_A} \leq \Pr\{\min(\boldsymbol{X}_A^N(\underline{0}), \boldsymbol{X}_A^N(\underline{1})) \geq \min_{\underline{v} \in \mathbb{F}_2^s / \{\underline{0},\underline{1}\}} \boldsymbol{X}_A^N(\underline{v})\}. \tag{22}$$

As the error probability P_{e_A} is a non-decreasing function of a, two extreme cases will be considered only. Hence, the following distinction is made: Either Z is randomly selected from Z^0, or Z is randomly selected from $Z^{n/2}$.

- Z^0 — all vectors $\underline{v} \in \mathbb{F}_2^s$ satisfy Equation (17), so that $P_{e_A} \to 0$ for sufficiently large N.

- $Z^{n/2}$ — all vectors $\underline{v} \in \mathbb{F}_2^s$ are equally likely (Corollary 5.4). Consequently, $P_{e_A} \to 1 - 2^{s-1}$ $(s > 1)$ for sufficiently large N.

Based on the results of Section 5, the following heuristics are used to obtain an upper bound on P_{e_A}. Define for all $\underline{u}, \underline{v} \in \mathbb{F}_2^s$ with $0 < w_H(\underline{v}) < s$:

$$\lambda_A(\underline{u}, \underline{v}) = \frac{P_{Z_A(\underline{u})}}{P_{Z_A(\underline{u})} + P_{Z_A(\underline{v})}} \simeq \frac{\mathrm{VC}_{A,\underline{u}}(Z^a)}{\mathrm{VC}_{A,\underline{u}}(Z^a) + \mathrm{VC}_{A,\underline{v}}(Z^a)} \tag{23}$$

and

$$N_A(\underline{u}, \underline{v}) = [P_{Z_A(\underline{u})} + P_{Z_A(\underline{v})}]N \simeq [\mathrm{VC}_{A,\underline{u}}(Z^a) + \mathrm{VC}_{A,\underline{v}}(Z^a)]\frac{N}{|Z^a|}. \tag{24}$$

The average error probability $P_{e_A}(\underline{v})$ that one will find less subvectors $\underline{v}$ than $\underline{0}$, or less subvectors $\underline{v}$ than $\underline{1}$, is given by

$$P_{e_A}(\underline{v}) = P_e(\lambda_A(\underline{0}, \underline{v}), N_A(\underline{0}, \underline{v})) = P_e(\lambda_A(\underline{1}, \underline{v}), N_A(\underline{1}, \underline{v})) \quad \text{with} \quad 0 < w_H(\underline{v}) < s. \tag{25}$$

The overall average error probability P_{e_A} can be upper bounded as follows

$$P_{e_A} < \sum_{0 < w_H(\underline{v}) < s} P_{e_A}^2(\underline{v}). \tag{26}$$

Let $a = 0$ and define $\lambda_i = \lambda_A(\underline{0}, \underline{v})$ and $L_i = N_A(\underline{0}, \underline{v})$, where $0 < i = w_H(\underline{v}) < s$. Next, substitute Equations (8) and (25) in Expression (26) to obtain the desired result

$$P_{e_A} < \sum_{i=1}^{s-1} \binom{s}{i} P_e^2(\lambda_i, L_i) < \sum_{i=1}^{s-1} \binom{s}{i} [4\lambda_i(1 - \lambda_i)]^{L_i}. \tag{27}$$

Hence, it is expected that $\mathrm{VC}_{A,\underline{v}}(Z)$ will be minimal for some vector $\underline{v}$, $(0 < w_H(\underline{v}) < s)$ with probability P_{e_A} which is upper bounded by (27). Due to the fact that the underlying code parameters $[n, k]$ are fixed, a cryptanalist is only 'free' in choosing s. If P_{e_A} is minimized according to the upper bound (27), then for all interesting cases the optimal value $s = 4$ is found.

In [4], codes of length 250 bits are suggested. For a binary [250, 226]-code, we find $P_{e_A} < 10^{-131}$ for $s = 4$.

7   Local Majority Voting

Let $C^{\underline{m}}$ be a set of distinct cryptograms of a message $\underline{m}$, i.e., $C^{\underline{m}} = \underline{m}E + \mathcal{Z}$, where $\mathcal{Z}$ is the corresponding set of error patterns. The set $C^{\underline{m}}$ is a translation of the set $\mathcal{Z}$ by a vector $\underline{m}E$, therefore

$$\mathrm{VC}_{A,(\underline{v}+(\underline{m}E)_A)}(C^{\underline{m}}) = \mathrm{VC}_{A,\underline{v}}(\mathcal{Z}). \tag{28}$$

For sufficiently large $|\mathcal{Z}|$, it is expected that $\mathrm{VC}_{A,\underline{v}}(\mathcal{Z})$ will be minimal for either $\underline{v} = \underline{0}$ or $\underline{v} = \underline{1}$. As a consequence, $\mathrm{VC}_{A,(\underline{v}+(\underline{m}E)_A)}(C^{\underline{m}})$ will be minimal for $(\underline{m}E)_A$ or $(\underline{m}E)_A + \underline{1}$ with probability $1 - P_{e_A}$. Let $\widehat{(\underline{m}E)}_A$ be a vector $\underline{w}$ for which $\mathrm{VC}_{A,\underline{w}}(C^{\underline{m}})$ is minimal.

Algorithm - LMV

Step 1.   Encipher a message $\underline{m}$ until L distinct cryptograms $\underline{c} \in C^{\underline{m}}$ have occurred. Let $C^{\underline{m}}$ be defined as $C^{\underline{m}} = \underline{m}E + \mathcal{Z}$, where $\mathcal{Z}$ is the corresponding set of L error patterns.

Step 2.   Define a subset $A = \{a_1, \ldots, a_s\} \subseteq \{1, \ldots, n\}$, where $s \leq n$. Set the estimate $\widehat{(\underline{m}E)}_A$ equal to a vector $\underline{v} \in \mathbb{F}_2^s$ for which $\mathrm{VC}_{A,\underline{v}}(C^{\underline{m}})$ is minimal, i.e.,

$$\widehat{(\underline{m}E)}_A \in_R \left\{ \underline{v} \in \mathbb{F}_2^s \mid \forall_{\underline{w} \in \mathbb{F}_2^s} \left[\mathrm{VC}_{A,\underline{v}}(C^{\underline{m}}) \leq \mathrm{VC}_{A,\underline{w}}(C^{\underline{m}}) \right] \right\}. \tag{29}$$

For L large enough, one of the following two possibilities will occur with overwhelming probability

$$\widehat{(\underline{m}E)}_A = (\underline{m}E)_A \quad \text{or} \quad \widehat{(\underline{m}E)}_A = (\underline{m}E)_A + \underline{1}. \tag{30}$$

Summary.   The LMV-algorithm finds an estimate $\widehat{(\underline{m}E)}_A$ of $(\underline{m}E)_A$ from L observations of $(\underline{m}E + \underline{Z})_A$. For sufficiently large L, (30) holds with overwhelming probability.

Remark.   All subvectors $\underline{v} \in \mathbb{F}_2^s$ can be considered as estimates. The best estimate for $(\underline{m}E)_A$, say est_1, equals $\min_{\underline{v}} \mathrm{VC}_{A,\underline{v}}(C^{\underline{m}})$. Accordingly, label all elements $\underline{v}$ such that

$$\mathrm{VC}_{A,est_1}(C^{\underline{m}}) \leq \mathrm{VC}_{A,est_2}(C^{\underline{m}}) \leq \ldots \leq \mathrm{VC}_{A,est_{2^s}}(C^{\underline{m}}). \tag{31}$$

As observed before, one expects to find

$$est_1 = est_2 + \underline{1}, \tag{32}$$

with high probability. If this is indeed the case, then accept est_1 as a correct estimate. If (32) does not hold, then one might decide not to trust est_1.

8   Global Majority Voting

Choose two integers r and s. Let $\mathcal{A}$ be a set of r different subsets $A \subseteq \{1, \ldots, n\}$ having cardinality s. Label the elements of the set $\mathcal{A}$, i.e., $\mathcal{A} = \{A_1, \ldots, A_r\}$. One can say that $\mathcal{A}$ determines r subvectors $\underline{z}_{A_i}$ of length s. The elements $A_i \in \mathcal{A}$ should satisfy

$$\left\{ \bigcup_{1 \leq i \leq r} A_i \right\} = \{1, \ldots, n\}; \tag{33}$$

$$\forall_{i,\ 1\leq i<r}\ \left\{A_i \bigcap A_{i+1}\right\} \neq \emptyset. \tag{34}$$

By (33) these r subvectors $\underline{z}_{A_i}$ cover a vector of length n. By (34) any subvector $\underline{z}_{A_i}$ has some positions in common with at least one other subvector $\underline{z}_{A_j}$. Consequently, we have $n \leq r(s-1)+1$. By choosing $\mathcal{A}$ such that A_i and A_{i+1} share more than one position, one might be able to detect wrong estimates.

Algorithm - GMV

Step 1. Perform r local majority votes (Algorithm LMV) to obtain r estimates $\widehat{(mE)}_{A_i}$ $(1 \leq i \leq r)$. For L large enough, one of the following two possibilities will occur with overwhelming probability for each i:

$$\widehat{(mE)}_{A_i} = (mE)_{A_i} \quad \text{or} \quad \widehat{(mE)}_{A_i} = (mE)_{A_i} + \underline{1}. \tag{35}$$

Step 2. Derive an estimate $\widehat{(mE)}$ of $\underline{m}E$ as follows. Let A_1 determine s symbols of $\widehat{(mE)}$. The subvectors determined by A_1 and A_2 have at least one position in common. Then, with overwhelming probability, one can set the positions of $\widehat{(mE)}$ determined by A_2 conform to the positions determined by A_1. By repeating this for all A_i $(1 \leq i \leq r)$, one is able to find an estimate for $\underline{m}E$. Again for L large enough, one of the following two possibilities will occur with overwhelming probability:

$$\widehat{(mE)} = \underline{m}E \quad \text{or} \quad \widehat{(mE)} = \underline{m}E + \underline{1}. \tag{36}$$

Summary. The GMV-algorithm performs r-times an LMV. The r obtained local estimates are synchronized conform the overlapping positions and in such a way that they cover the complete vector $\underline{m}E$. For L large enough, the GMV-algorithm provides an estimate for $\underline{m}E$, possibly corrupted with the all-one vector.

9 Extended Majority Voting Attack

Let $\underline{m}_\nu = \underline{m} + \underline{u}_\nu$, where $\underline{u}_\nu$ is the ν-th unit vector. The EMV-attack obtains k vectors $\widehat{(m_\nu E)}$, from which an estimate for the νth-row of E is computed, i.e., $\hat{\underline{e}}_\nu = \widehat{(mE)} + \widehat{(m_\nu E)}$. The estimates are obtained from a global majority vote (Algorithm GMV). Accordingly,

$$\hat{E} = E + A, \tag{37}$$

where A is a $k \times n$ matrix, and where the set of all rows of A consists of the all-one vector and the zero vector only. Depending on the structure of $\mathcal{Z}$, the matrix A can be removed by either defining an equivalent cryptosystem or removing the all-one vectors.

Algorithm - EMV

Step 1. Let $\underline{u}_\nu$ be the ν-th unit vector. Perform a GMV for the message $\underline{m}_\nu = \underline{m} + \underline{u}_\nu$ $(1 \leq \nu \leq k)$. For large L, the GMV-algorithm determines an estimate $\widehat{(m_\nu E)}$ for which with overwhelming probability it holds that

$$\widehat{(m_\nu E)} = \underline{m}_\nu E \quad \text{or} \quad \widehat{(m_\nu E)} = \underline{m}_\nu E + \underline{1}, \qquad 1 \leq \nu \leq k. \tag{38}$$

Step 2. Compute the estimate $\hat{e}_\nu$ of the ν-th row of the encryption matrix E as follows

$$\hat{e}_\nu = \widehat{(m_\nu E)} + \widehat{(mE)}, \qquad 1 \le \nu \le k. \tag{39}$$

Let $\hat{E} = (\hat{e}_1^T, \cdots, \hat{e}_k^T)^T$ be an estimate for the encryption matrix $\cdot E$.

Step 3. Let $\hat{Z} = m\hat{E} + C^m$.

- Case $\hat{Z} + \underline{1} = \hat{Z}$. The matrix $\hat{E}$ and the set $\hat{Z}$ together define an equivalent cryptosystem.

- Case $\hat{Z} + \underline{1} \ne \hat{Z}$. Compare the set $C^{m_\nu} + \widehat{e_\nu}$ with the set C^m. If the two sets are different, let $\widehat{e_\nu} = \widehat{e_\nu} + \underline{1}$. Repeat this for all row estimates $\widehat{e_\nu}$ $(1 \le \nu \le k)$ to obtain $\bar{E}$. Further, let $\bar{Z} = C^m + m\bar{E}$ be the set of corresponding error vectors. The matrix $\bar{E}$ and the set $\bar{Z}$ together define an equivalent cryptosystem.

Costs. In general, it suffices for the above attack to use L $(L \ll N)$ encryptions to perform a successful voting attack. In this paper we will only consider the (worst) case costs $L = N = |\mathcal{Z}|$, i.e., all distinct error vectors are used, like in [5]:

- $O(kN \log N)$ encryptions;

- $O(kN \log N)$ vector operations;

- $O(N)$ units of memory.

The overall error probability P_e for the above EMV-attack is upper bounded by

$$P_e < 1 - (1 - P_{e_A})^{r(k+1)} < r(k+1)P_{e_A}. \tag{40}$$

For a binary $[250, 226]$-code (see Section 6) we find: $P_e < 10^{-127}$ with $r = 83$, $N = 2^{24}$ and $kN \log N < 2^{37}$.

Remark. If all distinct error vectors $(L = N)$ have to be considered, then the above EMV-attack can be improved considerably. In this situation we are not interested how an estimate $(\widehat{mE})_{A_i}$ looks like, but to find an estimate for which the vector count is different from all other estimates, i.e.,

$$\forall_{\underline{v} \in \mathbb{F}_2^s / (\widehat{mE})_{A_i}} \quad \mathrm{VC}_{A_i, (\widehat{mE})_{A_i}}(C^m) \ne \mathrm{VC}_{A_i, \underline{v}}(C^m). \tag{41}$$

Next, find a corresponding estimate in C^{m_ν}, say $(\widehat{m_\nu E})_{A_i}$, for which it holds that

$$\mathrm{VC}_{A_i, (\widehat{m_\nu E})_{A_i}}(C^{m_\nu}) = \mathrm{VC}_{A_i, (\widehat{mE})_{A_i}}(C^m). \tag{42}$$

Obviously, the s positions A_i of the νth-row of the encryption matrix E follow from

$$(\underline{e}_\nu)_{A_i} = (\widehat{m\bar{E}})_{A_i} + (\widehat{m_\nu \bar{E}})_{A_i}. \tag{43}$$

In this way, we obtain $(\underline{e}_\nu)_{A_i}$ for $1 \le \nu \le k$.

By repeating this for all A_i $(1 \le i \le r = n/s)$ one is able to find $\underline{e}_\nu$ $(1 \le \nu \le k)$. Note, we do not need overlapping positions. This procedure will yield a correct encryption matrix E.

10 Conclusion

The object of this paper is to show that due to the imposed weight constraint the RN-system is vulnerable to an EMV-attack.

It has been shown that more than one condition exists for which an MV-attack can be averted. In particular, this condition can be

$$|w_H(z) - \frac{n}{2}| \leq a, \tag{44}$$

for any a.

An EMV-attack has been presented. It appears that the attack is essentially optimal if the error vectors satisfy Hamming weight $n/2$, that is, a weight constraint with $a = 0$. As a consequence, the choice of the weight constraint in [4] is essentially the worst one. The attack will not succeed when the weight constraint is relaxed to $a = n/2$, i.e., when it is dropped. Therefore, one should select the error vectors at random satisfying the syndrome property, but without any weight constraint. As a consequence, the error vector generating process will become much easier.

We conclude with the remark that even without a weight constraint the attacks presented in [1, 3] and [5] remain as powerful.

Acknowledgements

We are grateful to Peter de Rooij, René Struik and Henk van Tilborg for very helpful discussions concerning this work.

References

[1] R. Heiman, 'On the Security of Cryptosystems based on Linear Error-Correcting Codes', M.Sc. Thesis, Feinburg Graduate School, Weizmann Institute of Science, Rehovot, Israel, August 1987.

[2] R.J. McEliece, A Public-Key Cryptosystem Based on Algebraic Coding Theory, DSN Progress Report 42-44, Jet Propulsion Laboratory, Pasadena, pp. 114-116, 1978.

[3] J. Meijers and J. van Tilburg, 'On the Rao-Nam Private-Key Cryptosystem using Linear Codes', *Proceedings 1991 IEEE-ISIT*, p. 126, Budapest, Hungary, June 1991.

[4] T.R.N. Rao and K.H. Nam, 'Private-Key Algebraic-Code Encryptions', *IEEE Trans. Inform. Theory*, vol. IT-35, no. 4, pp. 829-833, July 1989.

[5] R. Struik and J. van Tilburg, 'The Rao-Nam Scheme is Insecure Against a Chosen-Plaintext Attack', in: *Advances in Cryptology - CRYPTO '87*, Carl Pomerance ed., Lecture Notes in Computer Science # 293, Springer-Verlag, pp. 445-457, 1988.

A SECURE ANALOG SPEECH SCRAMBLER USING THE DISCRETE COSINE TRANSFORM

B. Goldburg*, E. Dawson**, S. Sridharan***

*School of Electrical and Electronic Systems Engineering
Queensland University of Technology
GPO Box 2434
Brisbane Queensland 4001
Australia

**School of Mathematics and
Information Security Research Centre
Queensland University of Technology
GPO Box 2434
Brisbane Queensland 4001
Australia

***School of Electrical Engineering
University of Technology Sydney
Sydney New South Wales 2001
Australia

ABSTRACT

A secure technique for analog speech encryption is described which uses the Discrete Cosine Transform (DCT). The paper shows how the proposed technique overcomes weaknesses identified in other analog scramblers by using transform domain permutation and the insertion of dummy transform components. It is shown that the DCT scrambler with dummy spectrum insertion offers a similar level of security to a digital scrambler on a bandlimited channel while providing better quality of recovered speech.

1. INTRODUCTION

Many of the analog speech encryption devices once thought to be secure have been shown to have weaknesses which allow the recovery of intelligible speech [4]. Attacks are possible because the scrambling process fails to prevent access to the speech information contained in the scrambled speech. The hopping window scrambler [3], for example, permutes speech segments in time. This results in a distortion of the speech time envelope, reducing the intelligibility of the speech. For practical implementation reasons the speech segments have a duration greater than 32ms. These segments are left intact and contain a significant amount of speech information.

Some trained listeners are able to directly interpret speech which has been scrambled in this way. Failing this it has been shown [4] that it is possible to reorder the speech segments using the spectral information contained within the unaffected speech segments. A cryptanalyst can completely recover the original speech using this technique.

The goal of cryptographers who design analog speech scramblers is to distort the incoming speech such that no useful information can be obtained from the ciphertext (scrambled speech) alone.With the correct key and algorithm however it should be possible to recover the speech with a minimum degradation in quality. The quest for such a scheme has seen a progression from the time domain to the frequency domain in an attempt to disguise speech information. Although this step improved security to an extent, frequency domain techniques such as the bandsplitter have also proved to be vulnerable to attack [4].

Another analog speech encryption technique has become feasible due to the recent advances in microelectronics with the availability of general and special purpose digital signal processing devices. In this method speech samples are first converted to a transform domain where a permutation is applied to the transform coefficients. The permuted transform coefficients are converted back to the time domain and transmitted. The main attraction of this method is that it should result in significantly lower residual intelligibility provided the transformation removes speech redundancy. Transforms that realize a time to frequency domain transformation are particularly useful in this application since they allow distortion of the pitch and formant information directly. One transform that has been extensively researched in the past for encryption of speech is the Discrete Fourier Transform (DFT) (see [7], [10] and [11]).

This paper describes research that has been conducted on the use of Discrete Cosine Transform (DCT) for speech encryption. A description of the scrambling technique is given in Section 2. The DCT is of interest since it is known to approach the optimum performance of the Karhunen Loeve Transform in terms of decorrelation of information [1]. A major finding of this research is that the DCT offers higher quality output, lower residual intelligibility and greater cryptanalytic strength compared to a DFT based speech encryption system as will be demonstrated in Section 3. In Section 4 it will be shown how to prevent the ciphertext alone attacks on frequency domain scramblers from [4] in the case of the DCT scrambler by using dummy transform component insertion. In Section 5 a comparison will be made between the performance of the DCT scrambler and a digital scrambler on a bandlimited channel.

2. DESCRIPTION OF THE SCRAMBLING TECHNIQUE

Consider the DCT of a vector x of length N representing one frame of speech samples given by

$$u = Fx$$

where F denotes the transform matrix given by

$$F_{ij} = (c(j))/N \cos[(2i+1)\pi j/(2N)] \quad \text{for } i,j = 0,1,...,N-1;$$

$$c(j) = \sqrt{2} \quad \text{for } j = 0;$$

$$= 2 \quad \text{for } j = 1,2,...,N-1.$$

The scrambling is performed by an NxN matrix P applied to the DCT vector u to produce a vector v given by

$$v = Pu$$

The scrambled speech y in the time domain is obtained by applying the inverse transformation F^{-1} on v given by

$$y = F^{-1}v$$

Restrictions have to be imposed on P due to bandwidth limitations of speech. Speech for telephony is restricted in bandwidth to 300-3400 Hz. The components outside the desired bandwidth are set to zero and the remainder are permuted. In the case for N = 256 samples per analysis frame, M = 197 components are permuted. Methods for generating permutations so that all M! permutations are available have been addressed in [11]. This would imply for the proposed system, that 197! permutations are possible. Clearly, an exhaustive key search is not feasible.

3.   COMPARISON WITH DFT SCRAMBLER

Suppose that a DFT based scrambler is used on a channel limited to the bandwidth between 300-3400 Hz using N = 256 samples per analysis frame. In this case there are 97 spectral coefficients available for permutation. The DCT scrambler, as described in Section 2, has a much larger number of permutations available for use compared to the DFT scrambler. This gives the DCT scrambler superior security under an exhaustive key search attack.

A subjective listening test similar to that proposed in [6] was used to evaluate the performance of the DFT and DCT scramblers in terms of the residual intelligibility of the scrambled speech. Thirty subjects were asked to identify the digits 0 to 9 after being scrambled. Each person was asked to identify forty scrambled one digit numbers from each scrambler. For the DFT and DCT scrambler the average percentage of correct digits were 17.2% and 17.1% respectively. This result demonstrates the similar performance of the two scramblers in relation to the residual intelligibility of the scrambled speech. It should be noted that the same subjects were given the same test on scrambled speech from an eight segment hopping window scrambler. In this case the average percentage of correct digits was 47.7%. This, clearly, demonstrates the better performance of DFT and DCT scramblers over the hopping scrambler in lowering the residual intelligibility of the scrambled speech.

In order to measure the residual intelligibility of the scrambled speech and the quality of the recovered speech of a speech scrambler, objective distance measures can be used which quantify the similarities between two speech samples. Two such measures are the Frequency Variant Spectral Distance (FVSD) as described in [8] and the Segmental Spectral Signal to Noise Ratio (SSNR) as described in [7]. Table 1 contains the results of using each of these measures to compare the original and scrambled speech, and the original and recovered speech for both the DFT and DCT scrambler. This table indicates the lower residual intelligibility and the better quality in the recovered speech offered by the DCT scrambler.

	FVSD	SSNR(dB)
DFT (Original/Scrambled)	102.1	-2.2
DCT (Original/Scrambled)	99.9	-3.7
DFT (Original/Recovered)	22.7	16.6
DCT (Original/Recovered)	14.6	45.3

Table 1

Comparison of DFT and DCT Scramblers

The permutation of transform coefficients preserves the signal energy within a given frame. Talk spurt and intonation information is still recoverable from the scrambled speech. In addition all the coefficients contributing to the original spectral shape are still present in the scrambled spectrum. They need only be rearranged in order to recover intelligibility. As shown in [4] this information can be used to conduct a ciphertext only attack on a DFT scrambler. This technique uses a template consisting of a vector codebook that can be used in place of the plaintext spectrum. A brief description of this attack is as follows:

Recent work in speech coding (see [5]) has shown that the speech waveform can be represented by a finite number of feature vectors or a vector codebook. Consider a codebook in which the feature vectors are speech spectra obtained by training the codebook on a large set of speech data. Transform components in the scrambled speech are rearranged in order to minimise the mismatch between the decrypted spectrum and the codebook vector. Each vector in the codebook is examined in turn as the template necessary for pattern matching. The resulting decrypted spectrum giving the minimum mismatch with its respective codebook vector, is assumed to be the recovered spectrum. It is beneficial to accumulate permutation information over all available ciphertext frames in the fixed key system.

A DFT scrambler permuting 88 coefficients, with a frame size $N = 256$, was cryptanalysed using the procedure described above. The attack was performed on a system using a fixed permutation, as well as one using a constantly varying permutation. 300 frames or ten seconds of speech were used for the attack. Only a small number of coefficients were able to be positioned correctly, however in each case this was sufficient to recover intelligible speech. As expected, the fixed permutation attack gave more intelligibility. The attack was also applied to a DCT scrambler permuting 197 coefficients and with a frame size of 256 using a fixed permutation, as well as one using a constantly varying permutation. A much smaller percentage of components were able to be placed correctly with the DCT scrambler. A comparison of the results is given in Table 2.

Table 2 gives the percentage of components which were placed within two positions of their correct location for each case. It was found that as this percentage increased, there was a corresponding increase in the intelligibility of the recovered speech. Figures 1 to 4 show the distribution of the component distances from their original positions. It is clear from these figures that more intelligibility was able to be recovered from the attacks on the DFT scrambler. This indicates that the DFT scrambler is far less secure than the DCT scrambler.

	DCT	DFT
Original/Scrambled	2.5%	3.7%
Original/Recovered (Varying Permutation)	4.8%	8.9%
Original/Recovered (Fixed Permutation)	10.7%	23%

Table 2

Results of Attack on DCT and DFT Scramblers

4. ENERGY MODIFICATION USING INSERTION OF DUMMY COMPONENTS

One of the main problems associated with designing a speech scrambler is to conceal the energy variations in the scrambled speech. To overcome this problem it was proposed to use (see [7]) the substitution of dummy spectral components for a predefined block of components from the original speech spectrum. The magnitude of the components are chosen such that for any given frame the energy will be close to an established upper limit. This produces constant energy scrambled speech which resembles random noise. In [7] it is suggested that dummy components should be adaptively positioned so that significant transform components would not be discarded by the process. This requires that the components be recognizable as dummy components and hence run the risk of being detected and removed by a cryptanalyst.

In the proposed DCT based scrambler the location of the dummy components is fixed prior to permutation, but is chosen carefully to ensure that the insertion process has little effect on the recovered speech quality. Ten components with random amplitudes are used. They are scaled in order to maintain the desired constant energy limit. Following the scrambling operation the dummy components will be distributed throughout the spectrum. In this fashion these components should be undetectable due to their random nature.

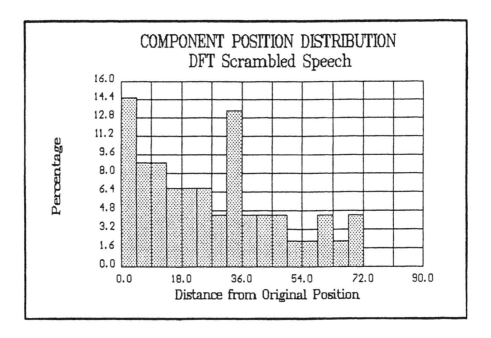

Figure 1

DFT Scrambled Speech

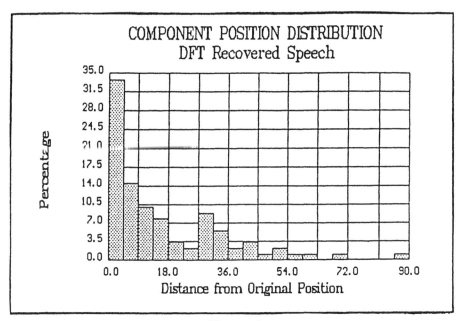

Figure 2

DFT Recovered Speech After Attack

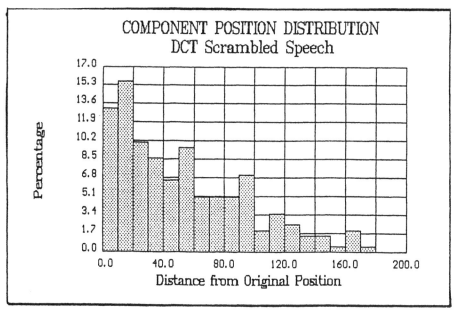

Figure 3

DCT Scrambled Speech

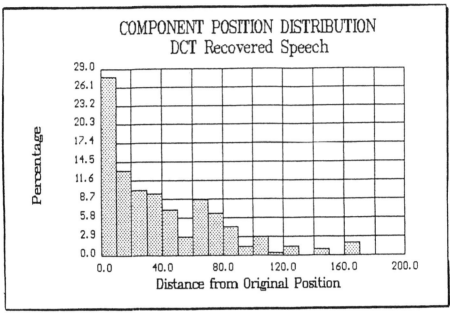

Figure 4

DCT Recovered Speech After Attack

Observe that for silent frames the ten dummy components will be very easily detected since they will be the only components in the spectrum with a significant amplitude. To overcome this problem silent frames are treated as a special case. When the energy of an input frame of speech falls below a predefined threshold it is said to be silent. In this case the entire spectrum is replaced by a dummy spectrum whose components have been selected randomly such that their magnitudes match the amplitude distribution of non-silent frames. One component in the spectrum is used to indicate that such a substitution has been made. The scrambling process will move this component to a position unknown to a cryptanalyst.

Following the descrambling process the receiver must determine whether the current frame was originally a silent frame. It does this by interrogating the component used to signal such an event. If the frame was originally silent then the entire spectrum is replaced with a silent frame. If not then only the ten dummy components are zeroed. If a sampling rate of 8 KHz is used, with a selection of N = 256 samples in each analysis frame, 197 spectral components lie within the usable bandwidth from 300 to

3400 Hz. Thus only six percent, (ten dummy components and one signalling component) of the usable spectral components are lost as a result of this process.

The codebook based attack described in Section 3 is possible due to the fact that all the original transform components are present in an unaltered form in the scrambled spectra. A useful side effect of the dummy component insertion technique is the introduction of high energy dummy transform components. This confuses the attack described in Section 3. A transform domain scrambler using this energy modification is secure under the above mentioned attack. The attack is unsuccessful in determining the original position of the transform components, since the dummy components are selected in their place. The component position distribution of Figure 5 demonstrates the poor performance of the attack.

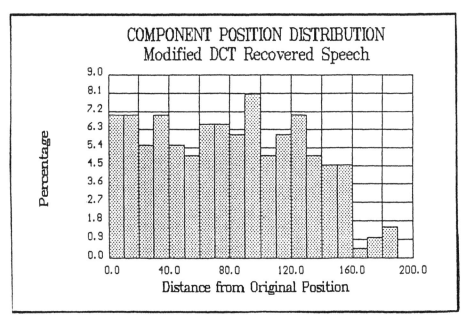

Figure 5

Modified DCT Recovered Speech After Attack

5.   COMPARISON WITH DIGITAL SPEECH SCRAMBLERS ON ANALOG CHANNELS

In digital speech encryption for the standard analog channel the analog signal is first digitized at a rate of 64 kbits/sec. The digital bits are compressed using a speech coder to generate a signal at a bit rate suitable for the limited bandwidth of the telephone channel. The bit stream is then encrypted using either a block cipher or a stream cipher. Current research and development work in this direction indicates that it is possible to design highly secure speech transmission systems using this approach with speech coders operating at 4.8 kbits /sec as described in [2] for transmission on the standard telephone line. It should be noted that there is a degradation in the quality of the recovered speech in such a system caused by the speech compression.

A comparison was made of the performance of the DCT scrambler which includes dummy spectrum insertion as described in Section 4 and a digital scrambler used on an analog channel which operates with a speech coder at 4.8 kbits/sec using a secure stream cipher such as described in [9]. The residual intelligibility of the scrambled speech and the quality of the recovered speech were used as criterion for the measure of performance. The comparison was made on the basis of the objective distance measures FVSD and SSNR. Table 3 contains measures resulting from a comparison of original and scrambled speech from the DCT scheme described above and the output of a 4.8 kbits/sec speech coder fed into a stream cipher. Both produced scrambled speech with very low residual intelligibility. The digital scrambler performs best according to its FVSD, however the SSNR would suggest that the DCT scheme gives lower intelligibility. To further demonstrate the low intelligibility of the scrambled speech in the DCT scrambler with dummy spectrum insertion the same thirty subjects, as used in the experimental results given in Section 3, were asked to identify the digits 0 to 9. In this case on the average only 9.4% of the digits were identified correctly. This result, corresponding to guessing odds, demonstrates the low intelligibility in the DCT scrambler with dummy spectrum insertion.

Table 3 contains an evaluation of the quality of the recovered speech resulting from the 4.8 k bits/sec of digital speech coders and the speech recovered after scrambling using the DCT scrambler. The analog scrambler by far has the best recovered speech quality according to the distance measures used. Informal listening tests verified these results. It should be noted that the development of the Integrated Services Digital Network (ISDN) will allow speech to be transmitted without the use of speech compression. The inherent loss of recovered speech quality will not be an issue in ISDN.

	FVSD	SSNR (dB)
DCT (Scrambled)	140.7	-22.3
4.8 kbit/sec coder (Scrambled)	152.2	-11.1
DCT (Recovered)	6.76	53.8
4.8 kbit/sec coder (Recovered)	30.03	24.4

Table 3

Comparison of DCT Scrambler and 4.8 kbits/sec Digital Scrambler

6. CONCLUSIONS

A secure speech scrambler using the DCT has been presented. It was shown that by using dummy spectrum insertion one can design a DCT scrambler which is secure from the vector codebook attack described in [4]. A comparison of the performance of the DCT scrambler and a digital speech scrambler was made. The residual intelligibility measures for each of the scramblers were very similar suggesting that no significant advantage would be gained by using the digital device in preference to the analog one. The recovered speech quality measures, however, revealed that the DCT scrambler is far superior to the digital schemes tested. The speech recovered from the DCT scrambler was very clear and natural.

It should be noted that the results described in this paper are based on computer simulation. Further investigation is needed into the actual hardware design of a DCT scrambler.

REFERENCES

1. N. Ahmed, T. Natarajan and K. Rao, "Discrete Cosine Transform", **IEEE Trans. on Computers,** 1974, pp. 90-93.

2. B. Atal, V. Cuperman and Gershoff (editors), **Advances in Speech Coding,** Kluwer Academic Press, 1990.

3. H. Beker and F. Piper, **Secure Speech Communications,** Academic Press, 1985.

4. E. Dawson, B. Goldburg and S. Sridharan, "The Automated Cryptanalysis of Analog Speech Scramblers" in **Abstracts of EUROCRYPT 91,** 1991, pp. 203-207.

5. R.M. Gray, "Vector Quantization", IEEE ASSP Mag., April 1984, pp 4-29.

6. N. Jayant, B. McDermott, S. Christensen and A. Quinn, "A comparison of four methods for analog speech privacy", IEEE Trans. on Communications, Vol. COM-29, No. 1, 1981, pp. 18-23.

7. A. Matsunaga, K. Koga, M. Ohkawa," An analog speech scrambling system using the FFT technique with high level security", IEEE Journal in Selected Areas in Communications, vol. 7, May 1989, pp. 540-547.

8. S.R. Quackenbush, T.P. Barnwell and M.A. Clements, Objective Measures of Speech Quality, Prentice Hall, 1988.

9. R.A. Rueppel, New Approaches to Stream Ciphers, PhD Thesis, Swiss Federal Institute of Technology, 1984.

10. K. Sakurai,T. Koga and T. Muratani, "A speech scrambler using Fast Fourier Transform Techniques", IEEE Selected Areas in Communications, vol SAC-2, No. 3, May 1984, pp. 434-442.

11. S. Sridharan, E. Dawson and B. Goldburg, "A Fast Fourier Transform Based Encryption System" to appear in IEE Proceedings Part I Communications, Speech and Vision.

An Oblivious Transfer Protocol and Its Application for the Exchange of Secrets

Lein Harn and Hung-Yu Lin
Computer Science Telecommunications Program
University of Missouri - Kansas City
Kansas City, MO 64110

<Abstract>
The oblivious transfer protocol is a powerful tool in the design of cryptographic applications, such as coin flipping by the telephone, exchanging secrets and sending certified mail. In this paper, for our purpose of extending the oblivious transfer to the exchange of secrets, we redefine a verifiable oblivious transfer protocol which has the three properties of fairness, verifiability and security. The structure of the protocols is similar to the original protocols proposed by Rabin and Blum. The major difference is that our protocols are based on the difficulty of the discrete logarithm.

1. Introduction

Oblivious transfer is a powerful tool in the design of protocols for the exchange of secrets between two mutually distrustful parties. In this paper, a verifiable oblivious transfer protocol is presented and is then modified for the exchange of secrets.

The problem of the exchange of secrets has been widely investigated in the past few years, but so far as we know, almost all of the proposed protocols are based on the difficulty of factoring or the quadratic residue assumption(QRA). In this paper, the security of the proposed protocol will be based on the difficulty of discrete logarithm and, with negligible restriction, the secrets in exchange can be chosen arbitrarily by the participating parties. In other papers, the secret is of a special form, for example, the factorization of a composite number.

It is obvious that the strings of bits can be exchanged by following the single bit exchange protocol. But it takes too much time because exchanging even a single bit constitutes many iterations. In this paper, each participant is allowed to choose his own secret bits and then embeds the secret in the exponent of a modulo exponentiation. Therefore, the secret to be exchanged is not of a special form and the protocol can proceed in a more efficient way, which will be shown in this paper.

All of the discussion should be in the probabilistic scenario, but for simplicity, we will present it in a deterministic way.

2. Related works

The problem of the exchange of secrets was first addressed by Rabin [11] and Blum [3,4]. Blum's paper [4] was probably the best known at first. In that paper, the secret is the factorization of some composite number instead of a single bit or a string of bits. Unfortunately, since the secret of the factorization of a composite number is indeed not a single bit and some linear relationship exists among the exchanged bits, it is shown by Hastad and Shamir [8] that this protocol is not so secure as was claimed. Later, Luby, Micali, and Rackoff [10] proposed a new protocol for the exchange of a single secret bit by flipping a symmetrically-biased coin; Tedrick improved the fairness by exchanging even half a bit [12] and reducing the computational advantage from 2-1 to 5-4 [13]. Recently Cleve [6] proposed his controlled gradual disclosure scheme to guarantee that one's confidence of a secret is steadily increasing towards 1 instead of drifting towards 1 by following a random walk.

In general, if the release of a secret is based on the result of flipping two independently identical coins, there may be significant difference in each opponent's knowledge of the other's secret due to the nature of random walk. However, if only one coin, which is biased according to the secrets from two parties, is flipped and the results are observed by two parties, one can infer his/her opponent's secrecy by watching the reaction of his/her opponent. But one thing in common for all of the above schemes is that the security is based on the difficulty of factoring or QRA. The first protocol based on the difficulty of discrete logarithm was proposed by Brickle, Chaum, Damgard, and Graaf [5]. Although, with their protocol one can convince the others that his secrecy of discrete logarithm is within some interval without revealing anything, the exact release of this secret is time consuming unless this problem is transferred to another problem in which security again is based upon factorization.

3. Verifiable Oblivious Transfer
3.1 Definition

According to the general definition of oblivious transfer, wherein Alice transfers a secret, which has been notarized by a third party who is trusted by both Alice and Bob, to Bob, the following two conditions need to be satisfied at the end of the protocol:

(a) With probability 1/2, Bob receives the secret, and with probability 1/2, Bob learns nothing about the secret.

(b) Alice does not know whether or not Bob receives the secret.

Many oblivious transfer protocols, either non-interactive [1,7] or interactive [2], have been implemented with different assumptions. The strongest one assumes that both players follows the protocol faithfully. However, in the real world this assumption is hardly acceptable between two mutually distrustful parties. The weakest one assumes nothing else except one cryptographic assumption, i.e., the hardness of factoring. Therefore, each message sent needs to be committed to prevent one from denying what he/she has sent to the other. By employing a trusted intermediary or using some cryptographic signature scheme, this commitment problem can be easily solved.

For our purpose of extending the oblivious transfer to the exchange of secrets, we now define a verifiable oblivious transfer protocol which has the three properties:

(1) Fairness: Conditions (a) and (b) hold true if both parties follow the protocol.

(2) Verifiability: Any cheating from Alice can be detected by Bob with probability almost equal to 1.

(3) Security: Bob cannot obtain Alice's secret with probability higher than 1/2 if he does not follow the protocol.

We ignore the possible situation where one denies what he/she has sent to the other in the process. In the following protocol we do not use signature or bit-commitment technique for the messages transferred between Alice and Bob.

3.2 Protocol

Before any data can be transferred between Alice and Bob, we assume that a large prime p, where $p = 4*p' + 1$, p' is also prime, and a primitive element, e, of G(p), where G(p) is the Galois Field of p, are known to both. Alice chooses her own secret α with $gcd(\alpha, p-1)=1$ and α is a quadratic non-residue of p, i.e., $\alpha \in QNR_p$, and submits $A_s = e^{\alpha} \pmod{p}$ and $A_{1-s} = \alpha^{\alpha} \pmod{p}$, where $s \in \{0,1\}$ and is known only to Alice, to a trusted third party for notarization. From now on, the "secret" in the following protocol refers to α in the notarized value. The problem of whether α, instead of some α', in this protocol is the real secret of Alice is the same as whether the factorization of N, instead of N' is the real secret of Bob in many other oblivious transfer protocols and this will not be discussed in this paper. After A_0 and A_1 are notarized, Bob starts the oblivious transfer by following the steps below:

Step 1. Bob randomly chooses a secret number b with gcd(b , p-1) = 1, and sends to Alice

$$C_1 = A_0{}^b \quad \text{mod p or } C_1 = A_1{}^b \quad \text{mod p.}$$

Step 2. Alice computes and sends to Bob

$$C_2 = C_1{}^{\alpha^{-1}} \quad \text{mod p.}$$

Step 3. Bob computes

$$C_3 = C_2{}^{b^{-1}} \quad \text{mod p.}$$

If $C_3 = e$, then Bob knows nothing about Alice's secret(i.e. Bob loses the game); otherwise Bob checks if

$$A_{s'} = e^{C_3} \quad \text{mod p for s' = 0 or s' = 1.}$$

If so, Bob knows Alice's secret $\alpha = C_3$ (i.e., Bob wins the game). Otherwise, Bob can charge Alice of cheating.

3.3 Protocol Analysis

Before starting to prove the properties as stated in the definition, we need to explain the assumptions made in this protocol:

(A1) Alice prefers not to reveal her secret to Bob and Bob prefers not to let Alice know what he has received.

Without this assumption, there are many ways for Alice to reveal her secret to Bob which violate the above definitions. For example, Alice can choose small α such that Bob can derive α with probability much higher than 1/2. This assumption is quite natural for those who want to play around with this protocol.

(A2) Given C and g where $C = g^x \pmod{p}$ and g is a generator, it is computationally infeasible to compute x.

This is exactly what the well known Diffie-Hellman assumption is based on. With certain modification, we make another assumption in this protocol.

(A2') Suppose that g_0 and g_1 are primitive elements. Given C_0, C_1, g_0, g_1, and p, with $C_i = g_0{}^x \pmod p$ and $C_{1-i} = g_1{}^x \pmod p$, where x is a random number and $i \in \{0,1\}$, it is computationally infeasible to decide either $C_0 = g_0{}^x \pmod p$ or $C_1 = g_0{}^x \pmod p$ without knowing x.

At first glance, this assumption seems to be much stronger than (A2) since the secret involved in (A2') is only one bit, while in (A2) the secret is x. But so far no one can prove that the problem in (A2) is harder than in (A2'). In fact, the two assumptions are believed to have the same complexity. Bellare and Micali's paper [1] implicitly made the same assumption. One potential problem of this assumption is that Bob may distinguish A_0 from A_1 using the quadratic residue property. By choosing α carefully we can remove this problem as shown below:

Lemma 1. By choosing α with $\alpha \in QNR_p$ and $\gcd(\alpha, p-1) = 1$, both A_0 and A_1 are in QNR_p.

<proof>: Since e is a primitive element of G(p) and $\gcd(\alpha, p-1) = 1$, so $e \in QNR_p$ [9] and α is odd. That is, both A_0 and A_1 are in QNR_p.

According to the above discussion, we know that the secret, α, needs to be in QNR_p with $\gcd[\alpha, p-1] = 1$. Now we would like to examine how many numbers within [1, p-1] satisfy these conditions.

Lemma 2. 1/4 of the numbers within [1, p-1] can be selected as the secret.

<Proof> We know that $L(1/p) = L(-1/p)$ if p is congruent to 1 mod 4, where L is the Legendre symbol. Therefore, we can conclude that if x is in QNR_p and x is even, then p-x is also in QNR_p and is odd. Since half of the numbers within [1, p-1] are in QNR_p, so 1/4 of the numbers within [1, p-1] are odd quadratic non-residue of mod p and can be chosen as the secret.

(A3) THe publication of $A_s = e^\alpha \pmod p$ and $A_{1-s} = \alpha^\alpha \pmod p$ does not reveal the secret, α.

Some efforts have been tried by the authors to prove the vulnerability of this assumption. So far none of these attacks seems to work and we encourage readers to prove it.

After explicitly explaining the assumptions and the size of the secret to be revealed in this protocol, we can now prove the fairness, verifiability, and security.

Theorem 1. This protocol is fair if Alice and Bob follow the protocol.

<Proof>: First we assume that Bob can obtain α with probability greater than 1/2. This implies that Bob can distinguish (α^α) from (e^α). Since both α and e are primitive elements of $G(p)$, it contradicts assumption (A2') (in fact without knowing α, it's much more difficult for Bob to make the distinction). Then we assume that Alice knows whether or not Bob receives α with probability greater than 1/2. This implies that Alice can distinguish $(\alpha^\alpha)^b$ from $(e^\alpha)^b$. Again, this contradicts assumption (A2'). So, this protocol is fair if Alice and Bob follow the protocol.

Theorem 2. This protocol is verifiable.

<Proof>: The only way Alice can cheat in this protocol is by sending back in Step 2 to Bob $\mathbb{C}$, instead of $C_2 = C_1^{\alpha^{-1}} \pmod p$. Suppose that Alice can cheat Bob without being detected. That is, Alice can choose $\mathbb{C}$ such that $\mathbb{C} = e^b$ or $\mathbb{C} = \alpha^b$ on her will. However, this means that either she can make the distinction between e^b and α^b or she knows b and therefore it contradicts assumption (A2) or (A2'). So, this protocol is verifiable.

Theorem 3. This protocol is secure.

<Proof>: If Bob deviates from the protocol, that is, in step 1 Bob sends to Alice $\mathbb{C}_1$ instead of $C_1 = A_0^b \pmod p$ or $C_1 = A_1^b \pmod p$, and after step 2 Bob receives $C_2 = \mathbb{C}_1^{\alpha^{-1}}$. Now Bob knows one more equation $\mathbb{C}_1 = C_2^\alpha$. However, from assumption (A2) we know this does not give Bob any advantage in obtaining Alice's secret. So, this protocol is secure.

4. The exchange of secrets
4.1 Problem description

In this section, the verifiable oblivious transfer in section 3 will be used as a constructing block to build up the protocol for the exchange of secrets. First we have to explain what we are trying to achieve in the exchange of secrets. Suppose Alice and Bob are managers of two competing companies who want to exchange secret market information over the computer network. The trading of the information is voluntary, that is, anyone can stop the exchanging of the secrets anytime during the process without any obligation. In such a case, the difference of chances between both parties in correctly guessing the other's secret should be negligible. In addition, any cheating in the process will be detected with very high probability - almost equal to 1. No intermediary is needed during the process of the exchange and they try not to go to court if possible, except when one obtains the other's secret while the other denies it.

4.2 Protocol

Both Alice and Bob will follow the same sequence of steps in turn, We will only talk about the actions in one side. Suppose that α and β are the secrets of Alice and Bob, respectively, which have the same properties as mentioned in Section 3.2. A_0, A_1 and B_0, B_1 are, respectively, the corresponding notarized information of Alice and Bob with $A_0 = e^\alpha \bmod p$, $A_1 = \alpha^\alpha \bmod p$, $B_0 = e^\beta \bmod p$, and $B_1 = \beta^\beta \bmod p$ (Note that Alice and Bob do not need to permute the notarized information as in Section 3.2). Alice starts the protocol first.

Step 1 For i=1 , 2,..., 100, Alice randomly chooses a_i with $\gcd(a_i , p-1) = 1$, and computes

$$\mathbb{B}_{i,s_i} = (B_0)^{a_i} \bmod p \text{ and } \mathbb{B}_{i,1-s_i} = (B_1)^{a_i} \bmod p,$$

where s_i is in $\{1,0\}$ and is known only to Alice. Then Alice sends these 100 pairs of ciphertext ($\mathbb{B}_{i,s_i}$, $\mathbb{B}_{i,1-s_i}$) to Bob and Alice will receives 100 pairs of ciphertext ($\mathbb{A}_{i,t_i}$, $\mathbb{A}_{i,1-t_i}$) from Bob, where t_i is known only to Bob.

Step 2. For each pair of ciphertext received, Bob computes

$$\mathbb{B}'_{i,0} = (\mathbb{B}_{i,s_i})^{\beta^{-1}} \quad \text{mod} \quad p \quad \text{and} \quad \mathbb{B}'_{i,1} = (\mathbb{B}_{i,1-s_i})^{\beta^{-1}} \quad \text{mod} \quad p.$$

Then Bob exchanges the 100 pairs of data with Alice one bit at a time, starting from the least significant bit, that is, 100 pairs of bits in each turn.

Step 3. For the i-th pair of bits Allice received, she checks, by knowing her own s_i and a_i, if the least significant bit of e^{a_i} mod p is equal to the corresponding least significant bit in $(\mathbb{B}'_{i,0}, \mathbb{B}'_{i,1})$. This implies that the other bit received is the least significant bit of β^{a_i}. Only when Bob performs the operation faithfully in Step 2 and returns the bits in the same order, Alice will continue to exchange second least significant bits, and so on. Otherwise, Bob cheated and Alice can stop the protocol.

Step 4. After both parties faithfully exchange all of the bits, Alice knows all the bits of β^{a_i} and e^{a_i} and she can derive β from β^{a_i} by computing

$$\beta = (\beta^{a_i})^{a_i^{-1}} \quad \text{mod} \quad p.$$

The analysis of this protocol is almost the same as the verifiable oblivious transfer protocol. We will only briefly discuss the difference.

From assumption (A2), we know in the above protocol that the chance of cheating without being detected is $1/2^{100}$. That is, any cheating will be detected with probability almost equal to 1. If Bob terminates the process for some reason at the middle of protocol, all the advantage he has over Alice is just one bit of α^{b_i} and this is negligible in comparison with the size of the secret, α.

5. Conclusion

We have proposed a verifiable oblivious transfer protocol and its extension for the exchange of secrets. The structure of the protocols is similar to that of the original protocols proposed by Rabin and Blum. The major difference is that our protocols are based on the difficulty of discrete logarithm. Because of this difference, the secret in our protocol can be chosen as a random number, while the secret in original protocols must be of a special form, i.e., the factorization of a composite number.

References

[1] Bellare, M., and Micali, S., Non-interactive oblivious transfer and applications, *Avances in Cryptology: CRYPTO '89*, pp. 547-5557.

[2] Berger, R., Peralta, R., and Tedric, T., A provably secure oblivious transfer protocol, *Avances in Cryptology: Proc. of EUROCRYPT '84*, pp. 379-386.

[3] Blum, M., Three applications of oblivious transfer: 1. Coin flipping by telophone, 2. How to exchange secrets, 3. How to send certified electronic mail, Dept. EECS, University of California, Berkeley, Calif., 1981.

[4] Blum, M., How to exchange (secret) keys, *ACM Transaction on Computer System, Vol. 1, No. 2*, May 1983, pp. 175-193.

[5] Brickle, E., Chaum, D., Damgard, I., and van de Graaf, J., Gradual and verifiable release of a secret, *Advances in Cryptology: CRYPTO '87*, pp. 156-166.

[6] Cleve, R., Controlled gradual disclosure schemes for random bits and their applications, *Avances in Cryptology: CRYPTO '89*, pp. 573-588.

[7] Harn, L., and Lin, H. Y., Non-interactive oblivious transfer, *Electronics Letters, Vol.26 , No. 10*, May 1990, pp. 635-636.

[8] Hastad, J., and Shamir, A., The cryptographic security of truncated linearly related variables, *Proc. of 17th STOC*, 1985, pp. 355-362.

[9] Knuth, D., The Art of Computer Programming, Vol. 2, Addison Wesley, Reading, MA, 1973.

[10] Luby, M., Micali, S., and Rackoff, C., How to simultaneously exchange a secret bit by flipping a symmetrically biased coin, *Proc, 22nd Ann. IEEE Symp. on Foundations of Computer Science*, 1983, pp. 11-21.

[11]. Rabin, M., How to exchange secret by oblivious transfer. Harvard Center for Research in Computer Technology, Cambridge, Mass., 1981.

[12] Tedric, T., How to exchange half a bit, *Advances in Cryptology: Proc. of CRYPTO' 83*, pp. 147-151.

[13] Tedric, T., Fair exchange of secrets, *Advances in Cryptology: Proc. of CRYPTO '84*, pp. 434-438.

4 Move Perfect ZKIP of Knowledge with No Assumption

Takeshi Saito & Kaoru Kurosawa
Dept. of Electrical and Electronic Eng.
Tokyo Institute of Technology
Tokyo, 152 Japan

Kouichi Sakurai
Computer & Information Systems Laboratory
Mitsubishi Electric Corporation
5-1-1 Ofuna, Kamakura, 247 Japan

Abstract

This paper presents a 4-move perfect ZKIP of knowledge with no cryptographic assumption for the random self reducible problems [TW87] whose domain is $NP \cap BPP$. The certified discrete log problem is such an example. (Finding a witness is more difficult than the language membership problem.) A largely simplified 4-move ZKIP for the Hamilton Circuit problem is also shown. In our ZKIP, a trapdoor coin flipping protocol is introduced to generate a challenge bit. P and V cooperatively generate a random bit in a coin flipping protocol. In a trapdoor coin flipping protocol, V who knows the trapdoor can create the view which he can later reveal in two possible ways: both as head and as tail.

1   Introduction

A zero knowledge interactive proof system (ZKIP) of knowledge consists of two polynomially bounded probabilistic Turing machines, a prover(P) and a verifier(V). P has a secret in his auxiliary input tape and convinces V that P knows the secret without leaking any additional information. A protocol (P,V) is zero knowledge if any $\tilde{V}$ can create by himself everything $\tilde{V}$ sees in the conversation with P (view). The atomic protocol of usual ZKIPs is as follows.

1. P puts something in an envelope and sends the envelope to V.

2. V sends a challenge bit to P.

3. P answers by revealing something related to the content of the envelope.

The usual ZKIPs repeat the atomic protocol polynomially many times in serial (unbounded moves). However, the parallel execution of the atomic protocol(3move) is not zero knowledge[GK90]. [FS89] showed 4-move ZKIPs of knowledge for the Hamilton Circuit(HC) problem relying on the chosen discrete log assumption(CDLA) and one way permutations. They used a trapdoor bit commitment scheme to make the envelope. P who doesn't know the trapdoor cannot change the content of the envelope later. $\tilde{V}$ who knows the trapdoor can create the envelope by himself which he can later reveal as anything.

This paper presents a 4-move perfect ZKIP of knowledge with no cryptographic assumption for the random self reducible problems [TW87] whose domain is $NP \cap BPP$.

The certified discrete log problem is such an example. (Finding a witness is more difficult than the language membership problem.) A largely simplified 4-move ZKIP for the HC problem is also shown. In our ZKIP, a trapdoor coin flipping protocol is introduced to generate a challenge bit. P and V cooperatively generate a random bit in a coin flipping protocol. In a trapdoor coin flipping protocol, V who knows the trapdoor can create the view which he can later reveal in two possible ways: both as head and as tail.

"V's challenge bits" are computed by "a trapdoor coin flipping protocol" in the proposed ZKIP while "P's first message(envelope)" is computed by "a trapdoor bit commitment" in [FS89]. If we assume only one way permutations, our ZKIP for the HC problem is much more efficient than that of [FS89].

(As to language membership, [BMO90] showed a 5 move perfect ZKIP for the random self reducible problems.)

2 Preliminaries

2.1 Interactive Proof System and Zero Knowledge [FS90]

Our model of computation is the probabilistic polynomial time interactive Turing machine (both for the prover P and for the verifier V). In this section, the common input is denoted by x, and its length is denoted by $\mid x \mid = n$. Each machine has an auxiliary input tape. P's auxiliary input is denoted by ω. V's auxiliary input is denoted by y. $\nu(n)$ denotes any function vanishing faster than the inverse of any polynomial. Formally:

$$\forall k \; \exists N \; s.t. \; \forall n > N \; \nu(n) < \frac{1}{n^k}$$

Negligible probability is probability behaving as $\nu(n)$. *Overwhelming* probability is probability behaving as $1 - \nu(n)$.

$A(x)$ denotes the output of a probabilistic algorithm A on input x. This is a random variable. $V_p(x)$ denotes V's output after interaction with P on common input x. $M(x; A)$(where A may be either P or V) denotes algorithm A as one computation step for M.

Definition 2.1: Let R be a relation $\{(x, \omega)\}$ testable in polynomial time, where $\mid x \mid = \mid \omega \mid$ (this restriction can be met by standard padding techniques for all relations of interest).

For any x, its *witness set* $\omega(x)$ is the set of ω such that $(x, \omega) \in R$.

Definition 2.2: An *interactive proof of knowledge* system for relation R is a pair of algorithms (P, V) satisfying:

1. *Completeness*: $\forall (x, \omega) \in R \ Prob(V_{P(x,\omega)}(x) accepts) > 1 - \nu(n)$

2. *Soundness* : $\exists M \ \forall P' \ \forall x \ \forall \omega'$
   $Prob(V_{P'(x,\omega)}(x) accepts) < Prob(M(x, \omega'; P') \in \omega(x)) + \nu(n)$

The probability is taken over the coin tosses of V, P' and M. The knowledge extractor M runs in expected polynomial time, and uses P' as a blackbox.

Remark: If $\omega(x)$ is empty, this definition implies that the probability that V accepts is negligible.

Definition 2.3: Proof system (P, V) is *zero knowledge* (*ZK*) over R if there exists a simulator M which runs in expected polynomial time, such that for any probabilistic polynomial time V', for any $(x, \omega) \in R$, and any auxiliary input y to V', the two ensembles $View_{PV'}(x, y)$ and $M(x, y; V')$ are polynomially indistinguishable. M is allowed to use V' as a subroutine.

$$View_{PV'}(x, y) = (\rho, a_1, b_1, \ldots)$$

where ρ is the random tape of V' and $a_i(b_i)$ is the i-th message of $P(V')$.

2.2   WI and WH [FS90]

Definition 2.4: Proof system (P, V) is *witness indistinguishable* (WI) if for any V', for any large enough input x, for any $\omega_a \in \omega(x)$ and $\omega_b \in \omega(x)$, and for any auxiliary input y for V', the ensembles, $V'_{P(x,\omega_a)}(x, y)$ and $V'_{P(x,\omega_b)}(x, y)$ generated as V's view of the protocol are indistinguishable.

Theorem 2.1: Let (P, V) be any zero knowledge protocol. Then the protocol is WI.

Theorem 2.2: WI is preserved under parallel composition of protocols.

The above two Theorems establish a methodology for constructing WI protocols. Take the basic step of a ZKIP. By theorem 2.1, it is WI.

Definition 2.5: G is a *generator* for relation R if on input 1^n it produces instances $(x, \omega) \in R$ of length n. G is an *invulnerable generator* if for any polynomial time non uniform *cracking* algorithm C, $prob((x, C(x)) \in R) < \nu(n)$, where $x = G(1^n)$. The probability is taken over the coin tosses of G and C.

Definition 2.6: Let (P, V) be a proof of knowledge system for relation R, and let G be a generator for this relation. (P, V) is *witness hiding* (WH) oñ (R, G) if there exists a witness extractor M which runs in expected polynomial time, such that for any non uniform polynomial time V'

$$Prob(V'_{P(x,\omega)}(x) \in \omega(x)) < Prob(M(x; V', G) \in \omega(x)) + \nu(n)$$

where $x = G(1^n)$. The probability is taken over the distribution of the inputs and witness, as well as the random tosses of P and M. The witness extractor is allowed to use V' and G as blackboxes.

What we need now is to establish a connection between WI and WH.

Given relation $R = \{(x, \omega)\}$, define R^2, where $(x_a, x_b, \omega) \in R^2$ iff $(x_a, \omega) \in R$ or $(x_b, \omega) \in R$. Given a generator G for R, obtain a generator G^2 for R^2, by applying G twice independently, and discarding at random one of the two witnesses.

Theorem 2.3: Let G be a generator for relation R. Let (P, V) be a proof of knowledge system for R^2 (P proves knowledge of a witness of one of two instances in R). Then if (P, V) is WI over R^2, then it is WH over (R^2, G^2).

If an NP problem has random self reducibility properties, it is easy to construct WI over R^2.

Let's consider the certified discrete log problem. That is, $R = \{(p, g, c, x), \omega\}$, where p is a prime, g is a generator for Z_p^*, c is the factorization of p-1 [P75], and $x = g^\omega \bmod p$. Then, the WH for (R^2, G^2) is given as follows.

Protocol Q

The common input is (p, g, c, x_a, x_b), where $x_a = g^{\omega_a} \bmod p$ and $x_b = g^{\omega_b} \bmod p$. P proves that he knows the discrete log of either x_a or x_b. The basic step of the protocol is:

(step 1) P chooses secretly, randomly and independently t_a, t_b, and computes $\phi_a = x_a g^{t_a} \bmod p$, $\phi_b = x_b g^{t_b} \bmod p$. P sends those two values in a random order to V.

(step 2) V replies by a challenge:0 or 1.

(step 3) If P receives 0, P reveals t_a and t_b, and V checks that y_a and y_b were constructed correctly (satisfy $\phi = xg^t \bmod p$). If P receives 1, P reveals the discrete log of *only one of* the ϕ's(which equals $\omega + t \pmod{p-1}$).

This basic step is executed $\log p$ independent times *in parallel*.

2.3 Trapdoor bit commitment

Definition 2.7[FS89]: A trapdoor *bit commitment scheme* consists of a *commit* stage and a *reveal* stage. The scheme must satisfy the following properties:

[Completeness]: Party P can commit to any bit b (either 0 or 1).

[Soundness]: P has negligible probability of constructing a commitment which he can later reveal in two possible ways: both as 0 and as 1.

[Security]: Party V has negligible probability of predicting the value of a committed bit.

[Trapdoor]: V (through some trapdoor information) can construct commitments, indistinguishable from P's commitments, which he can later reveal in two possible ways: both as 0 and as 1.

3 Trapdoor coin flipping

We introduce and define a trapdoor coin flipping protocol as follows.
Definition 3.1: A trapdoor coin flipping scheme is a protocol (P,V) which generate a random bit. It must satisfy the following properties:

[Completeness] The flipped coin becomes head or tail at random, when P and V are honest.

[Soundness] P has at most negligible probability of predicting or biasing the value of a flipped coin.

[Security] V has at most negligible probability of predicting or biasing the value of a flipped coin.

[Trapdoor] V (through some trapdoor information) can create the view, which he can later reveal in two possible ways: both as head and as tail.

A trapdoor coin flipping protocol is easily obtained by making use of the trapdoor bit commitment scheme as follows.

[step 1] P commits a random bit e_P by a trapdoor bit commitment scheme.

[step 2] V sends a random bit e_V.

[step 3] P reveals e_P.

[step 4] The flipped coin is $e = e_P + e_V \bmod 2$.

Example

We show an example based on the certified discrete log problem.

[step 1] V sends P (p, g, c, x_a, x_b) such that $x_a = g^{w_a}$ and $x_b = g^{w_b} \bmod p$, where c is the factorization of $p - 1$.

[step 2] V proves that he knows w_a or w_a by a WH protocol(protocol Q).

[step 3] P chooses a random bit e_P and computes

$$
\begin{cases}
y_a = x_a g^{r_a} & y_b = x_b g^{r_b} & \text{if } e_p = 0 \\
y_a = x_k g^{r_k} & y_b = g^l & \text{if } e_p = 1
\end{cases}
$$

where r_a, r_b, l are random numbers, k is a or b. P sends y_a and y_b in a random order.

[step 4] V sends a random bit e_V.

[step 5] P sends e_P and

$$
\begin{cases}
(r_a, r_b) & \text{if } e_p = 0 \\
l & \text{if } e_p = 1
\end{cases}
$$

[step 6] V checks what he received.

[step 7] The flipped coin is $e = e_P + e_V \bmod 2$.

4 4 move ZKIP with no assumption

4.1 Proposed ZKIP

We present a perfect 4-move ZKIP with "no assumption" for the certified discrete log problem. It is easily generalized to the random self reducible problems [TW87] whose domain is $NP \cap BPP$.

The common input is (p, g, c, y), where $y = g^s \bmod p$, p is a prime, g is a generator of Z_n^*, and c is the factorization of $p - 1$. P proves that he knows s. Let $n =| p |$. The ZKIP consists of 3 subprotocols.

Subprotocol 1 (move 1-3)

- V chooses ω_{ai}, ω_{bi} at random $(i = 1 \dots n)$
  and sends $x_{ai}(= g^{\omega_{ai}} \bmod p)$, $x_{bi}(= g^{\omega_{bi}} \bmod p)$.

- V proves that he knows ω_{ai} or ω_{bi} by protocol Q (see 2.2).
  (The roles are reversed. V is the prover and P is the verifier.)

Subprotocol 2 (coin flipping)

[move 4] P chooses random number $\{r_{ai}\}, \{r_{bi}\}, \{l_i\}$, random bits $\{e_{Pi}\}$ and computes

$$\begin{cases} y_{ai} = x_{ai}g^{r_{ai}} \quad y_{bi} = x_{bi}g^{r_{bi}} & \text{if } e_{Pi} = 0 \\ y_{bi} = x_{ki}g^{r_{ki}} \quad y_{bi} = g^{l_i} & \text{if } e_{Pi} = 1 \end{cases}$$

where k is a or b. P sends y_{ai} and y_{bi} in a random order.

[move 6] V sends random bits $\{e_{Vi}\}$ $(i = 1 \dots n)$

[move 7] P sends $\{e_{Pi}\}$ and

$$\begin{cases} (r_{ai}, r_{bi}) & \text{if } e_{Pi} = 0 \\ l_i & \text{if } e_{Pi} = 1 \end{cases}$$

(bit revealing)
V verifies what he received. ($e_{Pi} + e_{Vi} \bmod 2$ serves as the V's challenge bit.)

Subprotocol 3 (main part)

[move 5] P chooses random numbers $\{v_i\}$ and sends $\{z_i\}$ such that

$$z_i = g^{v_i} \bmod p \quad (i = 1 \ldots n)$$

[move 8] P sends $\{t_i\}$ such that

$$t_i = \begin{cases} v_i \bmod p & \text{If } e_{P_i} + e_{V_i} = 0 \bmod 2 \\ v_i + s \bmod p & \text{if } e_{P_i} + e_{V_i} = 1 \bmod 2 \end{cases}$$

V verifies.

Full protocol

The full protocol executes subprotocol 1,2 and 3 by regrouping the 8 moves into 4 super-moves; (1), $(2, 4, 5)$, $(3, 6)$ $(7, 8)$.

Remark

(y_{ai}, y_{bi}) is a trapdoor bit commitment of e_{P_i}. If one knows ω_{ai} or ω_{bi}, he can reveal the commital as both 0 and 1.

4.2 Correctness

Completeness: Trivial.

Soundness: We describe a knowledge extractor M, which stops in expected polynomial time, and the probability it outputs s is the same (up to negligible additive terms) as the probability that (a possibly cheating) P' convinces a truthful V.

[1] M chooses ω_{ai}, ω_{bi} at random and sets $x_{ai} = g^{\omega_{ai}} \bmod p$, $x_{bi} = g^{\omega_{bi}} y \bmod p$. M sends x_{ai} and x_{bi} to P' in a random order. After this, M executes the remained protocol(P', V) by faithfully simulating V's part. If V rejects, M stops and outputs nothing. Otherwise,

[2] M repeatedly resets P' to step 5 of the protocol, chooses new random challenges $\{e_{V_i}\}$, until P' again meets these challenges successfully.

[3] In the two successful executions, if the first $\{e_{Pi}\}$ and the second $\{e_{Pi}\}$ are the same at move 7 (P' reveals the commitals in the same way), M can derive s from these two executions because M gets v_i and $v_i + s$ at move 8.

[4] In the two successful executions, if the first $\{e_{Pi}\}$ and the second $\{e_{Pi}\}$ are different at move 7 (P' reveals the commitals in a different way), M can derive s from these two executions with overwhelming probability, because P' reveals $l_i = \omega_{bi} + s + r_{bi}$ and r_{bi} at move 7 for some i. (If P' can reveal the commitals in a different way, he must know s.)

The details will be given in the final paper.

Zero-knowledge: We describe a simulator S, which for any(possibly cheating) V' creates in expected polynomial time something indistinguishable from the view of V'. The simulator first performs P's part in moves 1-3. If V' does not complete this subprotocol successfully, S stops. Otherwise, S repeats move 2, each time with different randomly chosen challenges, until V' again successfully meets M's challenges. From the two successful executions S can find ω_{ai} or ω_{bi}. To guard against an infinite execution in case there is only one set of challenges that V answers correctly, S tries in parallel to find ω_{ai} or ω_{bi} by himself (using exhaustive search).

Once M finds a ω_{ai} or ω_{bi}, he can create instances of the trapdoor coin flipping scheme which he can open both as 0 and as 1. This allows him to carry out P's part in moves 5-8, without knowing s and without using resettable simulation.

The output of S is perfectly indistinguishable from V's view of the protocol when executed with a real P.

4.3   Theorem

[Main Theorem]
Any random self reducible problem whose domain is NP $\cap$ BPP problem has a 4 move perfect ZKIP of knowledge with no assumption.

4.4 Efficiency for NP problems

It is easy to apply our technique to the 4 move ZKIP of the Hamilton Circuit problem under the assumption of the existence of one way permutations. Let n be the number of nodes of the input graph. In the 4move ZKIP of [FS89], the prover commits the bits of the incidence matrices of n graphs by using a trapdoor bit commitment scheme. The number of bits to be trapdoor committed is n^3. In our ZKIP, the trapdoor commitment is used to generate V's challenge bits. The number is only n. When relying on the assumption that one way permutations exist, the bit length of the value of the trapdoor committed bit is quite large. Thus, the number of bits communicated in our ZKIP is only $1/n^2$ of that of [FS89].

References

[TW87] M.Tompa and H.Woll
"Random self-reducibility and zero knowledge interactive proofs for possession of information", Proc.28th FOCS, pp.472-482(1987).
[GK90] O.Goldreich and H.Krawczyk
"On the composition of zero knowledge proof systems", Proc.ICALP 90, pp.268-282(1990).
[FS89] U.Feige and A.Shamir
"Zero-Knowledge Proofs in Two Rounds", Advances in Cryptology-Crypt'89,Lecture Notes in Computer Science 435, Spring-Verlag, Berlin, pp.526-544(1990)
[FS90] U.Feige and A.Shamir
"Witness Indistinguishable and Witness", ACM Annual Symposium on Theory of Computing, pp.416-426(May 1990)
[BMO90] M.Bellare, S.Micali, and R.Ostrovsky
"The(True) Complexity of statistical Zero-Knowledge," ACM Annual Symposium on Theory of Computing,pp.494-502(May 1990)
[P75] V.Pratt
"Every prime has a succinct certificate," SIAM J. Computing 4 pp.214-220(1975)

On the Complexity of Constant Round ZKIP of Possession of Knowledge

Toshiya Itoh[*]   *Kouichi Sakurai*[†]

Abstract:

In this paper, we show that if a relation R has a three move blackbox simulation zero-knowledge interactive proof system of possession of knowledge, then there exists a probabilistic polynomial time algorithm that on input $x \in \{0,1\}^*$, outputs y such that $(x,y) \in R$ with overwhelming probability if $x \in$ dom R, and outputs "$\perp$" with probability 1 if $x \notin$ dom R. In the present paper, we also show that without any unproven assumption, there exists a four move blackbox simulation perfect zero-knowledge interactive proof system of possession of the prime factorization, which is *optimal* in the light of the round complexity.

1 Introduction

A notion of interactive proof systems (IP) was introduced by Goldwasser, Micali, and Rackoff [GMR1] to characterize a class of languages that are *efficiently* provable in an interactive manner. Informally, an interactive proof system (P, V) for a language L is an interactive protocol between a computationally unbounded probabilistic Turing machine P and a probabilistic polynomial time Turing machine V, in which when $x \in L$, V accepts x with at least *overwhelming* probability, and when $x \notin L$, V accepts x with at most *negligible* probability, where the probabilities are taken over all of the possible coin tosses of both P and V.

In [GMR1], they also introduced a notion of zero-knowledge (ZK) to capture total amount of knowledge released from a prover P to a verifier V during the execution of an interactive proof system (P, V). Goldreich and Oren [GO] investigated which properties (e.g., round complexity, randomness of provers or verifiers, etc.) are *essential* in zero-knowledge interactive proof systems (ZKIP) for languages, and showed several results on the triviality of ZKIP for languages, i.e., (1) if a language L has a Las Vegas GMR1 ZKIP, then $L \in \mathcal{RP}$; (2) if a language L has a GMR1 ZKIP with deterministic verifiers, then $L \in \mathcal{RP}$; (3) if a language L has a one move GMR1 ZKIP, then $L \in \mathcal{BPP}$; (4) if a language L has a two move auxiliary input ZKIP, then $L \in \mathcal{BPP}$; and (5) if a language L has an auxiliary input ZKIP with deterministic provers, then $L \in \mathcal{BPP}$. Furthermore, Goldreich and Krawczyk [GKr] showed that if a language L has a three move blackbox simulation private coin type ZKIP (or a constant round blackbox simulation public coin type ZKIP), then $L \in \mathcal{BPP}$.

From a somewhat practical point of view (e.g., secure identification protocols, etc.), Feige, Fiat, and Shamir [FFS] and Tompa and Woll [TW] formulated an alternative notion of "interactive proof systems of knowledge (or possession of knowledge)." Specifically, Tompa and

---

[*] Department of Information Processing, The Graduate School at Nagatsuta, Tokyo Institute of Technology, 4259 Nagatsuta, Midori-ku, Yokohama 227, Japan. (e-mail: titoh@nc.titech.ac.jp)

[†] Computer & Information Systems Laboratory, Mitsubishi Electric Corporation, 5-1-1 Ofuna, Kamakura 247, Japan. (e-mail: sakurai@isl.melco.co.jp)

Woll [TW] showed that for any random self-reducible relation R, there exists a (polynomially many move) blackbox simulation perfect ZKIP of possession of knowledge. In addition, they pointed out in [TW] that any ZKIP of possession of knowledge for a relation R is also a ZKIP for language membership on dom R.

The parallel execution of the protocol in [TW] is a three move interactive proof system of possession of knowledge for any random self-reducible relation R. Is such a three move protocol (blackbox simulation) zero-knowledge ? The triviality theorem by Goldreich and Krawczyk [GKr] immediately induces that the three move protocol is *never* blackbox simulation zero-knowledge when dom $R \notin \mathcal{BPP}$, while it provides nothing when dom $R \in \mathcal{BPP}$, e.g., certified discrete logarithm, etc. Intuitively, it may not be (blackbox simulation) zero-knowledge even when dom $R \in \mathcal{BPP}$, however, anything on this has not been known at all. One of our interests in this paper is to solve this open problem.

In this paper, we show several results on the triviality of ZKIP of possession of knowledge with respect to a relation R, some of which are stronger than those in [GKr], [GO], especially when dom $R \in \mathcal{BPP}$. More precisely, we show the following theorems:

1. (Theorem 3.1) if a relation R has a one move GMR1 ZKIP of possession of knowledge, then there exists a probabilistic polynomial time algorithm that on input $x \in \{0,1\}^*$, outputs y such that $(x,y) \in R$ with overwhelming probability if $x \in$ dom R, and outputs "$\perp$" with probability 1 if $x \notin$ dom R;

2. (Theorem 3.2) if a relation R has a two move auxiliary input ZKIP of possession of knowledge, then there exists a probabilistic polynomial time algorithm that on input $x \in \{0,1\}^*$, outputs y such that $(x,y) \in R$ with overwhelming probability if $x \in$ dom R, and outputs "$\perp$" with probability 1 if $x \notin$ dom R;

3. (Theorem 3.3) if a relation R has an auxiliary input ZKIP of possession of knowledge with deterministic provers, then there exists a probabilistic polynomial time algorithm that on input $x \in \{0,1\}^*$, outputs y such that $(x,y) \in R$ with overwhelming probability if $x \in$ dom R, and outputs "$\perp$" with probability 1 if $x \notin$ dom R;

4. (Theorem 3.6) if a relation R has a constant move blackbox simulation *public coin* type ZKIP of possession of knowledge, then there exists a probabilistic polynomial time algorithm that on input $x \in \{0,1\}^*$, outputs y such that $(x,y) \in R$ with overwhelming probability if $x \in$ dom R, and outputs "$\perp$" with probability 1 if $x \notin$ dom R;

5. (Theorem 3.7) if a relation R has a three move blackbox simulation *private coin* type ZKIP of possession of knowledge, then there exists a probabilistic polynomial time algorithm that on input $x \in \{0,1\}^*$, outputs y such that $(x,y) \in R$ with overwhelming probability if $x \in$ dom R, and outputs "$\perp$" with probability 1 if $x \notin$ dom R;

These theorems imply that for nontrivial relations R, the maximum lower bound of the round complexity of (blackbox simulation) ZKIP of possession of knowledge is at least "four."

Feige and Shamir [FS1] showed that any relation R whose domain is in $\mathcal{NP}$ has a four move blackbox simulation perfect ZKIP of possession of knowledge under the certified discrete logarithm assumption, however, the correctness (soundness) of proofs of possession of knowledge depends on the unproven assumption. Recently, Bellare, Micali, and Ostrovsky [BMO] showed that any random self-reducible relation R in a sense of [TW] has a five move blackbox simulation perfect ZKIP for language membership on dom R, and Sakurai and Itoh [SI] recently showed that without any unproven assumption, the protocol in [BMO] is also the ones of possession of knowledge with respect to any random self-reducible relation R. Thus in the case of blackbox simulation ZKIP of possession of knowledge for nontrivial relations R, the optimal round complexity is at least "four" but not greater than "five." It has been still open whether or not the optimal round complexity in this case is exactly "four."

In this paper, we also show that there exists a "four" move blackbox simulation perfect ZKIP of possession of the prime factorization. To the authors' best knowledge, this is the first concrete example of a four move blackbox simulation (perfect) ZKIP of possession of nontrivial knowledge without any unproven assumption, and solves the open problem on the optimality of the round complexity for (blackbox simulation) ZKIP of possession of knowledge.

2   Preliminaries

This section first presents notions of interactive proof systems for languages and of possession of knowledge, and then gives a definition of blackbox simulation perfect (statistical) (computational) zero-knowledge. Formal definitions for GMR1 and auxiliary input perfect (statistical) (computational) zero-knowledge can be found in [GMR2].

Definition 2.1 [GMR1], [GMR2]:   *An interactive protocol (P, V) is an interactive proof system for a language L iff*

- **Membership Completeness:** *For every $k > 0$ and every sufficiently large $x \in L$, (P, V) halts and accepts x with probability at least $1 - |x|^{-k}$, where the probabilities are taken over the coin tosses of both P and V;*

- **Membership Soundness:** *For every $k > 0$, every sufficiently large $x \notin L$, and any P^*, (P^*, V) halts and accepts x with probability at most $|x|^{-k}$, where the probabilities are taken over the coin tosses of both P^* and V.*

It should be noted that P's resource is computationally unbounded, while V's resource is bounded by probabilistic polynomial (in $|x|$) time.

Definition 2.2 [FFS]:   *An interactive protocol (P, V) is an interactive proof system of possession of knowledge for a relation R iff*

- **Possession Completeness:** *For every $k > 0$ and every sufficiently large x such that there exists a y with $(x, y) \in R$, $(P(y), V)$ halts and accepts x with probability at least $1 - |x|^{-k}$, where the probabilities are taken over the coin tosses of both P and V;*

- **Possession Soundness:** *For each $\ell > 0$, there exists an expected polynomial (in $|x|$) time Turing machine E such that for any $k > 0$, any $s \in \{0, 1\}^*$, and any P^*, if $(P^*(s), V)$ halts and accepts x with probability at least $|x|^{-\ell}$, then $(x, E^{P^*}(x)) \in R$ with probability at least $1 - |x|^{-k}$ for sufficiently large x, where E is allowed to have a blackbox access to P^* and the probabilities are taken over the coin tosses of both E and V.*

Note that both P's and V's resource are bounded by probabilistic polynomial time in $|x|$.

Let X and Y be finite (or countably infinite) sets, and let $R \subseteq X \times Y$ be a relation. Here we use dom R to denote "domain" of R, i.e., dom $R = \{x \in X \mid (x, y) \in R$ for some $y \in Y\}$.

The following are the definitions for GMR1, auxiliary input, and blackbox simulation perfect (statistical) (computational) zero-knowledge interactive proof systems of possession of knowledge with respect to a relation R.

Definition 2.3 [GMR1]: *An interactive proof system (P, V) of possession of knowledge with respect to a relation R is GMR1 perfect (statistical) (computational) zero-knowledge iff for all V^*, there exists a probabilistic polynomial time Turing machine M_{V^*} such that for any $(x, y) \in R$, the distribution ensembles $\{M_{V^*}(x)\}_{x \in \text{dom } R}$ and $\{\langle P(x,y), V^*(x)\rangle\}_{x \in \text{dom } R}$ are perfectly (statistically) (polynomially) indistinguishable.*

Definition 2.4 [GMR2]: *An interactive proof system (P, V) of possession of knowledge with respect to a relation R is auxiliary input perfect (statistical) (computational) zero-knowledge iff for all V^*, there exists a probabilistic polynomial time Turing machine M_{V^*} such that for any $(x, y) \in R$ and any $\mu \in \{0,1\}^*$, the distribution ensembles $\{M_{V^*}(x, \mu)\}_{x \in \text{dom } R}$ and $\{\langle P(x,y), V^*(x, \mu)\rangle\}_{x \in \text{dom } R}$ are perfectly (statistically) (polynomially) indistinguishable.*

Definition 2.5 [GO]: *An interactive proof system (P, V) of possession of knowledge with respect to a relation R is blackbox simulation perfect (statistical) (computational) zero-knowledge iff there exists a probabilistic polynomial time Turing machine M_U such that for for every polynomial Q bounding the running time of V^* and for any $(x, y) \in R$, the distribution ensembles $\{M_U^{V^*}(x)\}_{x \in \text{dom } R}$ and $\{\langle P(x,y), V^*(x)\rangle\}_{x \in \text{dom } R}$ are perfectly (statistically) (polynomially) indistinguishable even when the distinguishers are allowed to have a blackbox access to V^*.*

3 Triviality of ZKIP of Possession of Knowledge

3.1 Language Membership versus Possession of Knowledge

It is already known in [TW] that for any relation R, a ZKIP of possession of knowledge can be converted into the one for a language membership problem on dom R. Then some of the results on the triviality of ZKIP for a language shown in [GO], [GKr] can be directly applied to the triviality of dom R in the case of ZKIP of possession of knowledge with respect to a relation R. On the other hand, Sloan [S] observed that there are some distinctions between ZKIP for a language and ZKIP of possession of knowledge, i.e., there exist a language $L \in \mathcal{NP}$ that has a ZKIP (P, V) for a language L, but (P, V) is not a proof of possession of a witness w for $x \in L$, unless $\mathcal{NP} \cap \mathcal{BPP} = \mathcal{P}$. In general, the ZKIP of possession of knowledge restricts the type of protocols more strongly than the ZKIP for a language does.

3.2 Triviality Results on ZKIP of Possession of Knowledge

This subsection shows several results on the triviality of ZKIP of possession of knowledge with respect to a relation R, which are proven under the same condition as [GO], [GKr]. It should be noted that some of the results are stronger than the ones for language membership problem shown in [GO], [GKr], especially when dom $R \in \mathcal{BPP}$.

Throughout this paper, we use "move" to denote a (single) message transmission from a prover P (resp. a verifier V) to a verifier V (resp. a prover P).

Theorem 3.1: *If a relation R has a one move GMR1 perfect (or computational) ZKIP of possession of knowledge, then there exists a probabilistic polynomial (in $|x|$) time algorithm that on input $x \in \{0,1\}^*$, outputs y such that $(x, y) \in R$ with overwhelming probability if $x \in$ dom R, and outputs "$\perp$" with probability 1 if $x \notin$ dom R.*

Sketch of Proof: In any one move interactive proof system (P, V) of possession of knowledge on common input x, P sends to V a message α and checks whether or not $\rho_V(x, \alpha, r) = $ "accept," where r is the random coin toss of V and ρ_V is a polynomial (in $|x|$) time computable predicate for an honest verifier V. Consider the following interactive proof system (P^*, V): (1) on input x, P^* runs the simulator M_V for an honest verifier V to generate $\langle x, r', \alpha \rangle$, and sends to V a message α; and (2) V checks whether or not $\rho_V(x, r, \alpha) = $ "accept."

When $x \in$ dom R, P^* causes V to accept with overwhelming probability, because the underlying (P, V) is a GMR1 perfect (or computational) ZKIP of possession of knowledge with respect to the relation R. Then it follows from the definition of soundness (see Definition 2.2.) that there exists a probabilistic polynomial (in $|x|$) time algorithm for computing with overwhelming probability y such that $(x, y) \in R$. On the other hand, when $x \notin$ dom R, any y does not satisfy $(x, y) \in R$. Thus there exists a probabilistic polynomial (in $|x|$) time algorithm that on input $x \in \{0, 1\}^*$, outputs y such that $(x, y) \in R$ with overwhelming probability if $x \in$ dom R, and outputs "$\perp$" with probability 1 if $x \notin$ dom R. ∎

Theorem 3.2: *If a relation R has a two move auxiliary input perfect (or nonuniformly computational) ZKIP of possession of knowledge, then there exists a probabilistic polynomial (in $|x|$) time algorithm that on input $x \in \{0, 1\}^*$, outputs y such that $(x, y) \in R$ with overwhelming probability if $x \in$ dom R, and outputs "$\perp$" with probability 1 if $x \notin$ dom R.*

Sketch of Proof: In any two move interactive proof system (P, V) of possession of knowledge on common input x, V sends to V a message α followed by a message β with which P responds to V, and V checks whether or not $\rho_V(x, r, \alpha, \beta) = $ "accept," where r is the random coin toss of V and ρ_V is a polynomial (in $|x|$) time computable predicate for an honest verifier V. Consider the following interactive proof system (P^*, V): (1) V sends to P a message α; (2) P^* runs the simulator M_{V^*} on input x and α (auxiliary input) to generate $\langle x, (\alpha, r'), \alpha, \beta \rangle$, and sends it to P, and sends to V a message β, where V^* is a dishonest verifier that has an α as its auxiliary input; and (3) V checks whether or not $\rho_V(x, r, \alpha, \beta) = $ "accept."

When $x \in$ dom R, P^* causes V to accept with overwhelming probability, because the underlying (P, V) is an auxiliary input perfect (or nonuniformly computational) ZKIP of possession of knowledge with respect to the relation R. Then it follows from the definition of soundness (see Definition 2.2.) that there exists a probabilistic polynomial (in $|x|$) time algorithm for computing with overwhelming probability y such that $(x, y) \in R$. On the other hand, when $x \notin$ dom R, any y does not satisfy $(x, y) \in R$. Thus there exists a probabilistic polynomial (in $|x|$) time algorithm that on input $x \in \{0, 1\}^*$, outputs y such that $(x, y) \in R$ with overwhelming probability if $x \in$ dom R, and outputs "$\perp$" with probability 1 if $x \notin$ dom R. ∎

Theorem 3.3: *If a relation R has an auxiliary input perfect (or computational) ZKIP of possession of knowledge in which the prover is deterministic, then there exists a probabilistic polynomial (in $|x|$) time algorithm that on input $x \in \{0, 1\}^*$, outputs y such that $(x, y) \in R$ with overwhelming probability if $x \in$ dom R, and outputs "$\perp$" with probability 1 if $x \notin$ dom R.*

Sketch of Proof: We assume that on input $x \in \{0, 1\}^*$, the interactive proof system $(P, .V)$ of possession of knowledge with deterministic provers consists of $t(|x|)$ rounds and V moves first, where t is a polynomial in $|x|$. Since P is deterministic, P's i-th message α_i is uniquely determined by x and V's messages $\beta_1, \beta_2, \ldots, \beta_i$. After $t(|x|)$ rounds, V checks whether or not $\rho_V(x, r, \beta_1, \alpha_1, \beta_2, \alpha_2, \ldots, \beta_t, \alpha_t) = $ "accept," where r is the random coin toss of V and ρ_V is a polynomial (in $|x|$) time computable predicate for an honest verifier V. Consider the following interactive proof system (P^*, V): for each i $(1 \leq i \leq t)$, (1) V sends to P a message β_i; (2) when receiving the message β_i from V, P^* runs the simulator $M_{V_i^*}$ on input x and $\beta_1, \beta_2, \ldots, \beta_i$ to generate $\langle x, (\beta_1, \beta_2, \ldots, \beta_i), r', \beta_1, \alpha_1, \beta_2, \alpha_2, \ldots, \beta_i, \alpha_i \rangle$, and sends to V a message α_i, where V_i^* is a dishonest verifier that has $\beta_1, \beta_2, \ldots, \beta_i$ as its auxiliary input and sends β_j $(1 \leq j \leq i)$ to P in the j-th round; and (3) V checks whether or not $\rho_V(x, r, \beta_1, \alpha_1, \beta_2, \alpha_2, \ldots, \beta_t, \alpha_t) = $ "accept."

When $x \in$ dom R, P^* causes V to accept with overwhelming probability, because the underlying (P, V) is an auxiliary input perfect (or computational) ZKIP of possession of knowledge with deterministic provers. Then it follows from the definition of soundness (see Definition 2.2.) that there exists a probabilistic polynomial (in $|x|$) time algorithm for computing with overwhelming probability y such that $(x, y) \in R$. On the other hand, when $x \notin$ dom R, any y does not satisfy $(x, y) \in R$. Thus there exists a probabilistic polynomial (in $|x|$) time algorithm that on input $x \in \{0, 1\}^*$, outputs y such that $(x, y) \in R$ with overwhelming probability if $x \in$ dom R, and outputs "$\perp$" with probability 1 if $x \notin$ dom R. ∎

The proofs for the Theorems 3.1, 3.2, and 3.3 are almost the same as [GO], however, the results here are *stronger* than the ones in [GO] especially when dom $R \in \mathcal{BPP}$.

Theorem 3.4: *If a relation R has a three move blackbox simulation Arthur-Merlin (public coin) type ZKIP of possession of knowledge, then there exists a probabilistic polynomial (in $|x|$) time algorithm that on input $x \in \{0, 1\}^*$, outputs y such that $(x, y) \in R$ with overwhelming probability if $x \in$ dom R, and outputs "$\perp$" with probability 1 if $x \notin$ dom R.*

Sketch of Proof: In any three move Arthur-Merlin type interactive proof system (P, V) of possession of knowledge on common input x, P first sends to V a message α, V responds to P with a message β, and P sends to V a message γ. After the interactions above, V checks whether or not $\rho_V(x, \alpha, \beta, \gamma) = $ "accept," where ρ_V is a polynomial (in $|x|$) time computable predicate for an honest verifier V. Here we treat only a deterministic predicate, however, the analysis similar to this can be applied to the case of probabilistic predicate, i.e., V checks whether or not $\rho_V(x, r, \alpha, \beta, \gamma) = $ "accept," where r is the random coin tosses of V. Without loss of generality, we assume that there exists a polynomial ℓ such that $|\alpha| = |\beta| = \ell(|x|)$.

For the three move blackbox simulation ZKIP of possession of knowledge on common input x, the behavior of the simulator M with a blackbox access to a (dishonest) verifier V^* is *completely* determined by the common input x, the random coin tosses R_M of M, and the response from V^*. We assume that the running time of the simulator M is *strictly* bounded by some polynomial in $|x|$, and M resets V^* at most $t = t(|x|)$ times during the simulation process, where t is a polynomial in $|x|$. Let $|R_M| = q(|x|)$, where q is a polynomial bounding the number of random coin tosses used by M on input x.

Define the following procedure F that uses M as a subprocedure. For an input $x \in \{0, 1\}^*$, choose $R_M \in_R \{0, 1\}^{q(|x|)}$ and $\beta^{(i)} \in_R \{0, 1\}^{\ell(|x|)}$ $(1 \leq i \leq t)$, and run M on input x and R_M. If M first generate $\alpha^{(1)}$, then the procedure F responds with $\beta^{(1)}$. For the i-th different string $\alpha^{(i)}$ generated by M, F responds with $\beta^{(i)}$, and if the same string $\alpha^{(j)}$ $(1 \leq j \leq i)$ is generated by M, then F responds with $\beta^{(j)}$ as before. Thus $\alpha^{(i)}$ is *completely* determined by x, R_M, and $(i - 1)$ strings $\beta^{(1)}, \beta^{(2)}, \ldots, \beta^{(i-1)}$ in the procedure, i.e., there exists a deterministic function f_M such that $\alpha^{(i)} = f_M(x, R_M, \beta^{(1)}, \beta^{(2)}, \ldots, \beta^{(i-1)})$. We use $F_M(x, R_M, \beta^{(1)}, \beta^{(2)}, \ldots, \beta^{(t)}) = (x, \alpha, \beta, \gamma)$ to denote a conversation generated by the simulator M when activated with these parameters. It is easy to see that with overwhelming probability, there exists an index i $(1 \leq i \leq t)$ such that $\alpha = \alpha^{(i)}$ and $\beta = \beta^{(i)}$. We call a vector $(x, R_M, \beta^{(1)}, \beta^{(2)}, \ldots, \beta^{(t)})$ to be *successful* for M if $F_M(x, R_M, \beta^{(1)}, \beta^{(2)}, \ldots, \beta^{(t)}) = (x, \alpha, \beta, \gamma)$ is an accepting conversation for an honest verifier V, i.e., $\rho_V(x, \alpha, \beta, \gamma) = $ "accept," and we call a vector $(x, R_M, \beta^{(1)}, \beta^{(2)}, \ldots, \beta^{(t)})$ to be *successful* at i if it is successful for M and $\alpha = \alpha^{(i)}$ and $\beta = \beta^{(i)}$.

Goldreich and Krawczyk [GKr] showed that if $x \in$ dom R, there exists a nonnegligible fraction of all possible vectors $(x, R_M, \beta^{(1)}, \beta^{(2)}, \ldots, \beta^{(t)})$ that are successful for M. This implies that there exists an index i $(1 \le i \le t)$ such that a nonnegligible fraction of all possible vectors $(x, R_M, \beta^{(1)}, \beta^{(2)}, \ldots, \beta^{(t)})$ are successful at i, because t is a polynomial in $|x|$. Thus there exists a nonnegligible fraction of all possible prefices $(x, R_M, \beta^{(1)}, \beta^{(2)}, \ldots, \beta^{(i-1)})$, each of which has a nonnegligible fraction of all possible continuations $(\beta^{(i)}, \beta^{(i+1)}, \ldots, \beta^{(t)})$ being successful at the index i, i.e., the vector $(x, R_M, \beta^{(1)}, \ldots, \beta^{(i-1)}, \beta^{(i)}, \ldots, \beta^{(t)})$ is successful at the index i. For such an index i $(1 \le i \le t)$, define a set $B(x, \alpha^{(i)})$ to be

$$B(x, \alpha^{(i)}) = \left\{ \beta \in \{0,1\}^{\ell(|x|)} \mid \exists \gamma \; \rho_V(x, \alpha^{(i)}, \beta, \gamma) = \text{``accept.''} \right\}.$$

It follows from the observation above that the set $B(x, \alpha^{(i)})$ is a nonnegligible fraction of all possible $\beta \in \{0,1\}^{\ell(|x|)}$, i.e., there exists a polynomial p such that $\|B(x, \alpha^{(i)})\|/2^{\ell(|x|)} \ge 1/p(|x|)$. Here we call such a set $B(x, \alpha^{(i)})$ to be *prolific*.

For any successful vector $(x, R_M, \beta^{(1)}, \beta^{(2)}, \ldots, \beta^{(t)})$ at an index i $(1 \le i \le t)$, it is easy to check whether or not the set $B(x, \alpha^{(i)})$ is prolific. This can be done as follows:

(1) choose $s = |x| \cdot p(|x|)$ pairs $\langle \delta_k^{(i)}, \delta_k^{(i+1)}, \ldots, \delta_k^{(t)} \rangle \in_R \{0,1\}^{(t-i+1)\cdot\ell(|x|)}$ $(1 \le k \le s)$.

(2) for each k $(1 \le k \le s)$, compute $(x, \alpha, \beta_k, \gamma_k) = F_M(x, R_M, \beta^{(1)}, \ldots, \beta^{(i-1)}, \delta_k^{(i)}, \ldots, \delta_k^{(t)})$, and check whether or not $(x, R_M, \beta^{(1)}, \ldots, \beta^{(i-1)}, \delta_k^{(i)}, \ldots, \delta_k^{(t)})$ is successful at i.

If $B(x, \alpha^{(i)})$ is prolific, then there exists at least a single pair $\langle \delta_k^{(i)}, \delta_k^{(i+1)}, \ldots, \delta_k^{(t)} \rangle$ of $(t-i+1)$ strings such that the vector $(x, R_M, \beta^{(1)}, \ldots, \beta^{(i-1)}, \delta_k^{(i)}, \ldots, \delta_k^{(t)})$ is successful at the index i with overwhelming probability; otherwise there does not exist any such a successful pair of strings with overwhelming probability. Consider the following interactive proof system (P^*, V):

P1-1: P^* chooses $R_M \in_R \{0,1\}^{q(|x|)}$ and $\beta^{(j)} \in_R \{0,1\}^{\ell(|x|)}$ $(1 \le j \le t)$.

P1-2: P^* computes $(x, \alpha, \beta, \gamma) = F_M(x, R_M, \beta^{(1)}, \beta^{(2)}, \ldots, \beta^{(t)})$.

P1-3: If $\rho_V(x, \alpha, \beta, \gamma) = $ "accept," then go to step P1-1; otherwise continue.

P1-4: P^* determines with f_M an index i $(1 \le i \le t)$ such that $\alpha = \alpha^{(i)}$ and $\beta = \beta^{(i)}$.

P1-5: P^* checks whether or not the set $B(x, \alpha^{(i)})$ is prolific.

P1-6: If $B(x, \alpha^{(i)})$ is not prolific, then go to step P1-1.

$P \to V$: $\alpha = \alpha^{(i)}$

V1: V chooses $\beta \in_R \{0,1\}^{\ell(|x|)}$.

$V \to P$: β

P2-1: P^* checks whether or not $\beta \in B(x, \alpha)$.

P2-2: If $\beta \notin B(x, \alpha)$, then P^* halts (or gives up).

P2-3: P^* chooses $(t-i)$ strings $\delta^{(j)} \in_R \{0,1\}^{\ell(|x|)}$ $(i+1 \le j \le t)$ and computes $(x, \alpha, \beta, \gamma) = F_M(x, R_M, \beta^{(1)}, \ldots, \beta^{(i-1)}, \beta, \delta^{(i+1)}, \ldots, \delta^{(t)})$.

$P \to V$: γ

V2: V checks whether or not $\rho_V(x, \alpha, \beta, \gamma) = $ "accept."

Note that in step P2-1, P^* can check whether or not $\beta \in B(x, \alpha)$ in (probabilistic) polyno-mial (in $|x|$) time. When receiving $\beta \in_R \{0,1\}^{\ell(|x|)}$, P^* chooses $(t-i)$ strings $\delta^{(j)} \in_R \{0,1\}^{\ell(|x|)}$ $(i+1 \le j \le t)$, and computes $(x, \alpha, \beta, \gamma) = F_M(x, R_M, \beta^{(1)}, \ldots, \beta^{(i-1)}, \beta, \delta^{(i+1)}, \ldots, \delta^{(t)})$. It is easy to see that $\beta \in B(x, \alpha)$ iff $\rho_V(x, \alpha, \beta, \gamma) =$ "accept."

When $x \in$ dom R, P^* causes V to accept with overwhelming probability, because the underlying (P, V) is a three move blackbox simulation Arthur-Merlin type ZKIP of possession of knowledge with respect to the relation R. Then it follows from the definition of soundness (see Definition 2.2.) that there exists a probabilistic polynomial (in $|x|$) time algorithm for computing with overwhelming probability y such that $(x, y) \in R$. On the other hand, when $x \notin$ dom R, any y does not satisfy $(x, y) \in R$. Thus there exists a probabilistic polynomial (in $|x|$) time algorithm that on input $x \in \{0,1\}^*$, outputs y such that $(x, y) \in R$ with overwhelming probability if $x \in$ dom R, and outputs "$\perp$" with probability 1 if $x \notin$ dom R. $\blacksquare$

In [TW], Tompa and Woll showed that for any random self-reducible relation R, there exists a *polynomially* many round blackbox simulation perfect ZKIP of possession of knowledge. The triviality theorem on a three move ZKIP for languages [GKr] directly implies that the parallel execution of the protocol in [TW] is *never* a blackbox simulation perfect ZKIP unless dom $R \in \mathcal{BPP}$. The theorem above (Theorem 3.4) induces a stronger result than the one in [GKr] on the zero-knowledgeness of parallel execution of the protocol in [TW].

Corollary 3.5: *The parallel execution of the protocol for any random self-reducible relation R in [TW] is not blackbox simulation (perfect) ZKIP of possession of knowledge even when dom $R \in \mathcal{BPP}$, unless there exists a probabilistic polynomial (in $|x|$) time algorithm that for any $x \in$ dom R, outputs y such that $(x, y) \in R$ with overwhelming probability.*

The result above (Theorem 3.4) can be extended to more general cases, i.e., a constant move blackbox simulation Arthur-Merlin (public coin) type ZKIP of possession of knowledge and a three move blackbox simulation general (private coin) type ZKIP of possession of knowledge. More precisely, we can show the following theorems in almost the same way as as Theorem 3.4.

Theorem 3.6: *If a relation R has a constant move blackbox simulation Arthur-Merlin (public coin) type ZKIP of possession of knowledge, then there exists a probabilistic poly-nomial (in $|x|$) time algorithm that on input $x \in \{0,1\}^*$, outputs y such that $(x, y) \in R$ with overwhelming probability if $x \in$ dom R, and outputs "$\perp$" with probability 1 if $x \notin$ dom R.*

Sketch of Proof: Without loss of generality, we assume that any constant move Arthur-Merlin type interactive proof system (P, V) consists of $2k+1$ message exchanges for some constant $k > 0$. We use α_i (resp. β_i) to denote the i-th ($1 \le i \le k$) message sent by P (resp. V), and use γ to denote the last message sent by P.

In almost the same way as **Sketch of Proof** for Theorem 3.4, we can define a proce-dure F and a successful vector $(x, R_M, \beta^{(1)}, \beta^{(2)}, \ldots, \beta^{(t)})$ for M. Here we also define a vector $(x, R_M, \beta^{(1)}, \beta^{(2)}, \ldots, \beta^{(t)})$ to be *successful* at $\langle i_1, i_2, \ldots, i_k \rangle$, where $1 \le i_1 < i_2 < \ldots < i_k \le t$, if for each j ($1 \le j \le k$), $\alpha_j = \alpha^{(i_j)}$ and $\beta_j = \beta^{(i_j)}$. We can show in a way similar to **Sketch of Proof** for Theorem 3.4 that there exists a set of indices $\langle i_1, i_2, \ldots, i_k \rangle$ such that the set

$$B(x, \alpha^{(i_1)}, \ldots, \alpha^{(i_k)})$$
$$= \left\{ \langle \beta_1, \ldots, \beta_k \rangle \in \{0,1\}^{k \cdot \ell(|x|)} \mid \exists \gamma \; \rho_V(x, \alpha^{(i_1)}, \beta_1, \ldots, \alpha^{(i_k)}, \beta_k, \gamma) = \text{"accept."} \right\}$$

is *prolific*, and that a set of indices $\langle i_1, i_2, \ldots, i_k \rangle$ for $B(x, \alpha^{(i_1)}, \alpha^{(i_2)}, \ldots, \alpha^{(i_k)})$ to be prolific can be found in probabilistic polynomial (in $|x|$) time with overwhelming probability.

Recalling that the underlying (P,V) is a constant move blackbox simulation Arthur-Merlin type ZKIP of possession of knowledge withe respect to the relation R, we can show in a way similar to Theorem 3.4 that there exists a probabilistic polynomial (in $|x|$) time algorithm that on input $x \in \{0,1\}^*$, outputs y such that $(x,y) \in R$ with overwhelming probability if $x \in$ dom R, and outputs "$\bot$" with probability 1 if $x \notin$ dom R. ∎

Theorem 3.7: If a relation R has a three move blackbox simulation general (private coin) type ZKIP of possession of knowledge, then there exists a probabilistic polynomial (in $|x|$) time algorithm that on input $x \in \{0,1\}^*$, outputs y such that $(x,y) \in R$ with overwhelming probability if $x \in$ dom R, and outputs "$\bot$" with probability 1 if $x \notin$ dom R.

Sketch of Proof: Here we use the same notations as **Sketch of Proof** for Theorem 3.4. Let $r \in \{0,1\}^{\ell(|x|)}$ be random coin tosses of an honest V on input x, where ℓ is a polynomial in $|x|$. Without loss of generality, we assume that $|\alpha| = |r| = \ell(|x|)$.

The procedure F chooses $r^{(i)} \in_R \{0,1\}^{\ell(|x|)}$ ($1 \le i \le t$) instead of choosing $\beta^{(i)} \in_R \{0,1\}^{\ell(|x|)}$, and computes $\beta^{(i)} = V(x, r^{(i)}, \alpha^{(i)})$. Define a vector $(x, R_M, r^{(1)}, r^{(2)}, \ldots, r^{(t)})$ to be *successful* for M if $F_M(x, R_M), r^{(1)}, r^{(2)}, \ldots, r^{(t)}) = (x, r, \alpha, \beta, \gamma)$ is an accepting conversation for an honest verifier V, i.e., $\rho_V(x, r, \alpha, \beta, \gamma) = $ "accept," and a vector $(x, R_M, r^{(1)}, r^{(2)}, \ldots, r^{(t)})$ to be *successful* at i if $\alpha = \alpha^{(i)}$, $r = r^{(i)}$, and $\beta = V(x, r^{(i)}, \alpha^{(i)})$. In almost the same way as **Sketch of Proof** for Theorem 3.4, we can show that there exists an index i ($1 \le i \le t$) such that the set

$$B(x, \alpha^{(i)}) = \left\{ r \in \{0,1\}^{\ell(|x|)} \mid \exists \gamma \; \rho_V(x, r, \alpha^{(i)}, V(x, r, \alpha^{(i)}), \gamma) = \text{"accept."} \right\}$$

is *prolific*, and that an index i ($1 \le i \le t$) for $B(x, \alpha^{(i)})$ to be prolific can be found in probabilistic polynomial (in $|x|$) time with overwhelming probability.

Recalling that the underlying (P,V) is a three move blackbox simulation ZKIP of possession of knowledge with respect to the relation R, we can show in a way similar to Theorem 3.4 that there exists a probabilistic polynomial (in $|x|$) time algorithm that on input $x \in \{0,1\}^*$, outputs y such that $(x,y) \in R$ with overwhelming probability if $x \in$ dom R, and outputs "$\bot$" with probability 1 if $x \notin$ dom R. ∎

The proofs for the Theorems 3.4, 3.6, and 3.7 are almost the same as [GKr], however, the results here are *stronger* than the ones in [GKr] especially when dom $R \in \mathcal{BPP}$.

4 Does the Optimal ZKIP Lie in Four Moves ?

As shown in [GKr], any language that has a three move blackbox simulation ZKIP is in $\mathcal{BPP}$. The result similar to this can be shown for the ZKIP of possession of knowledge with respect to a relation R. (see Theorem 3.7 in subsection 3.2.) Then this implies that for nontrivial languages (resp. relations), the maximum lower bound of the round complexity of (blackbox simulation) ZKIP for languages (resp. of possession of knowledge) is at least "four."

The next natural question is how many the optimal round complexity of (blackbox simulation) ZKIP for nontrivial languages (resp. relations) is. Such a question is currently investigated independently by some researchers. Brassard, Crépeau, and Yung [BCY] presented a six move blackbox simulation perfect zero-knowledge interactive argument [BCC] for any language $L \in \mathcal{NP}$ under the *certified discrete logarithm assumption* (CDLA) (or more generally under the assumption that one-way group homomorphisms exist). In addition, Goldreich and Kahan [GKa] announced that any language $L \in \mathcal{NP}$ has a five move blackbox simulation computational ZKIP based on claw-free pairs of functions, and Naor and Yung [NY] showed that any language $L \in \mathcal{AM}$ has a bounded move blackbox simulation computational ZKIP (especially, any language $L \in \mathcal{NP}$ has a four move blackbox simulation computational ZKIP) based on a family of collision intractable hash functions. Recently, Feige and Shamir [FS1] proved that for any relation R whose domain is in $\mathcal{NP}$, there exists a four move blackbox simulation perfect ZKIP of possession of knowledge under the CDLA, however, the correctness (soundness) of the proof of possession of knowledge depends on the unproven assumption.

Our interest in this paper is to design a four move (blackbox simulation) ZKIP without any unproven assumption, because it enables us to capture the *intrinsic* properties of problems. Recalling that QNR [GMR1] and GNI [GMW], each of which seems to be not in $\mathcal{BPP}$, have four move blackbox simulation perfect ZKIP, the *possible* lower bound on the round of ZKIP for membership of the *nontrivial* language is exactly "four." Bellare, Micali, and Ostrovsky [BMO] showed that for any random self-reducible relation R in a sense of [TW] (e.g., QR, GI, etc.), there exists a five move blackbox simulation perfect ZKIP for language membership on dom R, and we can easily show that it is also the one of possession of knowledge without any unproven assumption. It is not known whether or nor the five move protocol in [BMO] is essentially optimal for any random self-reducible relation R.

From the result above on the five move perfect ZKIP [BMO] and the triviality theorem on three move blackbox simulation ZKIP (see Theorem 3.7.), it follows that in the case of blackbox simulation ZKIP of possession of knowledge, the optimal round complexity is at least "four" but not greater than "five." It has been open whether or not the optimal round complexity for blackbox simulation ZKIP of possession of knowledge is exactly "four."

5 Four Move ZKIP of Possession of Knowledge

The possession of the prime factorization of an odd composite x is polynomially equivalent to the ability of computing one of square roots of quadratic residues modulo x. Thus the four move perfect ZKIP of possession of the prime factorization of x is induced from that of ability of computing one of square roots of quadratic residues modulo x.

Here we use POI to denote a set of positive odd integers, and we use PF to denote a set of k ($k \geq 2$) tuples of odd prime powers. Define a relation $R \subseteq POI \times PF$ to be $(x, y) = (x, \langle p_1^{e_1}, p_2^{e_2}, \ldots, p_k^{e_k} \rangle) \in R$ iff $x = p_1^{e_1} p_2^{e_2} \cdots p_k^{e_k}$, p_i is an odd prime, and $e_i \geq 1$ ($1 \leq i \leq k$). Then dom R is a set of positive odd composites (not a prime power), and for any $x \in$ dom R, $y \in R(x)$ is the prime factorization of x.

Here we refer to the relation R as COMP. In the following, we show that there exists a four move perfect ZKIP of possession of knowledge for the (nontrivial) relation COMP. In order to do this, we need to present a technical lemma and a definition below.

Lemma 5.1 (Lemma 10 in [TW]): *Let x be an odd composite, and let SQRT be a probabilistic polynomial (in $|x|$) time algorithm that for a nonnegligible fraction of quadratic residues u in Z_x^*, outputs a single square root of u modulo x with nonnegligible probability. Then there exists an expected polynomial (in $|x|$) time algorithm FACT that, on input x, outputs the prime factorization of x with overwhelming probability.*

Definition 5.2 [FS1],[FS2]: *An interactive proof system (P, V) of possession of knowledge with respect to a relation R is said to be witness indistinguishable iff for any V^*, all sufficiently large $x \in$ dom R, any $y_1, y_2 \in R(x)$, and any auxiliary input ν, the two distribution ensembles $\{P(x, y_1), V(x, \nu)\}_{x \in \text{dom } R}$ and $\{P(x, y_2), V(x, \nu)\}_{x \in \text{dom } R}$ are indistinguishable.*

Feige and Shamir [FS2] observed that if an interactive proof system (P, V) of possession of knowledge is (perfect) zero-knowledge, then the parallelized protocol of (P, V) is (perfectly) witness indistinguishable. This fact plays an important role in designing a four move perfect ZKIP of possession of the prime factorization of an odd composite x.

Theorem 5.3: *There exists a four move perfect ZKIP of possession of the prime factorization of an odd composite (not a prime power) without any unproven assumption.*

Sketch of Proof: The following is a protocol that for all sufficiently large $x \in \text{dom } R$, the prover P demonstrates the verifier V his possession of the prime factorization of $x \in \text{dom } R$.

Four Move Perfect ZKIP of Possession of the Prime Factorization
common input: $x \in POI$ $(n = |x|)$.

V1: If x is an odd composite[1] (not a prime power), then V generates $r, r_i \in_R Z_x^*$, and computes $u \equiv r^2 \pmod{x}$ and $u_i \equiv r_i^2 \pmod{x}$ for each i $(1 \leq i \leq n)$; otherwise V immediately halts and rejects;

$V \to P$: $u, \langle u_1, u_2, \ldots, u_n \rangle$.

P1: P generates $s_i \in_R Z_x^*$, $e_i \in_R \{0, 1\}$, and computes $v_i \equiv s_i^2 \pmod{x}$ $(1 \leq i \leq n)$.

$P \to V$: $\langle v_1, v_2, \ldots, v_n \rangle$, $\langle e_1, e_2, \ldots e_n \rangle$.

V2: V generates $f_i \in_R \{0, 1\}$, and computes $w_i \equiv r^{e_i} r_i \pmod{x}$ $(1 \leq i \leq n)$.

$V \to P$: $\langle w_1, w_2, \ldots, w_n \rangle$, $\langle f_1, f_2, \ldots, f_n \rangle$.

P2: If $w_i^2 \equiv u^{e_i} u_i \pmod{x}$ for each i $(1 \leq i \leq n)$, then P computes one of the square roots s of u modulo x with the prime factorization of x, and evaluates $z_i \equiv s^{f_i} s_i \pmod{x}$ $(1 \leq i \leq n)$; otherwise P immediately halts and rejects.

$P \to V$: $\langle z_1, z_2, \ldots, z_n \rangle$.

V3: If $z_i^2 \equiv u^{f_i} v_i \pmod{x}$ for each i $(1 \leq i \leq n)$, then V immediately halts and accepts; otherwise V immediately halts and rejects.

(End of Protocol)

Possession Completeness: Trivial !

Possession Soundness: The following is a description of the construction for a probabilistic polynomial (in $|x|$) time Turing machine E. (see Definition 2.2.)

Construction for E
input: $x \in POI$ $(n = |x|)$.

1. Simulate step V1.

2. Run P^* on input u, u_i, f_i $(1 \leq i \leq n)$ to generate v_i, e_i $(1 \leq i \leq n)$.

3. Simulate step V2.

4. Run P^* on input w_i, f_i $(1 \leq i \leq n)$ to generate z_i $(1 \leq i \leq n)$.

5. Simulate step V3.

---

[1] This can be determined in zero error probabilistic polynomial (in $|x|$) time. (see, e.g., [AH], [R], [SS].)

(a) If V halts and rejects in step V3, then go to step 1.

6. Reset P^* to the state in step V2, and choose $\tilde{f}_i \in_R \{0, 1\}$ $(1 \leq i \leq n)$.

7. Run P^* on input w_i, $\tilde{f}_i$ $(1 \leq i \leq n)$ to generate $\tilde{z}_i$ $(1 \leq i \leq n)$.

8. Simulate step V3.

(a) If V halts and rejects in step V3, then go to step 6.

9. Search an index j $(1 \leq j \leq n)$ such that $f_j \neq \tilde{f}_j$.

(a) If such an index does not exist, then go to step 6.

10. Find $\tilde{s} \in Z_x^*$, one of the square roots of u modulo x, from z_j and $\tilde{z}_j$.

(a) If $\tilde{s} \equiv \pm r \pmod{x}$, then go to step 1.

11. Compute $y = \gcd(y, \tilde{s} \pm r)$, and output y.

(End of Construction for E)

Assume that $(P^*(s), V)$ halts and accepts with nonnegligible probability. Then $P^*(s)$ replies with appropriate responses for a nonnegligible fraction of all possible messages u, u_i, f_i $(1 \leq i \leq n)$ challenged by V. This implies that in step 9, E in expected polynomial (in $|x|$) time finds an index j $(1 \leq j \leq n)$ such that $f_j \neq \tilde{f}_j$ with overwhelming probability. Without loss of generality, we assume that $f_j = 0$ and $\tilde{f}_j = 1$. Then in step 10, E obtains both $z_j^2 \equiv v_j$ $\pmod{x}$ and $\tilde{z}_j^2 \equiv u v_j \pmod{x}$, and in expected polynomial (in $|x|$) time finds $\tilde{s} \equiv \tilde{z}_j z_j^{-1}$ $\pmod{x}$, one of the square roots of u modulo x. It should be noted that $\tilde{s} \not\equiv \pm r \pmod{x}$ with probability $1 - 2^{-k+1}$, because the part of V's proof of possession of the square root of u modulo x is *perfectly* witness indistinguishable [FS1], [FS2]. Thus it follows from Lemma 5.1 that E outputs in expected polynomial (in $|x|$) time the prime factorization of x with overwhelming probability by iterating the procedure above polynomially many times.

Zero-Knowledgeness: The following is a description of the construction for a probabilistic polynomial (in $|x|$) time Turing machine M. (see Definition 2.5.)

Construction for M

input: $x \in POI$ $(n = |x|)$.

1. Check whether or not x is an odd composite.

(a) If x is an odd composite (not a prime power), then M runs V^* on input x and r_{V^*} to generate u, u_i $(1 \leq i \leq n)$; otherwise M halts and outputs x.

2. Simulate step P1.

3. Run V^* on input v_i, e_i $(1 \leq i \leq n)$ to generate w_i, f_i $(1 \leq i \leq n)$.

4. Check whether or not $w_i^2 \equiv u^{e_i} u_i \pmod{x}$ for each i $(1 \leq i \leq n)$.

(a) If $w_j^2 \not\equiv u^{e_i} u_j \pmod{x}$ for some j $(1 \leq j \leq n)$, then M halts and outputs x, r_{V^*}, u, u_i, v_i, e_i, w_i, f_i $(1 \leq i \leq n)$.

(b) Define a set I to be $I := \{\langle e_1, e_2, \ldots, e_n \rangle\}$.

5. Reset V^* to the state in step P1, and choose $\tilde{e}_i \in_R \{0, 1\}$ $(1 \leq i \leq n)$.

6. Run V^* on input v_i, $\tilde{e}_i$ $(1 \leq i \leq n)$ to generate $\tilde{w}_i$, $\tilde{f}_i$ $(1 \leq i \leq n)$.

7. Check whether or not $\tilde{w}_i^2 \equiv u^{\tilde{e}_i} u_i \pmod{x}$ for each i $(1 \leq i \leq n)$.

   (a) If $\tilde{w}_j \not\equiv u^{\tilde{e}_j} u_j \pmod{x}$ for some j $(1 \leq j \leq n)$, then $I := I \cup \{\langle \tilde{e}_1, \tilde{e}_2, \ldots, \tilde{e}_n \rangle\}$.

   (b) If $\|I\| = 2^n$, then M exhaustively, i.e., in exponential (in $|x|$) time, searches $\tilde{s} \in Z_x^*$, one of the square roots of u modulo x, and go to step 9.

8. Search an index j $(1 \leq j \leq n)$ such that $e_j \neq \tilde{e}_j$.

   (a) If such an index j $(1 \leq j \leq n)$ does not exist, then go to step 5.

   (b) Find $\tilde{s} \in Z_x^*$, one of the square roots of u modulo x, from w_j and $\tilde{w}_j$.

9. Simulate step P2 by evaluating $z_i \equiv \tilde{s}^{f_i} s_i \pmod{x}$ $(1 \leq i \leq n)$.

10. Output x, r_{V^*}, u, u_i, v_i, e_i, w_i, f_i, z_i $(1 \leq i \leq n)$.

(End of Construction for M)

From the results in [AH], [R], [SS], it follows that step 1 runs in zero error probabilistic polynomial (in $|x|$) time. In the real interaction between P and V^*, if V^* does not know any square root of u modulo x, then P halts and rejects with overwhelming probability. Thus step 7-(b), which in general takes exponential (in $|x|$) time, is executed with negligible (exponentially small) probability. Without loss of generality, we assume that $e_j = 0$ and $\tilde{e}_j = 1$. Then in step 8-(b), M obtains both $w_j^2 \equiv u_j \pmod{x}$ and $\tilde{w}_j^2 \equiv u u_j \pmod{x}$, and in expected polynomial (in $|x|$) time finds $\tilde{s} \equiv \tilde{w}_j w_j^{-1} \pmod{x}$, one of the square roots of u modulo x. In general, $\tilde{s}$ is not necessarily identical to s that P possesses in the real interaction with V^*. However, M perfectly simulates the real interaction between P and V^*, because the part of P's proof of possession of the square root of u modulo x is *perfectly* witness indistinguishable [FS1], [FS2].

Hence the protocol above is a four move perfect ZKIP of possession of knowledge for the relation COMP without any unproven assumption. ∎

Remark 5.4: The protocol above is a *parallelized* and *interleaved* protocol of possession of the prime factorization of an odd composite x by Tompa and Woll. (see Theorem 11 in [TW].)

Theorem 5.5: *The four move perfect ZKIP of possession of knowledge for COMP is optimal with respect to the round complexity unless there exists an expected polynomial (in $|x|$) time algorithm that on input any odd composite x, outputs the prime factorization of x.*

The optimality of the theorem above is guaranteed by Theorem 3.7. It is worth noting that the result above (Theorem 5.5) can not be induced from the result by [GKr], because the set of odd composites can be recognized in zero error probabilistic polynomial time. (see, e.g., [AH], [R], [SS].) Thus the triviality theorem on ZKIP of possession of knowledge for a relation R is *nontrivial* especially when dom $R \in \mathcal{BPP}$.

6 Concluding Remarks

In this paper, we extended the triviality theorems on ZKIP for a language by Goldreich and Oren [GO], and Goldreich and Krawczyk [GKr] to the ones on ZKIP of possession of knowledge. These theorems for possession of knowledge are not necessarily induced from the ones for language membership, and some of them are stronger especially for a relation R whose domain is in $\mathcal{BPP}$. We also showed that there exists a four move blackbox simulation perfect ZKIP of possession of knowledge for a nontrivial relation. To the authors' best knowledge, this is the first concrete example of a four move blackbox simulation (perfect) ZKIP of possession of knowledge without any unproven assumption[2], and solves the open problem on the optimality of the round complexity for (blackbox simulation) ZKIP of possession of knowledge.

Our result on a four move ZKIP of possession of the prime factorization of x heavily depends on the random self-reducibility of quadratic residues modulo x, however, it is still open whether or not there exists a four move blackbox simulation (perfect) ZKIP of possession of square roots modulo x. Note that the possession of square roots modulo x is polynomial time reducible to that of the prime factorization of x. Thus it is somewhat contradictory, because the proof of possession of the prime factorization of x, which is enough to compute square roots modulo x, has a lower round complexity than that of square roots modulo x.

Goldreich and Oren [GO] showed that if a language L has a ZKIP with *perfect soundness*, then $L \in \mathcal{RP}$, however, we could not show a result similar to this, because of the *slight* difference of its definition for soundness.

- What is a (natural) definition for "perfect soundness" in interactive proof systems of possession of knowledge ?

The triviality theorems on ZKIP of possession of knowledge do not necessarily imply that any relation R has a four move blackbox simulation ZKIP of possession of knowledge without any unproven assumption, but only guarantee that if a relation R has a three move blackbox simulation ZKIP of possession of knowledge, then it is easy to find y such that $(x, y) \in R$ for almost all $x \in \text{dom } R$. It is worth noting that the parallel execution of the protocol in [TW] for QR can not be blackbox simulation zero-knowledge under the quadratic residuosity assumption, however, it can be *witness hiding* [FS1], [FS2] or *non-transferable* [FFS]. Thus there exists an interactive proof system of possession of knowledge for a relation R that is *not* blackbox simulation zero-knowledge but witness hiding.

In designing (secure) cryptographic protocols, it is often enough to be witness hiding (not necessarily zero-knowledge), however, the triviality theorems on witness hiding protocols are not well known like in the case of ZKIP for languages or of possession of knowledge.

- Is the *optimal* round complexity for witness hiding protocols "three" ?

Acknowledgments:

The authors wish to thank Osamu Watanabe, Kaoru Kurosawa, and Tatsuaki Okamoto for their several valuable comments on the earlier version of this work.

References

[AH] Adleman, L.M. and Huang, M.D.A., "Recognizing Primes in Random Polynomial Time," Proc. of STOC, pp.462-469 (May 1987).

[BCC] Brassard, G., Chaum, D., and Crépeau, C., "Minimum Disclosure Proofs of Knowledge," *JCSS*, Vol.37, No.2, pp.156-189 (October 1988).

---

[2] Just after our result on the four move blackbox simulation perfect ZKIP, Saitoh and Kurosawa [SK] showed that any random self-reducible relation in a sense of [TW] whose domain is in $\mathcal{BPP}$ has a four move blackbox simulation perfect ZKIP of possession of knowledge without any unproven assumption.

[BCY] Brassard, G., Crépeau, C., and Yung, M., "Everything in $\mathcal{NP}$ Can Be Argued in Perfect Zero-Knowledge in a Bounded Number of Rounds," Proc. of ICALP'89, LNCS 372, *Springer-Verlag*, Berlin, pp.123-136 (1989).

[BMO] Bellare, M., Micali, S., and Ostrovsky, R., "Perfect Zero-Knowledge in Constant Rounds," Proc. of STOC, pp.482-493 (May 1990).

[FFS] Feige, U., Fiat, A., and Shamir, A., "Zero-Knowledge Proofs of Identity," Proc. of STOC, pp.210-217 (May 1988).

[FS1] Feige, U. and Shamir, A., "Zero-Knowledge Proofs of Knowledge in Two Rounds," Proc. of Crypto'89, LNCS 435, *Springer-Verlag*, Berlin, pp.526-544 (1990).

[FS2] Feige, U. and Shamir, A., "Witness Indistinguishable and Witness Hiding Protocols," Proc. of STOC, pp.416-426 (May 1990).

[GKa] Goldreich, O. and Kahan, A., "Using Claw-Free Permutations to Constant Round Zero-Knowledge Proofs for $\mathcal{NP}$," in preparation (1989).

[GKr] Goldreich, O. and Krawczyk, H., "On the Composition of Zero-Knowledge Proof Systems," Proc. of ICALP'90, LNCS 443, *Springer-Verlag*, Berlin, pp.268-282 (1990).

[GMR1] Goldwasser, S., Micali, S., and Rackoff, C., "The Knowledge Complexity of Interactive Proof Systems," Proc. of STOC, pp.291-304 (May 1985).

[GMR2] Goldwasser, S., Micali, S., and Rackoff, C., "The Knowledge Complexity of Interactive Proof Systems," *SIAM J. Comput.*, Vol.18, No.1, pp.186-208 (February 1989).

[GMW] Goldreich, O., Micali, S., and Wigderson, A., "Proofs that Yield Nothing But Their Validity or All Languages in NP Have Zero-Knowledge Proofs," *Tech. Rep.* #544, Israel Institute of Technology, Department of Computer Science (March 1989).

[GO] Goldreich, O. and Oren, Y., "Definitions and Properties of Zero-Knowledge Proof Systems," *Tech. Rep.* #610, Israel Institute of Technology, Department of Computer Science (February 1990).

[NY] Naor, M. and Yung, M. "Universal One-Way Hash Functions and their Cryptographic Applications," Proc. of STOC, pp.33-43 (May 1989).

[R] Rabin, M.O., "Probabilistic Algorithm for Primality Testing," *Journal of Number Theory*, Vol.12, pp.128-138 (1980).

[S] Sloan, R., "All Zero-Knowledge Proofs are Proofs of Language Membership," *Tech. Memo.*, MIT/LCS/TM-385, MIT Laboratory for Computer Science (February 1989).

[SI] Sakurai, K. and Itoh, T., "Language Membership versus Possession of Knowledge in Constant Round ZKIP," *IEICE Trans.*, Vol.E74, No.8, pp.2118-2123 (August 1991).

[SK] Saitoh, T. and Kurosawa, K., "4-Move Perfect ZKIP of Knowledge with No Assumption," *these proceedings* (November 1991).

[SS] Solovay, R. and Strassen, V., "A Fast Monte Calro Test for Primality," *SIAM J. Comput.*, Vol.6, No.1, pp.84-85 (March 1977).

[TW] Tompa, M. and Woll, H., "Random Self-Reducibility and Zero-Knowledge Interactive Proofs of Possession of Information," Proc. of FOCS, pp.472-482 (October 1987).

On the Power of Two-Local Random Reductions

Lance Fortnow
University of Chicago
Computer Science Department
1100 E. 58th Street
Chicago, IL 60637
fortnow@cs.uchicago.edu

Mario Szegedy
AT& T Bell Laboratories
P.O. Box 636
Murray Hill, NJ 07974

ms@research.att.com

Abstract

We show that any language that has a two-locally-random reduction in which the target functions are boolean is in NP/poly∩co-NP/poly. This extends and simplifies a result by Yao.

1 Introduction

Suppose Frank wanted to access a database. Frank had access privledges to this database but for security reasons Frank could not reveal his question to this database. What can Frank learn under this requirement? What if Frank had access to several copies of the same database?

Abadi, Feigenbaum and Kilian [1] looked at the following game based on this scenario: Suppose a probabilistic polynomial-time player has access to a trustworthy oracle. This player wishes to use this oracle to determine the value of some complex function of some input but does not wish to reveal any information about the input besides its length. Abadi, Feigenbaum and Kilian [1] showed that any language reducible to an oracle in this fashion lies in NP/poly∩co-NP/poly.

Beaver and Feigenbaum [2] looked at the power of having the polynomial-time player query several separated oracles, i.e. oracles that can not communicate among themselves or listen to the conversation between a different oracle and the player. Beaver and Feigenbaum show the surprising result that, given $n+1$ different oracles, *any* function has such a *locally-random reduction*. Beaver, Feigenbaum, Kilian and Rogaway [3] improved this result to show that $n/(c \log n)$ oracles suffice for any positive constant c.

Virtually nothing was known about the complexity of two-locally-random reduction. Perhaps one could use two separate and trustworthy oracles to determine the value of any function without revealing more than the input length. Extending an idea of Yao [6], we give a partial negative result: Any language with a two-locally-random reduction with boolean oracles is in NP/poly∩co-NP/poly.

2   The Main Theorem

First we formally define local-random reductions:

Definition 1 *A function f has a k-locally-random reduction if there exist polynomial-time functions σ and ϕ and a polynomial $q(n)$ such that for input x and every r of length $q(|x|)$, there exists arbitrary oracle functions $g_1, \ldots, g_{k(n)}$ such that*

$$f(x) = \phi(x, r, g_1(\sigma(1, x, r)), \ldots, g_{k(n)}(\sigma(k(n), x, r)))$$

and for each i, $\sigma(i, x, r)$ and $\sigma(i, y, r)$ are distributed identically when $|x| = |y|$ and r is chosen uniformly at random over all strings of length $q(|x|)$.

We say a language has a k-locally-random reduction if its characteristic function has.

The outputs of the σ functions are the questions asked to the oracles, the r is the random coins of the querier, the g functions are the oracle responses and the ϕ function is the computation done after the response. If k is a constant, we will often use $\sigma_i(x, r)$ for $\sigma(i, x, r)$.

Locally-random reductions are a restriction of instance hiding schemes where the oracles can flip coins and more importantly interact with the polynomial-time player including having future answers depend on previous questions. See Beaver and Feigenbaum [2] for further details.

We can also look at *random-self reductions* as a restriction of local-random reductions by requiring $g_1 = \ldots = g_k = f$. For more precise definitions and theorems about random-self reductions see [1, 5, 4].

We can now state the main theorem:

Theorem 2 *If L has a two-local random reduction with oracles g_1 and g_2 where g_1 and g_2 output a single bit then L is in NP/poly$\cap$co-NP/poly.*

Our proof was inspired by a weaker result by Yao [6]: Any language with a two-local-random reduction with oracles that output only a single bit each is in PSPACE/poly. Besides obtaining stronger consequences our proof is also reasonably simpler than Yao's original proof.

Yap [7] shows that if NP$\subseteq$co-NP/poly then the polynomial-time hierarchy collapses to the third level. From this fact we immediately get

Corollary 3 *If SAT (or any other NP-hard language) has a two-local-random-reduction where the oracles only output single bit responses then the polynomial-time hierarchy collapses to the third level.*

3 Proof of Main Theorem

To prove Theorem 2, we need only show L is in NP/poly because by Definition 1 a language has a k-local-random reduction if and only if its complement also has one.

For every fixed n the characteristic function of the words of L having length n is a boolean function $f_n = f$.

Suppose f has a two-local-random reduction as required by the theorem. Consider the multisets $M_1 = \{\sigma_1(0,r) \mid r \in \{0,1\}^{q(n)}\}$, $M_2 = \{\sigma_2(0,r) \mid r \in \{0,1\}^{q(n)}\}$ (where 0 denotes the n bit string of zeros). The distributional equivalence of σ_1 and σ_2 imply that for every input x of f the multisets $\{\sigma_1(x,r) \mid r \in \{0,1\}^{q(n)}\}$ and $\{\sigma_2(x,r) \mid r \in \{0,1\}^{q(n)}\}$ can be identified with M_1 and M_2 respectively. The elements of these multisets will be called points.

Note 4 *We are dealing with multisets instead of sets to insure that $\sigma_i(x,r_1)$ and $\sigma_i(x,r_2)$ map to distinct points of M_i.*

We suppose that M_1 and M_2 are disjoint and introduce the following convention:

When talking about the value of g on a given point, we mean the value of g_1 if the point is in M_1 and the value of g_2 if the point is in M_2.

Let j be an element of $\{1,2\}$. We say that the values $x, r, g_j(\sigma_j(x,r))$ sets $f(x)$ if the value of $\phi(x,r,g_j(\sigma_j(x,r)), w)$ does not depend on w.

In the case $x, r, g_j(\sigma_j(x,r))$ does not set $f(x)$, we can obtain the value of $w = g_{3-j}(\sigma_{3-j}(x,r))$ from $x, r, g_j(\sigma_j(x,r)), f(x)$.

Definition 5 *For some x and r let $y = \sigma_j(x,r)$, $y' = \sigma_{3-j}(x,r)$. We say that $g_j(y)$ forces $g_{3-j}(y')$ through x and r if $x, r, g_j(\sigma_j(x,r))$ does not set $f(x)$.*

Forcing can be iterated. A sequence of length n points $y_0, \ldots, y_m$ is a *forcing path with respect to a subset of length-n inputs S* if for every i, $0 \le i < m$

1. i is even and there exists $x \in S$ and $r \in \{0,1\}^{q(|x|)}$ such that $y_i = \sigma_1(x,r)$ and $y_{i+1} = \sigma_2(x,r)$ and $g_1(y_i)$ forces $g_2(y_{i+1})$ through x and r.

2. i is odd and there exists $x \in S$ and $r \in \{0,1\}^{q(|x|)}$ such that $y_i = \sigma_2(x,r)$ and $y_{i+1} = \sigma_1(x,r)$ and $g_2(y_i)$ forces $g_1(y_{i+1})$ through x and r.

The *description* of the forcing path consists of the points $y_0, \ldots, y_m$ and the corresponding x's and r's used to force g_1 and g_2 along the path.

If the value of f is known for a subset of inputs, then the values of g along any forcing path with respect to this subset are forced by the value of the first point. These values can be computed in polynomial time if given the description.

From now on any forcing path will start at $\sigma_1(0,0)$.

The idea of the NP/poly algorithm is that the value of g at the point $\sigma_1(0,0)$ and the value of f on a small, but appropriate set of inputs will force enough values of g to compute f. Recall that with the help of ϕ we can compute $f(x)$ in polynomial time for an arbitrary choice of r from the values $g_1(\sigma_1(x,r))$ and $g_2(\sigma_2(x,r))$.

For some x the nondeterministic guess will include an r with the property that both $g_1(\sigma_1(x,r))$ and $g_2(\sigma_2(x,r))$ are forced (or in some cases only one of them) and the description of the corresponding forcing paths.

The case in which we need only one of the above values is when f is set by x, r and this value. In this case it is enough to give the forcing path to the corresponding point.

The polynomial advice of our NP/poly machine will contain:

1. The value $g_1(\sigma_1(0,0))$,

2. a polynomial length sequence of inputs $x_0, \ldots, x_m$,

3. the sequence of values of f at these points: $f(x_1), \ldots, f(x_m)$.

It remains to show that a small subset of inputs with the desired properties exists.

Lemma 6 *For every function f that has a 2-local-random reduction to g_1 and g_2 using random strings of length $q(n)$, there is a set of length n inputs $x_1, \ldots, x_m$ ($m \leq q(n)+1$) such that for every x there is an r with one of the following properties:*

1. $\sigma_1(x,r)$ and $\sigma_2(x,r)$ are both on a forcing path (with respect to $x_1,\ldots,x_m$) of length m.

2. There is a $j \in \{1,2\}$ such that $\sigma_j(x,r)$ is on a forcing path (with respect to $x_1,\ldots,x_m$) of length at most m and $x,r,g_j(\sigma_j(x,r))$ sets $f(x)$.

Proof of the lemma:

We construct an exponentially expanding set system $S_0 \subset S_1 \subset \cdots \subset S_m \subseteq M_1 \cup M_2$ and a set of inputs $x_1,\ldots,x_m$ ($m \leq l+1$) recursively such that S_i can be reached by a forcing path with respect to $x_1,\ldots,x_i$ of length at most i.

Moreover for every x there is an r such that one of the following cases holds:

1. both $\sigma_1(x,r)$ and $\sigma_2(x,r)$ are in S_m.

2. there is a $j \in \{1,2\}$ that $\sigma_j(x,r) \in S_m$ and $x,r,g_j(\sigma_j(x,r))$ sets the value of f.

$S_0 = \{\sigma_1(0,0)\}$. Suppose that $S_1,\ldots,S_i$ are already constructed, but S_i does not satisfy the properties required for S_m, i.e. there is an x that none of the conditions 1 and 2 hold for x.

Choose such an x for x_{i+1}.

For every pair $\sigma_1(x_{i+1},r), \sigma_2(x_{i+1},r)$ that coincides with one of the points of S_i (such a pair can never coincide with two points of S_i because otherwise condition 1 would hold for x_{i+1}), the value of the point of the pair that is outside S_i is forced by x_{i+1}, r and the value of g on the point that belongs to S_i.

Let the set S_{i+1} include exactly the points of the pairs that have one common point with S_i. The values of g at these points are forced in $i+1$ step by $x_0,\ldots,x_{i+1}$.

Observe that the size of S_{i+1} is exactly twice the size of S_i. This follows from the fact that for each point of S_i there is exactly one coinciding pair and that the second elements of the pairs are all distinct (see Note 4). The upper bound on m is now implied by $|M_1| = |M_2| = 2^{q(n)}$.

4 Final Comments

There is still a large gap in our knowledge of local-random reductions. Here are some open questions:

- What is the power of two-local-random reductions when the query to one depends on the answer given by the other?

- What is the power of two-local-random reductions when the oracles can output any number of bits?

- What is the minimum $k(n)$ such that any function has a $k(n)$-local-random-reduction.

5 Acknowledgments

We acknowledge Andrew Yao for several important ideas taken from his talk at the DIMACS workshop [6]. Our discussions with Joan Feigenbaum and Gábor Tardos contributed significantly to obtaining the result. We also thank Joan Feigenbaum for her comments on this exposition. Finally, we would like to thank the anonymous referee for several helpful comments. The first author is supported in part by NSF Grant CCR-9009936.

References

[1] M. Abadi, J. Feigenbaum, and J. Kilian. On Hiding Information from an Oracle, *J. Comput. System Sci.* 39 (1989), 21–50.

[2] D. Beaver and J. Feigenbaum. Hiding Instances in Multioracle Queries, *Proc. of the 7th STACS* (1990), Springer Verlag LNCS 415, 37–48.

[3] D. Beaver, J. Feigenbaum, J. Kilian, and P. Rogaway. Security with Low Communication Overhead, *Proc. of the 10th CRYPTO* (1990), Springer Verlag LNCS, to appear.

[4] J. Feigenbaum and L. Fortnow. On the Random-Self-Reducibility of Complete Sets, *Proc. of the 6th Structures* (1991), IEEE, 124–132.

[5] J. Feigenbaum, S. Kannan, and N. Nisan. Lower Bounds on Random-Self-Reducibility, *Proc. of the 5th Structures* (1990), IEEE, 100–109.

[6] A. Yao. An Application of Communication Complexity to Cryptography, *Lecture at DIMACS Workshop on Structural Complexity and Cryptography* (1990).

[7] C. Yap. Some Consequences of Nonuniform Conditions on Uniform Classes, *Theor. Comput. Sci.* 26 (1983), 287–300.

A Note On One-Prover, Instance-Hiding Zero-Knowledge Proof Systems

(Extended Abstract)

Joan Feigenbaum
AT&T Bell Labs, Rm 2C473
600 Mountain Avenue
Murray Hill, NJ 07974-0636
jf@research.att.com

Rafail Ostrovsky
MIT Lab for Computer Science
545 Technology Square
Cambridge, MA 02139
raf@theory.lcs.mit.edu

Abstract

In this note we study two cryptographic notions introduced in recent years: zero-knowledge proof systems [GMR] and instance-hiding schemes [AFK]. Addressing an open problem of [BFS] and extending the work in [AFK], we show that, if one-way permutations exist, the following two statements are equivalent for any function f:

(i) f has a one-prover, instance-hiding, zero-knowledge proof system.

(ii) f is computable in polynomial space and has an instance-hiding scheme that leaks at most the length of the input.

1 Introduction

In [GMR], the notion of *zero-knowledge* was introduced. Informally, a zero-knowledge proof system for f is a protocol between two players A (an infinitely powerful prover) and B (a probabilistic polynomial-time verifier), with a common input (x, y), in which A convinces B that $y = f(x)$ without revealing anything else. The intuition behind this definition is that A convinces a polynomially bounded B of the validity of a theorem, without revealing the proof.

In [AFK], the notion of *instance hiding* was introduced. Again informally, an instance-hiding scheme for the function f is a protocol between two players A (an infinitely powerful oracle) and B (a probabilistic polynomial-time querier) in which B, who has a private input x, uses the superior computing power of A to derive $f(x)$ without revealing to A anything about x except its length.[1] The intuition here is to allow a weak player B to utilize resources of a strong (and honest!) player A to help B in computing $f(x)$ without revealing his private data to A. Possible applications include financial transactions and shared use of supercomputers.

---

[1] The *term* "instance-hiding scheme" does not appear in [AFK]. There, these protocols are called "encryption schemes for functions."

The natural question of combining these two notions was addressed in [AFK, BFS]. Intuitively, this is the (realistic) scenario in which both players distrust each other and hence wish to conceal some information. For example, assume that the weak player B has a secret input x and that he is willing to pay for A's services to compute $f(x)$. Nevertheless, B does not want to tell A what x is. Moreover, B does not trust A to give a correct answer. Thus, B requires a proof from A that, indeed, A correctly helped B to compute $f(x)$, for some (unknown to A) x. On the other hand, A also has a legitimate concern that B will exploit the fact that A does not know what value x_i he is helping B with and that B will manage to compute $f(x_1)$ and $f(x_2)$ for the price of $f(x_1)$ only. Can both conditions (zero-knowledge and instance-hiding) be satisfied at the same time? Under what assumptions? For which functions? In [BFS], it was shown that, with multiple physically separated A's, the above requirements can be achieved for any f whose characteristic function is computable in nondeterministic exponential time. It was left open what can be done with just one A. Some specific examples of one-prover, instance-hiding, zero-knowledge proofs systems (based on standard number-theoretic examples) were given in [AFK]. In this note, we show:

MAIN THEOREM: *Assume that one-way permutations exist. Then a function f has a one-prover, instance-hiding, zero-knowledge proof system if and only if f is computable in polynomial space and has an instance-hiding scheme that leaks at most the length of the input.*

2 Preliminaries

Let V and P be a pair of interactive Turing Machines. As in [GMR], the verifier V is a probabilistic polynomial-time Turing Machine (with a private source of random coins) and the prover P is computationally unbounded. The notation V^* (resp. P^*) is used to denote a potentially cheating verifier (resp. prover); that is, V^* (resp. P^*) may follow the algorithm of the legitimate verifier V (resp. prover P), or it may deviate from the algorithm in an attempt to gain some advantage. As in [AFK], the input x is known only to the verifier. The output produced by V^* after interacting with P^* is an element of the set $\{0,1\}^* \bigcup \{reject\}$ and is denoted by $(V^*(x), P^*)$ – note that the output is a random variable; it depends on the coins of V^* and P^* as well as on the input. The *transcript* of messages sent between V^* and P^* on input x, denoted $Trans(V^*, P^*, x)$, is a random variable, and its distribution is induced by the random coin-tosses and the algorithms of both prover and verifier. V^*'s *view* of the protocol execution on input x is denoted $View_{(V^*, P^*)}(V^*, x)$ and consists of the transcript, together with V^*'s coin-tosses.

Similarly P^*'s view of the protocol execution on input x is denoted $View_{(V^*,P^*)}(P^*, x)$ and consists of the transcript, together with P^*'s coin-tosses.

Recall the definition of *instance-hiding scheme* given in [AFK]:

Definition 2.1 *The protocol (V, P) is an* **instance-hiding scheme for the function** f *if it satisfies the following properties.*

(A) For all sufficiently large x, $\text{Prob}((V(x), P) = f(x)) > 3/4$.

(B) For all inputs x and x' with $|x| = |x'|$, the random variables $\text{View}_{(V,P)}(P, x)$ and $\text{View}_{(V,P)}(P, x')$ are identically distributed.

Condition (B) captures the notion of instance-hiding – that is, the transcript leaks no more than the length of x. A more general definition of instance-hiding is given in [AFK]; if we restrict attention to the case in which the "leaking" function L is just $|x|$, then condition (B) is equivalent to the definition in [AFK].

Next, recall the definition of instance-hiding proof system given in [BFS]:

Definition 2.2 *The protocol (V, P) is an* **instance-hiding proof system for the** **function** f *if it satisfies the following properties.*

(i) For all sufficiently large x, $\text{Prob}((V(x), P) = f(x)) > 3/4$.

(ii) For all sufficiently large x, for all P^*, $\text{Prob}((V(x), P^*) \notin \{f(x), \text{reject}\}) < 1/4$.

(iii) For all P^*, for all x and x' with $|x| = |x'|$, the random variables $\text{View}_{(V,P^*)}(P^*, x)$ and $\text{View}_{(V,P^*)}(P^*, x')$ are identically distributed.

In both of these definitions, the success probability can be amplified to $1 - 2^{-poly(|x|)}$ straightforwardly, using sequential repetition and majority vote.

Next, we present a modification of the definition of zero-knowledge given in [GMR].

Definition 2.3 *An instance-hiding proof system (V, P) for the function f is* **zero-know-** **ledge** *if, for any probabilistic polynomial-time V^*, there is a probabilistic, expected-polynomial-time oracle machine M_{V^*} (called the* **simulator***) with the following property: During its execution on input x, M_{V^*} may make exactly one query x' to an f-oracle,*

and the query must have length $|x|$. *The distribution of the simulator's output* $M_{V^*}(x)$ *is* computationally indistinguishable from $\text{View}_{(V^*,P)}(V^*, x')$.

In the [GMR] definition of zero-knowledge, the simulator is given the value $f(x)$. The intuition behind that definition is that the prover prevents the verifier from learning anything except $f(x)$. We have modified the definition so that the simulator may obtain one value of the function, say $f(x')$, at an arbitrary input x' that has the same length as x. This is the right notion of zero-knowledge for our setting: In an instance-hiding proof system for f, the prover does not know x and cannot infer anything about it (except its size) from the messages sent by the verifier; hence the prover cannot hope to prevent the verifier from learning, say, $f(x')$ instead of $f(x)$, where $|x'| = |x|$. Instance-hiding, zero-knowledge proof systems were first defined in [BFS]; a more thorough discussion of this definition can be found there.

Note that, in the above definition, the zero-knowledge requirement must hold only with respect to a polynomially bounded verifier. Analogous definitions, in which no information is leaked in the information-theoretic sense, could be defined. In this paper, we concentrate on the computational notion of zero-knowledge.

If (P, V) is an interactive protocol satisfying any of the above definitions, then $V(x, r, p)$ denotes the move that V would make given input x, random string r, and (possibly empty) transcript prefix p. Similarly, $P(r', p)$ denotes the move that P would make given random input r' and (possibly empty) transcript prefix p. We stress that, because these are instance-hiding protocols, P cannot use x directly in computing his next move.

In our construction, we use *Combined Oblivious Transfer* (COT), introduced in [Y]. In this two-party protocol, player S (the sender) has a private input x, and player R (the receiver) has a private input y. They are also given a poly-size circuit $C(\cdot, \cdot)$. (In this note, we allow S to be infinitely powerful and require R to be polynomially bounded.) At the end of the protocol, R learns $C(x, y)$ (but does not learn anything about x that is not already revealed by $C(x, y)$) while S learns nothing. (We note that COT with reversed roles of S and R is also possible.) In [Y] an implementation of COT based on factoring is given. In [GMW], an implementation based on trapdoor permutations is presented. Finally, in [OVY], it is shown how to implement COT based any one-way function.

3 The construction

We separate the proof of our main result into two Lemmas. The assumption that one-way permutations exist is only needed for the second.

Lemma 3.1 *Suppose that f is computable in polynomial space and has an instance-hiding scheme. Then f has an instance-hiding proof system.*

Proof (sketch): Let (P, V) be an instance-hiding scheme for f. We use it to construct (P', V'), an instance-hiding proof system for f. Let $n = |x|$, and let $m = m(n)$ be a polynomial upper bound on the number of moves in $Trans(P, V, x)$; we can assume without loss of generality that m is even. Recall that x is the private input to V'. Let r be the private random string of V' and $l = l(n)$ be the (polynomial) length of r.

Players P' and V' interact to produce a transcript t of the instance-hiding scheme (P, V) for instances of length n. This transcript must have two properties, described informally as follows: It must be "instance-hiding," and, with high probability, it must be "correct for all x." We now describe these properties in more detail and show why P' can prove (interactively) to V' that t has them.

The instance-hiding property is just what we expect it to be: For any x_1 and x_2 of length n, the number of r's that cause P and V to produce t on input x_1 is the same as the number that cause P and V to produce t on input x_2. P' must prove that t has this property round by round, and he does so as follows. For his first move, V' simply computes the question $q_1 = V(x, r)$ and sends it to P'. Suppose that $p = (q_1, a_1, \ldots, q_{i-1}, a_{i-1})$ is the prefix of t that has been computed so far. That is, P' has just sent the answer a_{i-1} to V'. P' then proves (interactively) to V' that, for all x_1 and x_2 of length n, for all questions q_i, the number of r's, consistent with input x_1 and prefix p, such that $V(x_1, r, p) = q_i$ is the same as the number of r's, consistent with input x_2 and prefix p, such that $V(x_2, r, p) = q_i$. If, for any i, this interactive proof fails, V' just terminates the overall protocol and outputs "reject." These "subproofs" can be accomplished, because the statement "for all x_1 and x_2 of length n, for all questions q_i, the number of r's, consistent with input x_1 and prefix p, such that $V(x_1, r, p) = q_i$ is the same as the number of r's, consistent with input x_2 and prefix p, such that $V(x_2, r, p) = q_i$" is a PSPACE statement [LFKN, S]. V' (resp. P') computes the questions q_i (resp. a_i) exactly as V (resp. P') computes them.

Intuitively, t has the desired correctness property if it is very likely that, on all inputs x of length n, V gets the correct answer $f(x)$ if the transcript produced is t. We will make this notion more precise shortly, but this intuitive description suffices to finish the high level description of the proof of the lemma. We show that, if P' and V' behave correctly, i.e., if they produce transcripts according to the same distribution produced by P and V, then, with high probability, t has the correctness property. This in turn implies that, with high probability, V' will get the right answer $f(x)$. We also require P' to *prove* (interactively) to V' that t has the correctness property. This protocol takes place at the end, i.e., after P' sends his final answer $a_{m/2}$ and before V' computes his candidate for $f(x)$. Once again, it is possible to obtain such a protocol because the statement that t has the correctness property is a PSPACE statement; the fact that f is computable in polynomial space is needed here.

We now make the definition of correctness more precise. For any random string r of V, there are some x's for which V always gets the right answer on input x and random string r and some x's for which V can get the wrong answer on input x and random string r. Say that r is *bad* if there is at least one x for which V can get the wrong answer on input x and random string r. We can assume without loss of generality that an r chosen uniformly at random from $\{0,1\}^l$ is bad with probability at most 2^{-n}. If this is not true of the original instance-hiding scheme (P, V), we can replace it with another instance-hiding scheme that proceeds by running n^d independent copies of (P, V) and outputting the plurality of the answers. The existence of a suitable polynomial n^d has a standard proof using Chernoff bounds.

For any transcript t, there is a set $R(t)$ of random strings r that are consistent with t; these are the r's for which there exists at least one x such that t could be produced on input x and random string r. Say that t is bad if at least $1/10$ of the elements of $R(t)$ are bad. Formally, the transcript t has the correctness property if it is not bad. What can be shown is that, if P' and V' behave correctly, then the probability that they produce a t that has the correctness property is at least $9/10$. This in turn implies that, for any x, V' outputs the correct $f(x)$ with probability greater than $3/4$. To see that t has the correctness property with probability at least $9/10$, observe that, if t is bad with probability greater than $1/10$, then r is bad with probability at least $1/100$, which is a contradiction.

Note that the correctness property can be proven after a complete transcript t has been produced, but the instance-hiding property must be proven round by round.

Finally, note that we have argued that the *transcripts* of the proof system (P', V') will be identically distributed for all inputs of length n, while Definition 2.2 requires that the *views* be identically distributed. In fact, it can be shown that it is sufficient to argue that the transcripts are identically distributed. A proof of this fact will be given in the full version of the paper.

Lemma 3.2 *Assume that one-way permutations exist. Then any function f that has an instance-hiding proof system in fact has one that is zero-knowledge.*

Proof (sketch): Let (P, V) be an instance-hiding proof system for f. We use (P, V) to construct (P', V'), a zero-knowledge, instance-hiding proof system for f. The notation x, m, r, and l is as in the proof of Lemma 3.1. The proof system (P', V') consists of a set-up phase and an execution phase. In the set-up phase, P' first constructs a secure encryption function E and sends it to V'. Next, V' and P' each toss a sequence of l fair coins; call these sequences $s_1 = s_{11}s_{12} \ldots s_{1l}$ and $s_2 = s_{21}s_{22} \ldots s_{2l}$, respectively. V' then commits to s_1 and P' to s_2. In the execution phase of the protocol, the sequence $r = r_1 r_2 \ldots r_l$, where $r_i = s_{1i} \oplus s_{2i}$, plays the role of the verifier's random input in (P, V). Intuitively, the execution phase is constructed by replacing each move of V by a COT protocol and replacing each answer of P with an encrypted answer. Essentially, all of the above steps can be done, given a protocol to execute COT which is "simulatable" in zero-knowledge. The work of [OVY] provides such a protocol based on any one-way permutation. (Without the simulatability requirement, [OVY] provides a COT protocol based on any one-way function; however, to implement simulatability, permutations are needed.)

We now give more details. The first step of (P, V) is for V to compute the first question $q_1 = V(x, r)$ and send it to P. In the first step of (P', V'), the players execute a COT protocol. P' supplies the input s_2, and V' supplies the inputs x and s_1. The output of protocol is $E(q_1)$; both players are given this output. P' then decrypts the question q_1, computes the answer a_1 that would be given by P, and sends $E(a_1)$ to V'. More generally, suppose that $(E(q_1), E(a_1), \ldots, E(q_i), E(a_i))$ is the transcript prefix. The next step is a COT protocol to which P' supplies $(q_1, a_1, \ldots, q_i, a_i)$ and s_2, and V' supplies x and s_1. The output, which is given to both players, is $E(q_{i+1})$. P' then decrypts the question, computes a_{i+1} using the same algorithm as P, and sends $E(a_{i+1})$ to V'. The last step is a COT protocol in which P' supplies s_2 and the entire transcript $(q_1, a_1, \ldots, q_m, a_m)$, V' supplies x and s_1, and the output is $f(x)$. This time, of course, only V' gets the output. Notice that a simulator, after querying an f-oracle, can encrypt a random string, thus providing a run that is polynomial-time indistinguishable from the actual transcript.

Suppose that a function f satisfies Definition 2.2. We may conclude immediately that f satisfies Definition 2.1 and that f has a one-prover interactive proof system; it is shown in [LFKN, S] that the latter is equivalent to the conclusion that f is computable in polynomial space. Thus, combining Lemmas 3.1 and 3.2 yields our main result:

Theorem: *Assume that one-way permutations exist. Then f has a one-prover, instance-hiding proof system that is zero-knowledge if and only if f is computable in polynomial space and has an instance-hiding scheme.*

In the full version of the paper, we discuss how the assumption of the existence of a one-way permutation could be weakened further.

4 Acknowledgement

Part of the work of the second author was done at AT&T Bell Laboratories. The rest was supported by an IBM Graduate Fellowship.

References

[AFK] M. Abadi, J. Feigenbaum, and J. Kilian. On Hiding Information from an Oracle, *J. Comput. System Sci.* 39 (1989), 21–50.

[BFS] D. Beaver, J. Feigenbaum, and V. Shoup. Hiding Instances in Zero-Knowledge Proof Systems, *Proc. of the 10th CRYPTO* (Santa Barbara, CA; August, 1990), Springer Verlag LNCS 537, 1991, 326–338.

[GMR] S. Goldwasser, S. Micali, and C. Rackoff. The Knowledge Complexity of Interactive Proof Systems, *SIAM J. Comput.* 18 (1989), 186–208.

[GMW] S. Goldreich, S. Micali and A. Wigderson. How to Play ANY Mental Game, *Proc. of the 19th STOC* (New York, NY; May, 1987), ACM, 218–229.

[LFKN] C. Lund, L. Fortnow, H. Karloff, and N. Nisan. Algebraic Methods for Interactive Proof Systems, *Proc. of the 31st FOCS* (St. Louis, MO; October, 1990), IEEE, 2–10.

[OVY] R. Ostrovsky, R. Venkatesan, and M. Yung. Fair Games Against an All-Powerful Adversary, *Proc. of SEQUENCES '91* (Positano, Italy; June, 1991).

[S] A. Shamir. IP = PSPACE, *Proc. of the 31st FOCS* (St. Louis, MO; October, 1990), IEEE, 11–15.

[Y] A. C. Yao. How to Generate and Exchange Secrets, *Proc. of the 27th FOCS* (Toronto, Ontario; October, 1986), IEEE, 162–167.

AN EFFICIENT ZERO-KNOWLEDGE SCHEME FOR THE DISCRETE LOGARITHM BASED ON SMOOTH NUMBERS

Yvo Desmedt*
Department of EE & CS
University of Wisconsin — Milwaukee
P.O. Box 784, WI 53201 Milwaukee
U.S.A.

Mike Burmester†
Department of Mathematics
RHBNC - University of London
Egham, Surrey TW20 OEX
U.K.

Abstract. *We present an interactive zero-knowledge proof for the discrete logarithm problem which is based on smooth numbers. The main feature of our proof is its communication complexity (number of messages exchanged, number of bits communicated) which is less than that of competing schemes.*

1 Introduction

A great deal of theoretical research has recently been carried out in the area of zero-knowledge (*e.g.*, [BMO90,BGG90,Sha90]). Earlier on, research had focussed on making practical and efficient zero-knowledge schemes [FS87,CEvdG88,FFS88,GQ88, Bet88,Sch90]. Unfortunately some of these schemes were not proven to be zero-knowledge. An important aspect in the design of practical protocols is their communication complexity. Indeed while in the United States local phone calls are charged by the number of calls (or by a flat rate), in Japan and in many European countries these are charged by minute (or time unit), as long distance calls are charged in the United States. So the design of zero-knowledge protocols with low communication cost is an important aspect for many countries. Although some research has concentrated on reducing the communication complexity of zero-knowledge protocols [KMO89], the results are theoretical and focus on **NP**-complete languages. For practical cryptosystems **NP**-completeness is not used, but problems such as factoring and the discrete logarithm, which are believed not to be **NP**-complete are heavily utilized. The discrete logarithm is an important one-way function which is often used in modern cryptography. Many key exchange protocols [DH76,Kon81,MO] and the El Gamal signature scheme [ElG85] have been based on the discrete logarithm.

In this paper we present an efficient interactive zero-knowledge proof for the discrete logarithm problem. Our scheme is based on smooth numbers (see *e.g.*, [COS86]) and

---

*Supported in part by NSF Grant NCR-9004879 and NSF Grant NCR-9106327.
†Research carried out while visiting the University of Wisconsin – Milwaukee.

the Pohlig-Hellman algorithm [PH78].

The novelty of our proof is its high efficiency. If $|x|$ is the length of the input, then the number of rounds (messages exchanged) for our protocol is typically $\log^\epsilon |x|$ and the communication complexity (number of bits exchanged) is $\log^{1+\epsilon} |x|$ for the verifier and $|x| \log^\epsilon |x|$ for the prover. Overall, when compared to the Chaum–Evertse–van de Graaf protocol [CEvdG88] for the discrete logarithm, the communication complexity of our protocol is reduced by a factor $\log |x|$.

This paper is organized as follows. In Section 2 we overview the definitions and the mathematics and algorithms required to understand the text. In Section 3 we present our protocol for the discrete logarithm over Z_p^*, p prime, and in Section 4 we extend it.

2 Preliminaries

In this section we give the definitions and discuss some aspects of group theory and smooth numbers. Readers familiar with these topics may skip this section.

2.1 DEFINITIONS

In the next section we shall consider a zero-knowledge proof of membership in a language (set) L. Informally a proof is a two party protocol (P, V). P is the prover and V is the verifier. P wants to prove to V that the common input x belongs to L. Both the prover and the verifier use randomness (toss coins). (P, V) is an *interactive proof for L* [GMR89] if, for common input x,

- *Completeness:* For all k, for any sufficiently large x in L, V accepts with probability at least $1 - |x|^{-k}$ ($|x|$ is the binary length of x).

- *Soundness:* For all k, for any sufficiently large x not in L, for any prover P', V accepts with probability at most $|x|^{-k}$.

Informally, a proof is *zero-knowledge* if, when the input x is in L, then the 'view' of any verifier V' can be simulated in expected polynomial time in $|x|$. For a formal definition see [GMR89].

We will consider protocols for which the number of rounds is 'almost-constant'. This to a large extent will account for their efficiency. A function $t = t(n)$ of the natural numbers is *almost-constant* if it is an unbounded function which grows very slowly, e.g., significantly slower than $\log n$. For our purpose it is not necessary to specify such functions any further, but if it helps to have a particular function in mind then we could take $t(n) = \log^* n$ [BB88] (so $t(n)$ grows slower than any iterated logarithm).

2.2 GROUP THEORY

In a general context we will consider families of groups, but informally we will just speak about groups. We require the following notation and results from group theory.

Z_n is the set of residues of integers modulo n and Z_n^* is the multiplicative group of residues which are coprime to n. The *order of a group* is the number of its elements. The *order of an element* α, ord(α), is the smallest positive integer x for which $\alpha^x = 1$. We denote the group generated by β as $\langle \beta \rangle$. It is well known that for any divisor d of the order of a (finite) cyclic group there is exactly one cyclic subgroup which has this order [Jac85]. Furthermore, if k, l are complementary coprime factors of the order of an Abelian group G, then G is the direct product of a subgroup of order k and a subgroup of order l [Jac85].

2.3 THE DISCRETE LOGARITHM PROBLEM

We shall consider the *discrete logarithm* problem and the *discrete logarithm decision* problem. The first problem deals with *finding* or *computing* integers x such that $\alpha = \beta^x \pmod{n}$, where $\alpha, \beta \in Z_n^*$ and n is a natural number (β is not necessarily a generator when Z_n^* is cyclic). The second deals with recognizing the *existence* of such x's. So the discrete logarithm decision problem is concerned with recognizing the elements of the group $\langle \beta \rangle$ generated by β. Clearly the discrete logarithm decision problem can be reduced to the discrete logarithm problem. We are particularly interested in the case when $n = p$ is a prime. It is believed that the discrete logarithm decision problem is hard [PH78,Odl84,AM86,COS86].

2.4 SMOOTH NUMBERS AND THE POHLIG-HELLMAN ALGORITHM

We use the definition given in [COS86]. An integer z is *smooth with respect to* y if for each prime p_i which divides z we have $p_i \leq y$. If y is understood, we speak of z being smooth. If the order of $\beta \in Z_p^*$, p a prime, is smooth relative to $|p|^c$, c a constant, then the discrete logarithm of β^x can be calculated efficiently [PH78]. In particular the elements of $\langle \beta \rangle$ can be recognized in time polynomial in $|p|$.

Suppose that the order of $\beta \in Z_p$, ord(β), is $|p|$ smooth, where p is a prime. Then ord(β) can easily be calculated. Indeed, write $p - 1 = l_p \cdot k_p$, where l_p is $|p|$ smooth and k_p is the largest divisor of $p-1$ whose prime factors are greater than $|p|$. Observe that k_p and l_p can easily be calculated, e.g., using recursively the Euclidean algorithm, starting with checking $\gcd(2, p-1)$, etc. So $\beta^{l_p} \equiv 1 \pmod{p}$. When $l_p = p_1^{m_1} \cdots p_a^{m_a}$, where p_i are primes, then for any $p_i \leq |p|$ and $p_i \mid l_p$, the largest prime factor $p_i^{m_i'}$ of ord(β) is the smallest power of p_i for which $\beta^{l \cdot p_i^{m_i'}/p_i^{m_i}} \equiv 1 \pmod{p}$. So it is feasible to compute ord(β) when ord$(\beta) \mid l_p$. Observe that if the ord(β) is known then it is easy to check membership in $\langle \beta \rangle$: indeed $\alpha \in \langle \beta \rangle$ if and only if $\alpha^{\text{ord}(\beta)} = 1$. In this paper we only need this simplified Pohlig-Hellman algorithm, except in Section 4.

3 A z-k proof for discrete logarithm over Z_p^*, p a prime

We shall consider a proof of membership for the language (set) L of the discrete logarithm decision problem. The elements of L are the strings $x = (\alpha, \beta, p)$, where p is a prime, $\alpha, \beta \in Z_p^*$, and $\alpha\beta^s \equiv 1 \pmod p$ for some integer $s \in Z_{p-1}$ (so s is equal to minus the discrete logarithm). Consider the following protocol:

Protocol (P,V): Input $x = (\alpha, \beta, p)$

Step 0 V checks that p is a prime [SS77] and that $\alpha, \beta \in Z_p^*$. Then V computes $k = k_p$ where k_p is the largest divisor of $p - 1$ whose prime factors are larger than $|p|$ and computes $l = l_p = (p - 1)/k_p$. Then V checks that $\alpha^k \in \langle \beta^k \rangle$ (using the algorithm discussed in 2.4). If any one of these conditions fails then V halts and rejects. Otherwise the following Steps are repeated t times independently, where $t = t(|p|)$ is almost-constant.

Step 1 P sends to V: $z = \beta^{lr} \pmod p$, where $r \in_R Z_{p-1}$.[1]

Step 2 V sends to P: q, where $q \in_R Z_v$, $v = |p|$.

Step 3 P verifies that $q \in Z_v$, and if so, sends to V:

$$y \equiv r + qu \pmod{p-1}, \text{ where the } u \text{ is such that } \alpha^l \beta^{lu} \equiv 1 \pmod p.$$

Step 4 V verifies that $y \in Z_{p-1}$ and that $z \equiv \alpha^{lq} \beta^{ly} \pmod p$. If this check fails then the protocol is halted.

After t complete rounds V accepts (V is convinced).

We can replace $p - 1$ by k_p in Steps 1–4 of the protocol above. This makes the scheme more efficient without affecting the zero-knowledge proof aspect.

Theorem 1 *The protocol (P, V) is a perfect zero-knowledge proof for the language*
$L = \{(\alpha, \beta, p) \mid p \text{ is a prime}; \alpha, \beta \in Z_p^*; \exists s \in Z_{p-1} : \alpha\beta^s \equiv 1 \pmod p\}$.

Proof. In Step 0 the verifier V (who is a polynomially bounded in $|x|$) checks that $\alpha^k \in \langle \beta^k \rangle$. V can do this as explained in Section 2.4. Indeed, $(\beta^k)^l \equiv 1 \pmod p$, so that the order of β^k is a smooth number. Let now $\alpha^k \equiv \beta^{kw} \pmod p$, for $w \in Z_{p-1}$.

The main part of the protocol is a zero-knowledge proof that $\alpha^l \in \langle \beta^l \rangle$. That is that, $\alpha^l \equiv \beta^{-lu} \pmod p$, for some u. Since l and k are relative prime, there exist integers L, K, with $lL + kK = 1$. Combining these results we see that $\alpha = \alpha^{lL+kK} = (\alpha^l)^L (\alpha^k)^K \equiv \beta^{-luL} \beta^{kwK} \equiv \beta^{-luL+kwK} \pmod p$. Then $\alpha\beta^s \equiv 1 \pmod p$ for $s \equiv luL - kwK \pmod{p-1}$, and therefore $x \in L$. It remains to show that we have a zero-knowledge proof that $\alpha^l \in \langle \beta^l \rangle$.

Completeness: Suppose that there exists a u such that $\alpha^l \equiv \beta^{-lu} \pmod p$. Then $\alpha^{lq} \beta^{ly} \equiv \alpha^{lq} \beta^{l(r+qu)} \equiv \alpha^{lq} \beta^{lr} \beta^{lqu} \equiv (\alpha^l \beta^{lu})^q z \equiv z \pmod p$. So the verification in Step 4 always checks and therefore V always accepts.

---

[1] $a \in_R A$ means that the element a is selected randomly with uniform distribution from the set A.

Soundness: Suppose that $\alpha^l \notin \langle \beta^l \rangle$. We shall show that it is not possible, even for an infinitely powerful dishonest prover, to answer more than one of the queries q in Step 3 of any particular round.

Indeed suppose that for some $z \in Z_p^*$ there exist $q_1, q_2 \in Z_v$, $q_1 > q_2$, with $z \equiv \alpha^{lq_1} \beta^{ly_1} \equiv \alpha^{lq_2} \beta^{ly_2} \pmod{p}$. Then $\alpha^{lq} \beta^{ly} \equiv 1 \pmod{p}$, where $q = q_1 - q_2$ and $y = y_1 - y_2$. Now $\gcd(lq, k) = 1$, since $q < v$, since the prime factors of l are less than or equal to v, and since the smallest prime factor of k is greater than v. So $lqM + kK = 1$ for appropriate integers M, K. Then $\alpha = \alpha^{lqM+kK} \equiv \beta^{-lyM} \alpha^{kK} \equiv \beta^{kwK-lyM} \pmod{p}$, since $\alpha^k \equiv \beta^{kw} \pmod{p}$ from Step 0. So $\alpha \in \langle \beta \rangle$ and hence $\alpha^l \in \langle \beta^l \rangle$, which is a contradiction.

Therefore the chance of success in t consecutive rounds when $\alpha^l \notin \langle \beta^l \rangle$ is $v^{-t} = |p|^{-ct}$, which is negligible (in $|p|$).

Zero-knowledge: We use the technique of 'probing and resetting' the verifier as described in [GMR89]. Suppose that $x \in L$. The 'simulator' $M_{V'}$ (a probabilistic expected polynomial time Turing Machine) first prepares 'transcripts' of the form ($z \equiv \alpha^{lq} \beta^{lr} \pmod{p}$), q, $y = r$), where $q \in_R Z_v$ and $r \in_R Z_{p-1}$. Then $M_{V'}$ obtains the appropriate distribution by probing V'. The expected number of probings is $O(tv)$, which is polynomial in $|p|$. In this case we have *perfect* zero-knowledge because the distribution generated by $M_{V'}$ is identical to the actual distribution. □

The communication complexity of the above protocol is roughly $2|p| \log^\varepsilon |p|$. Due to the preprocessing (Step 0) its computational complexity is similar to [CEvdG88], but it is faster (less multiplications) for the prover.

4 Generalizations

The result above can easily be generalized. Our discussion will not be very formal, but can easily be formalized (for further generalizations consult [BDB]).

Corollary 1 *Let H_n be an Abelian group[2] for which testing membership and the group operation can be performed in polynomial time. Then the protocol in Section 3 can easily be adapted to obtain a perfect zero-knowledge proof for the discrete logarithm decision problem, provided that, among other modifications (see [BDB]), m', a multiple of the order of H_n, is given and that we replace in the protocol Z_p^* by H_n, and all modulo p operations by the group operation of H_n, and $p - 1$ by m'.*

Proof. This is similar to Theorem 1 and is obtained by combining the extension of the "original" Pohlig-Hellman algorithm to Abelian groups with [BDB]. □

---

[2] Formally, an infinite family of Abelian groups $\{H_n\}$.

In this generalization the original Pohlig-Hellman algorithm is used to check whether $\alpha^k \in \langle \beta^k \rangle$. The verifier must check that m' is a multiple of the order of H_n.

The relaxed discrete logarithm problem [CEvdG88] deals with finding integers $x_1, \ldots, x_h$, such that $\alpha = \beta_1^{x_1} \cdots \beta_h^{x_h}$, where α and β belong to the Abelian group H_n. The corresponding decision problem deals with the *existence* of such x_i. The corollary above can be extended to the relaxed discrete logarithm decision problem over any Abelian group H_n (formally: family of polynomial-time groups) with h a constant.

In this text we assumed that $v = |p|$. This can easily be generalized to $|p|^a \leq v \leq |p|^b$, where a, b are constants, $0 < a < b$.

5 Conclusion

Computational number theory has mostly been used to cryptanalyze cryptosystems (*e.g.*, [COS86]). Here it helps to make a cryptographic protocol more efficient. It has allowed us to present a communication efficient zero-knowledge protocol for the discrete logarithm decision problem. A natural open problem is whether other zero-knowledge protocols can be made more efficient by using computational number theory.

In this paper we have only discussed zero-knowledge proofs for the discrete logarithm decision problem. Proofs of knowledge [TW87,FFS88] with lower (than [CEvdG88]) communication cost for the discrete logarithm (search) problem are discussed in [BDB]. We have not included them here, because they do not rely on smooth number based preprocessing.

REFERENCES

[AM86]   L. M. Adleman and K. S. McCurley. Open problems in number theoretic complexity. In D. Johnson, T. Nishizeki, A. Nozaki, and H. Wilf, editors, *Discrete Algorithms and Complexity, Proceedings of the Japan-US Joint Seminar (Perspective in Computing series, Vol. 15)*, pp. 263–286. Academic Press Inc., Orlando, Florida, June 4–6, Kyoto, Japan 1986.

[BB88]   G. Brassard and P. Bratley. *Algorithmics — Theory & Practice*. Prentice Hall, 1988.

[BDB]   M. Burmester, Y. Desmedt, and T. Beth. Efficient zero-knowledge identification schemes for smart cards. Accepted for publication in special issue on Safety and Security, The Computer Journal, February 1992, Vol. 35, No. 1, pp. 21–29.

[Bet88]   T. Beth. A Fiat-Shamir-like authentication protocol for the El-Gamal-scheme. In C. G. Günther, editor, *Advances in Cryptology, Proc. of Euro-*

crypt '88 (Lecture Notes in Computer Science 330), pp. 77–84. Springer-Verlag, May 1988. Davos, Switzerland.

[BGG90]   M. Bellare, O. Goldreich, and S. Goldwasser. Randomness in interactive proofs. In *31th Annual Symp. on Foundations of Computer Science (FOCS)*, pp. 563–572. IEEE Computer Society Press, October 22–October 24, 1990. St. Louis, Missouri.

[BMO90]   M. Bellare, S. Micali, and R. Ostrovsky. Perfect zero-knowledge in constant rounds. In *Proceedings of the twenty second annual ACM Symp. Theory of Computing, STOC*, pp. 482–493, May 14–16, 1990.

[CEvdG88] D. Chaum, J.-H. Evertse, and J. van de Graaf. An improved protocol for demonstrating possession of discrete logarithms and some generalizations. In D. Chaum and W. L. Price, editors, *Advances in Cryptology — Eurocrypt '87 (Lecture Notes in Computer Science 304)*, pp. 127–141. Springer-Verlag, Berlin, 1988. Amsterdam, The Netherlands, April 13–15, 1987.

[COS86]   D. Coppersmith, A. Odlyzko, and R. Schroeppel. Discrete logarithms in $GF(p)$. *Algorithmica*, pp. 1–15, 1986.

[DH76]   W. Diffie and M. E. Hellman. New directions in cryptography. *IEEE Trans. Inform. Theory*, IT-22(6), pp. 644–654, November 1976.

[ElG85]   T. ElGamal. A public key cryptosystem and a signature scheme based on discrete logarithms. *IEEE Trans. Inform. Theory*, 31, pp. 469–472, 1985.

[FFS88]   U. Feige, A. Fiat, and A. Shamir. Zero knowledge proofs of identity. *Journal of Cryptology*, 1(2), pp. 77–94, 1988.

[FS87]   A. Fiat and A. Shamir. How to prove yourself: Practical solutions to identification and signature problems. In A. Odlyzko, editor, *Advances in Cryptology, Proc. of Crypto '86 (Lecture Notes in Computer Science 263)*, pp. 186–194. Springer-Verlag, 1987. Santa Barbara, California, U. S. A., August 11–15.

[GMR89]   S. Goldwasser, S. Micali, and C. Rackoff. The knowledge complexity of interactive proof systems. *Siam J. Comput.*, 18(1), pp. 186–208, February 1989.

[GQ88]   L.C. Guillou and J.-J. Quisquater. A practical zero-knowledge protocol fitted to security microprocessor minimizing both transmission and memory. In C. G. Günther, editor, *Advances in Cryptology, Proc. of Eurocrypt '88 (Lecture Notes in Computer Science 330)*, pp. 123–128. Springer-Verlag, May 1988. Davos, Switzerland.

[Jac85]   N. Jacobson. *Basic Algebra I*. W. H. Freeman and Company, New York, 1985.

[KMO89]   J. Kilian, S. Micali, and R. Ostrovsky. Minimum resource zero-knowledge proofs. In *30th Annual Symp. on Foundations of Computer Science (FOCS)*, pp. 474–479. IEEE Computer Society Press, October 30–November 1, 1989. Research Triangle Park, NC, U.S.A.

[Kon81]   A. Konheim. *Cryptography: A Primer*. John Wiley, Toronto, 1981.

[MO]   J. L. Massey and J. K. Omura. A new multiplicative algorithm over finite fields and its applicability in public-key cryptography. Presented at Eurocrypt 83, Udine, Italy.

[Odl84]   A. M. Odlyzko. Discrete logs in a finite field and their cryptographic significance. In N. Cot T. Beth and I. Ingemarsson, editors, *Advances in Cryptology, Proc. of Eurocrypt 84 (Lecture Notes in Computer Science 209)*, pp. 224–314. Springer-Verlag, 1984. Paris, France April 1984.

[PH78]   S. C. Pohlig and M. E. Hellman. An improved algorithm for computing logarithms over $GF(p)$ and its cryptographic significance. *IEEE Trans. Inform. Theory*, IT-24(1), pp. 106–110, January 1978.

[Sch90]   C. P. Schnorr. Efficient identification and signatures for smart cards. In G. Brassard, editor, *Advances in Cryptology — Crypto '89, Proceedings (Lecture Notes in Computer Science 435)*, pp. 239–252. Springer-Verlag, 1990. Santa Barbara, California, U.S.A., August 20–24.

[Sha90]   A. Shamir. IP=PSPACE. In *31th Annual Symp. on Foundations of Computer Science (FOCS)*, pp. 11–15. IEEE Computer Society Press, October 22–October 24, 1990. St. Louis, Missouri.

[SS77]   R. Solovay and V. Strassen. A fast Monte-Carlo test for primality. *SIAM Journal on Computing*, 6(1), pp. 84–85, erratum (1978), ibid, 7,118, 1977.

[TW87]   M. Tompa and H. Woll. Random self-reducibility and zero-knowledge interactive proofs of possession of information. In *The Computer Society of IEEE, 28th Annual Symp. on Foundations of Computer Science (FOCS)*, pp. 472–482. IEEE Computer Society Press, 1987. Los Angeles, California, U.S.A., October 12–14, 1987.

An Extension of Zero-Knowledge Proofs and Its Applications

Tatsuaki Okamoto

NTT Laboratories
Nippon Telegraph and Telephone Corporation
1-2356, Take, Yokosuka-shi, Kanagawa-ken, 238-03 Japan

Abstract

This paper presents an extension (or relaxation) of zero-knowledge proofs, called *oracle-simulation* zero-knowledge proofs. It is based on a new simulation technique, called *no-knowledge-release-oracle simulation*, in which, roughly speaking, the view of the history of an interactive proof is simulated by the poly-time machine (simulator) with the help of an oracle which does not release any knowledge to the simulator. We show that, assuming the existence of a secure bit-commitment, any *NP language* has a *three round oracle-simulation* zero-knowledge proof, which is obtained by combining a public-coin-type zero-knowledge proof and a coin-flip protocol. This result is very exciting given the previously known negative result on the conventional zero-knowledge proofs, such that only *BPP languages* can have *three round black-box-simulation* zero-knowledge proofs. We also show some applications of this notion to identification systems based on digital signature schemes.

1 Introduction

1.1 Problem

The concept "zero-knowledge interactive proofs" was introduced by Goldwasser, Micali and Rackoff [GMR]. An interesting question from both the theoretical and practical viewpoints concerning zero-knowledge proofs is the lower bound of the round complexity of the zero-knowledge proofs. Recently Goldreich and Krawczyk [GKr] have almost answered the problem as follows: They have proven that a language L has a *three-round*[1] black-box simulation zero-knowledge interactive proof if and only if $L \in$ BPP. They also show that a language L has a *constant-round* black-box simulation zero-knowledge *public-coin-type* interactive proof if and only if $L \in$ BPP.

---

[1] Here, one "round" means one message transmission. Note that "round" is used as a couple of message transmission cycle (send and return) in [FeS1, FeS2].

However, this result does not imply that we cannot construct constant round (more than three) zero-knowledge *private-coin-type* interactive proofs. Brassard, Crépeau, and Yung have constructed a six-round perfect zero-knowledge pseudo-proof (or argument) for all NP languages under an assumption [BCY], and Feige and Shamir have shown a four-round perfect zero-knowledge pseudo-proof for all NP languages under the same assumption [FeS1]. On the other hand, Bellare, Micali, and Ostrovsky have shown a five-round perfect zero-knowledge proof for all random self-reducible languages [BMO1].

We can summarlize these results as follows:

$$ZK[3]=BPP \quad \text{(from [GKr])}$$
$$ZKAM=BPP \quad \text{(from [GKr])}$$
$$PZK'[4]\supseteq NP \quad \text{(from [FeS1])}$$
$$PZK[5]\supseteq RSR \quad \text{(from [BMO1])}$$

Here, $(P)ZK[n]$ means the class of languages that have black-box simulation (perfect) zero-knowledge interactive proofs with n rounds. $(P)ZK'[n]$ means the class of languages that have black-box simulation (perfect) zero-knowledge interactive *pseudo-*proofs with n rounds. RSR means the random self-reducible language class. ZKAM means the class of languages that have black-box simulation zero-knowledge *public-coin-type* interactive proofs with *constant rounds*. Note that we need an assumption for the result to get the result in [FeS1], while there are no unproven assumptions for the others [GKr, BMO1].

The key idea of constructing the five-round zero-knowledge proof by [BMO1] is to prevent a verifier from generating his challenge bits so that they are dependent on the value of prover's first message. For this purpose, in their protocol, before the verifier receives prover's first message, verifier's challenge bits are sent to the prover; the bits are concealed by a bit-commitment.

Another intuitive measure to prevent a verifier from generating specific challenge bits based on prover's first message is to combine a coin-flip protocol and a public-coin-type zero-knowledge protocol. For example, let us consider the following protocol, which is an interactive proof for the graph Hamiltonicity:

Protocol: 1 (Three Round IP for HAMILTONIAN)

Step 1 Prover P sends verifier V the bit-commitment, $BC(u_i, r_i)$, of t random bits, u_i $(i = 1, \ldots, t)$, where t is the size of the input, $G \in HAMILTONIAN$. P also sends V the first message X_i $(i = 1, \ldots, t)$ of the parallel version of Blum's zero-knowledge protocol for Hamiltonisity [Bl].

Step 2 V sends t random bits, v_i $(i = 1, \ldots, t)$.

Step 3 P opens the bit-commitment to V, and sends V the third message Y_i $(i = 1, \ldots, t)$ of the parallel version of Blum's protocol, where the second message of Blum's protocol (public coin challenge by the verifier) is set to be $e_i = u_i + v_i \bmod 2$, $(i = 1, \ldots, t)$.

Step 4 V checks the validity of both the bit-commitment and Blum's protocol. If verification succeeds, V accepts the proof.

Remark: When $G \in HAMILTONIAN$ is an input, X_i and Y_i are

$$X_i = (\alpha_i, \beta_i), \ Y = \delta_i.$$

$\alpha_i, \beta_i, \delta_i$ are as follows:

- Graph $\widehat{G}_i = \pi_i(G)$ (π_i: a random permutation),

- An $t \times t$ matrix $\alpha_i = \{\alpha_{ijk} \mid j, k = 1, 2, \ldots, t\}$, where $\alpha_{ijk}=BC(v_{ijk})$, and $v_{ijk} = 1$ if edge jk is present in the $\widehat{G}_i$, and $v_{ijk} = 0$ otherwise, and

- $\beta_i = BC(\pi_i)$.

- If $e_i = 0$, $\delta_i =($ decommit of α_{ijk} and β_i); otherwise, $\delta_i = \{$decommit of $\alpha_{ijk} \mid$ edge jk is in a Hamiltonian path in $\widehat{G}_i\}$.

(See Section 2 for the bit-comittment function, BC.)

Although this protocol is considered to release no additional knowledge to the verifier, we cannot prove it to be black-box zero-knowledge by the result [GKr], unless NP=BPP, since the above protocol is three rounds.

The question then is whether we can construct a *three round* interactive proof that releases no additional knowledge than the membership of the input (or whether the above protocol releases no additional knowledge to the verifier). If it does not release any additional knowledge, how can we prove the fact without the notion of zero-knowledge? (Is there a reasonable alternative notion to zero-knowledge?)

1.2 Results

This paper presents an affirmative answer to this question. That is, we show that the above protocol reveals no additional knowledge in the sense of a reasonable alternative notion to zero-knowledge, *oracle-simulation* zero-knowledge, which is a natural extension of zero-knowledge definition and includes zero-knowledge definition as the most strict case.

In this paper, we show the following result:

- All languages in NP have oracle-simulation zero-knowledge public-coin-type interactive proofs with *three rounds*, if a secure (one-round) bit-commitment scheme exists [Theorem 4.1].

We can summarize these results as follows:

$$OZK[3] \supseteq OZKAM[3] \supseteq NP \quad [\text{Theorem 4.1}]$$

Here, OZK[n] means the class of languages that have oracle-simulation zero-knowledge interactive proofs with n rounds. OZKAM[n] means the class of languages that have oracle-simulation zero-knowledge public-coin-type interactive proofs with n rounds. Note that we need an assumption to prove the results, [Theorem 4.1].

1.3 Related Work

We will now describe some work related to our approach.

Feige, Fiat and Shamir have introduced the notion of *no-transferable-information*, and the parallel versions of their zero-knowledge identification systems have been shown to be provably secure [FFS]. They have generalized this notion as a practical extension of zero-knowledge, which consists of two notions, *witness hiding* and *witness indistinguishability* [FeS1, FeS2]. In [FeS1], they use witness hiding and witness indistinguishable protocols as components to construct zero-knowledge pseudo-proofs with four rounds. They have also shown some other applications of these notions [FeS2]. Bellare, Micali, and Ostrovsky [BMO2] have introduced the notion of *witness independence*, which is almost equivalent to witness indistinguishability.

Next, we address the relationship between our oracle-simulation zero-knowledge (OZK) and their witness hiding (WH) and witness indistinguishability (WI).

First, WH and WI are defined for proofs of *knowledge* [FFS], where the prover's possession of knowledge is demonstrated. Therefore, they cannot be directly applied to proofs of *membership*. On the contrary, OZK is effective in proofs of *membership* as well as proofs of *knowledge*, as ZK is.

Moreover, WI is too weak as an alternative to zero-knowledge, as also described in [FeS1, FeS2]. That is, WI only guarantees to hide which witness is used by the prover for the protocol with the verifier, when there exist multiple witnesses for one input. Therefore, when there is only one witness for an input (e.g., a language in UP), WI does not give any guarantee. Moreover, even if many witnesses exist for an input, WI does not guarantee that knowledge common among these witnesses cannot be revealed.

On the other hand, WH is also too weak as an alternative to zero-knowledge, because WH does not guarantee that any partial information of a witness cannot be released. Moreover, WH seems to be too restrictive. All protocols which has been proven to be WH in [FeS1, FeS2] are a specific type of WI protocols, and we have not known any effective measure to prove WH directly without the help of WI or ZK. Therefore, we do not seem to be able to prove the security of Protocol 1 above, even if we restrict the protocol to be the knowledge proof model.

We can conclude as follows:

- WH and WI are defined for proofs of *knowledge*, therefore they cannot be directly applied to proofs of *membership*, while OZK can be applied to proofs of *membership* as well as proofs of *knowledge*.

- WI and WH are *too weak* as alternatives to zero-knowledge.

Remark: As shown in [FeS1, FeS2, BMO2], WI and WH are useful for many applications. However, these notions seem to be somewhat ad-hoc and technical for these applications. In contrast, OZK is considered to be a more natural extension of zero-knowledge (ZK) (See Subsection 3.2). For example, OZK trivially includes ZK as the most strict case.

2 Preliminaries

If algorithm A is a probabilistic algorithm, then for input x, the notation $A(x)$ refers to the probability space that assigns to the string σ the probability that A, on input x, outputs σ.

Let $U = \{U(x)\}$ be a family of random variables taking values in $\{0,1\}^*$, with the parameter x ranging in $\{0,1\}^*$. $U = \{U(x)\}$ is called a poly-bounded family of random variables, if, for some constant $e > 0$, all random variables $U(x) \in U$ assign positive probability only to strings whose length is exactly $|x|^e$.

Let $\mathcal{D} = \{D_x\}$ be a poly-size family of Boolean circuits; that is, for some constants $c, d > 0$, all D_x have one Boolean output and at most $|x|^c$ gates and $|x|^d$ inputs. Let $p_D(U(x)) = \Pr(a \leftarrow U(x) : D_x(a) = 1)$.

(P, V) is an interactive pair of Turing machines, where P is a prover, and V is a verifier [GMR, TM]. $\mathbf{View}_{(P,V)}(x)$, V's view of the history of (P, V), is denoted as (x, ρ, m), where ρ is the finite prefix of V's random tape that was read, and m is the final content of the communication tape on which P writes. $\mathbf{View}_{(P,V)}(x, t)$, V's view of the history of (P, V), is denoted as (x, t, ρ, m), where x is the input to V, t is the auxiliary input to V, ρ is the finite prefix of V's random tape that was read, and m is the final content of the communication tape on which P writes. $\mathbf{Com}_{(P,V)}(x)$, (P, V)'s communication, is denoted as the final content of the communication tape on which P and V write, where x is the input to (P, V).

M^O denotes a machine M with oracle O such that M is a machine and M interactively inquires of an oracle O through the inquiry tape on which M writes and the oracle tape on which O writes. $\mathbf{View}_{(O,M)}(x)$, M's view of the history of the interaction between M and O, is denoted as (x, ρ, m), where x is the input to M, ρ is the finite prefix of M's random tape that was read, and m is the final content of the oracle tape on which O writes. $\mathbf{Com}_{(O,M)}(x)$, (O, M)'s inquiries and answers, is denoted as the final content of the inquiry tape on which M writes and the oracle tape on which O writes, where x is the input to M^O.

Next, we roughly introduce the notion of the bit-commitment (see the formal definition in [N]). The bit-commitment protocol between Alice (committer) and Bob (verifier) consists of two stages; the *commit stage* and the *revealing stage*. In the commit stage, after their exchanging messages, Bob has some information that represents Alice's secret bit b. In the revealing stage, Bob knows b. Roughly, after the commit stage, Bob cannot guess b, and Alice can reveal only one possible value. Here, when the protocol needs k rounds in the commit stage, we call it k *round* bit-commitment. We can construct *one round* bit-commitment using probabilistic encryption [GM], and *two round* bit-commitment using any one-way function [H, ILL, N]. In the commit stage of one round bit-commitment, Alice generates a random number r and sends $BC(b, r)$ to Bob, where b is a committed bit, and BC is a bit-commitment function. When Alice wishes to commit l bits, $b_1, b_2, \ldots, b_l$, she sends $BC(b_1, r_1), \ldots, BC(b_l, r_l)$ to Bob. Hereafter, we will simply write $BC(\bar{b}, \bar{r})$ as $(BC(b_1, r_1), \ldots, BC(b_l, r_l))$, where $\bar{b} = (b_1, \ldots, b_l)$ and $\bar{r} = (r_1, \ldots, r_l)$.

Definition 2.1 Let $L \subset \{0,1\}^*$ be a language. Two families of random variables

$\{U(x)\}$ and $\{V(x)\}$ are *perfectly indistinguishable on L* if

$$\sum_{\alpha \in \{0,1\}^*} |\Pr(U(x) = \alpha) - \Pr(V(x) = \alpha)| = 0,$$

for all sufficiently long $x \in L$.

Two families of random variables $\{U(x)\}$ and $\{V(x)\}$ are *statistically indistinguishable on L* if

$$\sum_{\alpha \in \{0,1\}^*} |\Pr(U(x) = \alpha) - \Pr(V(x) = \alpha)| < \frac{1}{|x|^c},$$

for all constants $c > 0$ and all sufficiently long $x \in L$.

Two families of poly-bounded random variables $\{U(x)\}$ and $\{V(x)\}$ are $\mathcal{D}$-*computationally indistinguishable on L* if

$$|p_D(U(x)) - p_D(V(x))| < \frac{1}{|x|^c}$$

for all circuits $D \in \mathcal{D}$, all constant $c > 0$ and all sufficiently long $x \in L$.

$\{U(x)\}$ and $\{V(x)\}$ are *computationally indistinguishable on L*, if $\{U(x)\}$ and $\{V(x)\}$ are $\mathcal{D}$-*computationally indistinguishable on L* and $\mathcal{D}$ includes any poly-size family of circuits.

3   Oracle-Simulation Zero-Knowledge Proofs

3.1   Definitions

Definition 3.1 *Let $\mathcal{D}$ be a poly-size family of Boolean circuits. Let $\{R(x)\}$ be a family of random variables which are uniformly distributed over $\{0,1\}^t$, where the size of variables, t, is determined by the size of x, that is, $t = |a \leftarrow R(x)| = f(|x|)$. $\mathcal{D}$ is insensitive to a family of random variables, $\{U(x)\}$, on a language L, if there exists $\{R(x)\}$ such that $\{U(x)\}$ and $\{R(x)\}$ are $\mathcal{D}$-computationally indistinguishable on L.*

Definition 3.2 *Let M^O be a machine M which interactively communicates with oracle O with respect to a language L. O is a no-knowledge-release-oracle to M with respect to a language L, if there exist probabilistic polynomial-time machines T and W such that $\mathrm{Com}_{(O,M)}(x)$ and $T^M(x)$ are computationally indistinguishable on L, and that $\mathrm{View}_{(O,M)}(x)$ and $W^M(x)$ are $\mathcal{D}^*$-computationally indistinguishable on L, where $\mathcal{D}^*$ is the family of all poly-size circuits which are insensitive to $\mathrm{Com}_{(O,M)}(x)$.*

Definition 3.3 *An interactive proof (P, V) is computationally black-box oracle-simulation zero knowledge with respect to a language L if there exists a probabilistic polynomial-time machine M such that for any probabilistic polynomial-time machine V', there exists an oracle $O_{V'}$ such that $\mathrm{View}_{(P,V')}(x)$ and $M^{O_{V'},V'}(x)$ are computationally indistinguishable on L, and that $O_{V'}$ is a no-knowledge-release-oracle to $M^{V'}$ with respect to $\{0,1\}^*$.*

Note that the above definition of *oracle-simulation* zero knowledge is based on the notion of *black-box-simulation* zero knowledge. Hereafter, we will simply call *black-box oracle-simulation* zero knowledge *oracle-simulation* zero knowledge, since we will prove our main result, Theorems 4.1, using the above definition. However, we can also define *oracle-simulation* zero knowledge, based on the notion of *auxiliary-input* zero knowledge as follows:

Definition 3.4 *An interactive proof* (P, V) *is computationally auxiliary-input oracle-simulation zero knowledge with respect to a language* L *if, for any probabilistic polynomial-time machine* V', *and any* t, *there exists a probabilistic polynomial-time machine* $M_{V'}$ *and oracle* $O_{V'}$ *such that* $\mathbf{View}_{(P,V')}(x,t)$ *and* $M_{V'}^{O_{V'}}(x,t)$ *are computationally indistinguishable on* L, *and that* $O_{V'}$ *is a no-knowledge-release-oracle to* $M_{V'}$ *with respect to* $\{0,1\}^*$.

Proposition 3.5 *A computationally black-box oracle-simulation zero knowledge interactive proof with respect to a language* L *is computationally auxiliary-input oracle-simulation zero knowledge with respect to a language* L.

3.2 Is OZK Reasonable?

In this subsection, we show that the above definition (oracle-simulation zero-knowledge) must be considered to be reasonable as an alternative definition to the zero-knowledge definition [GMR, Or].

It has recently been realized by some researchers that the simulation technique with an oracle is very useful to prove the security of complicated protocols. Micali and Rogaway [MR] have shown an elegant definition of secure computation based on the idea of simulating a complicated real multi-party protocol by using the help of a simple virtual (ideal) protocol as an oracle. In other words, they have shown that the security problem of a real complicated protocol can be converted into that of a simple virtual protocol among the oracle, simulator, and adversary.

On the other hand, Goldreich and Petrank [GP] have shown that the knowledge complexity (the amount of released knowledge) of a protocol can be also measured by the simulation technique with an oracle. Roughly speaking, they have defined that the knowledge complexity (oracle version) of a protocol is the amount of knowledge that the oracle releases to the simulator, when the simulator with the help of the oracle can simulate the view of the protocol. In other words, they have shown that the amount of knowledge released by the real protocol can be equated to that of a simple virtual protocol between the oracle and the simulator.

Historically, the idea of using an oracle to help the simulation was firstly introduced in [GHY]. In the minimum-knowledge proof of membership [GHY], P releases one bit (the membership of the input x; $x \in L$ or $x \notin L$) to V. Then, roughly, for any input x ($x \in L$ or $x \notin L$), simulator M with oracle O can simulate the view of (P, V'), where oracle O releases one bit knowledge, $x \in L$ or $x \notin L$, to simulator M.

The oracles used in the above techniques [MR, GP, GHY] release some knowledge to the simulator. On the contrary, in this paper, we have introduced a new type of oracle for a simulator, the *no-knowledge-release-oracle*, which does not release any knowledge to the simulator but still helps the simulation. Similarly to the previous results, in our technique (simulation with no-knowledge-release-oracle), the amount of knowledge released by the real protocol can be also equated to that of a simple virtual protocol between the oracle (no-knowledge-release-oracle) and the simulator. Therefore, roughly speaking, the oracle-simulation zero-knowledge proof guarantees that no verifier V' can obtain any additional knowledge from the conversation with prover P, since oracle $O_{V'}$ gives no knowledge to simulator M through the conversation with M.

In order to show that oracle $O_{V'}$ gives no knowledge to simulator M, in our definition, we use a specific type of simulation technique. Here, we show that $\mathbf{Com}_{(O_{V'},M)}(x)$ is simulatable (approximable), and that $\mathbf{View}_{(O_{V'},M)}(x)$ is simulatable (approximable) when the class of distinguishers is restricted to be insensitive to $\mathbf{Com}_{(O_{V'},M)}(x)$. This simulation technique guarantees that $\mathbf{Com}_{(O_{V'},M)}(x)$ releases no additional knowledge and that $\mathbf{View}_{(O_{V'},M)}(x)$ releases no knowledge that creates an additional poly-time relation except the trivial ones. The reason why we use this specific type of simulation is that we cannot simulate $\mathbf{View}_{(O_{V'},M)}(x)$ in a usual manner, since if it is possible, we can simulate $\mathbf{View}_{(P,V')}(x)$ without oracle $O_{V'}$. Here, note that M always behaves honestly to $O_{V'}$, because M and $O_{V'}$ are introduced virtually to simulate the real view of (P, V').

Moreover, the definition of oracle-simulation zero-knowledge is a natural extension of the usual zero-knowledge definition. If $O_{V'}$ is a specific type of no-knowledge-release-oracle that answers nothing to the simulator, then the oracle-simulation zero-knowledge is clearly exactly same as the usual zero-knowledge.

4 Three Round Oracle-Simulation Zero-Knowledge Proofs

Theorem 4.1 *All languages in NP have oracle-simulation zero-knowledge public-coin-type interactive proofs with three rounds, if a secure (one-round) bit-commiment scheme exists.*

Sketch of Proof:

It is sufficient to show that Protocol 1 described in Subsection 1.1 is a computationally oracle-simulation zero-knowledge interactive proof, assuming the existence of a secure (one-round) bit-commitment scheme.

First, clearly, this protocol satisfies the completeness condition. Second, it satisfies the soundness condition, assuming that the bit-commitment scheme used in Protocol 1 is secure [N].

Then, we will show that the protocol satisfies the (computationally black-box) oracle-simulation zero-knowledgeness. First, simulator (probabilistic polynomial time machine) M and oracle $O_{V'}$ are constructed as follows:

Algorithm M :
Input $x \in HAMILTONIAN$.

Output $M^{O_{V'},V'}(x)$.

Step 1　M flips a coin t times, and obtains t random bits e'_i $(i = 1, \ldots, t)$.

Step 2　M generates (X'_i, Y'_i) $(i = 1, \ldots, t)$ using the standard simulation technique of Blum's protocol, such that (X'_i, e'_i, Y'_i) $(i = 1, \ldots, t)$ satisfies the verification equation of Blum's protocol.

Step 3　M sends (e'_i, X'_i) $(i = 1, \ldots, t)$ to oracle $O_{V'}$. Then, $O_{V'}$ returns (u'_i, r'_i, v'_i) $(i = 1, \ldots, t)$ to M.

Step 4　M outputs $(BC(u'_1, r'_1), \ldots, BC(u'_t, r'_t), X'_1, \ldots, X'_t; v'_1, \ldots, v'_t; u'_1, r'_1, \ldots, u'_t, r'_t, Y'_1, \ldots, Y'_t)$.

Oracle $O_{V'}$:

Input　(e'_i, X'_i) $(i = 1, \ldots, t)$

Output　(u'_i, r'_i, v'_i) $(i = 1, \ldots, t)$ such that $V'(BC(u'_1, r'_1), \ldots, BC(u'_t, r'_t), X'_1, \ldots, X'_t)$ $= (v'_1, \ldots, v'_t)$, $v'_i = u'_i \oplus e'_i$, and $(u'_1, \ldots, u'_t; r'_1, \ldots, r'_t)$ is uniformly distributed in set $U_{(X'_1, \ldots, X'_t)}(e'_1, \ldots, e'_t) = \{(u'_1, \ldots, u'_t; r'_1, \ldots, r'_t) \mid V'(BC(u'_1, r'_1), \ldots, BC(u'_t, r'_t), X'_1, \ldots, X'_t) = (u'_1 \oplus e'_1, \ldots, u'_t \oplus e'_t)\}$. Here, $V'(a)$ denotes the message that is sent by V' in Step 2 of Protocol 1, when V' receives a from the prover in Step 1.

First, clearly, the running time of algorithm M is expectedly polynomial in the size of the input. Then, we show that $\textbf{View}_{(P, V')}(x)$ and $M^{O_{V'}, V'}(x)$ are computationally indistinguishable on $HAMILTONIAN$.

Assuming that the bit-commitment scheme used in Protocol 1 is secure, the distribution of $e_i = u_i \oplus v_i$ $(i = 1, \ldots, t)$ of $\textbf{View}_{(P, V')}(x)$ is statistically indistinguishable from the uniformly distributed random bits. The distribution of P's coin-flip part of $\textbf{View}_{(P, V')}(x)$ is equivalent to the distribution of the corresponding part of $M^{O_{V'}, V'}(x)$. The computational relationship among the variables of $\textbf{View}_{(P, V')}(x)$ is equivalent to the relationship among the variables of $M^{O_{V'}, V'}(x)$, except the relationship among x, X_i, and X'_i. Here, $\{X_i\}$ of $\textbf{View}_{(P, V')}(x)$ and $\{X'_i\}$ of $M^{O_{V'}, V'}(x)$ are computationally indistinguishable. Therefore, in total, $\textbf{View}_{(P, V')}(x)$ and $M^{O_{V'}, V'}(x)$ are computationally indistinguishable on $HAMILTONIAN$.

Next, we show that there exists a probabilistic polynomial-time machine T such that $\textbf{Com}_{(O_{V'}, M)}(x)$ and $T^{V'}(x)$ are statistically indistinguishable. (Note that $T'^{M^{V'}}(x)$ can be converted into $T^{V'}(x)$.) First, a probabilistic polynomial-time machine T is constructed as follows:

Algorithm T :

Input　$x \in HAMILTONIAN$.

Output　$T^{V'}(x) = (e''_1, \ldots, e''_t, X'_1, \ldots, X'_t; u''_1, \ldots, u''_t, r''_1, \ldots, r''_t, v''_1, \ldots, v''_t)$.

Step 1　T randomly generates (u''_i, r''_i) $(i = 1, \ldots, t)$, where $t = |x|$.

Step 2　T calculates $BC(u''_i, r''_i)$ $(i = 1, \ldots, t)$, and generates X'_i $(i = 1, \ldots, t)$ in the same manner as Step 2 of Algorithm M.

Step 3　T sends $(BC(u''_i, r''_i), X'_i)$ $(i = 1, \ldots, t)$ to V' as the first message to V' in Step 1. Then, V' returns v''_i $(i = 1, \ldots, t)$ to T as the message to P in Step 2.

Step 4　T calculates $e''_i = u''_i \oplus v''_i$ $(i = 1, \ldots, t)$.

Step 5 T outputs $(e_1'', \ldots, e_t'', X_1', \ldots, X_t'; u_1'', \ldots, u_t'', r_1'', \ldots, r_t'', v_1'', \ldots, v_t'')$.

Clearly, the running time of algorithm T is expectedly polynomial in the size of the input x. Then, we show that $\mathbf{Com}_{(O_{V'},M)}(x)$ and $T^{V'}(x)$ are statistically indistinguishable.

$\mathbf{Com}_{(O_{V'},M)}(x)$ is $(e_1', \ldots, e_t', X_1', \ldots, X_t'; u_1', \ldots, u_t', r_1', \ldots, r_t', v_1', \ldots, v_t')$ such that $(e_1', \ldots, e_t')$, $(u_1', \ldots, u_t')$ and $(r_1', \ldots, r_t')$ are uniformly distributed random bits, $e_i' = u_i' \oplus v_i'$ $(i = 1, \ldots, t)$, and $V'(BC(u_1', r_1'), \ldots, BC(u_t', r_t'), X_1', \ldots, X_t') = (v_1', \ldots, v_t')$.

The computational relationship among the variables of $\mathbf{Com}_{(O_{V'},M)}(x)$ and that of $T^{V'}(x)$ are exactly equivalent. The only difference between $\mathbf{Com}_{(O_{V'},M)}(x)$ and $T^{V'}(x)$ is the distributions of $\{e_i'\}$ and $\{e_i''\}$. That is, $\{e_i'\}$ is statistically indistinguishable from $\{e_i''\}$, since V' can guess the value u_i'' from $BC(u_i'', r_i'')$ with negligible probability. Therefore, $\mathbf{Com}_{(O_{V'},M)}(x)$ is statistically indistinguishable from $T^{V'}(x)$.

Next, we show that there exists a probabilistic polynomial-time machine W such that $\mathbf{View}_{(O_{V'},M)}(x)$ and $W^{V'}(x)$ are $\mathcal{D}^*$-computationally indistinguishable, where $\mathcal{D}^*$ is the family of all poly-size circuits which are insensitive to $\mathbf{Com}_{(O,M)}(x)$. (Note that $W'^{M^{V'}}(x)$ can be converted into $W^{V'}(x)$.)

First, a probabilistic polynomial-time machine W is constructed as follows:

Algorithm W :

Input $x \in HAMILTONIAN$.
Output $W^{V'}(x) = (Y_1' \ldots, Y_t', e_1', \ldots, e_t', X_1', \ldots, X_t'; u_1'', \ldots, u_t'', r_1'', \ldots, r_t'', v_1'', \ldots, v_t'')$.

Step 1 W randomly generates (u_i'', r_i'') $(i = 1, \ldots, t)$, where $t = |x|$.
Step 2 W calculates $BC(u_i'', r_i'')$ $(i = 1, \ldots, t)$, and generates X_i', Y_i', e_i' $(i = 1, \ldots, t)$ in the same manner as Step 2 of Algorithm M.
Step 3 W sends $(BC(u_i'', r_i''), X_i')$ $(i = 1, \ldots, t)$ to V' as the first message to V' in Step 1. Then, V' returns v_i'' $(i = 1, \ldots, t)$ to T as the message to P in Step 2.
Step 4 W outputs $(Y_1' \ldots, Y_t', e_1', \ldots, e_t', X_1', \ldots, X_t'; u_1'', \ldots, u_t'', r_1'', \ldots, r_t'', v_1'', \ldots, v_t'')$.

Clearly, the running time of algorithm W is expectedly polynomial in the size of the input x. Then, we show that $\mathbf{View}_{(O_{V'},M)}(x)$ and $W^{V'}$ are $\mathcal{D}^*$-computationally indistinguishable.

The computational relationship of X_i', Y_i', e_i' between $\mathbf{View}_{(O_{V'},M)}(x)$ and $W^{V'}$ is exactly same. The computational relationship of X_i', u_i', v_i', r_i' (and $X_i', u_i'', v_i'', r_i''$) between $\mathbf{View}_{(O_{V'},M)}(x)$ and $W^{V'}$ is exactly same. $\{r_i'\}$ and $\{r_i''\}$ are information-theoretically independent from the other variables.

Therefore, the only possible difference between $\mathbf{View}_{(O_{V'},M)}(x)$ and $W^{V'}$ except the difference between $\{e_i' = u_i' \oplus v_i'\}$ and $\{e_i' \neq u_i'' \oplus v_i''\}$ is the linkage of $\{Y_i'\}$, $\{u_i'\}$ and $\{v_i'\}$ and that of $\{Y_i'\}$, $\{u_i''\}$ and $\{v_i''\}$.

$\mathcal{D}^*$ is insensitive to whether $\{e_i' = u_i' \oplus v_i'\}$ and $\{e_i' = u_i'' \oplus v_i''\}$ hold or not. Hence, if there is a poly-size circuit $D^* \in \mathcal{D}^*$ that distinguishes $\mathbf{View}_{(O_{V'},M)}(x)$ and $W^{V'}$ with non-negligible probability, then it must distinguish the difference between the linkage of $\{Y_i'\}$, $\{u_i'\}$ and $\{v_i'\}$ and that of $\{Y_i'\}$, $\{u_i''\}$ and $\{v_i''\}$. However, these variables are related through the bit-commitment function BC. Hence, if D^* distinguishes the difference between these likages, then this contradicts the asumption of BC.

Thus, $\mathbf{View}_{(O_{V'},M)}(x)$ and $W^{V'}$ are $\mathcal{D}^*$-computationally indistinguishable on L.

¶

5 Applications

This section introduces some applications of the oracle-simulation zero-knowledge proofs.

5.1 Identification

First, we will show a *three round* oracle-simulation zero-knowledge identification scheme using any digital signature scheme that is secure against *passive* attacks and whose pair of message and signature can be created uniformly using only the public-key. Hereafter, we call this type of signature scheme USPA (uniformly creatable digital signature scheme secure against passive attacks). For example, the RSA scheme is USPA, if the RSA scheme is secure passive attacks, because (m, s) can be uniformly generated by selecting s uniformly, then computing $m = s^e \bmod n$, where m is a message, s is its signature and (e, n) is the public-key.

Then, we show the protocol of an identification scheme using a passively secure digital signature scheme.

Protocol: 2 (Identification scheme)

Step 1 (Key generation and registration) First, we assume the existence of a USPA. Then, Prover P generates USPA's secret key, a_P, and public key, b_P, and publishes the public key with his name. Here, the size of message of the signature scheme is t. This step is done only once when P joins the identification system.

Step 2 Prover P generates random bits u_i $(i = 1, \ldots, t)$, and random strings r_i $(i = 1, \ldots, t)$, and calculates $x_i = BC(u_i, r_i)$ $(i = 1, \ldots, t)$. P sends $\{x_i\}$ to verifier V. Here, we suppose that BC is a secure bit-commitment function.

Step 3 V sends random bits v_i $(i = 1, \ldots, t)$ to P.

Step 4 P calculates t bit string $m = (u_1 \oplus v_1, \ldots, u_t \oplus v_t)$. P generates P's signature s of m using P's secret key, a_P, and sends s and (u_i, r_i) $(i = 1, \ldots, t)$ to V.

Step 5 V checks whether $x_i = BC(u_i, r_i)$ $(i = 1, \ldots, t)$ holds and whether s is a valid signature of m using P's public key, b_P.

Theorem 5.1 *The above identification scheme is (computationally black-box) oracle-simulation zero-knowledge, if BC is a secure bit-commitment scheme.*

The proof can be completed in a manner similar to the proof of Theorem 1.

If we assume the existence of a digital signature scheme which is secure against chosen message attacks, we can trivially construct a *two round* identification scheme.

However, this scheme is considered inefficient, since usually the assumed digital signature scheme is inefficient [GMRi, NY].

On the other hand, although the Fiat-Shamir scheme [FiS] is a typical and practical identification scheme based on a perfect zero-knowledge proof, the above protocol implies another typical way of constructing a practical identification scheme from a digital signature scheme. In other words, we have shown a technique that converts a digital signature scheme to an identification scheme, while [FiS] showed a technique to convert their identification scheme to a digital signature scheme.

When a primitive modification of ESIGN [FOM], where a one-way hash function is not used (or $h(m)$ is replaced by m), is used as USPA, the identification scheme of Protocol 2 is very practical and secure in the sense of Theorem 5.1, assuming that ESIGN is secure against passive attacks. A more practical version based on Protocol 2 has been presented in [FOM].

5.2 OZK for Proving the Possession of Factors

We will also introduce an oracle-simulation zero-knowledge proof to prove the possession of factors of a composite number. The proof is much more efficient than the previously proposed zero-knowledge proofs for the same purpose [TW].

When a modification of the Rabin scheme based on the idea [OS] is used as USPA, the above identification protocol is considered to be an oracle-simulation zero-knowledge proof which proves the possession of factors of a composite number. We will show the detailed description in the final paper.

6 Conclusion

The notion, oracle-simulation zero knowledge, is considered to have a considerable impact on both the practical and theoretical fields of cryptography, since oracle-simulation zero knowledge can be used to guarantee the security of many applications which zero-knowledge and other notions [FeS1, FeS2] cannot.

On the other hand, it will be interesting to study how well oracle-simulation zero knowledge proofs can overcome the negative consequences of zero-knowledge proofs. Theorem 4.1 in this paper is the first result from this viewpoint.

Acknowledgments

We would like to thank Toshiya Itoh and Kouichi Sakurai for their invaluable discussions and suggestions on the preliminary version of this paper. Without their help, we would miss a serious flaw in the definition of the oracle-simulation-zero-knowldge. We would also like to thank Atsushi Fujioka for carefully checking the preliminary manuscript and useful suggestions.

References

[BCC] G.Brassard, D.Chaum and C.Crépeau, "Minimum Disclosure Proofs of Knowledge," Journal of Computer and System Sciences, Vol.37, pp.156-189 (1988)

[BCY] G.Brassard, C.Crépeau and M. Yung, "Everything in NP Can Be Argued in Perfect Zero-Knowledge in a Bounded Number of Round," Proc. Eurocrypt'89 (1989)

[Bl] M.Blum, "How to Prove a Theorem So No One Else Can Claim It," ISO/ TC97/ SC20/ WG2 N73 (1986)

[BMO1] M.Bellare, S.Micali and R.Ostrovsky, "Perfect Zero-Knowledge in Constant Rounds," Proc. STOC (1990)

[BMO2] M.Bellare, S.Micali and R.Ostrovsky, "The (True) Complexity of Statistical Zero-Knowledge," Proc. STOC (1990)

[FeS1] U.Feige and A.Shamir, "Zero-Knowledge Proofs of Knowledge in Two Rounds," Proc. Crypto'89 (1989)

[FeS2] U.Feige and A.Shamir, "Witness Indistinguishable and Witness Hiding Protocols," Proc. STOC (1990)

[FFS] U.Feige, A.Fiat and A.Shamir, "Zero Knowledge Proofs of Identity," Proc. STOC, pp.210-217 (1987)

[FiS] A.Fiat and A.Shamir, "How to Prove Yourself," The Proc. of Crypto'86, pp.186-199 (1986)

[FOM] A.Fujioka, T.Okamoto and S.Miyaguchi: "ESIGN: An Efficient Digital Signature Implementation for Smart Cards", Proc. of Eurocrypt'91 (1991)

[GHY] Z. Galil, S. Haber, and C. Yung, "Minimum-Knowledge Interactive Proofs for Decision Problems", SIAM Journal on Computing, Vol.18, No.4, pp.711–739 (1989).

[GKr] O.Goldreich and H.Krawczyk "On the Composition of Zero-Knowledge Proof Systems," Technical Report #570 of Technion (1989)

[GM] S.Goldwasser, S.Micali, "Probabilistic Encryption," JCSS, 28, 2, pp.270–299 (1984).

[GMR] S.Goldwasser, S.Micali and C.Rackoff, "The Knowledge Complexity of Interactive Proofs," SIAM J. Comput., 18, 1, pp.186-208 (1989). Previous version, Proc. STOC, pp.291–304 (1985)

[GMRi] S.Goldwasser, S.Micali and R.Rivest, "A Digital Signature Scheme Secure Against Adaptive Chosen-Message Attacks," SIAM J. Compt., 17, 2, pp.281–308 (1988)

[GMW] O.Goldreich, S.Micali, and A.Wigderson, "Proofs that Yield Nothing But their Validity and a Methodology of Cryptographic Protocol Design," Proc. FOCS, pp.174-187 (1986)

[GP] O.Goldreich and E.Petrank, "Quantifying Knowledge Complexity," Proc. FOCS (1991)

[H] J.Håstad, "Pseudo-Random Generators under Uniform Assumptions," Proc. STOC (1990)

[ILL] R. Impagliazzo, L. Levin and M. Luby, " Pseudo-Random Number Generation from One-Way Functions," Proc. STOC, pp.12–24 (1989)

[MR] S.Micali and P.Rogaway, "Secure Computation," Proc. Crypto'91, (1991)

[N] M.Naor, "Bit Commitment Using Pseudo-Randomness," Proc. Crypto'89 (1990).

[NY] M.Naor and M.Yung, "Universal One-Way Hash Functions and Their Cryptographic Applications," Proc. STOC, pp.33–43 (1989)

[Or] Y.Oren, "On the Cunning Power of Cheating Verifiers: Some Observations about Zero Knowledge Proofs," Proc. FOCS, pp.462–471 (1987)

[OS] T.Okamoto, and A.Shiraishi "A Digital Signature Scheme Based on the Rabin Cryptosystem," (in Japanese) Spring Conference of IEICE Japan, 1439 (1985)

[TW] M.Tompa and H.Woll, "Random Self-Reducibility and Zero Knowledge Interactive Proofs of Possession of Information," Proc. FOCS, pp472–482 (1987)

Any Language in IP Has a Divertible ZKIP

Toshiya Itoh

Department of Information Processing,
The Graduate School at Nagatsuta,
Tokyo Institute of Technology
4259 Nagatsuta, Midori-ku, Yokohama 227, Japan
e-mail: titoh@nc.titech.ac.jp

Kouichi Sakurai

Computer & Information Systems Laboratory,
Mitsubishi Electric Corporation
5-1-1 Ofuna, Kamakura 247, Japan
e-mail: sakurai@isl.melco.co.jp

*Hiroki Shizuya**

Department of Electrical Communications,
Tohoku University
Aramaki-Aza-Aoba, Aoba-ku, Sendai 980, Japan
e-mail: shizuya@jpntohok.bitnet

Abstract:

A notion of "divertible" zero-knowledge interactive proof systems was introduced by Okamoto and Ohta, and they showed that for any *commutative* random self-reducible relation, there exists a divertible (perfect) zero-knowledge interactive proof system of possession of information. In addition, Burmester and Desmedt proved that for any language $L \in \mathcal{NP}$, there exists a divertible zero-knowledge interactive proof system for the language L under the assumption that probabilistic encryption homomorphisms exist. In this paper, we classify the notion of divertible into three types, i.e., *perfectly* divertible, *almost perfectly* divertible, and *computationally* divertible, and investigate which complexity class of languages has a perfectly (almost perfectly) (computationally) divertible zero-knowledge interactive proof system. The main results in this paper are: (1) there exists a perfectly divertible *perfect* zero-knowledge interactive proof system for graph non-isomorphism (GNI) without any unproven assumption; and (2) for any language L having an interactive proof system, there exists a computationally divertible computational zero-knowledge interactive proof system for the language L under the assumption that probabilistic encryption homomorphisms exist.

---

* The current address is Département d'IRO, Université de Montréal, C.P. 6128, Succ. A, Montréal, Québec, Canada, H3C, 3J7. (e-mail: shizuya@iro.umontreal.ca)

1 Introduction

A notion of interactive proof systems (IP) was introduced by Goldwasser, Micali, and Rackoff [GMR] to characterize a class of languages that are *efficiently* provable, while Babai [Ba] independently defined a similar (but somewhat different) notion of interactive protocols, Arthur-Merlin game (AM), as a natural extension of NP proofs to the probabilistic proofs just like that of $\mathcal{P}$ to $\mathcal{BPP}$. Informally, an interactive proof system (P, V) for a language L is an interactive protocol between a computationally unbounded probabilistic Turing machine P (a prover) and a probabilistic polynomial time Turing machine V (a verifier), in which when $x \in L$, V accepts x with al least *overwhelming* probability, and when $x \notin L$, V accepts x with at most *negligible* probability, where the probabilities are taken over the coin tosses of both P and V.

In [GMR], they also introduced a notion of zero-knowledge (ZK) to capture total amount of knowledge released from a prover P to a verifier V during the execution of an interactive proof system (P, V). From a somewhat practical point of view (e.g., secure identification protocols), Feige, Fiat, and Shamir [FFS] and Tompa and Woll [TW] formulated a notion of "zero-knowledge interactive proof systems of knowledge (or possession of information)." Specifically, Tompa and Woll [TW] showed that for any random self-reducible relation R, there exists a *perfect* zero-knowledge interactive proof system (ZKIP) of possession of information without any unproven assumption, and in addition, they proved in [TW] that such a (perfect) zero-knowledge interactive proof system of possession of information can be converted into the one for a language membership on dom R, where dom R denotes the domain of the relation R. This immediately implies that for any random self-reducible relation R, there exists a (perfect) zero-knowledge interactive proof system for a language membership on dom R.

Okamoto and Ohta [OO] introduced a notion of *commutative* random self-reducible as a variant of [TW], and showed that for any commutative random self-reducible relation R, there exists a divertible (perfect) zero-knowledge interactive proof system of possession of information (or for a language dom R) without any unproven assumption. Let (P, W, V) be an interactive protocol between a computationally unbounded probabilistic Turing machine P (a prover) and probabilistic polynomial time Turing machines W (a warden) and V (a verifier), in which

- a prover P and a verifier V always communicate with each other through a warden W;

- a warden W acts as an *active* eavesdropper, but always follows a predetermined program;

- when receiving a message q from V, W transforms q to q' and sends q' to P;

- when receiving a message a from P, W transforms a to a' and sends a' to V.

Informally, an interactive protocol (P, W, V) is a *divertible* interactive proof system for a language L iff (1) when $x \in L$, V accepts x with *overwhelming* probability; (2) when $x \notin L$, no matter what P does, V accepts x with *negligible* probability; and (3) the conversations between P and W are "uncorrelated" to those between W and V in a statistical or computational sense. (A formal definition for divertible interactive proof systems will be given in section 2.)

Recently, Burmester and Desmedt [BD] showed that a graph isomorphism (GI) has a (perfectly) divertible (perfect) zero-knowledge interactive proof system without any unproven assumption. It is worth noting that GI is known to be not commutative random self-reducible [OO] but random self-reducible [TW]. Furthermore, they proved in [BD] that for any language $L \in \mathcal{NP}$, there exists a divertible zero-knowledge interactive proof system (or argument [BCC]) under the assumption that probabilistic encryption homomorphisms exist, by demonstrating such an interactive protocol for an $\mathcal{NP}$-complete language SAT.

In this paper, we classify the notion of divertible into three types, i.e., *perfectly* divertible, *almost perfectly* divertible, and *computationally* divertible, according to how two families of random variables are *uncorrelated* with each other, and show that (1) there exists a perfectly divertible *perfect* zero-knowledge interactive proof system for graph non-isomorphism (GNI) without any unproven assumption; and (2) for any language $L \in \mathcal{IP}$, there exists a computationally divertible computational zero-knowledge interactive proof system for the language L under the assumption that probabilistic encryption homomorphisms exist, where $\mathcal{IP}$ denotes a class of languages that have interactive proof systems.

The novelty of the first result (see Theorem 4.1.) is that every language known so far to have a perfectly divertible perfect zero-knowledge interactive proof system without unproven assumption is in $\mathcal{NP}$, while GNI is *seemingly* not in $\mathcal{NP}$. In addition, the second result (see Theorem 4.2.) extends the result in [BD], i.e., any language $L \in \mathcal{NP}$ has a perfectly divertible computational zero-knowledge interactive proof system under the assumption that probabilistic encryption homomorphisms exist, to the one (with slightly weaker constraint on divertibility) for a much larger class of languages under the same assumption, i.e., any language $L \in \mathcal{IP}$ has a computationally divertible computational zero-knowledge interactive proof system under the assumption that probabilistic encryption homomorphisms exist. Recalling the result shown by Shamir [Sha] that $\mathcal{IP} = \mathcal{PSPACE}$, we can regard the second result as the one that any language $L \in \mathcal{PSPACE}$ has a computationally divertible computational zero-knowledge interactive proof system under the assumption that probabilistic encryption homomorphisms exist.

2 Preliminaries

In this section, we give definitions for interactive proof systems (IP), perfect (statistical) (computational) zero-knowledge interactive proof systems (ZKIP). In addition, we formally present the notions of perfectly (almost perfectly) (computationally) divertible.

It is worth noting that we classify the notion of divertible into three types in a statistical (or computational) sense according to its level of divertibility.

Definition 2.1 [GMR]: Let $L \subseteq \{0,1\}^*$ be a language, and let P and V be a pair of interactive (probabilistic) Turing machines. An interactive protocol (P, V) is said to be an interactive proof system for a language L iff

> **Completeness:** For each $k > 0$ and for sufficiently large $x \in L$ given as input to (P, V), V halts and accepts x with probability at least $1 - |x|^{-k}$, where the probabilities are taken over the coin tosses of P and V;
>
> **Soundness:** For each $k > 0$, for sufficiently large $x \notin L$, and for any P^*, on input x to (P^*, V), V halts and accepts x with probability at most $|x|^{-k}$, where the probabilities are taken over the coin tosses of P^* and V.

Note that P's resource is computationally unbounded, while V's resource is probabilistic polynomial (in $|x|$) time bounded.

Definition 2.2 [GMR]: Let $L \subseteq \{0,1\}^*$ be a language, and let $\{U(x)\}$ and $\{V(x)\}$ be families of random variables. The two families of random variables $\{U(x)\}$ and $\{V(x)\}$ are said to be statistically indistinguishable on L iff

$$\mathrm{Dis}(U, V, x) = \sum_{\alpha \in \{0,1\}^*} |\mathrm{Pr}\{U(x) = \alpha\} - \mathrm{Pr}\{V(x) = \alpha\}| < |x|^{-k},$$

for any $k > 0$ and all sufficiently large $x \in L$.

Definition 2.3 [GMR]: Let $L \subseteq \{0,1\}^*$ be a language, and let $\{U(x)\}$ and $\{V(x)\}$ be families of random variables. The two families of random variables $\{U(x)\}$ and $\{V(x)\}$ are said to be perfectly indistinguishable on L iff $\text{Dis}(U,V,x) = 0$ for all $x \in L$.

Let $C = \{C_x\}$ be a family of Boolean circuits C_x with one Boolean output. The family C is said to be a *poly-size* family of circuits iff for some $c > 0$, all $C_x \in C$ have at most $|x|^c$ gates. To feed samples from the probability distributions to such circuits, we will consider only *poly-bounded* families of random variables, i.e., families $U = \{U(x)\}$ such that for some $d > 0$, all random variable $U(x) \in U$ assigns positive probability only to strings whose length are exactly $|x|^d$. If $U = \{U(x)\}$ is a poly-bounded family of random variables and $C = \{C_x\}$ is a poly-size family of circuits, we use $P(U,C,x)$ to denote the probability that C_x outputs 1 on input a random string distributed according to $U(x)$.

Definition 2.4 [GMR]: Let $L \subseteq \{0,1\}^*$ be a language, and let $\{U(x)\}$ and $\{V(x)\}$ be families of poly-bounded random variables. The two families of poly-bounded random variables $\{U(x)\}$ and $\{V(x)\}$ are said to be computationally indistinguishable on L iff

$$|P(U,C,x) - P(V,C,x)| < |x|^{-k},$$

for all poly-size families of circuits C, for all $k > 0$ and all sufficiently large $x \in L$.

From Definitions 2.2, 2.3, and 2.4 (see above.), the following defines a perfect (statistical) (computational) zero-knowledge interactive proof system for a language L, respectively.

Definition 2.5 [GMR]: Let $L \subseteq \{0,1\}^*$ be a language, and let (P,V) be an interactive proof for a language L. The interactive proof system (P,V) is said to be a perfectly (statistically) (computationally) zero-knowledge interactive proof system for a language L iff for all $x \in L$ and all V^*, there exists an expected polynomial (in $|x|$) time Turing machine M with blackbox access to V^* such that $\{\text{View}_{(P,V^*)}(x)\}$ and $\{M(x,V^*)\}$ are perfectly (statistically) (computationally) indistinguishable on L, where $\text{View}_{(P,V^*)}(x)$ is the view that V^* can see during the interaction with P on input $x \in L$.

To define a divertible IP (or ZKIP) for a language L, we need to formalize how two families of random variables $\{U(x)\}$ and $\{V(x)\}$ are "uncorrelated" to each other on L.

Definition 2.6: Let $L \subseteq \{0,1\}^*$ be a language, and let $\{U(x)\}$ and $\{V(x)\}$ be families of random variables. The two families of random variables $\{U(x)\}$ and $\{V(x)\}$ are said to be almost perfectly independent on L iff

$$\text{Ind}(U,V,x) = \sum_{\alpha,\beta\in\{0,1\}^*} |\Pr\{U(x) = \alpha \wedge V(x) = \beta\} - \Pr\{U(x) = \alpha\}\Pr\{V(x) = \beta\}| < |x|^{-k},$$

for any $k > 0$ and all sufficiently large $x \in L$.

Definition 2.7: Let $L \subseteq \{0,1\}^*$ be a language, and let $\{U(x)\}$ and $\{V(x)\}$ be families of random variables. The two families of random variables $\{U(x)\}$ and $\{V(x)\}$ are said to be perfectly independent on L iff $\text{Ind}(U,V,x) = 0$, for all $x \in L$.

Definition 2.8: Let $L \subseteq \{0,1\}^*$ be a language, and let $\{U(x)\}$ and $\{V(x)\}$ be families of poly-bounded random variables. The two families of poly-bounded random variables $\{U(x)\}$

and $\{V(x)\}$ are said to be *computationally independent* on L iff for all poly-size families of circuits D, all $k > 0$, and all sufficiently large $x \in L$,

$$|P(U \times V, D, x) - P(U, V, D, x)| < |x|^{-k},$$

where $P(U \times V, D, x)$ is the probability that $D_x \in D$ outputs 1 on input a random string (α, β) distributed according to $U(x) \times V(x)$, and $P(U, V, D, x)$ is the probability that $D_x \in D$ outputs 1 on input a random string α (resp. β) distributed according to $U(x)$ (resp. $V(x)$).

From Definitions 2.6, 2.7, and 2.8 (see above.), the following defines a perfectly (almost perfectly) (computationally) divertible interactive proof system for a language L, respectively.

Definition 2.9: Let $L \subseteq \{0, 1\}^*$ be a language, and let (P, W, V) be an interactive protocol. The interactive protocol (P, W, V) is said to be a perfectly (almost perfectly) (computationally) divertible interactive proof system for a language L iff both $(P, W^{\leftrightarrow V})$ and $(^{P \leftrightarrow}W, V)$ are interactive proof systems for the language L, and for all (sufficiently large) $x \in L$, two families of random variables $\{(P, W^{\leftrightarrow V})(x)\}$ and $\{(^{P \leftrightarrow}W, V)(x)\}$ are perfectly (almost perfectly) (computationally) independent on L. In addition, (P, W, V) is said to be a divertible perfect (statistical) (computational) ZKIP for a language L iff both $(P, W^{\leftrightarrow V})$ and $(^{P \leftrightarrow}W, V)$ are perfect (statistical) (computational) ZKIP for the language L.

Remark 2.10: In Definition 2.9, $(P, W^{\leftrightarrow V})$ (resp. $(^{P \leftrightarrow}W, V)$) denotes a pair of interactive Turing machines (P, W) (resp. (W, V)) in which W communicates with V (resp. P) while interacting with P (resp. V). In addition, $\{(P, W^{\leftrightarrow V})(x)\}$ (resp. $\{(^{P \leftrightarrow}W, V)(x)\}$) denotes the associated family of random variables, the interaction between P and V (resp. W and V), with respect to all (sufficiently large) $x \in L$ given to the protocol (P, W, V).

3 Known Results

In [BD], Burmester and Desmedt showed that for any language $L \in \mathcal{NP}$, there exists a perfectly divertible computational zero-knowledge interactive proof system under the assumption that probabilistic encryption homomorphisms exist. The proof in [BD] is to demonstrate such a divertible ZKIP for SAT by an excellent "swapping" technique. To understand the "swapping" technique [BD], here we show an alternative proof for any $L \in \mathcal{NP}$ to have a perfectly divertible ZKIP, by demonstrating such a divertible ZKIP for Hamiltonian Cycle (HC).

To do this, we present a well-known computational ZKIP for HC [Bl]. Let G be a Hamiltonian graph with n vertices, and let H be a Hamiltonian cycle in G. We use $\text{Sym}(V)$ to denote a set of permutations over the vertices of a graph $G = (V, E)$, and use $f(\cdot, \cdot)$ to denote a (secure) probabilistic encryption common to both P and V.

Computational ZKIP for HC
common input: graph G with n vertices;

Repeat the following steps (from P1 to P2) t times;

P1. P chooses a permutation $\pi \in_R \text{Sym}(V)$ and computes an adjacency matrix $A = (a_{ij})$ of $G' = \pi G$. P encrypts a_{ij} of πG by $b_{ij} = f(a_{ij}, r_{ij})$, and sends $B = (b_{ij})$ to V.

V1. V chooses a bit $q \in_R \{0, 1\}$ and sends q to P as a challenge.

P2. If P receives "0," then P reveals π and all r_{ij}. V can check that the edges revealed indeed corresponds to the graph πG. If P receives "1," then P reveals a Hamiltonian cycle H' in $G' = \pi G$. V can easily check that this is the case from the structure of the adjacency matrix.

If the steps above (from P1 to P2) are successfully executed t times, then V halts and accepts G; otherwise V halts and rejects G. **(End of Protocol)**

Definition 3.1 [BD]: *A probabilistic encryption function $f(\cdot, \cdot)$ is said to be a probabilistic encryption homomorphism iff for any $b, b' \in \{0,1\}$ and any $r, r' \in \{0,1\}^k$, there exists an $r'' \in \{0,1\}^k$ such that $f(b,r) \cdot f(b',r') = f(b \oplus b', r'')$, where k is the security parameter and r'' is computable in polynomial time in k from r, r', b, and b'.*

Then using a "swapping" technique, we present a perfectly divitible computational ZKIP for HC under the assumption that probabilistic encryption homomorphisms exist.

Perfectly Divitible Computational ZKIP for HC
common input: graph G with n vertices;

Repeat the following steps (from P1 to V3) t times;

P1. P chooses $\pi_k \in_R \mathrm{Sym}(V)$, and computes $G_k = \pi_k G$ $(0 \le k \le 1)$. P encrypts the adjacency matrices $A_k = (a_{ij}^{(k)})$ of $G_k = \pi_k G$ by $b_{ij}^{(k)} = f(a_{ij}^{(k)}, r_{ij}^{(k)})$ $(0 \le k \le 1)$, and sends $B_0 = (b_{ij}^{(0)})$, $B_1 = (b_{ij}^{(1)})$ to W.

W1. For each entry $b_{ij}^{(k)}$ of B_k $(0 \le k \le 1)$, W computes $C_k = (c_{ij}^{(k)})$ in a way that

$$c_{ij}^{(k)} = f(0, s_{ij}^{(k)}) \cdot b_{ij}^{(k)} = f(0, s_{ij}^{(k)}) \cdot f(a_{ij}^{(k)}, r_{ij}^{(k)}) = f(a_{ij}^{(k)}, t_{ij}^{(k)}),$$

where $t_{ij}^{(k)}$ is computable from $a_{ij}^{(k)}$, $s_{ij}^{(k)}$, and $r_{ij}^{(k)}$ $(0 \le k \le 1)$ in polynomial time. (see Definition 3.1.) W chooses a bit $\alpha \in_R \{0,1\}$ and two permutations $\phi_0, \phi_1 \in_R \mathrm{Sym}(V)$, and computes $D_0 = (d_{ij}^{(0)}) = \phi_0 C_\alpha$, $D_1 = (d_{ij}^{(1)}) = \phi_1 C_{\bar{\alpha}}$, where $\bar{\alpha}$ denotes the inversion of a bit α. W sends D_0, D_1 to V.

V1. When receiving D_0, D_1 from W, V chooses a bit $q \in_R \{0,1\}$ and sends q to W as a challenge.

W2. When receiving q from V, W computes $Q = \alpha \oplus q$ and sends Q to P as a challenge.

P2. When receiving Q from W, P reveals π_Q, all $r_{ij}^{(Q)}$, and a Hamiltonian cycle $H_{\bar{Q}}$ in $G_{\bar{Q}} = \pi_{\bar{Q}} G$ to W.

W3. When receiving π_Q, all $r_{ij}^{(Q)}$, and a Hamiltonian cycle $H_{\bar{Q}}$ in $G_{\bar{Q}} = \pi_{\bar{Q}} G$, W computes $\phi_{Q \oplus \alpha} \pi_Q$, all $t_{ij}^{(Q)}$ (see W1.), and a Hamiltonian cycle $H'_{\bar{q}} = \phi_{\bar{Q}} H_{\bar{Q}}$ in $\phi_{\bar{q}} \pi_{\bar{Q}} G$ and sends them to V.

V3. When receiving $\phi_{Q \oplus \alpha} \pi_Q$, all $t_{ij}^{(Q)}$, and a Hamiltonian cycle $H'_{\bar{q}} = \phi_{\bar{Q}} H_{\bar{Q}}$ in $\phi_{\bar{q}} \pi_{\bar{Q}} G$, V can check that $D_q = (d_{ij}^{(q)})$ is indeed an encryption of $\phi_{Q \oplus \alpha} \pi_Q G$ and that $H_{\bar{q}}$ is a Hamiltonian cycle in $\phi_{\bar{q}} \pi_{\bar{Q}} G$.

If the steps above (from P1 to V3) are successfully executed t times, then V halts and accepts G; otherwise V halts and rejects G. **(End of Protocol)**

For each pair of messages from P (or V), W randomly *swaps* the order of each entry of the pair, and sends it to V (or P). This is the main idea of "swapping" technique [BD].

The swapping technique and the property of homomorphism guarantee the protocol above to be *perfectly* divertible, while the property of probabilistic encryption guarantees it to be computationally zero-knowledge like in the case of a computational zero-knowledge interactive proof system for $\mathcal{NP}$-complete languages [Bl], [GMW].

4 Main Results

4.1 Perfectly Divertible Perfect ZKIP for GNI

As shown in [BD], any $L \in \mathcal{NP}$ has a *perfectly* divertible computational zero-knowledge interactive proof system under the assumption that probabilistic encryption homomorphisms exist. On the other hand, the results so far on divertible zero-knowledge interactive proof systems without any unproven assumption are only known for some languages in $\mathcal{NP}$.

In this subsection, we show that a graph non-isomorphism (GNI) has a *perfectly* divertible perfect ZKIP without any unproven assumption, by combining the *slightly* modified perfect ZKIP for GNI [GMW] and the *swapping* technique inspired in [BD]. It should be noted that GNI is known to be in $\mathcal{AM}$ but is conjectured to be not in $\mathcal{NP}$.

Theorem 4.1: *There exists a perfectly divertible perfect zero-knowledge interactive proof system for graph non-isomorphism (GNI) without any unproven assumption.*

Proof: Our proof is heavily based on the perfect ZKIP for GNI by Goldreich, Micali, and Wigderson [GMW]. Unfortunately, the original protocol for GNI seems not to be converted to a perfectly divertible one, then we need to modify the protocol for GNI as follows:

Perfect ZKIP for GNI (Modified)

common input: (G_0, G_1), where both $G_0 = (V_0, E_0)$ and $G_1 = (V_1, E_1)$ have n vertices.

Repeat the following steps (from V0 to V3) t times;

V0. V chooses a bit $\alpha \in_R \{0, 1\}$ and two permutations $\pi_0, \pi_1 \in_R \mathrm{Sym}(V_0)$, and sends H_0, H_1 to P, where $H_0 = \pi_0 G_\alpha$, $H_1 = \pi_1 G_{\overline{\alpha}}$, and $\overline{\alpha}$ denotes the inversion of a bit α.

Repeat the following steps (from V1 to P2) n^2 times;

V1. V chooses a bit $\gamma_i \in_R \{0, 1\}$ and two permutations $\tau_{i,0}, \tau_{i,1} \in_R \mathrm{Sym}(V_0)$ $(i = 0, 1)$, and computes $T_{i,j} = \tau_{i,j} G_{j \oplus \gamma_i}$ $(0 \leq j \leq 1)$. V sends $(T_{0,0}, T_{0,1})$, $(T_{1,0}, T_{1,1})$ to P.

P1. P chooses a bit $q \in_R \{0, 1\}$, and sends q to V.

V2. V responds with $\{\gamma_q, \tau_{q,0}, \tau_{q,1}\}$ and $\{\alpha \oplus \gamma_{\overline{q}}, \tau_{\overline{q}, \alpha \oplus \gamma_{\overline{q}}} \pi_0^{-1}, \tau_{\overline{q}, \overline{\alpha \oplus \gamma_{\overline{q}}}} \pi_1^{-1}\}$ to P.

P2. P checks for each $j = 0, 1$ whether $\tau_{q,j}$ is an isomorphism between $T_{q,j}$ and $G_{j \oplus \gamma_q}$, and for each $j = 0, 1$ whether $\tau_{\overline{q}, \alpha \oplus \gamma_{\overline{q}} \oplus j} \pi_j^{-1}$ is an isomorphism between $T_{\overline{q}, \alpha \oplus \gamma_{\overline{q}} \oplus j}$ and H_j. If either condition is violated, then P halts; otherwise P continues.

P3. If P has completed n^2 iterations of the steps above from V1 to P2, then P answers with $\beta \in \{0, 1\}$ such that H_0 (H_1) is isomorphic to G_β $(G_{\overline{\beta}})$.

V3. V checks whether or not $\alpha = \beta$. If $\alpha \neq \beta$, then V halts and rejects (G_0, G_1); otherwise V continues.

If the whole steps above (from V0 to V3) are successfully executed t times, then V halts and accepts (G_0, G_1); otherwise V halts and rejects (G_0, G_1).   **(End of Protocol)**

Informally, the protocol above can be regarded as sequential execution of the *paring trick* of the ZKIP protocol for GNI in [GMW]. It is easy to see that the protocol above satisfies the properties of *completeness* and *soundness*. The proof of *zero-knowledgeness* is as follows: Let i denote the i-th iteration of the loop from V1 to P2 and let $q^{(i)}$ denote the i-th challenge of the prover in P1. As in the proof of zero-knowledgeness for GNI in [GMW], we reset the simulator at most n^2 times on inputs $(q^{(1)}, q^{(2)}, \ldots, q^{(n^2)})$ and $(q^{(1)}, q^{(2)}, \ldots, \overline{q}^{(i)})$ for each i $(1 \leq i \leq n^2)$. This process of resetting simulator provides us the *knowledge* of the verifier with overwhelming probability, thus we can simulate the *real* conversation between the prover and the verifier in a perfect zero-knowledge manner.

Then we can convert the (modified) perfect ZKIP for GNI to the *perfectly* divertible one.

Perfectly Divertible Perfect ZKIP for GNI

common input: (G_0, G_1), where both $G_0 = (V_0, E_0)$ and $G_1 = (V_1, E_1)$ have n vertices.

W1. When receiving H_0, H_1 from V, W chooses two permutations ϕ_0, $\phi_1 \in_R \text{Sym}(V_0)$ and a bit $a \in_R \{0, 1\}$, and sends $I_0 = \phi_0 H_a$, $I_1 = \phi_1 H_{\overline{a}}$ to P. (see step V0.)

W2. When receiving $(T_{0,0}, T_{0,1})$, $(T_{1,0}, T_{1,1})$ from V, W chooses a bit $b \in_R \{0, 1\}$, a bit $c_i \in_R \{0, 1\}$, and two permutations $\chi_{i,0}$, $\chi_{i,1} \in_R \text{Sym}(V_0)$, and sends $(S_{0,0}, S_{0,1})$, $(S_{1,0}, S_{1,1})$ to P, where $S_{i,j} = \chi_{i,j} T_{i \oplus b, j \oplus c_i}$ $(0 \leq i, j \leq 1)$. (see step V1.)

W3. When receiving a bit q from P, W sends a bit $e = q \oplus b$ to V. (see step P1.)

W4. V replies with $\{\gamma_e, \tau_{e,0}, \tau_{e,1}\}$ and $\{\alpha \oplus \gamma_{\overline{e}}, \tau_{\overline{e}, \alpha \oplus \gamma_{\overline{e}}} \pi_0^{-1}, \tau_{\overline{e}, \overline{\alpha \oplus \gamma_{\overline{e}}}} \pi_1^{-1}\}$ to W. (see step V2.)

W5. When receiving $\{\gamma_e, \tau_{e,0}, \tau_{e,1}\}$ and $\{\alpha \oplus \gamma_{\overline{e}}, \tau_{\overline{e}, \alpha \oplus \gamma_{\overline{e}}} \pi_0^{-1}, \tau_{\overline{e}, \overline{\alpha \oplus \gamma_{\overline{e}}}} \pi_1^{-1}\}$ from V, W checks whether or not $\tau_{e,j}$ is an isomorphism between $T_{e,j}$ and $G_{j \oplus \gamma_e}$, and whether or not $\tau_{\overline{e}, \alpha \oplus \gamma_{\overline{e}} \oplus j} \pi_j^{-1}$ is an isomorphism between $T_{\overline{e}, \alpha \oplus \gamma_{\overline{e}} \oplus j}$ and H_j $(0 \leq j \leq 1)$. If all the conditions above hold, then W replies with $u = \{u_1, u_2, u_3\}$, $v = \{v_1, v_2, v_3\}$,

$$u = \left\{ c_q \oplus \gamma_e, \chi_{q,0} \tau_{e,c_q}, \chi_{q,1} \tau_{e,\overline{c_q}} \right\},$$
$$v = \left\{ \alpha \oplus \gamma_{\overline{e}} \oplus a \oplus c_{\overline{q}}, \chi_{\overline{q}, \alpha \oplus \gamma_{\overline{e}} \oplus a \oplus c_{\overline{q}}} \tau_{\overline{e}, \alpha \oplus \gamma_{\overline{e}} \oplus a} \pi_a^{-1} \phi_0^{-1}, \chi_{\overline{q}, \overline{\alpha \oplus \gamma_{\overline{e}} \oplus a \oplus c_{\overline{q}}}} \tau_{\overline{e}, \overline{\alpha \oplus \gamma_{\overline{e}} \oplus a}} \pi_{\overline{a}}^{-1} \phi_1^{-1} \right\}$$

to P; otherwise W halts. (see step V2.)

W6. When receiving u and v from W (see W5.), P checks whether or not u_2 (resp. u_3) is an isomorphism between $S_{q,0}$ (resp. $S_{q,1}$) and G_{u_1} (resp. $G_{\overline{u_1}}$) and whether or not v_2 (resp. v_3) is an isomorphism between $S_{\overline{q}, v_1}$ (resp. $S_{\overline{q}, \overline{v_1}}$) and I_0 (resp. I_1). If either condition is violated, then P halts; otherwise P continues. (see step V2.)

W7. If P has completed n^2 iterations of the steps above from W2 to W6, then P answers with $\beta \in \{0, 1\}$ such that I_0 (I_1) is isomorphic to G_β $(G_{\overline{\beta}})$ to W. (see step P3.)

W8. When receiving a bit β from P, W answers with a bit $\beta' = a \oplus \beta \in \{0, 1\}$ such that H_0 (H_1) is isomorphic to $G_{\beta'}$ ($G_{\overline{\beta'}}$) to V. (see step P3.)

W9. V checks whether or not $\alpha = \beta'$. If $\alpha \neq \beta'$, then V halts and rejects; otherwise V continues. (see step V3.)

If the whole steps above (from W1 to W9) are successfully executed t times, then V halts and accepts (G_0, G_1); otherwise V halts and rejects (G_0, G_1). **(End of Protocol)**

It is easy (but somewhat cumbersome) to show that the protocol above satisfies the properties of *completeness*, *soundness*, and *zero-knowledgeness* in a straightforward way.
Here we informally show that the protocol above satisfies the property of perfect *divertibility*.

(W1) Each graph H_0, H_1 is randomly permuted and placed in random order, thus (H_0, H_1) is *perfectly* independent of (I_0, I_1).

(W2) In a way similar to (W1), $\{T_{i,j}\}$ is *perfectly* independent of $\{S_{i,j}\}$.

(W3) Note that b is randomly chosen in step W2. Thus q is *perfectly* independent of e.

(W4-5) In almost the same way as the above, $\{\gamma_e, \tau_{e,0}, \tau_{e,1}\}$, $\{\alpha \oplus \gamma_{\overline{e}}, \tau_{\overline{e},\alpha \oplus \gamma_{\overline{e}}} \pi_0^{-1}, \tau_{\overline{e},\overline{\alpha \oplus \gamma_{\overline{e}}}} \pi_1^{-1}\}$ are *perfectly* independent of $\{u_1, u_2, u_3\}$, $\{v_1, v_2, v_3\}$.

Thus $\{(^{P \leftrightarrow} W, V)(x)\}$ is *perfectly* independent of $\{(P, W^{\leftrightarrow V})(x)\}$ for all $x \in L$, and hence the protocol above is a *perfectly* divertible perfect zero-knowledge interactive proof system for graph non-isomorphism (GNI) without any unproven assumption. ■

It is easy to see that the protocol above for GNI can be applied to quadratic non-residuosity (QNI) and the nonmembership of a language $L = \langle a \rangle_p$ for a prime p and $a \in Z_p^*$ in a straightforward manner, however, both languages are in $\mathcal{NP}$.

4.2 Computationally Divertible ZKIP for Any Language in $\mathcal{IP}$

In this subsection, we show that for any language $L \in \mathcal{IP}$, there exists a computationally divertible computational zero-knowledge interactive proof system for the language L under the assumption that probabilistic encryption homomorphisms exist. This result can be regarded as an extension of [BD], i.e., any language $L \in \mathcal{NP}$ has a perfectly divertible computational ZKIP under the assumption that probabilistic encryption homomorphisms exist, while slightly relaxing the constraint on its divertibility. As pointed out in the **Introduction**, combining the result here with the one by Shamir [Sha], we can show that any language $L \in \mathcal{PSPACE}$ has a computationally divertible computational zero-knowledge interactive proof system under the assumption that probabilistic encryption homomorphisms exist.

Theorem 4.2: *For any $L \in \mathcal{IP}$, there exists a computationally divertible computational zero-knowledge interactive proof system for the language L under the assumption that probabilistic encryption homomorphisms exist.*

Proof: Our proof is heavily based on the computational ZKIP for any language $L \in \mathcal{IP}$ by Ben-Or et al [Betal]. As shown by Goldwasser and Sipser [GS], we assume without loss of generality that any language $L \in \mathcal{IP}$ has an Arthur-Merlin game [Ba].
Let (P', V') be an Arthur-Merlin game for a language L, and let $f(\cdot, \cdot)$ be a probabilistic encryption homomorphism. On input $x \in L$, let (P', V') have $t(|x|)$ interactions, let each message

sent by V' to P' (resp. sent by P' to V') be of length $m(|x|)$, and let rang $f \subseteq \{0,1\}^{k(|x|)}$ for some polynomial $t(\cdot)$, $m(\cdot)$, and $k(\cdot)$. We assume that for the probabilistic encryption homomorphism $f(b, s)$, the seed s is randomly chosen from $\{0,1\}^{\ell(|x|)}$ for each $b \in \{0,1\}$ and some polynomial $\ell(\cdot)$. For simplicity, we use $f(c, s)$ to denote $f(c, s) = f(c_1, s_1) \# f(c_2, s_2) \# \cdots \# f(c_m, s_m)$ for $c = c_1 c_2 \ldots c_{m(|x|)} \in \{0,1\}^{m(|x|)}$ and $s = s_1 s_2 \ldots s_{m(|x|)} \in \{0,1\}^{m(|x|) \cdot \ell(|x|)}$. We simply write t, m, k, and ℓ instead of $t(|x|)$, $m(|x|)$, $k(|x|)$, and $\ell(|x|)$, respectively. In addition, we assume that for each i $(1 \leq i \leq t)$, P' computes $a_i = P'(x, R_{P'}; q_1, a_1, \ldots, q_{i-1}, a_{i-1}, q_i)$ with $R_{P'}$ and V' computes $q_i = V'(x, R_{V'}; q_1, a_1, \ldots, q_{i-1}, a_{i-1})$ with $R_{V'} = q_1 q_2 \ldots q_t$, and after t interactions, V' evaluates $\rho_{V'}(x; q_1, a_1, \ldots, q_t, a_t)$ to determine whether or not to accept $x \in \{0,1\}^*$, where $\rho_{V'}$ denotes a polynomial (in $|x|$) time computable predicate for an honest verifier V'.

In [Betal], Ben-Or et al showed a constructive way to convert an interactive proof system (P, V) for a language $L \in \mathcal{IP}$ to a (computational) zero-knowledge interactive proof system $(\overline{P}, \overline{V})$ for the language L. The outline of the proof in [Betal] is: (1) convert a given interactive proof system (P, V) for a language $L \in \mathcal{IP}$ to an Arthur-Merlin game (P', V') for the language L; (2) $\overline{V}$ runs V' to generate a question q_i to $\overline{P}$ in the i-th interaction; (3) when receiving q_i from $\overline{V}$, $\overline{P}$ runs P' on input $x \in L$, R'_P, $q_1, a_1, q_2, a_2, \ldots, q_{i-1}, a_{i-1}, q_i$, to generate the answer a_i for the question q_i, and sends an encryption of a_i, $e_i = f(a_i, r_i)$; (4) after t interactions, $\overline{P}$ and $\overline{V}$ agree on a common $\mathcal{NP}$-statement, i.e.,

$$\exists r_1, r_2, \ldots, r_t, a_1, a_2, \ldots, a_t \left[\left\{ \bigwedge_{i=1}^{t} e_i = f(a_i, r_i) \right\} \wedge \phi_{V'}(x; q_1, a_1, q_2, a_2, \ldots, q_t, a_t) \right],$$

and reduce this $\mathcal{NP}$-statement to some $\mathcal{NP}$-complete language (e.g., 3COL, SAT, HC, etc.); and (5) $\overline{P}$ and $\overline{V}$ execute a computational ZKIP for the $\mathcal{NP}$-complete language (e.g., a computational ZKIP for 3COL [GMW]).

To convert the protocol by Ben-Or et al [Betal] to the divertible one, some technical difficulties arise: (D1) if W would transform a question q_i (made by $\overline{V}$) to q'_i in a perfectly divertible manner, then W might have no way to transform the (encrypted) answer e_i (made by $\overline{P}$) for q'_i to the (encrypted) answer e'_i for q_i; (D2) if W would transform an encrypted answer e_i (made by $\overline{P}$) to e'_i using a probabilistic encryption homomorphism, then $\overline{P}$ and $\overline{V}$ could not agree on a common $\mathcal{NP}$-statement like the above. To solve the difficulty (D1), we make $\overline{V}$ ask his questions to W in an encrypted form together with their seeds in an encrypted form, while to solve the difficulty (D2), we make W send to $\overline{P}$ his each seed for the transform of the (encrypted) answer e_i (made by $\overline{P}$) to e'_i in an encrypted form.

Then we finally have the following computationally divertible computational zero-knowledge interactive proof system for any language $L \in \mathcal{IP}$.

Computationally Divertible Computational ZKIP for Any Language in $\mathcal{IP}$

common input: $x \in \{0,1\}^*$, where $|x| = n$.

Repeat the following steps (from V1 to W2) for $1 \leq i \leq t$.

V1. V runs V' to generate an m bit string q_i and computes $Q_i = f(q_i, \alpha_i)$, $D_i = f(d_i, \beta_i)$, and $A_i - f(\alpha_i, \gamma_i)$, where $d_i \in \{0,1\}^\ell$ (dummy string). V sends an $(m + \ell + m\ell)k$ bit string $v_i = Q_i \# D_i \# A_i$ to W.

W1. When receiving v_i from V, W discards D_i, and generates an ℓ bit string s_i. W computes $Q'_i = f(0^m, \alpha'_i) \cdot Q_i$, $D'_i = f(s_i, \beta'_i)$, and $A'_i = f(0^{m \cdot \ell}, \gamma'_i) \cdot A_i$, and sends an $(m + \ell + m\ell)k$ bit string $w_i = Q'_i \# D'_i \# A'_i$ to P.

P1. When receiving w_i from W, P decrypts Q_i', D_i', and A_i' with his infinite power to get an m bit string q_i, an ℓ bit string s_i, and an $m\ell$ bit string α_i. This enables P to recover $Q_i = f(q_i, \alpha_i)$. P runs P' on input $x \in L$, $R_{P'}$, $q_1, a_1, \ldots, q_{i-1}, a_{i-1}, q_i$ to generate an m bit string a_i, and encrypts a_i by $e_i = f(a_i, \delta_i)$. P computes an $m\ell$ bit string ε_i such that $f(a_i, \varepsilon_i) = f(0^m, s_i) \cdot e_i$, and sends e_i to W.

W2. When receiving e_i from P, W computes $e_i' = f(0^m, s_i) \cdot e_i$, and sends e_i' to V.

If the steps above (from V1 to W2) are successfully executed t times, then V halts and accepts x; otherwise V halts and rejects x.

At this stage, V must show to $^{P \leftrightarrow}W$ that he *really* has q_i and α_i ($1 \leq i \leq t$) in a divertible and zero-knowledge manner. Note that this can be done in a way similar to the divertible ZKIP for GNI (see subsection 4.1.), using the homomorphism property of $f(\cdot, \cdot)$. (see Definition 3.1.) Thus, in a way similar to [Beta1], P, W, and V agree on the following $\mathcal{NP}$-statement, i.e.,

$$\exists a_1, \varepsilon_1, q_1, \alpha_1, a_2, \varepsilon_2, q_2, \alpha_2, \ldots, a_t, \varepsilon_t, q_t, \alpha_t$$
$$\left[\left\{ \bigwedge_{i=1}^{t} [e_i' = f(a_i, \varepsilon_i) \wedge Q_i = f(q_i, \alpha_i)] \right\} \wedge \rho_{V'}(x; q_1, a_1, q_2, a_2, \ldots, q_t, a_t) \right].$$

Then P, W, and V reduce this $\mathcal{NP}$-statement to SAT, and follow the protocol of a perfectly divertible computational ZKIP for SAT [BD]. **(End of Protocol)**

It is easy to see that the protocol above is a computational ZKIP for a language $L \in \mathcal{IP}$. Here we informally show that the protocol above is computationally divertible computational ZKIP for a language $L \in \mathcal{IP}$. From the definition of a probabilistic encryption homomorphism (see Definition 3.1.), it follows that v_i is computationally independent of w_i in W1. In a way similar to this, e_i is also computationally independent of e_i' in W2. For the $\mathcal{NP}$-statement above, it is guaranteed in [BD] that there exists a perfectly divertible computational ZKIP for SAT assuming the existence of probabilistic encryption homomorphisms. Then the whole protocol above is a computationally divertible interactive proof system.

Thus for any $L \in \mathcal{IP}$, there exists a computationally divertible computational ZKIP under the assumption that probabilistic encryption homomorphisms exist. ∎

4.3 Robustness of Transformations

In subsections 4.1 and 4.2, we showed that (1) there exists a transformation from an interactive proof system for GNI to a perfectly divertible perfect ZKIP for GNI without any unproven assumption; and (2) there exists a transformation from an interactive proof system for any language $L \in \mathcal{IP}$ to a computationally divertible computation ZKIP for the language $L \in \mathcal{IP}$ under the assumption that probabilistic encryption homomorphisms exist.

The transformation for GNI consists of two parts: (P1) transform an interactive proof system for GNI to a perfect ZKIP for GNI (see the protocol of perfect ZKIP for GNI (modified) in subsection 4.1.) without any unproven assumption; and (P2) transform the perfect ZKIP for GNI to a perfectly divertible perfect ZKIP for GNI (see the protocol of perfectly divertible perfect ZKIP for GNI in subsection 4.1.) without any unproven assumption, and in this transformation, the power of the prover of the resulting proof system for GNI is the same (within a probabilistic polynomial time factor) as that of the original one. On the other hand, the transformation for any language $L \in \mathcal{IP}$ consists of three parts (see the protocol of computationally divertible computational ZKIP of any language $L \in \mathcal{IP}$ in subsection 4.2.): (P1') transform an interactive proof system for the language $L \in \mathcal{IP}$ to an Arthur-Merlin proof system for the language $L \in \mathcal{IP}$ without any unproven assumption; (P2') transform the Arthur-Merlin proof system for the language $L \in \mathcal{IP}$ to a computational ZKIP for the language $L \in \mathcal{IP}$ under the assumption that probabilistic encryptions exist; and (P3') transform the computational ZKIP for the language $L \in \mathcal{IP}$ to a computationally divertible computational ZKIP for the language $L \in \mathcal{IP}$ under the assumption that probabilistic encryption homomorphisms exist, and in this transformation, the prover must be *infinitely* powerful regardless of the complexity of the underlying language $L \in \mathcal{IP}$ to respond with appropriate answers to the verifier in part (P1') and to decrypt encrypted messages from the warden in part (P3').

In this subsection, we discuss on the *constructive* nature of divertible ZKIP for languages considered in this paper. In order to do this, we provide in the below a definition for the notion of *robustness* given by Kilian [K].

Definition 4.3 [K]: *Let $\phi : (P, V) \rightarrow (P', V')$ be a transformation that converts an interactive proof system (P, V) for any language $L \in \mathcal{IP}$ to an interactive proof system (P', V'). The transformation ϕ said to be robust iff*

1. *For all (P, V), $\phi(P, V) = (P', V')$ accepts the same language L as (P, V);*

2. *P' can be evaluated by a probabilistic polynomial time Turing machine M with access to a blackbox evaluator for P,*

where "blackbox evaluator for P" means a function that takes as input a partial conversation so far between P and V, and outputs a distribution equal to that of P.

Ben-Or et al. [Betal] showed a transformation converting an interactive proof system for any language $L \in \mathcal{IP}$ to a computational ZKIP for the language $L \in \mathcal{IP}$ under the assumption that (secure) probabilistic encryptions exist. The transformation however is not robust, because it includes (as a subtransformation) a *nonrobust* transformation converting an interactive proof system for any language $L \in \mathcal{IP}$ to an Arthur-Merlin proof system for the language $L \in \mathcal{IP}$. On the other hand, Kilian [K] showed a *robust* transformation that converts an interactive proof system for any language $L \in \mathcal{IP}$ to a computational (resp. statistical) ZKIP for the language $L \in \mathcal{IP}$ under the assumption that cryptographically (resp. information theoretically) secure circuit evaluation schemes exist.

It is easy to see from the definition that the transformation for GNI is robust, while that for any language $L \in \mathcal{IP}$ is not robust. It is worth noting that the transformation for QNR and $\overline{\langle a \rangle}_p$ are also robust, where $\langle a \rangle_p = \{y \in Z_p^* \mid \exists x \in Z_{p-1} \text{ s.t. } y \equiv a^x \pmod{p}\}$.

To make the transformation for any language $L \in \mathcal{IP}$ robust, the following two technical difficulties arise: (D1) in part (P1'), how to robustly transform an interactive proof system for any language $L \in \mathcal{IP}$ to an Arthur-Merlin proof system for the language $L \in \mathcal{IP}$; and (D2) in part (P3'), how to robustly transform a computational ZKIP for the language $L \in \mathcal{IP}$ to a computationally divertible computational ZKIP for the language $L \in \mathcal{IP}$. (It should be noted that the part (P2') of the transformation is robust.)

We can show (with a slightly weaker assumption) that there exists a robust transformation from an Arthur-Merlin proof system for any language $L \in \mathcal{IP}$ to a computationally divertible computational ZKIP for the language $L \in \mathcal{IP}$ under the assumption that *trapdoor* probabilistic encryption homomorphisms exist, because the prover possessing the trapdoor can decrypt encrypted messages from the warden in (probabilistic) polynomial time. In general, this is not the case for an interactive proof system for a language $L \in \mathcal{IP}$, because it is not known whether or not there exists a robust transformation from an interactive proof system for any language $L \in \mathcal{IP}$ to an Arthur-Merlin proof system for the language $L \in \mathcal{IP}$.

Recently, Lund et al. [Letal] showed that a language PERM, $\#\mathcal{P}$-complete, has an Arthur-Merlin proof system, and Shamir [Sha] showed that a language QBF, $\mathcal{PSPACE}$-complete, has an Arthur-Merlin proof system. Then we can show that there exists a robust transformation converting an interactive proof system for QBF (resp. PERM) to a computationally divertible computational ZKIP for QBF (resp. PERM) under the assumption that (not necessarily trapdoor) probabilistic encryption homomorphisms exist, because the prover with access to the language QBF (resp. PERM) as an oracle can decrypt encrypted messages $f(b, r)$ from the warden by reducing a language $L \in \mathcal{NP}(\subseteq \#\mathcal{P} \subseteq \mathcal{PSPACE})$,

$$ L = \left\{ (a, b) \mid \exists r \subset [0, 1]^k \text{ s.t. } a = f(b, r) \right\}, $$

to the $\mathcal{PSPACE}$-complete language QBF (resp. $\#\mathcal{P}$-complete language PERM).

5   Concluding Remarks

In this paper, we showed several complexity theoretic results on the divertibility of interactive proof systems, i.e., which class of languages has a perfectly (almost perfectly) (computationally) divertible perfect (statistical) (computational) zero-knowledge interactive proof system. More precisely, we showed in this paper that (1) there exists a language (GNI), which is *seemingly* not in $\mathcal{NP}$ but in $\mathcal{AM}$, has a perfectly divertible perfect zero-knowledge interactive proof system without any unproven assumption, while the results known so far are only for languages in $\mathcal{NP}$; and (2) any language $L \in \mathcal{IP}$ has a computationally divertible computational zero-knowledge interactive proof system under the assumption that probabilistic encryption homomorphisms exist, which extends the result in [BD] with a slightly weaker constraint on its divertibility.

It is worth noting that the transformation for GNI is robust, while that for any language $L \in \mathcal{IP}$ is not robust. This is because it is not known whether or not there exists a robust transformation from an interactive proof system for any language $L \in \mathcal{IP}$ to an Arthur-Merlin proof system for the language $L \in \mathcal{IP}$. In this paper, we also showed that for a $\mathcal{PSPACE}$-complete language QBF (resp. $\#\mathcal{P}$-complete language PERM), there exists a robust transformation from an Arthur-Merlin proof system for QBF (resp. PERM) to a computationally divertible computational ZKIP for QBF (resp. PERM) under the assumption that (not necessarily trapdoor) probabilistic encryption homomorphisms exist.

As an alternative notion of divertibility, Desmedt [D] introduced a notion of "abuse" in the light of *subliminal channel* [Sim]. It should be noted that the notion of divertible is closely related to (but completely different from) the notion of "abuse." Here we investigated the divertibility of interactive proof systems in a complexity theoretic framework, however, Desmedt studied the notion of abuse in a cryptographic (or practical) framework to design secure cryptographic protocols, i.e., abuse-free (or subliminal-free) cryptographic protocols.

Fortnow [F] showed that if a language L has a statistical ZKIP, then $\overline{L} \in \mathcal{AM}$. This provides a good evidence that any $\mathcal{NP}$-complete language seems not to have a statistical ZKIP, because if any language $L \in$ co-$\mathcal{NP}$ has a constant round interactive proof system (i.e., co-$\mathcal{NP} \subseteq \mathcal{AM}$), then $\mathcal{PH}$ collapses to the 2-nd level [BHZ]. As a complementary result of [F], Aiello and Håstad [AH] showed that if a language L has a statistical ZKIP, then $L \in \mathcal{AM}$. It follows from the results above that any language having a statistical ZKIP is in $\mathcal{AM} \cap$ co-$\mathcal{AM}$.

As known so far, there exists a perfectly divertible perfect ZKIP (without any unproven assumption) for any language with commutative random self-reducibility [OO], for graph isomorphism [BD], and for graph non-isomorphism (Theorem 4.1 in subsection 4.1). It should be noted that all of them heavily depend on the homomorphic property of the language itself or its complement. In this way, the property of perfect (or almost perfect) divertibility seems to restrict the class of languages that have divertible ZKIP, however, any complexity theoretic characterization on this aspect is not known without any unproven assumption.

- What class of languages has a perfectly divertible perfect (statistical) (computational) zero-knowledge interactive proof system without any unproven assumption ?

Acknowledgments:

The authors would like to thank Tatsuaki Okamoto, Kaoru Kurosawa, and John Leo for their several valuable comments on the earlier version of this work.

References

[AH] Aiello, W. and Håstad, J., "Statistical Zero-Knowledge Languages Can Be recognized in Two Rounds," *IEEE Annual Symposium on Foundations of Computer Science*, pp.439-448 (October 1987).

[Ba] Babai, L., "Trading Group Theory for Randomness," *ACM Annual Symposium on Theory of Computing*, pp.421-429 (May 1985).

[BCC] Brassard, G., Chaum, D., and Crépeau, C., "Minimum Disclosure Proofs of Knowledge," *Journal of Computer and System Sciences*, Vol.37, No.2, pp.156-189 (October 1988).

[BD] Burmester, M.D.V. and Desmedt, Y., "All Languages in $\mathcal{NP}$ Have Divertible Zero-Knowledge Proofs and Arguments Under the Cryptographic Assumptions," Advances in Cryptology – Eurocrypt'90, Lecture Notes in Computer Science 474, *Springer-Verlag*, Berlin, pp.1-10 (1991).

[Betal] Ben-Or, M., Goldreich, O., Goldwasser, S., Håstad, J., Kilian, J., Micali, S., and Rogaway, P., "Everything Provable is Provable in Zero-Knowledge," Advances in Cryptology – Crypto'88, Lecture Notes in Computer Science 403, *Springer-Verlag*, Berlin, pp.37-56 (1990).

[BHZ] Boppana, R., Håstad, J., and Zachos, S., "Does co-$\mathcal{NP}$ Have Short Proofs," *Information Processing Letters*, Vol.25, No.2, pp.127-132 (May 1987).

[Bl] Blum, M., "How to Prove a Theorem So No One Else Can Claim It," *International Congress of Mathematicians*, pp.1444-1451 (August 1986).

[D] Desmedt, Y,. "Making Conditionally Secure Cryptosystems Unconditionally Abuse-Free in a General Context," Advances in Cryptology – Crypto'89, Lecture Notes in Computer Science 435, *Springer-Verlag*, Berlin, pp.6-16 (1990).

[F] Fortnow, L., "The Complexity of Perfect Zero-Knowledge," *ACM Annual Symposium on Theory of Computing*, pp.204-209 (May 1987).

[FFS] Feige, U., Fiat, A., and Shamir, A., "Zero-Knowledge Proofs of Identity," *ACM Annual Symposium on Theory of Computing*, pp.210-217 (May 1988).

[GMR] Goldwasser, S., Micali, S., and Rackoff, C., "The Knowledge Complexity of Interactive Proof Systems," *SIAM J. Comput.*, Vol.18, No.1, pp.186-208 (February 1989).

[GMW] Goldreich, O., Micali, S., and Wigderson, A., "Proofs that Yield Nothing But Their Validity or All Languages in NP Have Zero-Knowledge Proofs," *Technical Report* #544, Technion – Israel Institute of Technology, Department of Computer Science, Haifa, Israel (March 1989).

[GS] Goldwasser, S. and Sipser, M., "Private Coins versus Public Coins in Interactive Proof Systems," *ACM Annual Symposium on Theory of Computing*, pp.59-68 (May 1986).

[K] Kilian, J., "Achieving Zero-Knowledge Robustly," Proc. of Crypto'90 (August 1990).

[Letal] Lund, C., Fortnow, L., Karloff, H., and Nisan, N., "Algebraic Methods for Interactive Proof systems" *IEEE Annual Symposium on Foundations of Computer Science*, pp.2-10 (October 1990).

[OO] Okamoto, T. and Ohta, K., "Divertible Zero-Knowledge Interactive Proofs and Commutative Random Self-Reducibility," Advances in Cryptology – Eurocrypt'89, Lecture Notes in Computer Science 434, *Springer-Verlag*, Berlin, pp.134-149 (1990).

[Sha] Shamir, A., "$\mathcal{IP} = \mathcal{PSPACE}$," *IEEE Annual Symposium on Foundations of Computer Science*, pp.11-15 (October 1990).

[Sim] Simmons, G.J., "The Subliminal Channel and Digital Signatures," Advances in Cryptology – Eurocrypt'84, Lecture Notes in Computer Science 209, *Springer-Verlag*, Berlin, pp.364-378 (1985).

[TW] Tompa, M. and Woll, H., "Random Self-Reducibility and Zero-Knowledge Interactive Proofs of Possession of Information," *IEEE Annual Symposium on Foundations of Computer Science*, pp.472-482 (October 1987).

A Multi-Purpose Proof System

——— for identity and membership proofs

Chaosheng SHU   Tsutomu MATSUMOTO   Hideki IMAI

Division of Electrical and Computer Engineering
YOKOHAMA NATIONAL UNIVERSITY
156 Tokiwadai, Hodogaya-ku, Yokohama 240, Japan

Summary

In this paper, we propose a multi-purpose proof system which allows a user to perform various proof protocols needing to remember only one piece of secret data. These proofs include identity proof, membership proof without revealing one's identity, and combined identity and membership proof. When a user participates in a group, he will obtain a secret witness corresponding to the group's name from some administrator of the group. Using the secret witness, the user can prove his membership in this group. Many secret witnesses can be combined into one piece of secret data. From the secret data, the user can obtain the secret witness of the group he participates in. If the user participates in a new group afterward, he can also easily update his secret data. But the size of the secret data is independent of the number of the groups in which the user participates. Our system satisfies other desirable properties which were not attained by the previously proposed systems.

1 Introduction

How to prove individual's identity and/or membership in some groups is a very important problem in the application areas of cryptography. Many protocols [FS 86] [C1 85] [C2 86] [K 89] [KMI1 89] [KMI2 89] [OOK 90] [SMI 90] for such kind of proofs have been presented. In these proof protocols, any verifier can judge that a user is valid in an off-line environment, if the user holds the secret data corresponding to some public data about his name or the group names. In order to perform various proofs, it may be necessary for a user to remember many secret data. This will be a burden for the user to keep and manage them. How can we construct an efficient proof system which allows a user to perform various proof protocols needing to remember only one piece of secret data. In this paper, we give an answer by proposing a multi-purpose proof system. Our system deals with three types of proofs: identity proof, membership proof without revealing one's identity, and combined identity and membership proof (or called identity & membership proof). That is, a user who wants to perform these proofs only needs to remember one piece of secret data regarding his identity number and the groups' names. When a user participates in a group, he will obtain a secret witness corresponding to the group's name from some administrator of the group. Using the secret witness, the user can prove his membership in this group. Many secret witnesses can be combined into one piece of secret data. From this secret data, the user can obtain the secret witness of the group he participates in. If the user participates in a new group afterward, he can easily update this secret data. Some aims of the membership proof systems proposed by [C1 85] [C2 86] [K 89] [KMI1 89] [KMI2 89] [OOK 90] [SMI 90] are the same as ours. But, it will be seen that our system satisfies more conditions than they do.

In section 2, we describe the model of the system which includes three parts: the registration part, the participation part, and the proof part. We show the conditions about the system, and compare the conditions for our system with those of [KMI 89] [KMI2 89] [OOK 90] [SMI 90]. We explain the registration part in section 3, and the participation part in section 4. In particular, we propose a protocol called "secret hiding participation". In section 5 we present various proof protocols for identity proof, membership proof, and identity & membership proof. We briefly analyze the security of the system in section 6.

2 Model and Conditions

Different models will produce different systems and they will affect the overall properties of the system. Our objective is to construct a system as practical as possible. Thus, a realistic model is necessary. The model that we consider includes a trustworthy center C, a number of group administrators (GA), a set of users (U) who may be members of several groups. The duty of the center is only to manage the registration of the user's name and the group's name, and not to directly accept users as the members of a particular group. A group administrator is a manager who can accept users to be the members of his group. The group administrator must register his group name in the center, and need not to be assumed trustworthy. A user is a person (or an equipment) who must at first register his name in the center and may participate in any group if the administrator of the group accepts him. A user may be a prover or a verifier. When a proof protocol is carried out, the prover proves his identity or membership to the verifier, and the verifier checks the validity of the prover.

If only one center were set to directly accept users as the members of a particular group, the center would bear a heavy burden in management when the number of groups increases. The purpose of having a number of group administrators is to disperse the center's burden in management. This kind of model is also well seen in the real society and it can be considered as a realistic model.

Our system consists of three parts; the registration part, the participation part, and the proof part. In the registration part, (1) a user should register himself to the center, (2) each group administrator should register his group to the center. The participation part is a part in which a user applies for the membership in a group to the group administrator. In this part, the user can obtain a secret witness for various proofs in the future. The proof part includes some proof protocols which allow a prover, remembering only one piece of secret data, to prove his identity or membership to any verifier. In the system, we treat three types of proofs that are identity proof, membership proof, and identity & membership proof. Membership proof, here, means that the user only proves his membership in some groups without disclosing his identity. The identity & membership proof means that the user's identity and membership are simultaneously proved. The system is shown in Fig.1.

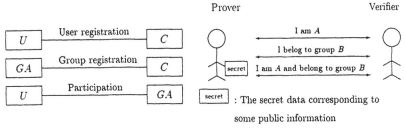

C: Center; GA: Group Administrator; U: User may be Prover or Verifier

Fig.1 The multi-purpose proof system

The main aim of our multi-purpose proof system is to enable a user, needing to remember only one piece of secret data, to execute various credential protocols as convenient as possible. Therefore, we demand that our system should satisfy the following conditions.

Multi-Purpose: A user who wants to execute various proof protocols only needs to remember one piece of secret data.

Autonomy: When a user wants to participate in a group, he must perform a application protocol with the group administrator. But it needs no communication with the center nor other group administrators both for the user and the group administrator. When a proof protocol is carried out, any verifier can independently verify the proofs without communicating with the group administrators nor the center.

Completeness: After execution of a proof protocol, a true prover is judged valid by any verifier with overwhelming probability.

Soundness: After execution of a proof protocol, any prover with false identity or false membership is undetected by the verifier with negligible probability.

Zero-knowledge: After execution of the proof protocol, no secret of the prover leaks to the verifier.

Hierarchy: If the center sets up some groups hierarchically, and a user is a member of a higher group, then the user is not only able to prove his higher membership, but also able to prove his lower membership without disclosing the verifier his higher membership.

Secret Isolation: The secrets of the center cannot be derived from the public information by the group administrators and the users. The group administrators and the users all have their own secrets, which cannot be derived by them with each other. The user's secret witness for proof of membership in a group is not known even to the group administrator of the same group.

Identity Connected: Any user cannot separate the secret witness for membership proof from the secret witness for identity proof.

Efficiency for Users: The size of the secret data is independent of the number of the groups in which the user participates.

Our conditions are different from those of [C1 85], [C2 86], [K 89], [KMI1 89], [KMI2 89], [OOK 90] and [SMI 90]. In [C1 85], [C2 86] and [K 89], the model of the systems itself is different from ours and their systems do not consider many conditions like ours. The main target of [KMI1 89], [KMI2 89], [OOK 90] is to construct an anonymous membership proof system without revealing one's identity. In [KMI1 89], four schemes were proposed. All of their schemes do not satisfy the autonomy condition because the process in which a user goes through to obtain the secret of a group requires him to go to the center. Scheme 1, 2, 3 of [KMI1 89] do not treat the membership proof for the multi-group situation. And Scheme 4 of [KMI1 89] does not satisfy the secret isolation condition, because the users are given the same secret for proof of membership of the same group. It is still not known whether the proof protocols of Scheme 1, 2, 4 have the property of zero knowledge. [KMI2 89] revised the schemes presented in [KMI1 89], and began to consider how to combine the secrets for proof of membership in plural groups into one piece of secret data. But the problem of the autonomy and the secret isolation is still not considered. [OOK 90] presented two schemes which were emphasized to be suitable for proof of membership in the groups in the hierarchy structure. However, both schemes do not satisfy the secret isolation condition because the users are given the same secret for proof of membership in the same group. Scheme 1 of [OOK 90] does not satisfy the autonomy condition either, because the process in which a user goes through to obtain the secret of a group requires him to go to the center.

On the other hand, the system of [SMI 90] cannot be used for group membership proof without disclosing one's identity.

Here, it is necessary to explain the role of the condition of "Identity-Connected". This is a condition which, to a certain extent, restricts the illegal action where a user gives his secret witness to others for anonymous membership proof. Assume some user A participates in a group GN; he holds a secret witness which he is able to use for membership proof. Now consider an illegal action in which A gives his secret witness to B who does not participate in group GN, and lets B perform a anonymous membership proof. Assume there are many members in GN. Since B uses A's secret witness to execute the anonymous membership proof, even if a verifier V knows of B's illegal action it is difficult for V to know who gave B the secret witness if B does not tell V. However, if this secret witness is also able to be used for A's identity proof, this illegal action will be dangerous for A, because A cannot guarantee that B will not use it to misrepresent A's identity. If verifier V

knows B's illegal action when B executes A's identity proof protocol, it is clear that A gave B the secret witness, and A will be punished. Later, we will see that in our system, any secret witness for membership proof can also be used for identity proof. In other words, a user cannot separate the secret witness for membership proof from the secret witness for identity proof.

3 Registration

The registration part includes two types of registration as follows.

1. User Registration A user registers a name to the center who gives the user some secrets originated from the identity.

2. Group Registration A group administrator registers his group to the center who gives the group administrator a secret corresponding to the name of the group.

We assume that a center is trustworthy. He should select some numbers and functions as follows.

$$p, q, L, n, n', g, a_0, \cdots, a_{t-1}, t, s, sig(), h().$$

Integers p, q are two large primes, $n = pq$ and $L = lcm(p - 1, q - 1)$. Integer n' is a prime such that $n' = 2n + 1$. This can be done as follows. The center first selects two large primes p, q such that $n = pq$, he computes $n' = 2n + 1$ and check whether n' is a prime. If n' is not a prime, he selects other p, q until n' is a prime. Then the center selects a g' which is a primitive element in $Z_{n'}^*$, and lets $g = (g')^2 \bmod n'$. In order to make the computation easier, p, q can be selected such that $p, q \equiv 3 \pmod 4$. Integers $a_0, \cdots, a_{t-1}$ are selected such that $1 \leq a_i < L$ for $0 \leq i \leq t - 1$, where t is called the security parameter. Positive integer s is a maximum number of groups that the center can manage. For ensuring the system secure, s should be selected as quite less than t. The center keeps p, q, L and $(a_0, \cdots, a_{t-1})$ secret, and makes n, n' and g public. Function $h()$ is a universal one-way hash function and is public to anyone.

This kind of hash function has the collision free property and can be constructed from any one-way function [Rom 90]. With the collision free property, it is difficult to find two different arguments for the same output of $h()$. Function $sig()$ is the center's public signature function whose outputs can be checked by anyone, and is assumed to be secure enough. In our system, we assume that it is difficult to extract mth roots $\bmod n$ of any integer without knowing p and q for any positive integer m and the discrete logarithm problem is hard.

3.1 User Registration

When a user U wants to register a name, he describes his identity number ID to the center. After checking the user's physical identity, the center computes $H = h(ID\|l)$, for some suitable number l. For different users U_i and U_j $i \neq j$ the center should adjust l_i, l_j so that H_i and H_j satisfies $gcd(H_i, H_j) = 1$, and $H_i = u_i^2 \bmod n$ for some u_i.

After computing H, u from ID and l for user U, the center signs two signatures $S_c = sig(H\|l)$ on H, l and $S_{cg} = sig(g^u \bmod n')$, on $g^u \bmod n'$. The center further computes

$$S_u = (u^{a_0} \bmod n, \cdots, u^{a_{t-1}} \bmod n)$$

and gives user U

$$(ID, u, S_u, H, l, S_c, S_{cg}).$$

After this stage, user U already has the ability to prove to any verifier his identity with ID, l, S_c, and the secret u. S_u is the secret with which the user can apply for membership in any group. Since the hash function h is collision free, there will not be more than one ID with the same H. S_{cg} will be used for membership proof.

3.2 Group Registration

When a new group administrator GA_i wishes to register a new group in the center, he should describe the center the group name GN_i. Given GN_i, the center chooses a suitable integer C_i such that $1 \leq C_i < L$, and computes a G_i which satisfies the following conditions,

$$f(C_i) = a_0 + a_1 C_i + \cdots + a_{t-1} C_i^{t-1} \mod L, \tag{1}$$

$$\gcd(f(C_i), L) = 1, \tag{2}$$

$$G_i = \frac{1}{f(C_i)} \mod L, \tag{3}$$

$$\gcd(G_i, G_j) = 1, for\ i \neq j. \tag{4}$$

This can be done as following. The center first evaluates polynomial (1), checks if (2) is satisfied. If (2) is satisfied, then the center computes (3) and checks if (4) is satisfied. If (2) or (4) is not satisfied, the center selects another C_i and do the same procedure above again until both the conditions are satisfied.

The center further produces a signature $S_{G_i} = sig(GN_i \| G_i)$ on GN_i and G_i, and makes (GN_i, G_i, S_{G_i}) public. This also means that GN_i has been admitted as a valid group name. The center gives C_i to group administrator GA_i as a secret. C_i will be used by GA_i for helping the users to compute the secret witnesses for their membership proof of the group GN_i. The output of $f(C_i)$ is kept as a secret by the center

4 Participation

In this section, we show the process how a user obtains the secret witness for his multi-purpose proofs in the future. When a user U wishes to participate in a group GN_i, he must execute a protocol with group administrator GA_i. Group administrator GA_i will help user U to obtain a secret witness about his group. The protocol achieves two targets. The first is that group administrator GA_i sees if U has registered his name in the center. This is accomplished by checking if U holds the correct S_u. The second is that GA_i helps U for enabling him to obtain a secret witness

$$P(G_i, u) = u^{\frac{1}{G_i}} \mod n$$

of GN_i, but GA_i cannot know the secret witness.

The protocol must satisfy the conditions below.

(1) A true user is always judged valid by the group administrator.

(2) The false user is always detected by the group administrator.

(3) The user's secrets u and S_u do not leak to the group administrator.

(4) The user can obtain the secret witness $P(G_i, u)$ without revealing it to the group administrator.

(5) The user can check if he obtains the correct $P(G_i, u)$.

(6) The group administrator does not reveal his secret C_i to the user.

We call this protocol "secret hiding participation", and show it as following Protocol 1. Let m denote a natural number determined before executing the protocol.

Protocol 1:

step 1-1 U picks $2m$ random numbers r_{1k}, r_{2k} ($\in_R Z_n^*$) for $1 \leq k \leq m$ such that $\gcd(r_{1k}, r_{2k}) = 1$, and computes

$$T_{1k} = H^{r_{1k}} \bmod n,$$
$$T_{2k} = H^{r_{2k}} \bmod n.$$

U gives GA_i
$$(ID, l, S_c, T_{11}, T_{21}, \cdots, T_{1m}, T_{2m}).$$

step 1-2 With ID and l, GA_i first computes $H = h(ID||l)$, and checks C's signature S_c. Then he sends a random binary vector $\mathbf{e} = (e_1, \cdots, e_m)$ to U.

step 1-3 If $e_k = 0$, U sets $Y_k = (r_{1k}, r_{2k})$ for $k = 1, \cdots, m$. If $e_k = 1$, U sets $Y_k = (S_u^{r_{1k}}, S_u^{r_{2k}})$, where

$$S_u^{r_{1k}} = (u^{a_0 \cdot r_{1k}} \bmod n, \cdots, u^{a_{t-1} \cdot r_{1k}} \bmod n),$$
$$S_u^{r_{2k}} = (u^{a_0 \cdot r_{2k}} \bmod n, \cdots, u^{a_{t-1} \cdot r_{2k}} \bmod n).$$

Then U sends $\mathbf{Y} = (Y_1, \cdots, Y_m)$ to GA_i.

step 1-4 GA_i checks if $\mathbf{Y}$ is correct for all k such that $1 \leq k \leq m$ as follows. If $e_k = 0$, GA_i checks if $H^{r_{1k}} \bmod n = T_{1k}$, $H^{r_{2k}} \bmod n = T_{2k}$ and $\gcd(r_{1k}, r_{2k}) = 1$. If $e_k = 1$, GA_i computes

$$P(G_i, u)^{r_{1k}} = \prod_{j=0}^{t-1} (u^{r_{1k} \cdot a_j})^{C_i^j} \bmod n = u^{f(C_i) \cdot r_{1k}} \bmod n = u^{\frac{1}{C_i} \cdot r_{1k}} \bmod n,$$

$$P(G_i, u)^{r_{2k}} = \prod_{j=0}^{t-1} (u^{r_{2k} \cdot a_j})^{C_i^j} \bmod n = u^{f(C_i) \cdot r_{2k}} \bmod n = u^{\frac{1}{C_i} \cdot r_{2k}} \bmod n.$$

GA_i further checks if

$$P(G_i, u)^{r_{1k} \cdot 2G_i} \equiv u^{\frac{1}{C_i} \cdot r_{1k} \cdot 2G_i} \equiv T_{1k} \pmod{n},$$

$$P(G_i, u)^{r_{2k} \cdot 2G_i} \equiv u^{\frac{1}{C_i} \cdot r_{2k} \cdot 2G_i} \equiv T_{2k} \pmod{n}.$$

step 1-1 – step 1-4 should be iterated λ times (λ is a suitable number). If each iteration is successful, GA_i judges that U is a valid user and continues to do step 1-5, otherwise he refuses to give U anything, but a message of rejection.

step 1-5 GA_i selects a

$$Y_k = (S_u^{r_{1k}}, S_u^{r_{2k}})$$

for some k at some iteration. He gives U

$$(P(G_i, u)^{r_{1k}}, \ P(G_i, u)^{r_{2k}}),$$

and corresponding to (T_{1k}, T_{2k}).

step 1-6 U checks if

$$P(G_i, u)^{r_{1k} \cdot G_i} \bmod n = u^{r_{1k}} \bmod n$$

$$P(G_i, u)^{r_{2k} \cdot G_i} \bmod n = u^{r_{2k}} \bmod n.$$

Then U computes $P(G_i, u)$ from $P(G_i, u)^{r_{1k}}, P(G_i, u)^{r_{2k}}$ with the Euclid's algorithm. If this step is successful, the participation is finished.

If user U participated in several groups (say $\mathbf{GN} = \{GN_1, \cdots, GN_w\}$, for some number w, $\mathbf{GN}$ denotes the set of groups), this means U is holding $u^{\frac{1}{G_1}} \bmod n, \cdots, u^{\frac{1}{G_w}} \bmod n$. From $\gcd(G_1, \cdots, G_w) = 1$, there are w integers x_i $(1 \le i \le w)$ which satisfy

$$G_1 x_1 + \cdots + G_w x_w = 1$$

and can be calculated by the extended Euclid's algorithm. From $u^{\frac{1}{G_i}} \bmod n, \cdots, u^{\frac{1}{G_w}} \bmod n$, user U can compute the master secret witness

$$PM(\mathbf{GN}, u) = u^{\prod_{GN_i \in \mathbf{GN}} \frac{1}{G_i}} \bmod n$$

like the method of [OOK 90]. This also means when user U participates in a new group afterward, he can easily update his master secret witness. From $PM(\mathbf{GN}, u)$, user U can extract u and any secret witness of one group or secret witnesses of several groups. This is why memorizing such a piece of secret data $PM(\mathbf{GN}, u)$ is enough for a user to perform multi-purpose proof protocols.

We can see among all of the secret witnesses, u is a basic secret witness to the other secret witnesses for membership proof. However, u is also the most important secret witness for user U. This is because u connects to the identity of user U. In the proof part, we will see that for user U, he can not separate the secret witness for membership proof from his secret number u. So, if a user gives others his secret witnesses to let them perform multi-purpose proofs, this illegal action is dangerous to the user, because the secret witness can also be used to prove the user's identity.

Note: Although S_u is also a secret of user U, it is only used for participation not for multi-purpose proof protocols. It is unnecessary for U to memorize S_u when he executes the multi-purpose proof protocols. We can see that the size of $PM(\mathbf{GN}, u)$ is independent of the number w of the groups in which the user participates. If n is selected such as $|n| = 512$ *bits*, then the size of $PM(\mathbf{GN}, u)$ is also upper bounded by 512 *bits*.

Now, let us check the conditions of Protocol 1. We at first see Conditions (1) and (2). From the protocol, it is clear that if user U correctly performs the protocol with the correct S_u, he is always judged valid by group administrator GA_i. While if U does not know the secret S_u or deviates from the protocol to give group administrator GA_i wrong $S_u^{r_{1k}}$, $S_u^{r_{2k}}$ for $1 \leq k \leq m$ in order to attempt to draw out some information about C_i, or $f(C_i)$, then his illegal actions is always detected by group administrator GA_i. This is because in step 1-2, group administrator GA_i, after receiving T_{1k} and T_{2k}, will query a random binary vector $\mathbf{e}$. It is not easy for U to give out a correct answer both to $e_k = 0$ and $e_k = 1$ for each k in step 1-3. The probability of the user succeeding in passing the group administrator's check at each iteration is only $1/2^m$. So with the iterations of step 1-1 – step 1-4, the probability of these undetected cheats can be decreased to any small number.

Next, we discuss Condition (3). Throughout the protocol, only in step 1-1 and step 1-3, user U gives group administrator GA_i some messages. In step 1-1, each pair T_{1k} and T_{2k} is only two random numbers for group administrator GA_i because r_{1k} and r_{2k} are randomly selected from Z_n^*. In step 1-3, when the query $e_k = 0$, user U only gives group administrator GA_i two random numbers r_{1k} and r_{2k} without including any information of S_u and u. When $e_k = 1$, U must send $S_u^{r_{1k}}$ and $S_u^{r_{2k}}$ to GA_i But since r_{1k} and r_{2k} are not simultaneously known to GA_i, it seems very difficult for GA_i to compute S_u and u from $S_u^{r_{1k}}$, $S_u^{r_{2k}}$ and T_{1k}, T_{2k}. $S_u^{r_{1k}}$, $S_u^{r_{2k}}$ are only two random numbers to group administrator GA_i too. So user U can maintain his secrets secure without revealing them to group administrator GA_i while executing the protocol.

Condition (4) is satisfied with reasons below. In step 1-1, although user U selects each pair r_{1k}, r_{2k} such that $\gcd(r_{1k}, r_{2k}) = 1$, this never changes the randomness of T_{1k}, T_{2k} and $S_u^{r_{1k}}$, $S_u^{r_{2k}}$, because anyone can select them uniformly from Z_n^*. While knowing r_{1k}, r_{2k}, user U can compute $P(G_i, u)$ with the Euclid's algorithm. But for group administrator GA_i, $P(G_i, u)^{r_{1k}}$ and $P(G_i, u)^{r_{2k}}$ are also random numbers without knowing r_{1k} and r_{2k}, and so he is impossible to obtain $P(G_i, u)$. This means it is difficult for GA_i to pretend to be even a user of his member.

We note that the property of $f(C_i) = \frac{1}{G_i} \bmod L$, enables U to check whether GA_i gives him correct $P(G_i, u)$. However from $P(G_i, u)$, computing $f(C_i)$ for U is infeasible, because this is the same as breaking RSA. This also implies that it is infeasible for U to compute C_i. So it is difficult for U to substitute for the role of GA_i. Therefore, Conditions (5) and (6) are satisfied.

Protocol 1 is different from the blind signature of [C1 85] because even group administrator GA_i does not know $\frac{1}{G_i} \bmod L$ corresponding to the public information G_i. Only the center knows $\frac{1}{G_i} \equiv f(C_i) \pmod{L}$. We can see that in the participation it needs no communication with the center nor other groups both for the user and the group administrator.

5   Multi-Purpose Proofs

In this section, we present various proof protocols for various purposes. The simplest proof is identity proof. When a prover U wants to prove to any verifier his identity, he should first compute u from $PM(\mathbf{GN}, u)$, then utilizes the protocol of [FS 86] according to the public information H, l and S_c. Here we omit the identity proof protocol. (The reader can refer to [FS 86] for details.) In the following we show Protocol 2, which enables a prover to prove to any verifier his membership in a group without revealing his identity. Denote the center by C, the prover by U, and any verifier by V. The name of the group is assumed to be GN_i, and can be checked if it is valid from the public information of (GN_i, G_i, S_{G_i}). Prover U first computes the secret witness $P(G_i, u) = u^{\frac{1}{G_i}} \bmod n$ for proof of membership in GN_i, from $PM(\mathbf{GN}, u)$. In the following other proof protocols, we assume that the verifier has made sure of the validity of the group's name, before the protocols are carried out.

Protocol 2: The following process must be repeated λ times (λ is a suitable number).

step 2-1 U picks a random number $r \in_R Z_n^*$, calculates $x = g^{ur^{G_i}} \bmod n'$, and sends S_{cg}, $g^u \bmod n'$, x to V.

step 2-2 V first checks C's signature S_{cg} with $g^u \bmod n'$. Then he picks a bit $e = 0$ or 1, and sends e to U.

step 2-3 If $e = 0$, U gives $Y = r$ to V. If $e = 1$, U gives $Y = ru^{\frac{1}{G_i}} \bmod n$ to V.

step 2-4 V first checks whether $\gcd(Y, n) = 1$. Then, if $e = 0$, V checks whether $g^{uY^{G_i}} \bmod n' = x$, and if $e = 1$, V checks whether $g^{Y_i^{G}} \bmod n' = g^{ur^{G_i}} \bmod n' = x$.

If each iteration is successful, V accepts the proof, otherwise he rejects.

In step 2-4, checking whether $\gcd(Y, n) = 1$ is necessary in this kind of proof protocols [BD 89], otherwise, the soundness is not satisfied.

Assume prover U is a member in $\mathbf{GN} = \{GN_1, \cdots, GN_w\}$. So, he knows the secrets $PM(\mathbf{GN}, u)$ $= u^{\prod_{GN_i \in \mathbf{GN}} \frac{1}{G_i}} \bmod n$. The following protocol 3 is a protocol for U to convince to V that he is a membership of $\mathbf{GN}$.

Protocol 3: The following process must be repeated λ times (λ is a suitable number).

step 3-1 U picks a random number $r \in_R Z_n^*$, calculates $x = g^{ur^{G_1 \cdots G_w}} \bmod n'$, and sends S_{cg}, g^u, x to V.

step 3-2 V first checks C's signature S_{cg} with g^u. Then he picks a bit $e = 0$ or 1, and sends e to U.

step 3-3 If $e = 0$, U gives $Y = r$ to V. If $e = 1$, U gives $Y = ru^{\frac{1}{G_1 \cdots G_w}} \bmod n$ to V.

step 3-4 V first checks whether $\gcd(Y, n) = 1$. Then, if $e = 0$, V checks whether $g^{uY^{G_1 \cdots G_w}} \bmod n' = x$, if $e = 1$, V checks whether $g^{Y^{G_1 \cdots G_w}} \bmod n' = g^{ur^{G_1 \cdots G_w}} \bmod n' = x$.

If each iteration is successful, V accepts the proof, otherwise he rejects.

Protocol 2 is a special case of Protocol 3, and the prover's identity does not need to be revealed. Both protocols are clear the zero-knowledge proof protocols. If U wants to prove the membership in only some subset of groups (say $\mathbf{GN}' \subset \mathbf{GN}$), U should first calculate $PM(\mathbf{GN}', u) = u^{\prod_{GN_i \in \mathbf{GN}'} \frac{1}{G_i}}$ $\bmod n$ from $PM(\mathbf{GN}, u)$, then performs the proof using Protocol 3. If we set $PM(\mathbf{GN}, u)$ to be a secret witness of a higher group, and $PM(\mathbf{GN}', u)$ to be of a lower membership, then using Protocol 3 the prover can prove his lower membership without disclosing to the verifier his higher membership.

The following Protocol 4 is an identity & membership proof protocol which enables a prover to simultaneously prove his identity and membership in a group. Assume the name of the group is GN_i.

Protocol 4: The following process must be repeated λ times (λ is a suitable number).

step 4-1 U picks a random number $r \in_R Z_n^*$, calculates $x = r^{2G_i} \bmod n$, and sends S_c, ID, l, x to V.

step 4-2 With ID and l, V first computes $H = h(ID\|l)$, and checks C's signature S_c. Then he picks a bit $e = 0$ or 1, and sends e to U.

step 4-3 If $e = 0$, U gives $Y = r$ to V. If $e = 1$, U gives $Y = rP(G_i, u) \bmod n$ to V.

step 4-4 V first checks whether $\gcd(Y, n) = 1$. Then, if $e = 0$, V checks whether $Y^{2G_i} \bmod n = x$, if $e = 1$, V checks whether $Y^{2G_i} \bmod n = x * H \bmod n$.

If each iteration is successful, V accepts the proof, otherwise he rejects.

Protocol 4 is also a zero-knowledge proof protocol, and we can see that the prover's identity is connected with the membership in the group. Protocol 5 for simultaneous proof of membership in several groups and identity is presented below. Assume prover U holds a secret witness which is $PM(\mathbf{GN}, u) = u^{\prod_{GN_i \in \mathbf{GN}} \frac{1}{G_i}} \bmod n$, $\mathbf{GN} = \{GN_1, \cdots, GN_w\}$.

Protocol 5: The following process must be repeated λ times (λ is a suitable number).

step 5-1 U picks a random number $r \in_R Z_n^*$, computes $x = r^{2G_1 \cdots G_w} \bmod n$, and sends S_c, ID, l, x to V.

step 5-2 With ID and l, V first computes $H = h(ID\|l)$, and checks C's signature S_c. Then he picks $e = 0$ or 1, and sends e to U.

step 5-3 If $e = 0$, U gives $Y = r$ to V. If $e = 1$, U gives to V

$$Y = rPM(\mathbf{GN}, u) \bmod n = ru^{\frac{1}{G_1 \cdots G_w}} \bmod n.$$

step 5-4 V first checks whether $\gcd(Y, n) = 1$. Then, if $e = 0$, V checks whether $Y^{2G_1 \cdots G_w} \bmod n = x \bmod n$, if $e = 1$, V checks whether

$$Y^{2G_1 \cdots G_w} \bmod n = x * H \bmod n.$$

If each iteration is successful, V accepts the proof, otherwise he rejects.

If U wants to prove the membership in only some subset of groups (say $\mathbf{GN'} \subset \mathbf{GN}$), U also can calculate $PM(\mathbf{GN'}, u) = u^{\prod_{GN_i \in \mathbf{GN'}} \frac{1}{G_i}} \bmod n$ from $PM(\mathbf{GN}, u)$, then performs the proof using Protocol 5. Protocol 4 is a special case of Protocol 5. If we set $PM(\mathbf{GN}, u)$ to be a secret witness of a higher group, and $PM(\mathbf{GN'}, u)$ to be of a lower membership, then using Protocol 5 the prover also can prove his lower membership without disclosing to the verifier his higher membership. Protocol 4 and 5 are zero-knowledge proof protocols and can be proved like [FFS 88] or [OO 88].

It can be seen that in all the protocols, the verifier independently verifies the proofs without communicating with the group administrators nor the center.

6 Security Analysis

Our system is constructed by the registration part, the participation part, and the proof part, so the security of the system should be considered from them. In the proof part, the proposed proof protocols are zero-knowledge proof protocols. These guarantee that the secrets of the users do not leak to any verifier after the execution of the proofs. However, since we assume only the center is trustworthy, the group administrators and users may illegally extract the secrets of others or may cheat to substitute for the role of others. For example, although the group administrator GA_i knows C_i, he cannot know $f(C_i)$ because there is no known efficient way to compute $f(C_i)$ without knowing $(a_0, \cdots, a_{t-1})$. However, if enough number of group administrators collude (say, they are the group administrators of the groups $GN_0, GN_2, \cdots, GN_t$), from the simultaneous polynomial equations (5),

$$
\begin{cases}
(a_0 + a_1 C_0 + \cdots + a_{t-1} C_0^{t-1}) G_0 \equiv 1 \pmod{L} \\
(a_0 + a_1 C_1 + \cdots + a_{t-1} C_1^{t-1}) G_1 \equiv 1 \pmod{L} \\
\vdots \\
(a_0 + a_1 C_t + \cdots + a_{t-1} C_t^{t-1}) G_t \equiv 1 \pmod{L},
\end{cases}
\tag{5}
$$

they can deduce a multiple of L, and thus p and q.

The method to deduce a multiple of L and p, q is as follows

Letting $\lambda_{ij} = G_i C_i^j$, we can get simultaneous polynomial equations (6) from (5).

$$
\begin{cases}
\lambda_{00} a_0 + \cdots + \lambda_{0(t-1)} a_{t-1} \equiv 1 \pmod{L} \\
\vdots \\
\lambda_{(t-1)0} a_0 + \cdots + \lambda_{(t-1)(t-1)} a_{t-1} \equiv 1 \pmod{L} \\
\lambda_{t0} a_0 + \cdots + \lambda_{t(t-1)} a_{t-1} \equiv 1 \pmod{L}
\end{cases}
\tag{6}
$$

From the first t equations of (6), we can write down equation (7) with matrices.

$$
\begin{bmatrix}
\lambda_{00} & \cdots & \lambda_{0(t-1)} \\
\vdots & \ddots & \vdots \\
\lambda_{(t-1)0} & \cdots & \lambda_{(t-1)(t-1)}
\end{bmatrix}
\begin{bmatrix}
a_0 \\
\vdots \\
a_{t-1}
\end{bmatrix}
\equiv
\begin{bmatrix}
1 \\
\vdots \\
1
\end{bmatrix}
\pmod{L}
\tag{7}
$$

Now let us define

$$
\Gamma =
\begin{bmatrix}
\lambda_{00} & \cdots & \lambda_{0(t-1)} \\
\vdots & \ddots & \vdots \\
\lambda_{(t-1)0} & \cdots & \lambda_{(t-1)(t-1)}
\end{bmatrix}.
$$

If $|\Gamma| \neq 0$, then Γ^{-1} exists, so from (7) we can get equation (8)

$$
\begin{bmatrix}
a_0 \\
\vdots \\
a_{t-1}
\end{bmatrix}
\equiv \Gamma^{-1}
\begin{bmatrix}
1 \\
\vdots \\
1
\end{bmatrix}
\pmod{L}
\tag{8}
$$

Also the last $((t+1)$th) equation of (6) can be written down as follows.

$$
[\lambda_{t0} \cdots \lambda_{t(t-1)}]
\begin{bmatrix}
a_0 \\
\vdots \\
a_{t-1}
\end{bmatrix}
\equiv 1 \pmod{L}.
\tag{9}
$$

By substituting (8) for (9), we get equation (10).

$$[\lambda_{t0} \cdots \lambda_{t(t-1)}] \Gamma^{-1} \begin{bmatrix} 1 \\ \vdots \\ 1 \end{bmatrix} \equiv 1 \pmod{L}. \tag{10}$$

Then equation (11) exists

$$[\lambda_{t0} \cdots \lambda_{t(t-1)}] \Gamma^{-1} \begin{bmatrix} 1 \\ \vdots \\ 1 \end{bmatrix} - 1 \equiv 0 \pmod{L}. \tag{11}$$

Since λ_{ij} and Γ^{-1} can be computed, it means that a multiple of L is obtained.

Now consider how to compute p ,q from the multiple of L. Let $\phi = bL$ for some integer b. It is well known that for some integer g such that $1 \leq g < n$ the following equations hold

$$g^\phi \equiv 1 \pmod{p},$$

$$g^\phi \equiv 1 \pmod{q},$$

$$g^\phi \equiv 1 \pmod{n}.$$

Let $R = \gcd(g^\phi - 1, n)$, from the equations above, it is clear that $R = p$ or q or n. Someone can repeat to select suitable g, and compute $R = \gcd(g^\phi - 1, n)$ until $R \neq n$. This time $R=p$ or q. $\square$

So, to avoid the possible collusion, the center must limit the number of groups to at most s which should be selected as quite less than t. Without enough number of the simultaneous polynomial equations, computing a multiple of L is considered to be impossible. In the system, the bigger the constant t is set, the more groups the center can manage. However, the amount of computation for registration and participation will increase. In the system of [KMI1 89], [KMI2 89], [OOK 90], there is not such upper bound in the number of groups the center can manage.

We have seen that in all the protocols about membership proof, user U should use $PM(\mathbf{GN}, u)$ to execute protocols. Even U proves his membership in only one group, he should use secret witness like $u^{\frac{1}{c_i}} \bmod n$ for some group GN_i. However, from $PM(\mathbf{GN}, u)$ or $u^{\frac{1}{c_i}} \bmod n$, whoever can extract secret number u and can prove U's identity. This is why it is dangerous to the user who gives others his secret witnesses to let them perform multi-purpose proofs.

We can note $g^u \bmod n'$ is independent of H. After user U proves his identity, the verifier will know U's H. However, since $u^2 \equiv H \pmod{n}$, the verifier can not compute u, and so he can not compute $y''\bmod n'$. Therefore, when user U gives the same verifier $g^u \bmod n'$ for membership proof, the verifier can not track U's identity. Inversely, when $g^u \bmod n'$ is given, since it is impossible for the verifier to compute u directly from $g^u \bmod n'$ because of the discrete logarithm problem, the verifier also can not compute H to track U's identity. It seems no possibility to compute U's secret u even with $g^u \bmod n'$ and other public information.

7 Conclusion

We have set up a model and constituted a system which allows a user to perform three types of proofs which are identity proof: membership proof without revealing the identity; and identity & membership proof: needing to remember only one piece of secret data. Not having to remember many different secrets to perform multiple protocols increases efficiency.

In this paper, a method called "secret hiding participation" is also proposed. It isolates the secrets of the users and of the group administrator. This makes that even the group administrator cannot pretend to be the users who belong to his group. This technique is believed to be useful for other applications. The secret data for multi-purpose proofs can also be considered to be applied for constructing a signature scheme, a key sharing scheme and etc.. The detailed security analysis for the whole system is still left to further study.

Acknowledgement The authors thank an anonymous reviewer for his comments which were helpful to improve a previous version of this paper.

References

[BD 89]   M.V.D.Burmester, Y.G.Desmedt "Remarks on soundness of proofs ", Electronics Letters, Vol. 25, No. 22, pp.1509–1511, 1989.

[C1 85]   D.Chaum, "Security without identification: Transaction systems to make big brother obsolete", Comm. of the ACM, Vol. 24, No. 10, pp.1030–1044, 1985.

[C2 86]   D.Chaum, "Showing credentials without identification:
Signatures transferred between unconditionally unlinkable pseudonyms", Advances in Cryptology, Eurocrypt'85, Springer-Verlag, pp.241–244, 1986.

[EH 90]   J.H.Evertse, E.van Heyst "Which new RSA signatures can be computed from some given RSA signatures ?", Advances in Cryptology-Eurocrypt'90, Springer-Verlag, pp.83–97, 1991.

[FFS 88]   U.Feige, A.Fiat, A.Shamir, "Zero knowledge proofs of identity", Journal of Cryptology, Vol.1 pp.77–94, 1988.

[FS 86]   A.Fiat, A.Shamir, "How to prove yourself : practical solutions to identification and signature problems", Advances in Cryptology-CRYPTO'86, Springer-Verlag, pp.186–194, 1987.

[K 89]   K.Koyama, "Demonstrating membership of a group using the Shizuya-Koyama-Itoh(SKI) protocol", Proc. SCIS'89, 1989.

[KMI1 89]   M.Kurosaki, T.Matsumoto, H.Imai,
"Simple Methods for Multipurpose Certification", Proc. SCIS'89, 1989. (in Japanese)

[KMI2 89]   M.Kurosaki, T.Matsumoto, H.Imai,
            "Methods to individually prove each membership for several groups", Tech. Rep. of
            IEICE, ISEC89-18, Japan, 1989. (in Japanese)

[OO 88]   K.Ohta, T.Okamoto, "A modification of the Fiat-Shamir scheme", Advances in
            Cryptology-CRYPTO'88, Springer-Verlag, pp.232–243, 1989.

[OOK 90]   K.Ohta, T.Okamoto, K.Koyama "Membership authentication for hierarchy multi-
            groups using the extended Fiat-Shamir Scheme", Advances in Cryptology-
            Eurocrypt'90, Springer-Verlag, pp.446–457, 1991.

[Rom 90]   J.Rompel, "One-way function are necessary and sufficient for signatures," Proc. 22nd
            STOC, pp.387–394, 1990.

[SMI 90]   C.Shu, T.Matsumoto, H.Imai, "How to simultaneously prove yourself and your
            membership", Tech. Rep. of IEICE, ISEC90-11, Japan, 1990.

Formal Verification of Probabilistic Properties in Cryptographic Protocols
(Extended Abstract)

Marie-Jeanne Toussaint
Université de Liège, Institut Montefiore, B28, Sart Tilman
B 4000 Liège (Belgium)
toussain@montefiore.ulg.ac.be

Abstract

We introduce an original method to verify probabilistic properties in cryptographic protocols. This method uses the representation of participants' knowledge
that we presented at CRYPTO'91. The modelization is based on the assumption
that the underlying cryptographic system is perfect and is an extension of the "Hidden Automorphism Model" introduced by Merritt.

1   Introduction

So far, a lot of effort has been devoted to the study of cryptographic systems. The
goal has been to develop cryptographic systems that are as secure as possible. But
this is not sufficient: the logic of the protocol has also to be correct. Like Merritt in
[Mer83], we study the problem of reasoning about cryptographic protocols, assuming the
underlying cryptographic system to be perfect. In [Mer83] and [MW85], a cryptosystem
is represented by an algebra (called the *crypto-algebra*) and its perfection is modeled by
the fact that the crypto-algebra is isomorphic to the free algebra of the same type.

In [TW91], we introduced a new representation of the participants' knowledge. The
messages and keys whose a participant knows the meaning are represented in his state
of knowledge by constants of the free-algebra whereas the ones whose he does not know
the meaning are represented by variables defined on the free-algebra. At CRYPTO'91
([Tou91a]), we presented a refinement of this representation by modeling the complete
knowledge that a participant is able to obtain (by computations and inferences).

In this paper, we present a method using this representation of participants' knowledge for proving probabilistic properties of cryptographic protocols. Some of these properties were already intuitively verified on an example in [TW91]; here we formally develop
and generalize the ideas of [TW91]. The possible global states of a protocol are represented in a tree called '*the protocol execution tree*'. The protocol execution tree has often
a finite length but an infinite width. It can be reduced to be finitely represented: the
resulting tree is simply called '*the reduced protocol execution tree*'. The branches of this

tree represent the different finite choices of the participants and are balanced by the corresponding probabilities of choice. The nodes of the reduced protocol execution tree are characterized by the values of the 'meaningful' variables of the protocol.

To model participants' knowledge about the protocol, we have to explicitly link participants' knowledge about the cryptosystem and the protocol. For each participant, we define a function called 'possible-instantiation' which associates with each cryptographic variable the set of values representing the possible instantiations of that variable given the participant's state of knowledge.

We define then a possibility relation[1] between the states of the protocol execution tree. Briefly, a protocol state is *possible* for a participant in a given protocol state if the instantiations of the variables known by the participant are identical in the two states and the participant's knowledge has the same representation in the two states. This possibility relation is an equivalence relation. But, in fact, in any current state of a protocol, a participant makes no distinction between two nodes 1 and 2 of the reduced tree if he estimates that these two nodes are equally likely i.e. that the probability that the current state belongs to node 1 is the same as the probability that this state belongs to node 2. We define the notion of 'probabilistically indistinguishable' nodes: two nodes 1 and 2 of the reduced tree are *probabilistically indistinguishable* for a participant in a given protocol state CS if the probabilistic measures of the sets of the protocol states of these two nodes possible in CS are identical. Using this model, we are able to verify probabilistic properties of cryptographic protocols.

The main originalities of our method are

- its generality (we can apply this method to protocols using public or private key cryptography, preserving the secret of data, to authentication protocols, signature schemes,... (see [Tou91b])),

- the ability to verify probabilistic properties of protocols,

- and the use of a model based on the assumption of a 'perfect' cryptosystem.

This paper is organized as follows. In Section 2, we present our representation of knowledge of the participants in a cryptographic protocol. In Section 3, we model the protocol by the protocol execution tree and we construct the reduced tree. Then, in Section 4, we define the function *possible − instantiation* which links the representation of knowledge of Section 2 and the protocol model of Section 3. Afterwards (Section 5), we define possibility relations between protocol states and the notion of 'probabilistically indistinguishable nodes'.

---

[1] i.e. a relation which for each state and each participant gives the states that this participant considers as possible from that state.

2 Modeling States of Knowledge

2.1 Modeling the Cryptosystem

A cryptosystem is seen as an algebraic system called the *crypto-algebra* C. Here, M denotes the set of (clear or enciphered) messages and K the set of keys. The operators of the crypto algebra C are usually the enciphering function D and the deciphering function D which are linked by some relations.

Following [MW85], we consider an idealization of the cryptosystem. It is the quotient by the relations between E and D of the free algebra of the same type as the crypto-algebra. This free-algebra is denoted here by F and its operators by e and d. The 'perfection' of the cryptosystem is modeled by assuming that the crypto-algebra is isomorphic to the free-algebra: the isomorphism between the free and crypto-algebras is denoted by 'φ'.

2.2 Modeling Participants' Knowledge

As in [TW91], we assume that a state of knowledge about the cryptosystem is a subset of $F \times C$ which can be partitioned in three special finite sets F, V and SV.

1. F (for *Fixed*) is formed by pairs (a, b) which define a one to one mapping from a subset of generators of F to a subset of generators of C. That corresponds to generators of the crypto-algebra that the participant has seen or that he knew at the start of the protocol and that he can label by a fixed element of the free-algebra. The set F will be described by specifying the free-algebra component of the pairs.

2. V (for *Variables*) is formed by pairs (x, y) where x is a fixed generator of the free-algebra whereas y ranges over a subset of the generators of the crypto-algebra. The pairs of this type correspond to generators of the crypto-algebra that the participant has not seen but the existence of which he is aware of. A pair (x, y) of V will be represented by a variable denoted by a tilded symbol, $\tilde{x}$, ranging over the image under the isomorphism φ^{-1} of the domain of y.

3. SV (for *Semi-Variables*) is formed by pairs (z, a) where a is a fixed element of the crypto-algebra and z belongs to a finite subset of $Cl(F \cup V) \setminus (Cl(F) \cup V)$ where $Cl(X)$ (for any subset X of F) denotes the closure of X under the enciphering and deciphering operators. These pairs correspond to elements that the participant has seen but is unable to label. A pair (z, a) of SV is represented by a variable (denoted z^*) on the free-algebra with an inclusion constraint on this variable in

the corresponding subset of $Cl(F \cup V) \setminus (Cl(F) \cup V)$ and occasionally inequality relations between variables.

The participants make some computations on their state of knowledge. We overvalue them by assuming that each participant is able to compute the closure $Cl(F \cup SV)$: this closure is called the *seen fraction* of the participant. The participants are also able to draw some inferences from their computations. All these inferences are modeled in [Tou91a] by unifications (i.e. restrictions of the domain of the variables so that two expressions become equal) and contra-unifications (i.e. eliminations of some values from the domain of the variables to prevent two expressions from becoming equal).

Example 2.1 In this paper, we consider as example an *oblivious transfer protocol* described in [Mer83].

Definition 2.1 *An oblivious transfer protocol has to have two properties (as stated in [BPT85]) :*

1. *The receiver B has a chance 1/2 of obtaining a message m.*
2. *The probability that A correctly guesses whether or not B has obtained m, is 1/2.*

Our example can be described as follows.

1. A(lice) chooses at random two different keys k_1 and k_2 and sends the two messages $ek1m = E(k_1, m)$ and $ek2m = E(k_2, m)$ to B(ob) (m denotes the information A wishes to transfer to B).

2. B chooses y at random in $\{ek1m, ek2m)\}$ and a key k_B; he sends $ekBy = E(k_B, y)$ to A.

3. A selects k_A at random in $\{k_1, k_2\}$ and sends $dkAekBy = D(k_A, E(k_B, y))$ to B; B computes $D(k_B, dkAekBy)$ to see if it is meaningful.

The above protocol is only valid if the cryptosystem is a secret key system such that the function D is left inverse of E. Moreover, the different enciphering (or deciphering) transformations have to be commutative.

Let us consider the states of knowledge of A and B at different moments of the execution.

- Just before the choice of k_A, A's knowledge state is $K(A, 2) = F(A, 2) \cup V(A, 2) \cup SV(A, 2)$:

$$F(A, 2) = \{m, k_1, k_2\}; \quad V(A, 2) = \{\widetilde{k_B}\}; \quad SV(A, 2) = \{ekBy^*\};$$

where $\widetilde{k_B} \in \mathcal{K}$; $ekBy^* \in \{e(\widetilde{k_B}, e(k_1, m)), e(\widetilde{k_B}, e(k_2, m))\}$

($K(X, n)$ denotes the knowledge state of participant X at Step n and $F(X, n)$, $V(X, n)$ and $SV(X, n)$ the sets composing this state of knowledge).

- At the end of the protocol, B's knowledge state is

  - either, in case $(y = ek1m$ and $k_A = k_1)$ or $(y = ek2m$ and $k_A = k_2)$,

    $$F(B, 3) = \{k_B, m\}, \quad V(B, 3) = \{\widetilde{k_1}, \widetilde{k_2}\}, \quad SV(B, 3) = \{ek1m^*, ek2m^*\}$$

    where $\widetilde{k_1}, \widetilde{k_2} \in \mathcal{K}$ $: \widetilde{k_1} \neq \widetilde{k_2}$; $ek1m^* = e(\widetilde{k_1}, \widetilde{m})$; $ek2m^* = e(\widetilde{k_2}, \widetilde{m})$

  - or, in case $y = ek1m$ (resp. $y = ek2m$) and $k_A = k_2$ (resp. $k_A = k_2$),

    $$F(B, 3) = \{k_B\}; \quad V(B, 3) = \{\widetilde{m}, \widetilde{k_1}, \widetilde{k_2}\};$$

    $$SV(B, 3) = \{ek1m^*, ek2m^*, dkAekBy^*\}$$

    where $\widetilde{m} \in \mathcal{M}$; $\widetilde{k_1}, \widetilde{k_2} \in \mathcal{K} : \widetilde{k_1} \neq \widetilde{k_2}$; $ek1m^* = e(\widetilde{k_1}, \widetilde{m})$; $ek2m^* = e(\widetilde{k_2}, \widetilde{m})$; $dkAekBy^* = d(\widetilde{k_2}, e(k_B, e(\widetilde{k_1}, \widetilde{m})))$(resp. $= d(\widetilde{k_1}, e(k_B, e(\widetilde{k_2}, \widetilde{m})))$.

    We see that the participants' knowledge about the cryptosystem depends on the current state of the protocol (i.e. here on the current values of y and k_A).

3 Modeling the Protocol

We want to use the representation of knowledge introduced in Section 2 to reason about the correctness of the protocol, but we first need a complete representation of what can happen when the protocol is executed.

3.1 The Protocol Execution Tree

A protocol is usually seen as a (finite) set of communicating processes. Here the communicating processes are called the *participants*. Intuitively, a cryptographic protocol is a protocol where some of the messages being exchanged are computed by using the enciphering and deciphering transformations of a cryptosystem (see [Bie89]). We give a more precise definition in terms of the crypto-algebra representing the cryptosystem being used.

We call *cryptographic variable* a protocol variable whose domain is a subset of the crypto-algebra and *cryptographic constant* a protocol constant which corresponds to a fixed element of the crypto-algebra: this element is thus identical for every execution of the protocol and is assumed to be known by each participant. A *cryptographic protocol* is then a protocol where the messages being exchanged correspond to cryptographic variables or cryptographic constants.

A cryptographic protocol is modeled by the *protocol execution tree*: it is the tree of the possible global states of the protocol. This tree can be seen as the unwinding of the partially synchronized product (defined for example in Chapter 4 of [TGH+89]) of states machines defining the communicating processes (see [MCF87] and [Var89]).

Example 3.1 The protocol execution tree of our oblivious transfer example is

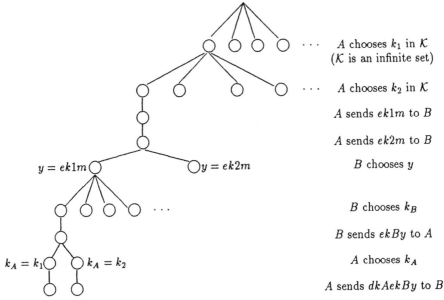

A chooses k_1 in $\mathcal{K}$
($\mathcal{K}$ is an infinite set)

A chooses k_2 in $\mathcal{K}$

A sends $ek1m$ to B

A sends $ek2m$ to B

B chooses y

B chooses k_B

B sends $ekBy$ to A

A chooses k_A

A sends $dkAekBy$ to B

where, for the clarity of the drawing, we have only ramified the leftmost node of each level of the tree. ∎

3.2 Cryptographic Protocol States

A cryptographic protocol state is the instantiation of some cryptographic variables.

Definition 3.1 *A cryptographic protocol state for a protocol with a set CV of cryptographic variables is a function CS from a subset of CV to the crypto-algebra.*

We can number the different actions of a protocol; this corresponds to numbering the levels of the protocol execution tree. We thus obtain a very simple discrete notion of time. For simplicity, we assume that the set of the cryptographic variables instantiated at time t is the identical for every execution of the protocol: we denote this set by 'CV_t'. (We will discuss this hypothesis in Remark 3.1). A cryptographic protocol state at time t ($t \leq 0$) is then a function from CV_t to C. We assume also that each cryptographic variable is instantiated only once during each run of the protocol. It seems indeed natural that distinct elements of the crypto algebra be represented by distinct cryptographic variables. (If the description of a protocol contains loops, we can consider that each execution of the loops defines new cryptographic variables denoted in a generic way.) We have thus $CV_{t'} \supseteq CV_t$ if $t' > t$.

With each protocol state CS and participant, we associate the knowledge state of that participant in that state. This state of knowledge is represented as explained in Section 2.

3.2.1 Characterization of a Cryptographic Protocol State

We could represent a cryptographic protocol state by the corresponding pairs "(variable, image of the variable in the crypto-algebra)" i.e. pairs of the form "(variable, sequence of bits)". But we would rather to emphasize the relation between the instantiations of the different cryptographic variables in a protocol state.

Basic Cryptographic Variables
In the description of a cryptographic protocol, some variables can be expressed in terms of other ones. In our oblivious transfer example, $ek1m, ek2m, y, ekBy, k_A$, and $dkAekBy$ can be expressed in terms of k_1, k_2, m, and k_B. We would like to determine the minimum number of cryptographic variables necessary for expressing the instantiations of every other variable in terms of these variables, the cryptographic constants, and the enciphering and deciphering operators. For this, we need some additional definitions.

Definition 3.2 *Let CS be a cryptographic protocol state and $cc_1, \ldots, cc_n$ the values of the cryptographic constants.*

- *The cryptographic variables $cv_1, \ldots cv_m$ (for any $m > 1$) belonging to the domain of CS are dependent under CS if $\exists i \in [1, m]$ such that*

  $$CS(cv_i) \in Cl(\{cc_1, \ldots cc_n, CS(cv_1), \ldots, CS(cv_{i-1}), CS(cv_{i+1}), \ldots, CS(cv_m)\})$$

  where $Cl(X)$ (for any subset X of the crypto-algebra) denotes the closure of X under the enciphering and deciphering operations of the crypto-algebra.
  If it is not the case $cv_1, \ldots, cv_m$ are said to be independent under CS.

- A basis in CS *is a maximal set of cryptographic variables that are independent under CS and the instantiations of which are primitive elements of the crypto-algebra C.*

There can be several bases in a protocol state CS. We choose the basis in CS such that the sum of the times of instantiation of all its cryptographic variables is minimal. This basis is called *the fundamental basis in CS*. (Under our hypotheses, the fundamental basis can be assumed identical in all the protocol states of a given level of the protocol execution tree.) In order to differentiate the cryptographic variables belonging to the fundamental basis from the others, we call the former ones *'basic'* and the others *'nonbasic'*. For example, at the end of the oblivious transfer example, k_1, k_2, k_B are basic and $ek1m, ek2m, y, ekBy, k_A, dkAekBy$ are non basic.

The instantiations of the nonbasic variables in a protocol state CS can be expressed in terms of the cryptographic constants, other cryptographic variables and the enciphering and deciphering operators. These expressions are called *relative values* of the corresponding variables in the protocol state CS. A basic cryptographic variable has an infinite number of possible values whereas a nonbasic cryptographic variable has a finite number of possible relative values. The *relative domain* of a cryptographic variable cv in a protocol state CS is the domain of cv in CS (i.e. the set of all the possible values of cv in CS) whose elements are represented by their relative value when cv is nonbasic. We define a function *'relative-domain(CS, .)'* which associates with every cryptographic variable its relative domain in CS. A protocol state CS can be characterized by the relative domain of the cryptographic variables i.e. by the function $relative - domain(CS, .)$.

3.3 Reduced Protocol Execution Tree

A protocol execution tree has a finite length but it can be very long (a level for each action of the protocol). The nodes of the protocol execution tree where there is no nondeterminism can be merged with their parent. We can also group nodes which correspond to infinite choices.

Example 3.2 We obtain in the case of our oblivious transfer example.

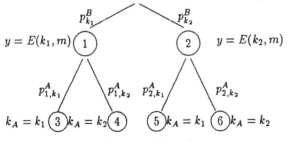

where we have assigned probabilities to each branch.

We have thus merged the states which differ only by the instantiations of the basic cryptographic variables.

Definition 3.3 *A* node of the reduced protocol execution tree *is the set of all the cryptographic protocol states CS which have identical functions relative* − *domain*(*CS*, .).

When no confusion is possible, we call the nodes of the reduced protocol tree the '*protocol nodes*' whereas the nodes of the (nonreduced) protocol execution tree are called the '*protocol states*'. *A* *protocol node at time t* is the set of all the corresponding protocol states at time *t*.

Remark 3.1 The hypothesis that the cryptographic variables are always instantiated at the same time of the execution of a protocol is not really necessary but simplifies the notations. If we do not assume that hypothesis, all the protocol states at a given time have not the same fundamental basis and we can only merge protocol states with the same fundamental basis. The number of protocol nodes is then larger and not necessarily finite. ∎

4 Participants' Knowledge about the Current Protocol State

In Section 2, we have represented the participants' knowledge about the cryptosystem being used. We would like to deduce from that the participants' knowledge about the protocol. We need to link the states of knowledge and the protocol. For each protocol state *CS* and participant *A*, we define a function which associates to each cryptographic variable *cv* the set of values that *A* considers as possible for the instantiation of *cv*. This function is denoted '*possible*−*instantiation*(*A*, *CS*)'. If we adopt a representation similar to the one of Section 2, we remark that the function *possible* − *instantiation*(*A*, *CS*) associates with each cryptographic variable a set of values which can be represented by a finite subset of $Cl(K(A, CS))$ where $K(A, CS)$ denotes *A*'s knowledge state associated with *CS*. *possible* − *instantiation*(*A*, *CS*) can thus be considered as a function from the domain of *CS* to $2^{Cl(K(A,CS))}$.

In any protocol state *CS* and for each cryptographic variable *cv* and participant *A*, the function *possible* − *instantiation*(*A*, *CS*) associates to *cv* a subset of its relative domain whose elements are represented by expressions in terms of the free-algebra component of elements of $F(A, CS)$, the variables of $V(A, CS)$ and $SV(A, CS)$ and the free-algebra operators. Note that if cv is basic, its image under $possible - instantiation(A, CS)$ is a singleton containing an element of $F(A, CS)$ or a variable of $V(A, CS)$ depending on whether A knows or not the instantiation of cv.

Example 4.1 In Example 2.1, we have built A's knowledge state at the end of Step 2 of our oblivious transfer example; the corresponding function $possible - instantiation$ is defined for every protocol state CS at the end of Step 2 by

$$
\begin{aligned}
possible - instantiation(A, CS)(m)(\text{resp.}(k_1), (k_2)) &= \{m\}(\text{resp.}\{k_1\}, \{k_2\}) \\
possible - instantiation(A, CS)(ek1m)(\text{resp.}(ek2m)) &= \{e(k_1, m)\}(\text{resp.}\{e(k_2, m)\}) \\
possible - instantiation(A, CS)(y) &= \{e(k_1, m), e(k_2, m)\} \\
possible - instantiation(A, CS)(k_B)(\text{resp.}(ekBy)) &= \{\widetilde{k_B}\}(\text{resp.}\{ekBy^*\}).
\end{aligned}
$$

Remark 4.1 For the syntax that we have defined in Chapter 4 of [Tou91b], we know that for every protocol state CS and participant A, each singleton containing an element v of $V(A, CS)$ (resp. of $SV(A, CS)$) is the image of one and only one (resp. of at least one) basic (resp. non-basic) cryptographic variable under the function $possible - instantiation(A, CS)$. In [Tou91b], we have also proved that, in our model, participants' knowledge about the cryptosystem and about the protocol is monotonously increasing during the execution of a protocol.

The states of a protocol node at time t differ only by the values of the basic cryptographic variables. The specific values in the crypto-algebra are of no interest for our analysis. It is thus natural that for any participant A, A's knowledge states '$K(A, CS)$', as well as the functions '$possible - instantiation(A, CS)$', have identical representations for all the protocol states of a protocol node CN at time t. These representations are associated to the node CN at time t and denoted respectively by '$K(A, CN, t)$' and '$possible - instantiation(A, CN, t)$'.

5 Defining Possibility Relations

At a given time of the execution of the protocol, a participant analyzes the protocol execution tree to reject all the protocol states which are not compatible with his knowledge and to discover (if it is possible) the current protocol state. The problem is to know which protocol states are possible for a participant in the current protocol state.

A participant A is assumed to be able to build the protocol execution tree and to know the representation of his knowledge in each protocol state i.e. the representation of $K(A, *)$ and *possible - instantiation*$(A, *)$ associated with all the protocol states. A protocol state CS' will be possible for a participant A in a given protocol state CS if the values of the basic cryptographic variables in CS' are possible for A in CS and if A's knowledge is represented in the same way in CS and CS'. We formalize that in the following definition where we use the usual notations.

Definition 5.1 *Let CS (resp. CS') be a cryptographic protocol state at time t (resp. t' ($t' \geq t$)).*

If $t' = t$, CS' is said possible *for A in CS if*

1. for every basic cryptographic variable $cv \in CV_t$

$$(cv, CS'(cv)) \in possible - instantiation(A, CS)(cv); \qquad (1)$$

2. $K(A, CS)$ and $K(A, CS')$, as well as $possible-instantiation(A, CS)$ and $possible-instantiation(A, CS')$, have identical representations.

When $t' > t$, the protocol state CS' is said possible *for A in CS if there is a protocol state CS'' at time t such that CS' is descendant of CS'' and CS'' is possible for A in CS.*

We will prove in the complete paper that the relation of possibility considered between cryptographic protocol states of a given time is an equivalence relation.

Afterwards, we define a possibility relation of a protocol node for a participant in a protocol state.

Definition 5.2 *A cryptographic protocol node is* possible *for a participant A in a protocol state CS if at least one protocol state belonging to this node is possible for A in CS.*

We can directly define the possibility of a node without considering Definition 5.1.

Proposition 5.1 *Let CS be a cryptographic protocol state at time t and CN a cryptographic protocol node.*
CN is said possible *for A in CS if and only if one of the two following conditions is satisfied*

- *if CN can be considered at time t and $K(A, CS)$ and $K(A, CN, t)$, as well as 'possible $-$ instantiation(A, CS)' and ' possible $-$ instantiation(A, CN, t)', have identical representations;*

- *or if there is a cryptographic protocol node CN' at time t such that CN is descendant of CN' and CN' is possible for A in CS.*

In [Tou91b], we define also a possibility relation between two protocol nodes: this relation is an equivalence one.

The protocol states of a node at a given time vary only on the instantiations of the basic cryptographic variables and are thus in infinite number. We could like to define on the set of protocol states a uniform probabilistic measure in order to reflect the fact that we view all instantiations of the basic variables as equally possible. If the set of these instantiations was finite, defining a uniform measure would be immediate. But it is countably infinite and such a uniform measure does not exist. In [TW91] and [Tou91b], we define a measure that approximates uniformity on infinite sets. We will give it in the complete paper. Here we assume that the possible protocol states are distributed under this probabilistic measure (when they are in infinite number). We can then define the possibility of a protocol node with a probability as follows.

Definition 5.3
A cryptographic protocol node CN at time t is possible with a probability α ($\alpha \in [0,1]$) for a user A in a protocol state CS

- *if CN is possible for A in CS and*

- *the set of the protocol states of CN at time t and possible for A in CS has a probabilistic measure equal to α in the set of all the protocol states at time t possible for A in CS.*

    Two cryptographic protocol nodes at time t are probabilistically indistinguishable for A in a protocol state CS if they are possible with the same probability.

Example 5.1 If we analyze the protocol execution tree of our oblivious transfer example (given in Subsection 3.3), we see that just before the choice of the key k_A, (i.e. at the end of Step 2), nodes 1 and 2 are possible for A in any protocol state at the end of Step 2. In the complete paper, we will apply the measure that approximates uniformity and we will prove that nodes 1 and 2 are possible with probability $1/2$ for A in any protocol state at the end of Step 2. A is thus unable to distinguish between nodes 1 and 2 and we have

$$p_{1,k_1}^A = p_{2,k_1}^A = p_{k_1}^A \quad \text{and} \quad p_{1,k_2}^A = p_{2,k_2}^A = p_{k_2}^A;$$

the probabilities that B obtains or not m at the end of the protocol are then the same if and only if

$$(p_{k_1}^B - p_{k_2}^B)(p_{k_1}^A - p_{k_2}^A) = 0.$$

Thus no participant can cheat the other since it is sufficient for one participant to choose his probabilities evenly to guarantee a fair outcome of the protocol. ∎

6 Conclusions

We have presented a method for formally proving probabilistic properties of cryptographic protocols. We have applied this method to verify the wanted probabilistic properties of an oblivious transfer protocol. But this proof is only valid if each participant sends only allowed messages because in the protocol execution tree only the actions specified by the protocol are represented. However, if we analyze the considered oblivious transfer protocol, we see that the probabilistic properties are no longer verified if the participants send unallowed messages.

In [Tou91b] and in a paper in preparation, we extend the method to consider the cheatings by sending unallowed messages or by intercepting or replacing some messages. However, in this extended abstract, we have preferred to emphasize the probabilistic aspect because this aspect (I think) had never been considered in the formal existing verification methods.

Acknowledgements

I would like to thank M. Merritt., J.-J. Quisquater, and P. Wolper for many helpful discussions about this work. I also address my thanks to F.N.R.S. and S.P.P.S. (Belgium) for financial support.

References

[BAN89]   M. Burrows, M. Abadi, and R. Needham. A Logic of Authentication. Technical Report 39, Digital — Systems Research Center (SRC), 1989.

[Bie89]   P. Bieber. *Aspects Epistémiques des Protocoles Cryptographiques*. PhD thesis, Université Paul-Sabatier de Toulouse (Sciences), October 1989.

[BM84]   M. Blum and S. Micali. How to Generate Cryptographically Strong Sequences of Pseudo-Random Bits. *SIAM Journal on Computing*, 13(4):850–864, 1984.

[BPT85]   R. Berger, R. Peralta, and T. Tedrick. A Provably Secure Oblivious Transfer Protocol. In T. Beth, N. Cot, and I. Ingemarsson, editors, *Lecture Notes in*

Computer Science. Advances in Cryptology, Proceedings of EUROCRYPT'84 #209, pages 379–386. Springer-Verlag, 1985.

[GMR89]   S. Goldwasser, S. Micali, and C. Rackoff. The Knowledge Complexity of Interactive Proof-Systems. *SIAM Journal on Computing*, 18(1):186–208, 1989.

[GNY90]   L. Gong, R. Needham, and R. Yahalom. Reasoning about Belief in Cryptographic Protocols. In *Proceedings of the 1990 IEEE Computer Society Symposium on Research in Security and Privacy*, pages 234–248. IEEE Computer Society Press, 1990.

[Kem89]   R. A. Kemmerer. Analyzing Encryption Protocols Using Formal Verification Techniques. *IEEE Journal on Selected Areas in Communications*, 7(4):448–457, 1989.

[MCF87]   J. K. Millen, S. C. Clark, and S. B. Freedman. The Interrogator: Protocol Security Analysis. *IEEE Transactions on Software Engineering*, 13(2):274–288, 1987.

[Mea90]   C. Meadows. Representing Partial Knowledge in an Algebraic Security Model. In *Proceedings of the Computer Security Foundations Workshop III*, pages 23–31. IEEE Computer Society Press, 1990.

[Mer83]   M. J. Merritt. *Cryptographic Protocols*. PhD thesis, Georgia Institute of Technology, 1983.

[MW85]   M. Merritt and P. Wolper. States of Knowledge in Cryptographic Protocols (extended abstract). Unpublished Manuscript, 1985.

[Syv91]   P. Syverson. The Use of Logic in the Analysis of Cryptographic Protocols. In *Proceedings of the 1991 IEEE Symposium on Research in Security and Privacy*, pages 156–170. IEEE Computer Society Press, 1991.

[TGH+89]   A. Thayse, P. Gribomont, G. Hulin, A. Pirotte, D. Roelants, D. Snyers, M. Vauclair, P. Gochet, P. Wolper, E. Grégoire, and P. Delsarte. *Approche logique de l'intelligence artificielle - 2. De la logique modale à la logique des bases de données*. Dunod, Bordas, Paris, 1989.

[Tou89]   M-J. Toussaint. Reasoning about Probabilistic Properties of Cryptographic Protocols (extended abstract). Abstract of the talk at the F.N.R.S. day on Computer Security, May 1989.

[Tou91a]   M-J. Toussaint. Deriving the Complete Knowledge of Participants in Cryptographic Protocols (Extended Abstract). in the proceedings of CRYPTO'91, August 1991.

[Tou91b] M.-J. Toussaint. *Verification of Cryptographic Protocols*. PhD thesis, Université de Liège (Belgium), 1991.

[TW91] M-J. Toussaint and P. Wolper. Reasoning about Cryptographic Protocols (Extended Abstract). In Joan Feigenbaum and Michael Merritt, editors, *Distributed Computing and Cryptography (October 1989)*, pages 245–262. DIMACS - Series in Discrete Mathematics and Theoretical Computer Science (AMS - ACM), 1991. Volume 2.

[Var89] V. Varadharajan. Verification of Network Security Protocols. *Computers & Security*, 8(8):693–708, 1989.

Cryptography and Machine Learning

Ronald L. Rivest*
Laboratory for Computer Science
Massachusetts Institute of Technology
Cambridge, MA 02139

Abstract

This paper gives a survey of the relationship between the fields of cryptography and machine learning, with an emphasis on how each field has contributed ideas and techniques to the other. Some suggested directions for future cross-fertilization are also proposed.

1 Introduction

The field of computer science blossomed in the 1940's and 50's, following some theoretical developments of the 1930's. From the beginning, both cryptography and machine learning were intimately associated with this new technology. Cryptography played a major role in the course of World War II, and some of the first working computers were dedicated to cryptanalytic tasks. And the possibility that computers could "learn" to perform tasks, such as playing checkers, that are challenging to humans was actively explored in the 50's by Turing [46], Samuel [39], and others. In this note we examine the relationship between the fields of cryptography and machine learning, emphasizing the cross-fertilization of ideas, both realized and potential.

The reader unfamiliar with either of these fields may wish to consult some of the excellent surveys and texts available for background reading. In the area of cryptography, there is the classic historical study of Kahn [20], the survey papers of Diffie and Hellman [11] and Rivest [37], and Simmons [44], as well as the texts by Brassard [8], Denning [10], and Davies and Price [9], among others. The CRYPTO and EUROCRYPT conference proceedings (published by Springer) are also extremely valuable sources. In the area of machine learning, there are standard collections of papers [29, 30, 23] for "AI" style machine learning, the seminal paper of Valiant [47] for the "computational learning theory" approach, the COLT conference proceedings (published by Morgan Kaufmann) for additional material of a theoretical nature, and the NIPS conference proceedings (also

---

*Supported by NSF grant CCR-8914428, ARO grant N00014-89-J-1988, and the Siemens Corporation. email address: `rivest@theory.lcs.mit.edu`

published by Morgan Kaufmann) for many interesting papers. The ACM STOC and the IEEE FOCS conference proceedings also contain many key theoretical papers from both areas. The Ph.D. thesis of Kearns [21] is one of the first major works to explore the relationship between cryptography and machine learning, and is also an excellent introduction to many of the key concepts and results.

2 Initial Comparison

Machine learning and cryptanalysis can be viewed as "sister fields," since they share many of the same notions and concerns. In a typical cryptanalytic situation, the cryptanalyst wishes to "break" some cryptosystem. Typically this means he wishes to find the secret key used by the users of the cryptosystem, where the general system is already known. The decryption function thus comes from a known family of such functions (indexed by the key), and the goal of the cryptanalyst is to exactly identify which such function is being used. He may typically have available a large quantity of matching ciphertext and plaintext to use in his analysis. This problem can also be described as the problem of "learning an unknown function" (that is, the decryption function) from examples of its input/output behavior and prior knowledge about the class of possible functions.

Valiant [47] notes that good cryptography can therefore provide examples of classes of functions that are hard to learn. Specifically, he references the work of Goldreich, Goldwasser, and Micali [14], who demonstrate (under the assumption that one-way functions exist) how to construct a family of "pseudo-random" functions $F_k : \{0,1\}^k \to \{0,1\}^k$ for each $k \geq 0$ such that (i) each function $f_i \in F_k$ is described by a k-bit index i, (ii) there is a polynomial-time algorithm that, on input i and x, computes $f_i(x)$ (so that each function in F_k is computable by a polynomial-size boolean circuit), and (iii) no probabilistic polynomial-time algorithm can distinguish functions drawn at random from F_k from functions drawn at random from the set of all functions from $\{0,1\}^k$ to $\{0,1\}^k$, even if the algorithm can dynamically ask for and receive polynomially many evaluations of the unknown function at arguments of its choice. (It is interesting to note that Section 4 of Goldreich et al. [14] makes an explicit analogy with the problem of "learning physics" from experiments, and notes that their results imply that some such learning problems can be very hard.)

We now turn to a brief comparison of terminology and concepts, drawing some natural correspondences, some of which have already been illustrated in above example.

Secret Keys and Target Functions

The notion of "secret key" in cryptography corresponds to the notion of "target function" in machine learning theory, and more generally the notion of "key space" in cryptography corresponds to the notion of the "class of possible target functions." For cryptographic (encryption) purposes, these functions must also be efficiently invertible, while no such requirement is assumed in a typical machine learning context. There is another aspect of this correspondence in which the fields differ: while in cryptography it is common to assume that the size of the unknown key is known to the cryptanalyst (this usually falls under the general assumption that "the general system is known"), there is much interesting research in machine learning theory that assumes that the complexity (size) of the target hypothesis is *not* known in advance. A simple example of this phenomenon is the problem of fitting a polynomial to a set of data points (in the presence of noise), where the degree of the true polynomial is not known in advance. Some method of trading off "complexity of hypothesis" for "fit to the data," such as Rissanen's Minimum Description Length Principle [36], must be employed in such circumstances. Further research in cryptographic schemes with variable-length keys (where the size of key is not known to the cryptanalyst) might benefit from examination of the machine learning literature in this area.

Attack Types and Learning Protocols

A critical aspect of any cryptanalytic or learning scenario is the specification of how the cryptanalyst (learner) may gather information about the unknown target function.

Cryptographic attacks come in a variety of flavors, such as *ciphertext only, known plaintext (and matching ciphertext), chosen plaintext,* and *chosen ciphertext.* Cryptosystems secure against one type of attack may not be secure against another. A classic example of this is Rabin's signature algorithm [35], for which it is shown that a "passive" attack—forging a signature knowing only the public signature verification key—is provably as hard as factorization, whereas an "active attack"—querying the signer by asking for his signature on some specially constructed messages—is devastating and allows the attacker to determine the factorization of the signer's modulus—a total break.

The machine learning community has explored similar scenarios, following the pioneering work of Angluin [2, 3]. For example, the learner may be permitted "membership queries"—asking for the value of the unknown function on some specified input—or "equivalence queries"—asking if a specified conjectured hypothesis is indeed equivalent to the unknown target hypothesis. (If the conjecture is incorrect, the learner is given a "counterexample"—an input on which the conjectured and the target functions give different results.) For example, Angluin [2] has shown that a polynomial number of membership and equivalence queries are sufficient to exactly identify any regular set (finite automaton), whereas the problem of learning a regular set from random examples is NP-complete [1].

Even if information is gathered from random examples, cryptanalytic/learning scenarios may also vary in the prior knowledge available to the attacker/learner about the distribution of those examples. For example, some cryptosystems can be successfully attacked with only general knowledge of the system and knowledge of the language of the plaintext (which determines the distribution of the examples). While there is some work within the machine learning community relating to learning from known distributions (such as the uniform distribution, or product distributions [40]), and the field of "pattern recognition" has developed many techniques for this problem [12], most of the modern research in machine learning, based on Valiant's PAC-learning formalization of the problem, assumes that random examples are drawn according to an arbitrary but fixed probability distribution that is unknown to the learner. Such assumptions seem to have little relevance to cryptanalysis, although techniques such as those based on the "theory of coincidences" [25, Chapter VII] can sometimes apply to such situations. In addition, we have the difference between the two fields in that PAC-learning requires learning for all underlying probability distributions, while cryptographic security is typically defined as security no matter what the underlying distribution on messages is.

Exact versus Approximate Inference

In the practical cryptographic domain, an attacker typically aims for a "total break," in which he determines the unknown secret key. That is, he exactly identifies the unknown cryptographic function. Approximate identification of the unknown function is typically not a goal, because the set of possible cryptographic functions used normally does not admit good approximations. On the other hand, the theoretical development of cryptography has focussed on definitions of security that exclude even approximate inference by the cryptanalyst. (See, for example, Goldwasser and Micali's definitions in their paper on probabilistic encryption [15].) Such theoretical definitions and corresponding results are thus applicable to derive results on the difficulty of (even approximately) learning, as we shall see.

The machine learning literature deals with both exact inference and approximate inference. Because exact inference is often too difficult to perform efficiently, much of the more recent research in this area deals with approximate inference. (See, for example, the key paper on learnability and the Vapnik-Chervonenkis dimension by Blumer et al. [7].) Approximate learning is normally the goal when the input data consists of randomly chosen examples. On the other hand, when the learner may actively query or experiment with the unknown target function, exact identification is normally expected.

Computational Complexity

The computational complexity (sometimes called "work factor" in the cryptographic literature) of a cryptanalytic or learning task is of major interest in both fields.

In cryptography, the major goal is to "prove" security under the broadest possible definition of security, while making the weakest possible complexity-theoretic assumptions. Assuming the existence of one-way functions has been a common such weakest possible assumption. Given such an assumption, in the typical paradigm it is shown that there is no polynomial-time algorithm that can "break" the security of the proposed system. (Proving, say, exponential-time lower bounds could presumably be done, at the expense of making stronger initial assumptions about the difficulty of inverting a one-way function.)

In machine learning, polynomial-time learning algorithms are the goal, and there exist many clever and efficient learning algorithms for specific problems. Sometimes, as we shall see, polynomial-time algorithms can be proved not to exist, under suitable cryptographic assumptions. Sometimes, as noted above, a learning algorithm does not know in advance the size of the unknown target hypothesis, and to be fair, we allow it to run in time polynomial in this size as well. Often the critical problem to be solved is that of finding a hypothesis (from the known class of possibile hypotheses) that is consistent with the given set of examples; this is often true even if the learning algorithm is trying merely to approximate the unknown target function.

For both cryptanalysis and machine learning, there has been some interest in minimizing space complexity as well as time complexity. In the cryptanalytic domain, for example, Hellman [18] and Schroeppel and Shamir [42] have investigated space/time trade-offs for breaking certain cryptosystems. In the machine learning literature, Schapire has shown the surprising result [41, Theorem 6.1] that if there exists an efficient learning algorithm for a class of functions, then there is a learning algorithm whose space complexity grows only logarithmically in the size of the data sample needed (as ϵ, the approximation parameter, goes to 0).

Unicity Distance and Sample Complexity

In his classic paper on cryptography [43], Shannon defines the "unicity distance" of a cryptosystem to be $H(K)/D$, where $H(K)$ is the entropy of the key space K (the number of bits needed to describe a key, on the average), and where D is redundancy of the language (about 2.3 bits/letter in English). The unicity distance measures the amount of ciphertext that must be intercepted in order to make the solution unique; once that amount of ciphertext has been intercepted, one expects there to be a unique key that will decipher the ciphertext into acceptable English. The unicity distance is an "information-theoretic" measure of the amount of data that a cryptanalyst needs to succeed in exactly identifying the unknown secret key.

Similar information-theoretic notions play a role in machine learning theory, although there are differences arising from the fact that in the standard PAC-learning model there may be infinitely many possible target hypotheses, but on the other hand only an approximately correct answer is required. The Vapnik-Chervonenkis dimension [7] is a key concept in coping with this issue.

Other differences

The effect of *noise* in the data on the cryptanalytic/learning problem has been studied more carefully in the learning scenario than the cryptanalytic scenario, probably because a little noise in the cryptanalytic situation can render analysis (and legitimate decryption) effectively hopeless. However, there are cryptographic systems [48, 28, 45] that can make effective use of noise to improve security, and other (analog) schemes [49, 50] that attempt to work well in spite of possible noise.

Often the inference problems studied in machine learning theory are somewhat more general than those that occur naturally in cryptography. For example, work has been done on target concepts that "drift" over time [19, 24]; such variability is rare in cryptography (users may change their secret keys from time to time, but this is dramatic change, not gradual drift). In another direction, some work [38] has been done on learning "concept hierarchies"; such a framework is rare in cryptography (although when breaking a substitution cipher one may first learn what the vowels are, and then learn the individual substitutions for each vowel).

3 Cryptography's impact on Learning Theory

As noted earlier, Valiant [47] argued that the work of Goldreich, Goldwasser, and Micali [14] on random functions implies that even approximately learning the class of functions representable by polynomial-size boolean circuits is infeasible, assuming that one-way functions exist, even if the learner is allowed to query the unknown function. So researchers in machine learning have focussed on the question of identifying which simpler classes of functions are learnable (approximately, from random examples, or exactly, with queries). For example, a major open question in the field is whether the class of boolean functions representable as boolean formulas in disjunctive normal form (DNF) is efficiently learnable from random examples.

The primary impact of cryptography on machine learning theory is a natural (but negative) one: showing that certain learning problems are computationally intractable. Of course, there are other ways in which a learning problem could be intractable; for example, learning the class of all boolean functions is intractable merely because the required number of examples is exponentially large in the number of boolean input variables.

A certain class of intractability results for learning theory are *representation dependent*: they show that given a set of examples, finding a consistent boolean function *represented in a certain way* is computationally intractable. For example, Pitt and Valiant [32] show that finding a 2-term DNF formula consistent with a set of input/output pairs for such a target formula is an NP-complete problem. This implies that learning 2-term DNF is intractable (assuming $P \neq NP$), but only if the learner is required to produce his answer in the form of a 2-term DNF formula. The corresponding problem for functions representable in 2-CNF (CNF with two literals per clause, which properly contains the set of functions representable in 2-term DNF) *is* tractable, and so PAC-learning the class of functions representable in 2-term DNF is possible, as long as the learner may output his answers in 2-CNF. Similarly, Angluin [1] has shown that it is NP-complete to find a minimum-size DFA that is consistent with a given set of input/output examples, and Pitt and Warmuth [33] have extended this result to show that finding an *approximately* minimum-size DFA consistent with such a set of examples is impossible to do efficiently. These representation-dependent results depend on the assumption that $P \neq NP$.

In order to obtain hardness results that are *representation-independent*, Kearns and Valiant [22, 21] turned to cryptographic assumptions (namely, the difficulty of inverting RSA, the difficulty of recognizing quadratic residues modulo a Blum integer, and the difficulty of factoring Blum integers). Of course, they also need to explain how learning could be hard in a representation-independent manner, which they do by requiring the learning algorithm not to output a hypothesis in some representation language, but rather to *predict* the classification of a new example with high accuracy.

Furthermore, in order to prove hardness results for learning classes of relatively simple functions, Kearns and Valiant needed to demonstrate that the relevant cryptographic operations could be specified within the desired function class. This they accomplished by the clever technique of using an "expanded input" format, wherein the input was arranged to contain suitable auxiliary results as well as the basic input values. They made use of the distribution-independence of PAC-learning to assume that the probability distribution used would not give any weight to inputs where these auxiliary results were incorrect.

By these means, Kearns and Valiant were able to show that PAC-learning the following classes of functions is intractable (here $p(n)$ denotes some fixed polynomial):

1. "Small" boolean formulae: the class of boolean formula on n boolean inputs whose size is less than $p(n)$.

2. The class of deterministic finite automata of size at most $p(n)$ that accept only strings of length n.

3. For some fixed constant natural number d, the class of threshold circuits over n variables with depth at most d and size at most $p(n)$.

They show that if a learning algorithm could efficiently learn any of these classes of functions, in the sense of being able to predict with probability significantly greater than

1/2 the classification of new examples, then the learning algorithm could be used to "break" one of the cryptographic problems assumed to be hard.

The results of Kearns and Valiant were also based on the work of Pitt and Warmuth [34], who develop the notion of a "prediction-preserving reducibility." The definition implies that if class A is so reducible to class B, then if class B is efficiently predictable, then so is class A. Using this notion of prediction-preserving reducibility, they show a number of classes of functions to be "prediction-complete" for various complexity classes. In particular, the problem of prediction the class of alternating DFAs is shown to be prediction-complete for P, and ordinary DFAs are as hard to predict as any function computable in log-space, and boolean formula are prediction-complete for NC^1. These results, and the notion of prediction-preserving reducibility, were central to the work of Kearns and Valiant.

The previous results assumed a learning scenario in which the learner was working from random examples of the input/output behavior of the target function. One can ask if cryptographic techniques can be employed to prove that certain classes of functions are unlearnable even if the learner may make use of queries. Angluin and Kharitonov [5] have done so, showing that (modulo the usual cryptographic assumptions regarding RSA, quadratic residues, or factoring Blum integers), that there is no polynomial-time prediction algorithm, *even if membership queries are allowed*, for the following classes of functions:

1. Boolean formulas,

2. Constant-depth threshold circuits,

3. Boolean formulas in which every variable occurs at most 3 times,

4. Finite unions or intersections of DFAs, 2-way DFAs, NFAs, or CFGs.

These results are based on the public-key cryptosystem of Naor and Yung [31], which is provably secure against a chosen-ciphertext attack. (Basically, the queries asked by the learner get translated into chosen-ciphertext requests against the Naor-Yung scheme.)

4   Learning Theory's impact on Cryptography

Since most of the negative results in learning theory already depend on cryptographic assumptions, there has been no impact of negative results on learning theory on the development of cryptographic schemes. Perhaps some of the results and concepts and Pitt and Warmuth [34] could be applied in this direction, but this has not been done.

On the other hand, the positive results in learning theory are normally independent of cryptographic assumptions, and could in principle be applied to the cryptanalysis of relatively simple cryptosystems. Much of this discussion will be speculative in nature, since there is little in the literature exploring these possibilities. We sketch some possible approaches, but leave their closer examination and validation (either theoretical or empirical) as open problems.

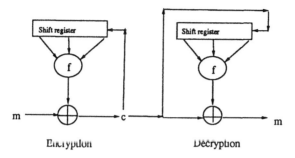

Figure 1: In cipher-feedback mode, each plaintext message bit m is encrypted by exclusive-oring it with the result of applying the function f to the last n bits of ciphertext, where n is the size of the shift register. The ciphertext bit c is transmitted over the channel; the corresponding decryption process is illustrated on the right.

Cryptanalysis of cipher-feedback systems

Perhaps the most straightforward application of learning results would be for the cryptanalysis of nonlinear feedback shift-registers operating in cipher-feedback mode. See Figure 1. The feedback function f is known only to the sender and the receiver; it embodies their shared "secret key."

If the cryptanalyst has a collection of matching plaintext/ciphertext bits, then he has a number of corresponding input/output pairs for the unknown function f. A learning algorithm that can infer f from such a collection of data could then be used as a cryptanalytic tool. Moreover, a chosen-ciphertext attack gives the cryptanalyst the ability to query the unknown function f at arbitrary points, so a learning algorithm that can infer f using queries could be an effective cryptanalytic tool.

We note that a definition of learnability that permits "approximate" learning is a good fit for this problem: if the cryptanalyst can learn an approximation to f that agrees with f 99% of the time, then he will be able to decrypt 99% of the plaintext.

Suppose first that we consider a known plaintext attack. Good cryptographic design principles require that f be about as likely to output 0 as it is to output 1. This would typically imply that the shift register contents can be reasonably viewed as a randomly drawn example of $\{0,1\}^n$, *drawn according to the uniform distribution*. (We emphasize that it is this assumption that makes are remarks here speculative in nature; detailed analysis or experimentation is required to verify that this assumption is indeed reasonable in each proposed direction.) There are a number of learning-theory results that assume that examples are drawn from $\{0,1\}^n$ according to the uniform distribution. While this assumption seems rather unrealistic and restrictive in most learning applications, it is a perfect match for such a cryptographic scenario. What cryptographic lessons can be drawn from the learning-theory research?

Schapire [40] shows how to efficiently infer a class of formula he calls "probabilistic read-once formula" against product distributions. A special case of this result implies that a formula f constructed from AND, OR, and NOT gates can be exactly identified (with high probability) in polynomial time from examples drawn randomly from the uniform distribution. (It is an open problem to extend this result to formula involving XOR gates in a useful way; some modification would be needed since the obvious generalization isn't true.) Perhaps the major lesson to be drawn here is that in the simplest formula for f each shift register bit should be used several times.

Linial, Mansour, and Nisan [27] have shown how to use spectral (Fourier transform) techniques to learn functions f chosen from AC^0 (constant depth circuit of AND/OR/NOT gates having arbitrarily large fan-in at each gate) from examples drawn according to the uniform distribution. Kushelevitz and Mansour [26] have extended these results to the class of functions representable as decision trees. Furst, Jackson, and Smith [13] have elaborated and extended these results in a number of directions. We learn the lesson that the spectral characteristics of f need to be understood and controlled.

Hancock and Mansour [17] have similarly shown that monotone $k\mu$ DNF formulae (that is, monotone DNF formulae in which each variable appears at most k times) are learnable from examples drawn randomly according to the uniform distribution. Although monotone formula are not really useful in this shift-register application, the negation of a monotone formula might be. We thus have another class of functions f that should be avoided.

When chosen-ciphertext attacks are allowed, function classes that are learnable with membership queries should be avoided. For example, Angluin, Hellerstein, and Karpinski [4] have shown how to exactly learn any μ-formula using members and equivalence queries. Similarly, Hancock [16] shows how to learn 2μ DNF formula (not necessarily monotone), as well as $k\mu$ decision trees, using queries.

We thus see a potential "pay-back" from learning theory to cryptography: certain classes of inferrable functions should probably be avoided in the design of non-linear feedback shift registers used in cipher-feedback mode. Again, we emphasize that verifying that the proposed attacks are theoretically sound or empirically useful remains an open problem. Nonetheless, these suggestions should provide some useful new guidelines to those designing such cryptosystems.

Other possibilities

We have seen some successful applications of continuous optimization techniques (such as gradient descent) to discrete learning problems; here the neural net technique of "back propagation" comes to mind. Perhaps such techniques could also be employed successfully in cryptanalytic problems.

Another arena in which cryptography and machine learning relate is that of *data compression*. It has been shown by Blumer et al [6] that pac-learning and data compression are essentially equivalent notions. Furthermore, the security of an encryption scheme is often enhanced by compressing the message before encrypting it. Learning theory may conceivably aid cryptographers by enabling ever more effective compression algorithms.

Acknowledgments
I would like to thank Rob Schapire for helpful comments.

References

[1] Dana Angluin. On the complexity of minimum inference of regular sets. *Information and Control*, 39:337–350, 1978.

[2] Dana Angluin. A note on the number of queries needed to identify regular languages. *Information and Control*, 51:76–87, 1981.

[3] Dana Angluin. Queries and concept learning. *Machine Learning*, 2(4):319–342, April 1988.

[4] Dana Angluin, Lisa Hellerstein, and Marek Karpinski. Learning read-once formulas with queries. Technical report, University of California, Report No. UCB/CSD 89/528, 1989.

[5] Dana Angluin and Michael Kharitonov. When won't membership queries help? In *Proceedings of the Twenty-Third Annual ACM Symposium on Theory of Computing*, pages 444–454, New Orleans, Louisiana, May 1991.

[6] Anselm Blumer, Andrzej Ehrenfeucht, David Haussler, and Manfred K. Warmuth. Occam's razor. *Information Processing Letters*, 24:377–380, April 1987.

[7] Anselm Blumer, Andrzej Ehrenfeucht, David Haussler, and Manfred K. Warmuth. Learnability and the Vapnik-Chervonenkis dimension. *Journal of the ACM*, 36(4):929–965, 1989.

[8] Gilles Brassard. *Modern Cryptology*. Springer-Verlag, 1988. Lecture Notes in Computer Science Number 325.

[9] D. W. Davies and W. L. Price. *Security for Computer Networks: An Introduction to Data Security in Teleprocessing and Electronic Funds Transfer*. John Wiley and Sons, New York, 1984.

[10] D. E. Denning. *Cryptography and Data Security*. Addison-Wesley, Reading, Mass., 1982.

[11] W. Diffie and M. E. Hellman. Privacy and authentication: An introduction to cryptography. *Proceedings of the IEEE*, 67:397–427, March 1979.

[12] Richard O. Duda and Peter E. Hart. *Pattern Classification and Scene Analysis*. Wiley, 1973.

[13] Merrick Furst, Jeffrey Jackson, and Sean Smith. Improved learning of AC^0 functions. In *Proceedings of the Fourth Annual Workshop on Computational Learning Theory*, pages 317–325, Santa Cruz, California, August 1991.

[14] O. Goldreich, S. Goldwasser, and S. Micali. How to construct random functions. In *Proceedings of the 25th IEEE Symposium on Foundations of Computer Science*, pages 464–479, Singer Island, 1984. IEEE.

[15] S. Goldwasser and S. Micali. Probabilistic encryption. *JCSS*, 28(2):270–299, April 1984.

[16] Thomas Hancock. Learning 2μ DNF formulas and $k\mu$ decision trees. In *Proceedings of the Fourth Annual Workshop on Computational Learning Theory*, pages 199–209, Santa Cruz, California, August 1991.

[17] Thomas Hancock and Yishay Mansour. Learning monotone $k\mu$ DNF formulas on product distributions. In *Proceedings of the Fourth Annual Workshop on Computational Learning Theory*, pages 179–183, Santa Cruz, California, August 1991.

[18] M. E. Hellman. A cryptanalytic time-memory trade off. *IEEE Trans. Inform. Theory*, IT-26:401–406, 1980.

[19] David P. Helmbold and Philip M. Long. Tracking drifting concepts using random examples. In *Proceedings of the Fourth Annual Workshop on Computational Learning Theory*, pages 13–23, Santa Cruz, California, 1991. Morgan Kaufmann.

[20] D. Kahn. *The Codebreakers*. Macmillian, New York, 1967.

[21] Michael Kearns. *The Computational Complexity of Machine Learning*. PhD thesis, Harvard University Center for Research in Computing Technology, May 1989. Technical Report TR-13-89. Also published by MIT Press as an ACM Distinguished Dissertation.

[22] Michael Kearns and Leslie G. Valiant. Cryptographic limitations on learning boolean formulae and finite automata. In *Proceedings of the Twenty-First Annual ACM Symposium on Theory of Computing*, pages 433–444, Seattle, Washington, May 1989.

[23] Yves Kodratoff and Ryszard S. Michalski, editors. *Machine Learning: An Artificial Intelligence Approach*, volume III. Morgan Kaufmann, Los Altos, California, 1990.

[24] Anthony Kuh, Thomas Petsche, and Ronald L. Rivest. Learning time-varying concepts. In *Proceedings 1990 Conference on Computation Learning and Natural Learning (Princeton)*, 1990. To appear.

[25] Solomon Kullback. *Statistical Methods in Cryptanalysis*. Aegean Park Press, 1976.

[26] Eyal Kushilevitz and Yishay Mansour. Learning decision trees using the fourier spectrum. In *Proceedings of the Twenty-Third Annual ACM Symposium on Theory of Computing*, pages 455–464, New Orleans, Louisiana, May 1991. ACM.

[27] Nathan Linial, Yishay Mansour, and Noam Nisan. Constant depth circuits, fourier transform, and learnability. In *Proceedings of the Thirtieth Annual Symposium on Foundations of Computer Science*, pages 574–579, Research Triangle Park, North Carolina, October 1989.

[28] R. J. McEliece. *A Public-Key System Based on Algebraic Coding Theory*, pages 114–116. Jet Propulsion Lab, 1978. DSN Progress Report 44.

[29] Ryszard S. Michalski, Jaime G. Carbonell, and Tom M. Mitchell, editors. *Machine Learning: An Artificial Intelligence Approach*, volume I. Morgan Kaufmann, Los Altos, California, 1983.

[30] Ryszard S. Michalski, Jaime G. Carbonell, and Tom M. Mitchell, editors. *Machine Learning: An Artificial Intelligence Approach*, volume II. Morgan Kaufmann, Los Altos, California, 1986.

[31] Moni Naor and Moti Yung. Public-key cryptosystems provably secure against chosen ciphertext attack. In *Proceedings of the Twenty-Second Annual ACM Symposium on Theory of Computing*, pages 427–437, Baltimore, Maryland, 1990. ACM.

[32] Leonard Pitt and Leslie G. Valiant. Computational limitations on learning from examples. *Journal of the ACM*, 35(4):965–984, 1988.

[33] Leonard Pitt and Manfred K. Warmuth. The minimum DFA consistency problem cannot be approximated within any polynomial. In *Proceedings of the Twenty-First Annual ACM Symposium on Theory of Computing*, Seattle, Washington, May 1989.

[34] Leonard Pitt and Manfred K. Warmuth. Prediction preserving reducibility. *Journal of Computer and System Sciences*, 41:430–467, 1990.

[35] M. Rabin. Digitalized signatures as intractable as factorization. Technical Report MIT/LCS/TR-212, MIT Laboratory for Computer Science, January 1979.

[36] Jorma Rissanen. *Stochastic Complexity in Statistical Inquiry*, volume 15 of *Series in Computer Science*. World Scientific, 1989.

[37] Ronald L. Rivest. Cryptography. In Jan van Leeuwen, editor, *Handbook of Theoretical Computer Science (Volume A: Algorithms and Complexity)*, chapter 13, pages 717–755. Elsevier and MIT Press, 1990.

[38] Ronald L. Rivest and Robert Sloan. Learning complicated concepts reliably and usefully. In *Proceedings AAAI-88*, pages 635–639, August 1988.

[39] A. L. Samuel. Some studies in machine learning using the game of checkers. *IBM Journal of Research and Development*, 3:211–229, July 1959. (Reprinted in *Computers and Thought*, (eds. E. A. Feigenbaum and J. Feldman), McGraw-Hill, 1963, pages 39–70).

[40] Robert E. Schapire. Learning probabilistic read-once formulas on product distributions. In *Proceedings of the Fourth Annual Workshop on Computational Learning Theory*, pages 184–198, Santa Cruz, California, August 1991.

[41] Robert Elias Schapire. *The Design and Analysis of Efficient Learning Algorithms*. PhD thesis, MIT EECS Department, February 1991. (MIT Laboratory for Computer Science Technical Report MIT/LCS/TR-493).

[42] R. Schroeppel and A. Shamir. A $TS^2 = O(2^n)$ time/space tradeoff for certain NP-complete problems. In *Proc. 20th Annual IEEE Symposium on Foundations of Computer Science*, pages 328–336, San Juan, Puerto Rico, 1979. IEEE.

[43] Claude Shannon. Communication theory of secrecy systems. *Bell System Technical Journal*, 28:656–715, October 1949.

[44] G. J. Simmons. Cryptology. In *The New Encyclopædia Brittanica*, pages 860–873. Encyclopædia Brittanica, 1989. (Volume 16).

[45] N. J. A. Sloane. Error-correcting codes and cryptography. In D. Klarner, editor, *The Mathematical Gardner*, pages 346–382. Wadsworth, Belmont, California, 1981.

[46] A. M. Turing. Computing machinery and intelligence. *Mind*, 59:433–460, October 1950. (Reprinted in *Computers and Thought*, (eds. E. A. Feigenbaum and J. Feldman), McGraw-Hill, 1963, pages 11–38).

[47] Leslie G. Valiant. A theory of the learnable. *Communications of the ACM*, 27(11):1134–1142, November 1984.

[48] A. D. Wyner. The wire-tap channel. *Bell Sys. Tech. J.*, 54:1355–1387, 1975.

[49] A. D. Wyner. An analog scrambling scheme which does not expand bandwidth, part 1. *IEEE Trans. Inform. Theory*, IT-25(3):261–274, 1979.

[50] A. D. Wyner. An analog scrambling scheme which does not expand bandwidth, part 2. *IEEE Trans. Inform. Theory*, IT-25(4):415–425, 1979.

Speeding up Prime Number Generation

Jørgen Brandt, Ivan Damgård and Peter Landrock

Aarhus University, Mathematical Institute,

Ny Munkegade, DK 8000 Aarhus C, Denmark

Abstract

We present various ways of speeding up the standard methods for generating provable, resp. probable primes. For probable primes, the effect of using test division and 2 as a fixed base for the Rabin test is analysed, showing that a speedup of almost 50% can be achieved with the same confidence level, compared to the standard method. For Maurer's algorithm generating provable primes p, we show that a small extension of the algorithm will mean that only one prime factor of $p - 1$ has to be generated, implying a gain in efficiency. Further savings can be obtained by combining with the Rabin test. Finally, we show how to combine the algorithms of Maurer and Gordon to make "strong provable primes" that satisfy additional security constraints.

1   Introduction

Nearly all the public-key systems proposed in the literature so far make use of large, random prime numbers. This is certainly the case for RSA, where the primes constitute the secret key. But also the best known implementations of Diffie-Hellman key exchange and El Gamal signatures make use of large primes. Thus fast algorithms for prime number generation is of obvious interest.

The simplest method for generating a prime number is to choose numbers at random and test for primality, until we find a prime. Usually, one would like the prime p produced to satisfy some extra conditions, most notably that $p - 1$ should have at least one large prime factor. Therefore one will constrain the search to a set of numbers that is guaranteed to have the desired property (see Gordon [Go] for example). To do the testing, a probabilistic test, usually the Miller-Rabin test [Ra], is used. Such tests do not always give correct answers, but are much more efficient than tests that do [AdHu]. Thus, such an algorithm leads to numbers that are *probable primes*: we can only assert with some probability that the output really is a prime. In [DaLa], Damgård and Landrock obtain some concrete estimates for the error probability of prime generation with this method.

In contrast with this, other authors (Maurer [Ma]) have proposed methods for generating *provable primes*. This is based on the fact that for a prime p, if a large enough part of the factorization of $p - 1$ is known, primality of p can be proved with no probability of error.

In the following, we study some of the known methods for prime generation, and analyse some efficiency improvements, some new and others well known. We also discuss

which of the well known security constraints are relevant for RSA applications, and how the primes generated can be guaranteed to satisfy these constraints.

2 Probable Primes

In this section, we will concentrate on methods that use the Rabin test for primality. This test may output "composite" or "probably prime". Let the test algorithm be denoted by *RaTest*. We then have

$$Prob(RaTest(n) = "probably\ prime" \mid n\ is\ composite) \leq 1/4$$

$$Prob(RaTest(n) = "probably\ prime" \mid n\ is\ prime) = 1$$

The Rabin test works by choosing at random a number modulo n, the so called *base*, and checking that it satisfies certain criteria. Of course, the natural idea to increase the degree of certainty, if we are told that n is probably prime, is to repeat the test some number of times with independently chosen bases. However, as pointed out by Brassard et al. in [BBCGP], the above probabilities are not immediately relevant when we are to assess the error probability that results from a given number of iterations. What we observe is the test result, so we should be interested in the probability that n is composite given that $RaTest(n) = "probably\ prime"$. This makes no sense, if we do not specify a distribution of the number n.

The standard distribution that is considered is a uniform choice of odd numbers in some interval. For the interval $[0, x]$, [EdPo] obtained a result for 1 iteration of the test, which was true already for the weaker primality test that only checks the Fermat congruence. Their result holds asymptotically as x goes to infinity. In [KiPo] weaker, but explicit and therefore computationally useful results were obtained. Recently in [DaLa], these results were improved and extended to any number of iterations, for the interval $[x/2, x]$. This interval was fixed mainly for practical reasons: if we choose $x = 2^k$, then the choice of interval means that we are looking for a prime of bit length exactly k, which is advantageous in many applications.

These results suggest the following algorithm for generating a large, random prime:

Algorithm 1

1. choose an odd n uniformly at random in $[x/2, x]$

2. If t independent iterations of *RaTest* return "probably prime", output n and stop, otherwise go to 1.

However, hardly anyone would use this algorithm in practice! For example, from en efficiency point of view, we are not interested in choosing basis uniformly at random. The Rabin test involves exponentiating the basis to the odd part of $n-1$, and the test spends nearly all its time doing this. Hence the best possible basis from this point of view would be 2, as this number is much faster to exponentiate than a random one: considering the standard square and multiply algorithm, it is clear that the multiplications reduce to 1-bit shifts. There is not even any reason to subtract the modulus because the subsequent modular squaring will bring down the size of the intermediate result anyway.

Furthermore, no practical test will apply Rabin's algorithm directly to a candidate. In stead, one would first use test division by small primes, which is much more efficient. Thus our algorithm would look as follows:

Algorithm 2

1. Choose an odd n uniformly at random from $[x/2, x]$.

2. Use test division on n by all primes less than r.

3. If the number passes step 2, do $RaTest$ with 2 as basis.

4. If step 3 outputs "probably prime", do at most t iterations of the Rabin test with independent random bases. If an iteration returns "composite", go to 1. If all of them return "probably prime", output n and stop.

Note that the estimates in [DaLa] on the error probability are still valid for this algorithm. To see this, let Y_t be the event that a number passes the t rabin tests, X the event that a composite number is chosen, and Z the event that a composite is chosen that passes the inital screening done by Algorithm 2. Then $Prob(Z|Y_t)$ is the probability that Algorithm 2 returns a composite, and since clearly

$$Prob(X|Y_t) = Prob(Z|Y_t) + Prob(X \setminus Z|Y_t) \geq Prob(Z|Y_t),$$

the error probability of Algorithm 2 is at most the error probability of Algorithm 1.

A final variation replaces the independent random choice of candidates by incremental search from a random starting point, such that a maximum of s candidates are examined:

Algorithm 3

1. Choose an odd n_0 uniformly at random from $[x/2, x]$. Put $n = n_0$.

2. Use test division on n by all primes less than r.

3. If n passes step 2, do $RaTest$ with 2 as basis.

4. If step 3 outputs "probably prime", do at most t iterations of the Rabin test with independent random bases.

   If an iteration returns "composite", put $n = n+2$ and go to step 2, unless s candidates have now been tested, in which case we output "fail" and stop.

   If all t iterations return "probably prime", output n and stop.

One can prove that if s is not too large, say a small constant times $log(x)$, the error probability of Algorithm 3 is not much larger than that of Algorithm 2. A proof will be included in the final version of this paper. Also included will be a discussion of the value of s to choose, such that the algorithm is almost certain not to output "fail".

The advantage of Algorithm 3 is that we can do the test division without having to divide: in an initialization stage, compute the residue modulo all the small primes of n_0. Then each time n is increased by 2, add 2 to the current value modulo each prime and test that no residue becomes 0. In practical experiments, we found that this optimized version of Algorithm 3 is about 25% faster than Algorithm 2, even if we use $s = \infty$ (and optimal values of r in both cases).

2.1 Efficiency

In this section we analyse the efficiency of Algorithm 3. We will assume that s is chosen large enough, such that except with negligible probability, the algorithm outputs a probable prime. We also assume that the optimized test division described above is used.

Recall that a composite n is called a pseudoprime base 2, if $2^{n-1} = 1 \bmod n$, and that $RaTest(n) = probably\ prime$ implies that n is a pseudoprime base 2. Pomerance [Po] has proved an upper bound on the number of pseudoprimes base 2: for all large enough x,

$$P(x) \leq x/\sqrt{L(x)},$$

where $P(x)$ is the number of pseudoprimes base 2 less than x, and $L(x) = x^{logloglogx/loglogx}$. He also conjectures that in fact $P(x)$ tends to $x \cdot L(x)^{-1+o(1)}$ for large x.

Although this is an asymptotic result, it suggests very strongly that the probability that a random composite passes the base-2 test is negligible. We shall assume this in the rest of this section. Note that this assumption only has to do with the analysis of the running time - the estimate of the error probability depends only on the results from [DaLa].

The assumption means that except with negligible probability, the algorithm will examine some composite candidates, reject some by test division, the rest by the base 2 test, and will accept the first prime it finds.

The rest of this section will be devoted to finding approximations for the optimal number of small primes to use for test division and for the expected running time that results from using the optimal r.

The test division will exclude all but some fraction $\alpha(r)$ of the possible n, where $\alpha(r)$ is easy to compute from r, assuming that candidate numbers are uniformly distributed. In fact, $\alpha(r) = (1 - 3^{-1}) \cdots (1 - p_j^{-1})$, where p_j is the largest prime number less than r. By Mertens's theorem $\alpha(r) \approx 2e^{-\gamma}/log(r) \approx 1/log(r)$, where γ is Eulers constant.

Let D be the time spent on test dividing one number by one small prime, R_2 the time for a Rabin test with base 2, and R the time for a Rabin test with random base.

If the primes were completely regularly distributed, we would expect to examine $log(x)/4$ candidates before finding a prime (note that we only look at odd numbers). Primes are of course not at all regularly distributed, and this will tend to make the search time for an incremental search larger. Practical evidence suggests that the average number of examined candidates is in fact $log(x)/2$, but given the current state of our knowledge about the distribution of primes, an actual proof of this seems to be hard to give. We will assume a search length of $log(x)/2$ in the following - but note that the expected search length does not influence the value of the optimal r.

We make the heuristic assumption that out of the $log(x)/2$ candidates, on average, a fraction $\alpha(r)$ will not be rejected by test division (this is only provably true if candidates are uniformly chosen). Thus, we spend $log(x)\alpha(r)R_2$ time units on doing base-2 tests. The time spent on test division is $log(x)Dr/log(r)$. Thus to find the best value of r, we must minimize the total expected running time

$$log(x)/2\left(\frac{Dr}{log(r)} + \frac{R_2}{log(r)}\right) + tR$$

as a function of r. Putting $f(r) = (Dr + R_2)/log(r)$, we find that $f(r)$ has a minimum when r satisfies that

$$rlog(r) - r = \frac{R_2}{D}$$

Using the prime number theorem and the fact that the r'th prime is about $rlog(r)$, it can be seen that a reasonable approximation to the optimal r is

$$r = \frac{R_2}{Dlog(R_2/D)}$$

We note that in [Ma], Maurer finds that a different value, namely $r = R_2/D$ is optimal. This is because he considers Algorithm 2, where one will of course not testdivide by all small primes, but reject a candidate, as soon as some small prime is found to divide it.

The result on r can be used for various purposes:

- if we assume that R_2 is cubic in $log(x)$, and D is constant, then it follows that the total running time is $O(log(x)^4/loglog(x))$, ignoring various low order terms. The same asymptotic running time was found by Maurer for Algorithm 2 in [Ma].

- Inserting the optimal value of r in the expresion for the running time shows that the time spent on rejecting composite candidates is approximately proportional to R_2. Concrete experiments have shown that R_2 can be as small as $R/2$, and thus Algorithm 2 is faster by a factor of 2 in this first phase, compared to an algorithm that uses random bases in all Rabin tests. Since by [DaLa], quite moderate values of t, e.g. 5-8 are sufficient for a very small error probability, the first phase is by far the most time consuming in a practical application.

- Given a concrete implementation, the formulas give a direct way of computing the optimal value of r and the resulting running time.

On an IBM PS/2 model 80 (16MHz 80386), generation of a probable 256 bit prime number p using the above methods takes about 8 sec. The program uses a variant of Gordon's method to ensure among other things that $p \pm 1$ has a large prime factor (about 120 bits in our implementation). Without these extra conditions, the program is quite a bit faster, about 4.5 seconds per prime. We use $t = 5$. Experiments indicate that the optimal number of small primes to use for test division in this implementation and with this output size is about 420. Based on measurements of R_2 and D, the analysis above would predict an optimal value of 430. Thus the validity of the analysis has been confirmed in practice.

It should be noted that measuring the optimal value is quite hard because the running time varies only very slightly in a large interval around the optimal value (less than 10% in the interval from 280 to 450).

Note also that the optimal test divide limit is very machine- and implemetation dependent: if for example we go from an 80386 machine to a 16-bit CPU like the 80286, the time for a test division increases by much less than that of an exponentiation, and so the optimal r value will be much larger for such a machine.

3 Provable Primes

In this section, we concentrate on variations of the method suggested by Maurer [Ma] for generating provable primes. This method is based on the following result by Pocklington:

α	$\rho(\alpha)$
1.5	0.59453 48919
2.0	0.30685 28194
2.5	0.13031 95618
3.0	0.04860 83883
3.5	0.01622 95932
4.0	0.00491 09256
4.5	0.00137 01177
5.0	0.00035 47247
6.0	0.00001 96497
7.0	0.00000 08746
8.0	0.00000 00323
9.0	0.00000 00010

Table 1: Distribution of the largest prime factor.

Theorem 2

Let $n - 1 = FR$, and let $q_1, ... q_t$ be the distinct prime factors of F. If there exists a number a such that $a^{n-1} = 1 \; mod \; n$ and for all $i = 1..t$, $(a^{(n-1)/q_i} - 1, n) = 1$, then any prime factor in n is congruent to 1 modulo F.

In particular, if $F > \sqrt{n}$, then trivially n is a prime. This suggests a straightforward algorithm for generating a random prime in some interval $[low, high]$: first generate recursively $q_1, q_2, ..$ until F, the product of the q's is larger than $\sqrt{high}$. Then choose random even R-values such that $n = FR + 1$ is in the right interval, until an n-value can be proved prime.

Maurer shows that if the q's are large, nearly any choice of a will suffice for proving primality of n (provided n really is prime!), so we are not likely to miss any primes, even if we only try once for each candidate. Furthermore, it is shown that if the number e is used to prove primality of the prime factors of $p - 1$ and $q - 1$, then the resulting RSA system with e as public exponent will not be easy to break by repeated encryption.

If the goal is to generate a prime uniformly chosen from the interval, then we should know something about the distribution of the prime factors of $n - 1$, in particular the distribution of their sizes is necessary. Fortunately, the distribution of the size of the largest prime factor of a number is well known. More precisely, for large N, one can compute $\rho(\alpha)$, the fraction of numbers x less than N whoose largest prime factor is less than $x^{1/\alpha}$. In Table 1, sample values of this function is given. From heuristic arguments, this distribution function seems to be also applicable if we add the condition that the number we are looking at is a prime minus 1.

What can be done to speed up this algorithm? First of all, we should of course use test division by small primes on a candidate before going into expensive exponentiations. Maurer suggests that since all candidates are of the form $n = FR + 1$ for fixed F, one can translate the condition that none of the small primes divide n into a condition on R. This will be faster to check, since R is usually much smaller than n (and certainly less than $\sqrt{n}$).

However, even if a candidate passes the test division, there is still no need to start application of the theorem above. We propose instead to do a Rabin test with base 2. This test will accept all the prime candidates, by the results of [Po], it will reject virtually all composite candidates, and it will be significantly faster than the exponentiations with random base required by Theorem 2. In a practical experiment, this use of the Rabin test reduced the average time for generation of one prime from 19.7 sec. to 17.5 sec. for generation of the same set of 20 256 bit primes. Furthermore, in a long series of tests, we have never seen a candidate accepted by the Rabin test that could not be proved prime (except for very small numbers). Another slight improvement can be obtained by observing that if we use $a = 2$ when applying Theorem 2, the fact that the candidate n passed the base 2 Rabin test implies that $a^{n-1} \bmod n = 1$, so the first exponentiation required by Theorem 2 can be discarded.

We then turn to a different issue: is it really necessary to generate prime factors of F, until we are above the square root of the final output? by the following result of Brillhart, Lehmer and Selfridge [BFL], the answer is no:

Theorem 3

Suppose $n = FR + 1$ satisfies the conditions of Theorem 2. Let R' be the odd part of R, and $F' = (n - 1)/R'$. Let r, s be defined by $R' = 2F's + r$, where $1 \leq r < 2F'$. Suppose $F' > \sqrt[3]{n}$. Then n is prime if and only if $s = 0$ or $r^2 - 8s$ is not a square.

This refined condition may seem to be too computationally costly to verify. However, this is not so in practice, at least not if we look at the variation using the Rabin test. For this variation, experience shows that the final part of Theorem 3 will only be used on the final candidate, and the extra computation required is only some trivial manipulations to find F', R', s, r, and perhaps a square root computation, which takes time negligible compared to the exponentiations.

Furthermore, the distribution of the largest prime factor shows that only 5% of the numbers x are expected to have all prime factors less than $x^{1/3}$. We suggest that we can easily live without these 5%, in which case we never have to generate more than 1 prime factor of $n - 1$. Hence this will simplify the code and save time compared to Maurer's original version for the approximately 30% of the numbers that have their largest prime factor less than $x^{1/2}$.

Unfortunately, these 30% of the cases also give rise to a theoretical problem, since they represent the cases where $F = q_1 < R$, whence R might contain a prime factor larger than q_1. But q_1 was chosen according to the distribution of the *largest* prime factor, so this will introduce a deviation from a uniform choice of a prime in the given interval. The problem only occurs, however, in approximately 12% of all cases, so we think that this deviation is not a serious one. It is certainly not something that would be a security problem for application to e.g. RSA since, compared to a uniform choice, our variation will generate primes p that tend to have larger prime factors of $p - 1$.

Using the methods outlined above, a provable 256 bit prime can be generated on an IBM PS/2 model 80 (16 MHz 80386) in about 18 sec. If one is willing to live with further deviations from a uniform choice, this can be reduced significantly: suppose we change the distribution with which the size of the prime q_1 is chosen such that it tends to be smaller than predicted by Table 1. This will of course make generation of q_1 faster. As an example, if all choices of relative length that are larger than 0.8 is reduced to 0.8, the time for generation of a prime drops to about 15 sec.

3.1 Security Constraints

If the primes generated are to be used for the RSA system, one usually demands that the primes satisfy some extra conditions that are designed to prevent a break of the resulting system by known attacks. Most important of these conditions are:

1. $p - 1$ should have a large prime factor.

2. $p + 1$ should have a large prime factor.

3. p should be constructed to avoid attacks by repeated encryption.

For probable primes, Gordon has shown one way to ensure that these conditions are satisfied. For provable primes, the situation is less clear.

Condition 3 is taken care of in Maurer's paper, and condition 1 is more or less automatically satisfied, if we use the variation of the algorithm suggested above, since it guarantees that there is a prime factor in $p - 1$ larger than $\sqrt[3]{p}$. Of course the precise meaning of "large" is not easy to define, as it depends on the computing power available to an attacker. Judging from the current state of the art, a large prime factor should be of length more than 50 bits at the very least. In fact, the ISO 11166-2 draft standard demands a length of at least 100 bits.

Condition 2 is not mentioned in [Ma]. One reason for arguing that it can be ignored is the recent cyclotomic polynomial factoring method by Bach and Shallit [BaSh]. They show that in fact there is an infinite number of conditions that had better be satisfied, if the primes are used in an RSA system: let P be any cyclotomic polynomial. Then a number n is easy to factor if it has a prime factor p for which $P(p)$ is a number with only small prime factors. The polynomials $X - 1$, $X + 1$ are examples of this, other examples are $X^2 + 1$, $X^2 + X + 1$. In fact, cyclotomic polynomials exist of any degree, so we get an infinite set of conditions. Given this fact, one might argue that none of these conditions are better than the others, and one should simply try to generate primes as uniformly chosen as possible, and hope that all the conditions are satisfied with large probability.

This is probably true if are looking at large enough numbers: Table 1 suggests very strongly that numbers of length significantly larger than 500 bits are very unlikely to have only small prime factors (although strictly speaking the table cannot be simultaneously used on all numbers of the form $P(p)$ because of the dependencies involved). However, the situation is different, if we talk about primes in the range 256-300 bits, which is a size suggested for many practical RSA applications. For such numbers, all prime factors of $p + 1$ can be expected to be smaller than 50 bits with probability about 0.03%. This is small, but certainly not negligible, if we are to generate keys for thousands of users. If instead we take numbers $P(p)$ where $deg(P) > 1$, $P(p)$ will have at least 512 bits, whence the probability drops below 10^{-11}. Thus for the RSA numbers that are realistic today, there is in fact good reason to take special care of $p + 1$, but no need to worry about the higher degree conditions.

How can this be done? a simple way out is to combine the algorithms of Maurer and Gordon:

1. Using your favorite prime generation program, generate an F with known factorization, and a large prime s, relatively prime to F.

2. Using Chinese Remaindering, find an odd p_0, such that $p_0 = 1 \bmod F$ and $p_0 = -1 \bmod s$.

3. Choose random values of L in some appropriate interval, until a number of the form $p = 2LFs + p_0$ can be proved prime by Maurer's method (or Theorem 3).

p_0 is likely to have about the same bit length as Fs, so since s must be "large", this puts a limit to how large F can be, relative to $p-1$. This introduces a slight deviation from the uniformity of primes otherwise produced by Maurer's method. Table 1 shows that for 256 bit primes and a 50 bit s, we loose about 10% of the primes this way.

A final observation is that the above algorithm may also be used to generate probable primes: suppose s and the prime factors of F are generated by e.g. Algorithm 3 from Section 2. Then what step 3 proves is that p is prime if the factors of F are prime. Thus our confidence in p is translated to confidence in the prime factors of F. Depending on the size chosen for these prime factors, this method may be more efficient than generating p by Algorithm 3 directly. This is because the exponentiations needed for the Rabin test are cubic in software, and therefore very rapidly become cheaper as the size of input numbers decreases.

4 Conclusion: are Provable Primes better than Probable?

We have shown a number of tricks that can speed up significantly the standard methods for generating both probable and provable primes. These ideas were shown not to affect the error probability of the algorithms, and their effect on the running time has been analysed and tested in practice.

We have also presented an algorithm that combines the well-known criteria for strong primes with Maurers algorithm for provable primes.

This still seems to leave open the question: should one choose provable or probable primes in a practical application?

One might say that provable primes are better, since they have zero error probability. Note, however, that in any cryptographic application one has to live with non-zero error probabilities anyway, for example the probability that a key is guessed by chance. Therefore, the question rather is whether one can efficiently bring down the error probability of probable primes to an acceptable level (for a system that uses RSA to exchange DES keys, an acceptable level might be the probability of guessing a random DES key).

This question was considered by Maurer [Ma], who in the first version of his paper arrived at the conclusion that if bounds in the order of 2^{-50} are placed on the error probability, probable primes become much less efficient the provable ones. But this was based on an overly pessimistic estimate of the error probability, as can be seen from the subsequent results in [DaLa]. In fact, our experiments indicate that the two methods have running times in the same order of magnitude, and that probable primes are a bit faster.

On top of this, the stack management necessitated by the recursive nature of provable prime generation means that the demand on data and program RAM is quite a bit larger for provable primes.

We therefore conclude that in the practical applications we are aware of, there is no reason to choose provable rather than probable primes.

References

[AdHu] L.Adleman and M.-D.Huang: *Recognizing Primes in Random Polynomial Time*, Proc. of STOC 1987, 462-469.

[BaSh] E.Bach and J.Shallit: *Factoring with Cyclotomic Polynomials*, Math. Comp. (1989), 52: 201-219.

[BBCGP] P.Beauchemin, G.Brassard,C.Crépeau, C.Goutier and C.Pomerance: *The Generation of Numbers that are Probably Prime*, J.Cryptology (1988) 1:53-64.

[BLS] J.Brillhart, D.H.Lehmer and J.L.Selfridge: *New Primality Criteria and Factorizations of* $2^m \pm 1$, Math. Comp. (1975), 29: 620-647.

[DaLa] Damgård and Landrock: *Improved Bounds for the Rabin Primality Test*, to appear.

[EdPo] P.Erdös and C.Pomerance: *On the Number of False Witnesses for a Composite Number*, Math. Comp. (1986), 46: 259-279.

[Go] J.Gordon: *Strong Primes are Easy to find*, Proc. of Crypto 84.

[Gu] R.K. Guy: *How to Factor a Number*, Proc. of the 5'th Manitoba Conference on Numerical Mathematics, 1975, University of Manitoba, Winnipeg.

[KiPo] S.H. Kim and C. Pomerance: *The Probability that a Randomly Probable Prime is Composite*, Math. Comp. (1989), 53: 721-741.

[Ma] U.Maurer: *The Generation of Secure RSA Products With Almost Maximal Diversity*, Proc. of EuroCrypt 89 (to appear).

[Po] C.Pomerance: *On the Distribution of Pseudoprimes*, Math. Comp. (1981), 37: 587-593.

[Ra] M.O. Rabin: *Probabilistic Algorithm for Testing Primality*, J.Number Theory (1980), 12: 128-138.

Two Efficient Server–Aided Secret Computation Protocols Based on the Addition Sequence

Chi–Sung Laih[*], Sung–Ming Yen[*] and Lein Harn[**]

[*]Department of Electrical Engineering,
National Cheng Kung University,
Tainan, Taiwan, Republic of China
TEL : 886–6–2361111 Ext–508
FAX : 886–6–2345482
E–mail: Laihcs@sun4.ee.ncku.edu.tw

[**]Computer Science Telecomunications Program
University of Missoouri–Kansas City
Kansas City, MO 64110
TEL : (816) 235–2367

Abstract

A server–aided secret computation protocol (SASC) is a method that allows a client (e.g. smart card) to compute a function efficiently with the aid of a powerful server (e.g. computer) without revealing the client's secrets to the server. Matsumoto et al. proposed a solution to the problem which is suitable for the RSA cryptosystem. Kawamura et al. have shown that a client, with a 10^5 times more powerful server's aid, can compute an RSA signature 50 times faster than the case without a server if the communication cost can be ignored. In this paper, we propose two SASC protocols based on the addition sequence to improve the efficiency. In the first protocol, since the addition sequence is determined by the server, it can improve the computational efficiency of the server only and it is suitable for the low speed communication link (e.g. 9.6 Kbps). It is expected that a client, with an 8982 times more powerful server's aid, can compute an RSA signature 50 times faster than the case without a server. In the second protocol, since the addition sequence is determined by the client, it can improve the computational efficiency of the client and server simultaneously but takes more communication time and it is suitable for the high speed communication link (e.g. above 10 Mbps). It is expected that a client, with a 3760 times more powerful server's aid, can compute an RSA signature 200 times faster than the case without a server.

I. Introduction

Smart cards are beginning to replace the conventional plastic cards in a variety of applications now. Since smart card has its own computational power and memory, it is very suitable for some applications which require computations on stored secret numbers. Under current technologies, the computational power of the smart card is still not so powerful to do very complicated computations that are required for most public key cryptosystems, e.g., RSA cryptosystem [1]. Recently, Matsumoto et al. [2] consider the approach by using a powerful untrusted auxiliary device (a server) to help a smart card (a client) to compute a secret function efficiently. Matsumoto et al. [3] and some other researchers [4–6] have proposed several protocols to solve the problem. Such protocols are referred to as the server–aided secret computation (SASC) protocols. The performance analysis of SASC protocols for the RSA cryptosystem is given by Kawamura et al. [7]. They have shown that a client, with a 10^5 times more powerful server's aid, can compute an RSA signature 50 times faster than in the case without a server if the communication cost can be ignored.

In this paper, we propose two protocols based on the concept of addition sequence to improve the efficiency. In the first protocol, since the addition sequence is determined by the server, it can only reduce the computational load for the server and is suitable for the low speed communication link (e.g., telephone lines). It is expected that a client, with an 8982 times more powerful server's aid, can compute an RSA signature 50 times faster than the case without a server. In the second protocol, since the addition sequence is determined by the client, it can reduce the computational load not only for the server, but also for the client and it is suitable for the high speed communication link (e.g., optical fibers). It is expected that a client, with a 3760 times more powerful server's aid, can compute an RSA signature 200 times faster than the case without a server.

The paper is organized as follows. In section II, we review the server–aided secret computation protocol proposed by Matsumoto et al. [3] and the concept of addition sequence. In section III we propose two improved protocols, one for low speed links and the other for high speed links. The performance analysis and some comparisons are given in section IV.

II. Review of Server–Aided SecretComputation Protocols and the Concept of Addition Sequence

A. Server–aided secret computation protocol [3]

Problem description:

A client has a pair of public keys (e, n) and a secret key d, where n is the product of two secret strong primes p and q, (i.e., n=pq, and ed≡1 mod ($\phi(n)$), where $\phi(n)$ = (p−1)(q−1) is the Euler's totient function of n). The client wishes to compute the signature S of a message M according to Eq.(1) with the aid of a server but reveals the client's secret d to the server with the probability less than or equal to the security level ϵ.

$$S \equiv M^d \pmod{n} .\tag{1}$$

Preparation:

The client splits his secret exponent d into an integer vector $D = (d_1, d_2, ..., d_m)$ and two binary vectors $F = (f_1, f_2, ..., f_m)$, and $G = (g_1, g_2, ..., g_m)$ such that

$$d \equiv \sum_{i=1}^{m} f_i \, d_i \pmod{p-1}\tag{2}$$

$$d \equiv \sum_{i=1}^{m} g_i \, d_i \pmod{q-1}\tag{3}$$

where the Hamming weights of W(F) and W(G) are less than or equal to L, and L<m (where $W(F) = \sum_{i=1}^{m} f_i$, $W(G) = \sum_{i=1}^{m} g_i$). The client keeps F and G as the secrets and D can be public.

Protocol:

Step 1: The client sends M,n and D to the server.

Step 2: The server computes the integer vector $Y = (y_1, y_2, ..., y_m)$ such that

$$y_i \equiv M^{d_i} \pmod{n}, \quad \text{for all i,}\tag{4}$$

and sends Y back to the client.

Step 3. The client computes Z_p and Z_q from F, G and Y as follows:

$$Z_p \equiv \prod_{i=1}^{m} y_i^{f_i} \pmod{p} \tag{5}$$

$$Z_q \equiv \prod_{i=1}^{m} y_i^{g_i} \pmod{q} \tag{6}$$

$$S \equiv Z_p \pmod{p} \tag{7}$$

$$S \equiv Z_q \pmod{q}. \tag{8}$$

Using the Chinese Remainder Theorem, the client can obtain the signature S from Z_p and Z_q [8].

Discussions:

(1) In this protocol, the client can obtain S with at most (L–1) multiplications mod p, (L–1) multiplications mod q, and one additional operation for computing the Chinese Remainder Theorem, where the lengths of p and q are about half of the length of n.

(2) The server needs to compute m exponentiations mod n.

(3) The cost of communication is 2(m+1) integers and each integer has at most $\log_2 n$ bits.

(4) Under the assumption that the RSA cryptosystem is secure, this protocol's security relies on searching the secret d via the domain of

$$\sum_{i=1}^{L} C\binom{m}{i}^2 . \tag{9}$$

If the security level ϵ is 10^{-30} (i.e., computationally secure), then m=50 and L=25 satisfy the condition.

Kawamura et al. have shown that this protocol allows a client to compute a signature 50 times faster, with a 10^5 times more powerful server's aid, than in the case without a server provided that the communication cost can be ignored. In the next section, we will proposed two protocols based on the addition sequence to improve the efficiency. We next review the concept of addition sequence.

B. Reviews of the addition sequence

The addition chain has been widely considered as an efficient method to compute the exponentiation and the evaluation of a polynomial since it requires the minimum number of multiplications [9]. An addition chain for a given number n is a sequence $\{a_0, a_1,..., a_r\}$ such that

(1) $a_0 = 1$, $a_r = n$,
(2) $a_i = a_j + a_k$ for $i>j \geq k$.

An addition sequence for a given increasing sequence $B = \{b_1, b_2, ...,b_t\}$ is a sequence $A = \{a_0, a_1,..., a_r\}$ such that

(1) $a_0 = 1$, $a_r = b_t$,
(2) $a_i = a_j + a_k$ for $i>j \geq k$,
(3) $B \subseteq A$.

Given a large number n, finding the shortest addition chain for n has been shown to be an NP–complete problem [10]. Similarly, computing the shortest addition sequence for a given increasing sequence B is also an NP–complete problem. This is due to the fact that the special sequence, $B = \{n\}$, includes the addition chain problem as its special case. Fortunately, we do not need to find the shortest addition sequence in the proposed protocols. Yao [11] estimates an upper bound of the length of an addition sequence:

$$\ell(b_1, b_2,...b_t) \leq \log_2 b_t + c t \log_2 b_t / \log_2 \log_2 b_t, \tag{10a}$$

where $\ell(b_1, b_2,..., b_t)$ is the length of a shortest addition sequence containing $(b_1, b_2,..., b_t)$ and

$$c = 2 + 4/\sqrt{\log_2 b_t}. \tag{10b}$$

Eq. (10) can be used to estimate the computational complexity of the proposed protocols. However, the simulated results show that the required length of an addition sequence is less than the estimated result by using Eq (10).

III. The Proposed SASC Protocols for the RSA Cryptosystem

In this section, we propose two SASC protocols based on the concept of addition sequence. Our protocols have much better performance than Matsumoto et al's protocol. Protocol–1 is very suitable for the low speed communication link (e.g., telephone lines) and protocol–2 is suitable for the high speed communication link (e.g., optical fibers). All the symbols we use in this section is the same as we used in section II.

Protocol–1:

This protocol is very similar to the protocol described in section II. The main difference is that the server can use an addition sequence to speedup the computations. We can use the following step 2' to replace step 2 in the previous protocol.

Step 2': The server constructs an addition sequence $A = \{a_0,...,a_r\}$ for the vector D received from the client and computes an integer vector $X = (x_1, x_2,.., x_r)$ with the aid of addition sequence A such that

$$x_i = M^{a_i} \bmod n, \quad \text{for } 1 \leq i \leq r .$$

Since A is an addition sequence, the server can compute X using at most r multiplications. Once X is obtained, since $D \subset A$, the server can directly obtain the integer vector $Y = (y_1, y_2,..., y_m)$ from X, where $y_i \equiv M^{d_i} \pmod{n}$, for $1 \leq i \leq m$, and sends Y back to the client.

Protocol–2:

Preparation:

In this protocol, the client will construct an addition sequence to speedup the computations for the client and the server simultaneously. The client splits the secret exponent d into two integer vectors $D_p = (d_{p1}, d_{p2},..., d_{pt})$ and $D_q = (d_{q1}, d_{q2},..., d_{qt})$ such that

$$d \equiv \sum_{i=1}^{t} d_{pi} \pmod{p-1} \tag{11}$$

$$d \equiv \sum_{i=1}^{t} d_{qi} \pmod{q-1} . \tag{12}$$

The client then constructs an increasing vector $D = (d_1, d_2,..., d_{2t}, K)$ from vectors

D_p, D_q and a random number $K > d_{2t}$, where

$$D_p \subset D \text{ and } D_q \subset D.$$

Finally, the client constructs an addition sequence $A = \{a_0, a_1, ..., a_r\}$ for the increasing vector D. The addition sequence A can be public but the vector D must be kept secret.

Protocol:

Step 1: The client sends M, n and A to the server.

Step 2: The server computes the integer vector $Y = (y_1, y_2, ..., y_r)$ such that

$$y_i \equiv M^{a_i} \pmod{n}, \quad \text{for } 1 \leq i \leq r \tag{13}$$

and sends Y back to the client. Since A is an addition sequence, the server can compute all y_i's using at most r multiplications.

Step 3: The client computes Z_p and Z_q from Y as follows:

$$Z_p \equiv \prod_{a_i \in D_p} y_i \pmod{p} \tag{14}$$

$$Z_q \equiv \prod_{a_i \in D_q} y_i \pmod{q} \tag{15}$$

$$S \equiv Z_p \pmod{p} \tag{16}$$

$$S \equiv Z_q \pmod{q}. \tag{17}$$

Using the Chinese Remainder Theorem, the client can obtain the signature S from Z_p and Z_q.

IV. Performance analysis and comparisons

In this section, we discuss the number of computations required for the client and the server in our protocols. The security level is 10^{-30} and the exponent d and the modulus n are 512–bit long.

As specified in [7], in protocol−1, m=50, L=25 satisfies the security condition. In step 2', the upper bound of the length of addition sequence A is

$$r = \log_2 M + (Cm\log_2 M)/(\log_2\log_2 M)$$
$$= 3856.$$

where M is the maximum element in D.

The simulated results show that the length of A is 1720 on average.

In protocol−2, under the same assumption that the RSA cryptosystem is secure, this protocol's security relies on searching the secret d via the domain of

$$C(\mathfrak{l})^2 . \tag{18}$$

If $C(\mathfrak{l})^2 > 10^{30}$,it will satisfy the same security condition,

$$\text{where} \quad r = \log_2 K + C(2t+1)\log_2 K/\log_2\log_2 K . \tag{19}$$

From Eq.(19) we know that r = 1336 and t = 6 satisfy the required security condition. However, the simulated results show that r = 720 and t = 7 satisfy the required security condition. Therefore, for the client, it needs 2(t−1) multiplications for computing Z_p and Z_q ((Eq.(14) and (15)) and one additional operation for computing the Chinese Remainder Theorem for evaluating S. For the server, it needs r multiplications. The required communication cost is 2(r+1) blocks of messages which is about $\frac{2(r+1)}{2(m+1)}$ times larger than the protocol described in section II, and each block has 512 bits.

Table 1 shows the required number of operations for the client and the server respectively and the communication costs in our protocols and the protocol proposed by Matsumoto et al.

From Table 1, the simulated results show that in protocol−1 the server can be 11.1 times faster than the server in the protocol proposed in [3] and the client has the same performance. The communication cost between these two protocols is the same. In protocol−2 the client (server) can be 4 (26.6) times faster than the client (server) in the protocol proposed in [3]. However, the required communication cost in our protocol is about 14.1 times more than that in their protocol. One thing we want to point out here is that in the protocol−2 it needs to construct an addition sequence. Since the exponent d is fixed all the time, we need only to construct the sequence once and store it for the rest of applications.

V. Conclusion

Two efficient SASC protocols based on the addition sequence are proposed in this paper. The first protocol is suitable for the applications with low speed communication links. The second protocol is suitable for the applications with high speed communication links. Performance analysis and some comparisons between proposed protocols and the existing one are also given. Results show that there is a trade–off between the communication cost and the efficiency.

REFERENCES

[1] R.L. Rivest, A. Shamir, and L. Adleman, "A method for obtaining digital signatures and public–key cryptosystem," Commun. ACM, Vol. 21, pp.120–126, Feb. 1978.

[2] T.Matsumoto and H. Imai, "How to use servers without releasing Privacy–Making IC cards more powerful," IEICE Technical Report, Rep. ISEC88–33. (May, 1988).

[3] T.Matsumoto and H. Imai, "Speeding up secret computations with insecure auxiliary devices," Proc. of CRYPTO'88, pp.497–506, 1988.

[4] S. Kawamura and A. Shimbo, "Computation methods for RSA with the aid of powerful terminals," 1989 Sym.on Cryptography & Inf. Security, Gotemba, Japan (Feb. 2–4 1989).

[5] S. Kawamura and A. Shimbo, "A method for computing an RSA signature with the aid of an auxiliary termimal," 1989 IEICE Autumn Natl. Conv. Rec. A–105.

[6] J.J. Quisquater and M. De Soete, "Speeding up smard card RSA computations with insecure coprocessors," Proc. SMART CARD 2000. Amsterdam (Oct. 1989).

[7] S. Kawamura and A. Shimbo, "Performance analysis of Server–Aided Secret Computation protocols for the RSA cryptosystem," The Trans. of the IEICE, vol. E73, No. 7, pp. 1073–1080, Jul. 1990.

[8] J.J. Quisquater and C. Couvreuer, "Fast decipherment algorithm for RSA public–key cryptosystem," Electron. Lett. 18, 21. pp. 905–907 (Oct. 1982).

[9] D.E. Knuth, The art of computer programming, Vol. II: Seminumerical algorithms. Reading, Addison Wesley, 1969.

[10] P. Downey and B. Leony and R. Sethi, "Computing sequences with addition chains," Siam Journ. Comput. 3 (1981) pp.638–696.

[11] Andrew Yao, "On the evaluation of powers," Siam. J. Comput. 5, (1976).

Table I Comparisons between our protocols and the protocol
proposed by Matsumoto et al.[3]

	The number of multiplications for the client (each with 256-bit modulus)	The number of multiplications for the server (each with 512-bit modulus)	Communication cost in blocks (each with 512 bits)
Protocol in [3]	48+ 1 CRT **	19150 *	102
Protocol-1	48+ 1 CRT	1720 †	102
		3856 ‡	
Protocol-2	12+ 1 CRT †	720 †	1442 †
	10+ 1 CRT ‡	1336 ‡	2674 ‡

* Our estimation is based on the "square and multiply" method;
  a 256-bit modulo exponentiation requires 383 multiplications
  on average.
** CRT : The Chinese Remainder Theorem.

† The simulated value.

‡ Value calculated by Eq.(10)

On Ordinary Elliptic Curve Cryptosystems

Atsuko Miyaji

Matsushita Electric Industrial Co., LTD.
1006, KADOMA, KADOMA-SHI, OSAKA, 571 JAPAN
miyaji@isl.mei.co.jp

Abstract

Recently, a method, reducing the elliptic curve discrete logarithm problem(EDLP) to the discrete logarithm problem(DLP) in a finite field, was proposed. But this reducing is valid only when Weil pairing can be defined over the m-torsion group which includes the base point of EDLP. If an elliptic curve is ordinary, there exists EDLP to which we cannot apply the reducing. In this paper, we investigate the condition for which this reducing is invalid. We show the next two main results.
(1) For any elliptic curve E defined over F_{2^r}, we can reduce EDLP on E, in an expected polynomial time, to EDLP that we can apply the MOV reduction to and whose size is same as or less than the original EDLP. (2) For an ordinary elliptic curve E defined over F_p (p is a large prime), EDLP on E cannot be reduced to DLP in any extension field of F_p by any embedding. We also show an algorithm that constructs such ordinary elliptic curves E defined over F_p that makes reducing EDLP on E to DLP by embedding impossible.

1 Introduction

Koblitz and Miller described how the group of points on an elliptic curve over a finite field can be used to construct public key cryptosystems([Mil], [Ko1]). The security of these cryptosystems is based on the elliptic curve discrete logarithm problem(EDLP). The best algorithm that has been known for solving EDLP is only the method of Pohlig-Hellman([Ko2]). Since it doesn't work for the elliptic curve over a finite field whose order is divided by a large prime, some works on the implementation of elliptic curve cryptosystems have been done ([Me-Va], [Be-Ca]). Recently Menezes, Vanstone and Okamoto([MOV]) proposed a noble method to reduce EDLP on an elliptic curve E defined over a finite field F_q to the discrete logarithm problem(DLP) in a suitable extension field of F_q. Using their method, H. Shizuya, T. Itoh and K. Sakurai([SIS]) gave a characterization for the intractability of EDLP from a viewpoint of computational complexity theory. T. Beth and F. Schaefer discussed the case where the extension degree of a finite field, in which EDLP is reduced to DLP, is lager than a constant. In this paper, we call their method ([MOV]) the MOV reduction.

The MOV reduction is constructed by a pairing defined over a m-torsion subgroup of an elliptic curve. It is called the Weil pairing. If an elliptic curve is supersingular, the Weil pairing is defined over any m-torsion subgroup of it. If an elliptic curve is ordinary (non-supersingular), there exists a m-torsion subgroup of it that the Weil pairing can't be defined over. Our main motivation for this work is to study EDLP on such m-torsion group of an ordinary elliptic curve.

Our result of this paper is following. For any elliptic curve E defined over $F_{q'}$, we can reduce EDLP on E to EDLP applied the MOV reduction in an expected polynomial time (Theorem 1). For a certain ordinary elliptic curve E defined over F_p (p is a large prime), we cannot reduce EDLP on E to DLP in any extension field of F_p by any embedding (Theorem 2).

Section 2 contains brief facts of the elliptic curves that we will need later. Section 3 explains the MOV reduction. Section 4 studies the case where we cannot apply the MOV reduction. Subsection 4-1 discusses how we can extend the MOV reduction to EDLP on any ordinary elliptic curve E defined over $F_{q'}$. Subsection 4-2 shows why we cannot reduce EDLP on an ordinary elliptic curves E defined over F_p to DLP in any extension field of F_p by embedding. Section 5 constructs ordinary elliptic curves E defined over F_p that makes reducing EDLP on E to DLP by embedding impossible.

Notation

p	: a prime
r	: a positive integer
q	: a power of p
F_q	: a finite field with q elements
K	: a field (include a finite field)
ch(K)	: the characteristic of a field K
K^*	: the multiplicative group of a field K
$\overline{K}$	: a fixed algebraic closure of K
E	: an elliptic curve
	If we remark a field of definition K of E, we write E/K.
#A	: the cardinality of a set A
o(t)	: the order of an element t of a group
Z	: the ring of integers

2 Background on Elliptic Curves

We briefly describe some properties of elliptic curves that we will use later For more information, see[Sil]. In the following, we denote a finite field F_q by K .

Basic Facts

Let E/K be an elliptic curve given by the equation, called the Weierstrass equation,

$$E : y^2 + a_1xy + a_3y = x^3 + a_2x^2 + a_4 + a_6 \qquad (a_1, a_3, a_2, a_4, a_6 \in K).$$

The j-invariant of E is an element of K determined by a_1, a_3, a_2, a_4 and a_6. It has important properties as follows.

(j-1) Two elliptic curves are isomorphic (over $\overline{K}$) if and only if they have the same j-invariant.

(j-2) For any element $j_0 \in K$, there exists an elliptic curve defined over K with j-invariant equal to j_0. For example, if $j_0 \neq 0, 1728$, we let

$E: y^2 + xy = x^3 - 36/(j_0 - 1728)x - 1/(j_0 - 1728)$. Then j-invariant of E is j_0.

The Group Law

A group law is defined over the set of points of an elliptic curve, and the set of points of an elliptic curve forms an abelian group. We denote the identity element ∞. After this, for $m \in \mathbb{Z}$ and $P \in E$, we let

$mP = P + \ldots\ldots + P$ (m terms) for $m > 0$,

$0P = \infty$, and

$mP = (-m)(-P)$ for $m < 0$.

The set of K-rational points on the elliptic curve E, denoted $E(K)$, is

$$E(K) = \{(x,y) \in K^2 \mid y^2 + a_1xy + a_3y = x^3 + a_2x^2 + a_4x + a_6\} \cup \{\infty\}.$$

$E(K)$ is a subgroup of E and a finite abelian group. So we can define the descrete logarithm problem over it.

Twist of E/K

A twist of E/K is an elliptic curve E'/K that is isomorphic to E over $\overline{K}$. We identify two twists if they are isomorphic over K.

Example Two elliptic curves E/K and E_1/K given below are twists each other.

$$E : y^2 = x^3 + a_4x + a_6$$
$$E_1: y^2 = x^3 + a_4c^2x + a_6c^3$$

$(a_4, a_6 \in K$, c is any non-quadratic residue modulo p$)$.

The Weil pairing

For an integer $m \geq 0$, the m - torsion subgroup of E, denoted $E[m]$, is the set of points of order m in E,

$E[m] = \{P \in E \mid mP = \infty\}$.

We fix an integer $m \geq 2$, which is prime to $p = ch(K)$. Let μ_m be the subgroup of the mth roots of unity in $\overline{K}$.

The Weil e_m-Pairing is a pairing defined over $E[m] \times E[m]$

$$e_m: E[m] \times E[m] \quad \rightarrow \quad \mu_m.$$

For a definition of the Weil e_m-pairing, see [Sil]. We list some useful properties of the Weil e_m-pairing.

For $E[m] \ni S, T, S_1, S_2, T_1, T_2,$

(e-1) Bilinear :

$$e_m(S_1+S_2, T)=e_m(S_1, T)e_m(S_2, T)$$
$$e_m(S, T_1+T_2)=e_m(S, T_1)e_m(S, T_2);$$

(e-2) Alternating :

$$e_m(S, T)=e_m(S, T)^{-1};$$

(e-3) Non-degenerate :

If $e_m(S, T)=1$ for all $S \in E[m]$, then $T=\infty$;

(e-4) Identity :

$e_m(S, S)=1$ for all $S \in E[m]$.

Number of Rational Points

As for $\#E(K)$, the following Hasse's theorem gives a bound of the number of rational points of an elliptic curve.

Theorem ([Sil]) Let E/K be an elliptic curve . Then $|\#E(K)-q-1| \leq 2q^{1/2}$.

Let $\#E(K)=q+1-a_q$. If $K=F_p$, we further have the next theorem by Deuring.

Theorem ([Deu]) Let a_p be any integer such that $|a_p| \leq 2p^{1/2}$. Letting $k(d)$ denote the Kronecker class number of d, there exist $k(a_p^2-4p)$ elliptic curves over F_p with number of points $p+1-a_p$, up to isomorphisms.

3 Reducing EDLP to DLP in a finite field

In this section, we briefly describe the MOV reduction of EDLP via Weil pairing. For more information, see [MOV].

First we give the definition of EDLP .

EDLP([Ko2])

Let E/F_q be an elliptic curve and P be a point of $E(F_q)$. Given a point $R \in E(F_q)$, EDLP on E to the base P is the problem of finding an integer $x \in Z$ such that $xP=R$ if such an integer x exists.

Next we mention about embedding the subgroup $<P> \subset E(K)$ generated by a point P into the multiplicative group of a finite extension field of K. This embedding is constructed via Weil pairing. It is the essence of the MOV reduction. In the following, we denote a finite field F_q by K and fix an elliptic curve E/K and a point $P \in E(K)$. We further assume that $o(P) = m$ is prime to $p=ch(K)$.

Embedding

Let Q be another point of order m such that $E[m]$ is generated by P, Q. Let K' be an extension field of K containing μ_m. We can define a homomorphism

$$f : <P> \rightarrow K'^*$$

by setting

$$f(nP) \quad = \quad e_m(nP, Q).$$

From the definition of Weil pairing, it follows easily that f is an injective homomorphism from $<P>$ into K'^*. As $K' \supset \mu_m$, the subgroup $<P>$ of E is a group isomorphism to the subgroup μ_m of K'^*.

Summary of the MOV reduction

We summarize the MOV reduction of EDLP, which finds an integer x such that $R = xP$ for a given $R \in E(K)$, with the above embedding.

We can check in probablistic polynomial time whether $R \in <P>$ or not. So we assume that $R \in <P>$. Since m is prime to p, we can construct an injective homomorphism f from $<P>$ into K'^* as stated above. Then the problem is equal to find an integer x such that $f(R) = f(P)^x$ for a given $f(R), f(P) \in K'$. In this way, we can reduce EDLP to DLP in an extension field K' of K.

Note that this reducing is invalid if m is divisible by $p = \text{ch}(K)$ because the above injective homomorphism cannot be defined in the case. The next section investigates this case.

4 Inapplicable case

Definition   Let E/F_q be an elliptic curve. If E has the properties $E[p^t] = \{\infty\}$ for all integer $t \geq 1$, then we say that E is supersingular. Otherwise we say that E is ordinary.

Remark   Let E be a supersingular elliptic curve. The definition of supersingular says that $o(T)$ is prime to $\text{ch}(K) = p$ for all $T \in E(K)$.

In the following, we denote a finite field F_q by K and fix an elliptic curve E/K and a point $P \in E(K)$. We further assume that $o(P) = m$ is divisible by $p = \text{ch}(K)$. From the above remark, it follows that E is ordinary. We will describe EDLP on such a point of an ordinary elliptic curve in the next two subsections.

4-1 Ordinary elliptic curves over F_{2^r}

In this subsection, we investigate the case of $q = 2^r$. Let m be expressed by $m = 2^t k$ (k is an integer prime to 2, t is a positive integer). And EDLP on E to the base P is finding an integer x such that $R = xP$ for given $R \in E(K)$ (section 2).

As we assume that $\gcd(m, 2) \neq 1$, we can't apply the MOV reduction directly to this case. So we extend the MOV reduction as follows.

The extended reducing method

If all the prime factors of k are small, then we can solve this problem with Pohlig-Hellman's method ([Ko2]). So we assume that k has a large prime factor.

Let $P' = 2^t P, R' = 2^t R$. Then in a probablistic polynomial time, we can check whether $R' \in <P'>$ or not ([MOV]). If $R' \notin <P'>$, then $R \notin <P>$. So we assume that $R' \in <P'>$. Since $o(P') = k$ is prime to 2, we can apply the MOV reduction ([MOV]) to this case. Namely, we can work in a suitable extension field of K and find an integer x' such that $R' = x'P'$. Then we get $2^t(R - x'P) = \infty$. If we assume that $R \in <P>$, we get $(R - x'P) \in <P>$. From the group theory, it follows easily that a finite cyclic group $<P>$ has only one subgroup whose order devides $m = \#<P>$. So we get $(R - x'P) \in <kP>$. Now we change the base P of EDLP into kP, then we have only to find an integer x" such that $R - x'P = x''(kP)$. Since $\#<kP>$ is 2^t, we can easily find an integer x" with Pohlig-Hellman's method ([Ko2]). So we can find an integer x by setting $x \equiv x' + x''k \pmod{m}$.

Now we summarize the extended reducing method as follows.

Condition : Find an integer x such that R=xP for given $R \in E(K)$. Let $m=o(P)$ be expressed
by $m = 2^t k$ (k is an integer prime to 2, t is a positive integer).

Method : (1)Find a non-trivial subgroup $<2^t P> \subset <P>$ whose order is prime to $p = ch(K)$.

(2)Embed $<2^t P>$ into the multiplicative group of a suitable extension field of K via an injective homorphism constructed by Weil pairing.

(3)Change EDLP on E to the base P into EDLP on E to the base kP. (Since all of the prime factors of $\#<kP>$ are small, we can easily solve such EDLP.)

The above discussion completes the proof of the following.

Theorem1   For any elliptic curve E/F_f and any point $P \in E(F_f)$, we can reduce EDLP on E (to the base P), in an expected polynomial time, to EDLP that we can apply the MOV reduction to and whose size is same as or less than the original EDLP.

Remark   We proved Theorem1 for a field F_f. We can extend the theorem to a field $F_{f'}$ if we can generate the tables of the discrete logarithm at most in a polynomial time in the element size.

4-2 Ordinary elliptic curves over F_p

In this subsection, we investigate the case of q=p. Let p be a large prime and m be expressed by $m = p^t k$ (k is an integer prime to p, t is a positive integer). From Hasse's theorem (section 2), there is a bound of $\#E(K)$. So the integer m must satisfy that $(m - p - 1) \leq 2p^{1/2}$.

The next result is easy to prove.

Lemma   Let p be a prime more than 7 and E/F_p be an ordinary elliptic curve. We assume that there is a point $P \in E(K)$ whose order is divisible by p. Then the point P has exactly order p. Furthermore $E(K)$ is a cyclic group generated by P.

So we try to solve EDLP on the above ordinary elliptic curve, namely an elliptic curve generated by a point of order p. Then non-trivial subgroup of $E(K)$ is only itself and p is a large prime. So we cannot apply the extended reducing method in section 4-1 to it.

We assume that $E(K) = <P>$ can be embedding into the multiplicative group of a suitable extension field K' of K via any way instead of Weil pairing. At this time we can reduce EDLP on E (to the base P) to DLP on K'. But, for any integer r, there is no any subgroup of K'^r, whose order is p. So we cannot embed $<P>$ into the multiplicative group of any extension field of K.

The next result follows the above discussion.

   Theorem2   For an elliptic curve E/F_p such that $\#E(F_p) = p$ and any point $P \neq \infty$ of $E(F_p)$, we cannot reduce EDLP on E (to the base P) to DLP in any extension field $F_p r$ of F_p by any embedding $<P>$ into the multiplicative group of $F_p r$.

5 Constructing elliptic curves

In this section, we describe the method of constructing elliptic curve E/F_p with p elements. In the following, let p be a large prime. We get the next result by Hasse's theorem and Deuring's theorem (section2).

   Lemma   Let $k(d)$ denote the Kronecker class number of d. There exist $k(1-4p)$ elliptic curves E/F_p with p elements, up to isomorphism.

Because of $k(1-4p) \geq 1$, we get that there exists at least one elliptic curve E/F_p with p elements for any given prime p. From the prime distribution, it follows easily that, for primes of $O(p)$, the number of elliptic curves E/F_p with p elements is at least $O(p/\log(p))$ ([Ri]). Now we mention how to construct such an elliptic curve E/F_p. Original work concerning this was done by Deuring ([La2], [At-Mo], [Mo]). In the following, we explain the essence of his work.

Let d be an integer such that $4p-1=b^2d$ (b is an integer). Then there is a polynomial $P_d(x)$ called class polynomial. For a definition of the class polynomial, see [La2], [At-Mo].

The class polynomial $P_d(x)$ has the following properties.

(c-1) $P_d(x)$ is a monic polynomial with integer coefficients.

(c-2) The degree of $P_d(x)$ is the class number of an order O_d of an imaginary quadratic field. (For a definition of the order, see [Sil] and for the class number, see [La1].)

(c-3) $P_d(x)=0$ splits completely modulo p.

Let j_0 be a root of $P_d(x)=0$ (modulo p). Then j_0 gives the j-invariant of an elliptic curve E/F_p with p elements. So we make an elliptic curve E/F_p with j-invariant j_0 as we mentioned in section2, and one of twists of E/F_p is an elliptic curve with p elements. Next we discuss how to find such curves among all twists in a practical way.

Decide which twist of E/F_p has an order p

For any twist E_t of E/F_p with j-invariant j_0, fix any point $X_t \neq \infty$ of $E_t(F_p)$ and calculate pX_t. If $pX_t = \infty$, then $E_t(F_p)$ has exactly p elements. This follows the section 4-2. For any given elliptic curve E/F_p , there are at most six twists modulo F_p-isomorphism. So we can decide which twist of E/F_p has an order p in a polynomial time of the element size.

Good d and good p

For a given large prime p, we can construct an elliptic curve E/F_p as we mentioned above. What prime p and integer d such that $4p-1=b^2d$ (b is an integer) are good for constructing such an elliptic curve? We will find a prime p and an integer d such that the order O_d has a small class number. Because if the order O_d has a large class number, the degree of $P_d(x)$ is large and it is cumbersome to construct $P_d(x)$.

Procedure for constructing an elliptic curve

We can construct an elliptic curve by the following algorithm.

Algorithm

(p-1) Choose an integer d such that the order O_d has a small class number from a list ([Ta]).

(p-2) Find a large prime p such that $4p-1=b^2d$ for an integer b.

(p-3) Calculate a class polynomial $P_d(x)$.

(p-4) Let $j_0 \in F_p$ be one root of $P_d(x)=0$ (modulo p).

(p-5) Construct an elliptic curve E/F_p with j-invariant j_0.

(p-6) Construct all twists of E/F_p.

(p-7) For any twist E_t of E/F_p, fix any point $X_t \neq \infty$ of $E_t(F_p)$ and calculate pX_t. If $pX_t = \infty$, then $E_t(F_p)$ has exactly p elements.

Remarks For a fixed integer d and any integer b, how many primes p satisfy the condition such that $4p-1=b^2d$? This is a problem to be solved.

Example We construct an elliptic curve by the above algorithm.

(p-1) Let d=19 then O_{19} has a class number 1.

(p-2) Let p=235208607464683519348911841623 then $4p-1=19*(1451*48496722383)^2$.

(p-3) Calculate a class polynomial $P_{19}(x)$ then we get $P_{19}(x)=x+884736$.

(p-4) Let $j_0 = -884736$.

(p-5) Let $E: y^2 = x^3 + a*x + b$

with a=185691005893171199485988223307, b= 99035203143024639725860038632.

(p-6) Twist of E/F_p is E_1 , where E_1 is as following,

$E_1: y^2 = x^3 + a_1 * x + b_1$

with a_1 = 185691005893171199485988223307 , b_1 = 136173404321658879623058029911 .

(p-7) Let E (F_p) $\ni$ X be (1, 12834397719522088187599559212) and E_1 (F_p) $\ni$ X_1 be (0, 2251799813687456). Calculate pX, pX$_1$ and we get pX $=\infty$, pX$_1 \neq \infty$. So E/F_p is generated by X , which has an order p.

Remark   In (p-2), we choose a prime p that is congruent 3 modulo 4. Because then we can find points of E (F_p) and E_1 (F_p) easily.

Using the above E/F_p and X , we construct EDLP on E to the base X. Then up to the present, the best algorithms that are known for solving this problem are only the method of Pohlig-Hellman.

We end this section by the next conclusion.

Conclusion   With the above algorithm, we can construct EDLP on E/F_p such that we cannot reduce it to DLP in any extension field by embedding.

6 Final remarks

For an ordinary elliptic curve E defined over F_p (p is a large prime), we showed in theorem 2 that there exists EDLP on E that cannot be reduced to DLP in any extension field of F_p by any embedding. What is the relation between such EDLP and DLP? It is an open problem to be solved. Which of such elliptic curve cryptosystems is good for implementation? It is another problem to be considered.

7 Acknowledgements

I wish to thank Makoto Tatebayashi for his helpful advice. I am grateful to Tatsuaki Okamoto for his teaching me about the computational complexity theory. I express my gratitude to Hiroki Shizuya for his arising my further research. I would like to thank Yoshihiko Yamamoto for his teaching me about the class polynomial. I am also grateful to Francois Morain for sending me his paper [At-Mo].

References

[At-Mo] A. O. L. Atkin and F. Morain, "Elliptic curves and primality proving", Research Report 1256, INRIA, Juin 1990. Submitted to Math. Comp.

[Be-Ca] A. Bender and G. Castagnoli, "On the implementation of elliptic curve cryptosystems", Advances in Cryptology - Proceedings of Crypto '89, Lecture Notes in Computer Science, 435 (1990), Springer-Verlag, 186-192.

[Be-Sc] T. Beth and F. Schaefer, "Non supersingular elliptic curves for public key cryptosystems", Abstracts for Eurocrypto 91 , Brighton, U.K. 155-159.

[Deu] M. Deuring, "Die Typen der Multiplikatorenringe elliptischer Funktionenkörper", Abh. Math. Sem. Hamburg 14 (1941), 197-272.

[Ko1] N. Koblitz, "Elliptic curve cryptosystems", Math. Comp. 48(1987), 203-209.

[Ko2] N. Koblitz, "A course in Number Theory and Cryptography", GTM114, Springer-Verlag, New York(1987).

[La1] S. Lang, "Algebraic Number Theory", GTM110, Springer-Verlag, New York(1986).

[La2] S. Lang, "Elliptic Functions", Addison-Wesley, 1973.

[Mil] V. S. Miller, "Use of elliptic curves in cryptography", Advances in Cryptology-Proceedings of Crypto'85, Lecture Notes in Computer Science, 218 (1986), Springer-Verlag, 417-426.

[Me-Va] A. Menezes and S. Vanstone, "The implementaion of elliptic curve cryptosystems", Advances in Cryptology - Proceedings of Auscrypt'90, Lecture Notes in Computer Science, 453(1990), Springer-Verlag, 2-13.

[Mo] F. Morain, "Building cyclic elliptic curves modulo large primes", Abstracts for Eurocrypto91 , Brighton, U.K. 160-164.

[MOV] A. Menezes, S. Vanstone and T. Okamoto, "Reducing elliptic curve logarithms to logarithms in a finite field", to appear in Proc. STOC'91.

[Ri] P. Ribenboim, "The book of prime number records", Springer-Verlag, New-York, 1988.

[Sil] J. H. Silverman, "The Arithmetic of Elliptic Curves", GTM106, Springer-Verlag, New York, 1986

[SIS] H. Shizuya, T. Itoh and K. Sakurai, "On the Complexity of Hyperelliptic Discrete Logarithm Problem", Proc. Eurocrypt'91, Lecture Notes in Computer Science, Springer-Verlag (to appear).

[Ta] T. Takagi, "Syotou seisuuronn kougi", Kyouritu Syuppan.

Cryptanalysis of another knapsack cryptosystem

Antoine Joux   Jacques Stern

December 12, 1991

Abstract

At the last Eurocrypt meeting, a cryptosystem based on modular knapsacks was proposed (see [11]). We show that this system is not secure, and we describe two different ways of breaking it using the LLL algorithm. This is one more example of a cryptosystem that can be broken using this powerful algorithm (see [1, 13, 14]). For more details, the reader should refer to [4].

1   The proposed cryptosystem

We briefly describe this knapsack based system. Every computation in this system are done in $\mathbb{Z}_p$, where p is a fixed prime, and an absolute value must be defined in $\mathbb{Z}_p$: $|a|$ is the smallest of the least positive remainders of a and $-a$. Given a number k, we say that a is k-small if $|a| \leq k$, and that a is k-large if $|a| \geq p/2 - k$. Let C, D, S be $n \times n$ matrices of small numbers, Δ be a diagonal matrix of large numbers and R be a $n \times n$ invertible matrix, and compute $A = R^{-1}(\Delta - SC)$, $B = -R^{-1}SD$ and $E = (A, B)$. Encryption of a bit vector x is done by computing $c = Ex$ and, if p is suffciently large, then decryption can be computed using the secret key R by the following formulas. First let $l = Rc$, the first half of x can then be obtained from the first half of l simply by translating small numbers into 0 and large numbers into 1. The remaining half of x can then be obtained by methods of linear algebra. The proofs can be found in [11].

The author of the cryptosystem then restricts himself to the case $S = I$, where I denotes the identity matrix and claims that in this case $p > 4kn$, is sufficent to ensure decryption. He also claims that $k = 1$ is a good choice. We present here two different methods to break the cryptosystem with $S = I$, $k = 1$ and with p being the smallest prime greater than $4n+1$. One of our method is key oriented and allows us to compute the secret key given the public one, the other is message oriented and gives the original message back without using the secret key. A sligthly different version of this second attack was found independently by an other researcher (cf [5]), and has already been exposed at Crypto 91. Thus, we won't give any details concerning this second attack. From a theoretical point of view, the attacks don't need any modification to perform with different values for S and k, however, we haven't made any practical tests in these cases.

2 Our attack on R

In this section, we show how, given the public key E, we can compute the private key R using the LLL algorithm. We run this algorithm n times, one for each line of R.

Let us describe how this is done. We first choose a large even scaling factor λ and we then define:

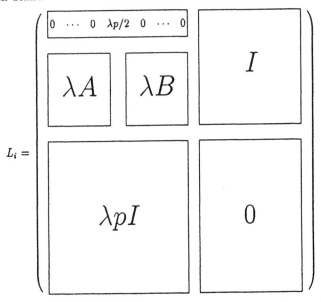

where $\lambda p/2$ on the first line of L_i is at the i-est position. L_i is a $(3n+1) \times (3n+1)$ matrix. We then use the LLL algorithm to find a reduced basis of the lattice generated by the <u>rows</u> of L_i, and we call M_i the resulting matrix. Let us now discuss what kind of short vectors the matrix M_i will presumably contain. First of all, there are some very short (and useless) vectors in the lattice L_i, they can be obtained by multiplying the first row of L_i by 2 or a single of the following n rows by p, and by reducing the $2n$ first components of the resulting vector modulo p using the $2n$ last rows of L_i. Using this method we can obtain $n + 1$ independent vectors of the lattice, these vectors have norm 2 or p, and are so short that they will surely appear in M_i. Moreover, it is easy to see that if we denote by $x_1, x_2, \cdots, x_{3n+1}$ the canonical basis of the vector space $\mathbb{R}^{3n+1}$, the $n + 1$ vectors we just found span the same subspace than $x_{2n+1}, x_{2n+2}, \cdots, x_{3n+1}$. We can now consider the other vectors of M_i; they must be linearly independant of the $n + 1$ vectors found so far. Thus their first $2n$ components can't be all zero. We can remark that since λ is large, in any of those vectors, the weight of the last $n + 1$ components will be negligible compared to that of the first $2n$ components. In order to simplify the argument, we can now discard the last components, and then divide everything by the scaling factor λ. It also easy to note that using the last rows of L_i to reduce it modulo p, any vector can be replaced by a vector having all his components in the range $-(p-1)/2, \cdots, (p-1)/2$. Thus to make a rough estimate of the size of an ordinary vector in M_i, it is sufficient to compute the average size of a random vector whose $2n$ components are in $\{-(p-1)/2, \cdots, -1, 0, 1, \cdots, (p-1)/2\}$. Before doing this computation, let us recall that $1 + 2 + \cdots + n = n(n+1)(2n+1)/6$ and compute the mean value of x^2 where x is a random value in $\{-(p-1)/2, \cdots, -1, 0, 1, \cdots, (p-1)/2\}$. We can write:

$$
\begin{aligned}
< x^2 > &= \frac{((p-1)/2)^2 + \cdots + 1^2 + 0 + 1^2 + \cdots + ((p-1)/2)^2}{p} \\
&= \frac{2(p-1)(p+1)p}{4p} \\
&= \frac{p^2 - 1}{2}
\end{aligned}
$$

Thus the average size of a random vector of $2n$ components each reduced modulo p is $n(p^2 - 1)$. Since $p = 4nk$, this average size is approximately $16k^2n^3$. We can now describe the short vector that is interesting and provides the ith line of R. If we let X_i denotes the ith line of a matrix X, we can write $R_i(A, B) = (\delta_i - C_i, -D_i)$, since $S = I$. Thus $R_i(A\ B)$ has all its components small except the ith which is large, if we now add (or substract since we work modulo p), $p/2$ from the ith component it will also become small, more precisely, it will be $2k$-small whereas the other components are k small. It is clear that we can lift these computation to L_i and obtain a vector whose components are small numbers scaled by λ followed by ± 1 and by R_i. To estimate the size of this vector we drop the $n+1$ last components and divide by λ and we find a maximal size of $(2n+3)k^2$. We can compute the ratio of the size of an ordinary vector and of the size of this particular vector; we find $16k^2n^3/((2n+3)k^2) \approx 8n^2$. This ratio is a good way to know if we can hope LLL to find a certain vector, the highest the ratio, the more certain we are the vector will be found. A ratio of $8n^2$ is not enough to ensure that the vector will be found: the theory only proves that with a ratio of $2^{n/2}$ the vector will be found. But LLL is usually much more efficient than it has been proven to be, as will be shown in the section discussing numerical results.

3 Our attack on a given message

In this section, we only give a lattice permitting to retrieve the original message x, given an encoded message c and the public key E. We first let λ be a large scaling factor, and we then define:

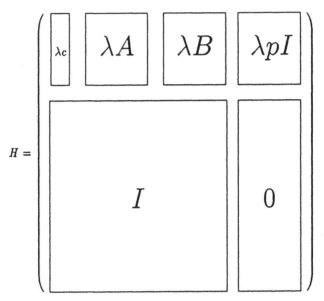

$$H = \begin{pmatrix} \lambda c & \lambda A & \lambda B & \lambda pI \\ & & & \\ & I & & 0 \\ & & & \end{pmatrix}$$

The lattice to reduce, in order to retrieve x, is the lattice generated by the <u>**columns**</u> of H.

4 First numerical results

Since we make an extensive use of the LLL algorithm in large dimensions ($3n + 1$ for both methods) we use a high speed implementation of the LLL algorithm which was written by the first named author in June 1990, or more precisely, two different implementations since, in the message oriented attack, we were able to use a faster but less numerically stable version. We run all our numerical experiments on a Sun 4/20 machine. Let us now describe a few experiments to show what can be done with the simple methods that have just been described. All the experiments are with $k = 1$, and unless otherwise stated with a LLL constant of 0.95 (see [10] for a definition of the LLL constant).

We first give a table of results concerning the message oriented attack, with the user CPU time. This is not very accurate, but it is sufficient to get an idea of the actual time needed.

dimension n	first try	second try	third try
10	1'58	1'49	1'56
15	14'27	13'56	14'00
20	55'57	56'04	56'22
25	160'15	159'53	160'55
30	372'37	379'39	375'10

We can see in this table that this attack works only for dimensions lower than 35 . We will soon explain how to go further. We now give a similar table for the key oriented attack.

dimension n	first try	second try	first try, new program
5	0'08	0'12	0'02
10	4'22	5'45	0'23
15	35'13	45'36	3'03
20	202'14	184'46	11'12
25	503'35	509'14	31'39
30	1189'48	1217'37	78'19
35	2387'52	2249'09	161'58

It is important to remember that the amounts of time[1] given here correspond to the computation needed to retrieve one line of the key. Thus the time needed to retrieve the whole is much longer, but we will see in the following section that partial knowledge about one message can be used to reduce the lattice dimension in the message oriented attack. For exemple, in dimension 30 it requires about 25 days to retrieve the whole key, but every passing day gives us about one line of the key, and thus ease the needed computation to decode incoming messages.

5 How to extend the attack or the art of semi-exhaustive search

We explain how the LLL algorithm can be coerced into making certain "choices". There are several ways of doing it depending on the kind of lattice we are working with.

We now describe the easiest case, the coertion of the message oriented attack. Let us suppose, for a moment, that we know a few bits of x; in this case, we try to solve a shorter system obtained by deleting the columns corresponding to known coefficients, and by substracting the columns corresponding to 1 from the codeword column c. This reduce the number of basis vector of the lattice by the number of known bits, and we can also reduce the length of the vectors by the same quantity by deleting the corresponding lines from the identity submatrix. Since we decreased the size of the knapsack without changing the value of p, it will be easier for LLL to find a meaningful solution. Now if we don't know any bit of x we can make an exhaustive search concerning a few bits. This will involve running the LLL algorithm many times, but we will find x in most cases. We call this kind of approach a semi-exhaustive search.

Such a semi-exhaustive search can also be done in the case of the key oriented attack. The method is more complicated and more expensive in computing time. We will describe the method in the full paper.

---

[1]Since the Asiacrypt conference, we have writen a new version of our program performing LLL reduction, this new version is much faster than the old one. As an example, we have added the time for the new program in the table of times for the key oriented attack.

6 Numerical result for the extended attack

We first give a table giving for each dimension the time needed for one run of the LLL algorithm and the number of bits affected by the semi-exhautive seach, in the case of the message oriented attack.

dimension n	number of bits	time
35	1	failed
35	2	702'13
40	1	1293'47
40	2	1253'39

This table shows that using 4 Sun 4/20, we can find most messages in dimension 40 within 21 hours. It is slow, but as we will see in the conclusion, it is much faster than exhaustive search. The reader might think than the number of results in our table is too small to validate the extension of our attack, but we have omitted the successful runs in dimension 35 and 40 with a number of searched bits greater than 2. We have also omitted another group of results with a LLL constant of 0.75, with such a constant, direct attack was not sufficient for dimension 25 or more, but the extended scheme did work well, of course with a greater number of searched bits.

7 Conclusion

In the original article, the only reference to the security of this system is that, in order to exclude exhautive search attacks, one should take n sufficiently large. From this remark, the author concludes that $n = 100$ seems to be suitable. This suggests several remarks. Firstly, if one only wants to exclude exhaustive search, $n = 40$ is truly sufficient since, if 1000 linear equations systems in dimension 40 can be solved every second, it will take more than 34 years to exhaust all the possibilities, and the average breaking time will be of 17 years. In most applications, even $n = 35$ should suffice since it will take a year or so to exhaust all possibilities. This is the reason which led us to conduct numerical experiment only for $n \leq 40$. A second remark is that for $n = 100$ with $k = 1$ and $p = 251$ (an example the author claims to be practical, although he cannot formally prove that decryption is ensured with such a p), the public key will use up 20000 bytes of memory, and the decoding time, consisting of a matrix product followed by linear algebra will certainly not permit to achieve a high speed cryptographic channel.

We have thus proven in this article that any reasonable implementation of the cryptosystem could be broken in a reasonable amount of time. Since the message oriented attack doesn't really use any knowledge about the cryptosystem apart from the trivial fact that it is a knapsack, we can add that the whole family of modular knapsack based cryptosystems can be broken using the same method as long as p is "sufficiently" large compared to n.

References

[1] L. Adleman. On breaking the iterated Merkle-Hellman public key cryptosystem. *Proceedings of the ACM Symposium on the Theory of Computing* (1982) 402-412.

[2] E. Brickell. Solving low density knapsacks. *Proceedings of Crypto 83.*

[3] E. Brickell. Breaking Iterated Knapsacks. *Proceedings of Crypto 84.* Lecture Notes in Computer Science 196.

[4] E. Brickell and A. M. Odlyzko. Cryptanalysis: A survey of recent results. *Proceedings IEEE* 1988.

[5] Y. M. Chee. The Cryptanalysis of a New Public-Key Cryptosystem based on Modular Knapsacks. *Proceedings of Crypto 91.* Lecture Notes in Computer Science, to appear.

[6] M. J. Coster, A. Joux, B. A. LaMacchia, A. Odlyzko, C. P. Schnorr and J. Stern. Improved Low-Density Subset Sum Algorithms. To appear.

[7] R. Kannan. Improved algorithms for integer programming and related lattice problems. *Proceedings of the ACM Symposium on the Theory of Computing.* (1983), 193-206.

[8] R. Kannan, A. K. Lenstra and L. Lovàsz. Polynomial factorisation and nonrandomness of bits of algebraic and some transcendental numbers, Carnegie-Mellon University. Computer Science Department Technical Report (1984).

[9] J. C. Lagarias and A. M. Odlyzko. Solving low-density subset sum problems. *Proceedings of IEEE symposium on the foundations of Computer Science.* (1983) 1-10.

[10] A. K. Lenstra, H. W. Lenstra and L. Lovàsz. Factoring polynomials with rational coefficients. *Math. Annalen* 261 (1982) 515-534.

[11] V. Niemi. A new trapdoor in knapsacks. *Advances in Cryptography - Proceedings of EUROCRYPT 90*, Lecture Notes in Computer Science, to appear.

[12] A. Shamir. A polynomial-time algorithm for breaking the basic Merkle-Hellman cryptosystem. *Proceedings of th IEEE symposium on the foundations of Computer Science.* (1982) 145-152.

[13] J. Stern. Secret linear congruential generators are not cryptographically secure. *Proceedings of the IEEE symposium on the foundations of Computer Science.* (1987) 421-426.

[14] J. Stern and P. Toffin. Crypanalysis of a public-key cryptosystem based on approximations by rational numbers. *Advances in Cryptography - Proceedings of EUROCRYPT 90*, Lecture Notes in Computer Science 473.

Collisions for Schnorr's Hash Function FFT-Hash Presented at Crypto '91

Joan Daemen, Antoon Bosselaers,
René Govaerts and Joos Vandewalle

Katholieke Universiteit Leuven, Laboratorium ESAT,
Kardinaal Mercierlaan 94, B–3001 Heverlee, Belgium.

Abstract

A method is described to generate collisions for the hash function FFT-Hash that was presented by Claus Schnorr at Crypto '91. A set of colliding messages is given that was obtained by this method.

1 Introduction

In the Rump Session of Crypto '91 Claus Schnorr presented FFT-Hash. This is a function that hashes messages of arbitrary length into a 128 bit hash value. It consists of two rounds, where every round is the combination of a Fast Fourier Transform over $GF(2^{16}+1)$ and a nonlinear recursion. It was claimed that producing a pair of messages that yield the same hashvalue is computationally infeasible. We have written a program that outputs a set of 384 bit messages that all have the same hash value for FFT-Hash. The CPU-time consumed is of the order of a few hours. An optimized version of the program is expected to take only a few minutes on a modern PC. The first collision was produced on October 3rd '91.

2 Description of FFT-Hash

Padding: The message is padded with a single "1" followed by a suitable number of "0" bits followed by the binary representation of its original length. The padded message can then be seen as the concatenation of a number of 128-bit blocks: $M_0 \parallel M_1 \ldots \parallel M_{n-1}$.

Algorithm for the hash function h: $H_i = g(H_{i-1} \parallel M_{i-1})$ for $i = 1, \ldots, n$. $H_i \in \{0,1\}^{128}$ and initial value $H_0 = 0123\ 4567\ 89ab\ cdef\ fedc\ ba98\ 7654\ 3210$ (hex.). The output of $h(M) = H_n$.

Algorithm for the function g: Let $p = 2^{16}+1$. The input to g is split up into 16 components $(e_0, \ldots, e_{15})$ with each component e_i consuming 16 bits. These e_i are treated as representations of integers modulo p. Define the FFT-transformation $FT_8(a_0, \ldots, a_7) = (b_0, \ldots, b_7)$ as

$$b_i = \sum_{j=0}^{7} 2^{4ij} a_j \bmod p \quad \text{for} \quad i = 0, \ldots, 7 \tag{1}$$

1. $(e_0, e_2, \ldots, e_{14}) = FT_8(e_0, e_2, \ldots, e_{14})$ This step is called a *FFT-step*

2. FOR(i=0 ; i<16 ; i++) $e_i = e_i + e_{i-1}e_{i-2} + e_{e_{i-3}} + 2^i \bmod p$
   All indices are taken modulo 16. This step is called a *recursion step*.

3. Second round: repeat step 1 and 2

The output of g is the 128-bit string $e_8 \parallel e_9 \ldots \parallel e_{15}$ where all occurrences of $p - 1 = 2^{16}$ are substituted by 0.

3   Weaknesses of FFT-Hash

1. The FFT step only affects the components with even index. For odd-indexed components no diffusion takes place.

2. The linearity of the FFT step can be used to impose certain values upon a number of output components. If for certain subsets of no more than 8 components, belonging to either the output or the input, the values are fixed, values for the remaining components can be computed such that equation 1 holds. This computation involves linear algebra alone.

3. The diffusion resulting from the recursion step can be completely eliminated by imposing 0 values to certain components. Suppose $(e_0, \ldots, e_{15})$ is the 16-tuple that has just undergone a recursion step. Suppose $e_5 = e_7 = 0$. Suppose also that e_6 was never addressed in the indirect indexing term $e_{e_{i-3}}$, hence $e_{i-3} \neq 6 \pmod{16}$ for all i at the moment they are used. Then the 12 MSB bits of e_6 only appear in the calculation for the new value of e_6. This can easily be seen because when $e_5 = e_7 = 0$ a product term $e_{i-1}e_{i-2}$ containing e_6 must be zero. Because the 12 MSB bits of e_6 can be altered without affecting the outcome of other components when the recursion is applied, e_6 will be called *isolated*. This can be applied to any component. Hence isolation of a component in a recursion step requires that the two neighboring components are 0 *and* that it is not addressed in the term $e_{e_{i-3}}$ for any i.

4 The Attack

The attack is based on the fact that it is possible to isolate a component during all four steps of g. The colliding messages consist of 3 blocks: M_0, M_1 and M_2. All effort goes into the search for appropriate M_1 and M_2 values. The attack is probabilistic. A subset of messagebits are given random values thereby fixing the remaining bits through a number of imposed relations. Starting from $H_1 = g(H_0 \parallel M_0)$ we have:

1. Calculation of M_1. The values are chosen in a way that the second component of M_1 ($= e_9$) has a maximum probability of staying isolated throughout the calculation of g. Certain changes in the 12 MSB bits of this component affect the intermediate hash value H_2 only in the second component. On the average 2^{23} different H_1, obtained by trying different M_0 values have to be tested. Only about 2^{11} of these survive a first check. For each of these remaining M_1 values 2^{15} trials have to be performed by varying ϕ (see figure).

2. Calculation of M_2. The values of M_2 are chosen in such a way that the second component of H_2 ($= e_1$) has maximum probability of being isolated and thus does not affect H_3. About 2^{22} different values of ϕ_1 and ϕ_2 have to be tried.

The figure illustrates the internal relations during the hashing process of the colliding messages. Q indicates the component that is isolated throughout the whole calculation.

The first result obtained by this method was a set of 805 colliding messages (in hexadecimal notation)

```
00a1 0000 0000 0000 0000 0000 000c 5b18
9156 XXXd 9e89 67e8 35f8 e2b0 12ec 26c0
570b 06ee ba21 8da5 6ec4 c27e 5d5d e6be
```

where XXX ranges over 1b5 to 4d9 that all hash to

```
527d c019 d8cb 1d92 162b f04c cfff 26c6
```

References

[1] C Schnorr, FFT-Hash, An Efficient Cryptographic Hash Function, *Rump Session Crypto '91.*

An arrow from e_i to e_j means $e_{e_i} = e_j$ or $e_i = j$ (mod 16)

Boxes containing a constant indicate the value that is imposed upon the component

Boxes containing a greek letter indicate variables that are isolated (denoted by ■) until they are used (as indicated in the down left corner) to impose a certain value to a component

A ⋆ in the down left corner indicates that we depend upon luck (prob: 2^{-16})

□ indicates that the component is fixed by an FFT relation

An empty box denotes a component that is fixed by initial values and/or internal relations

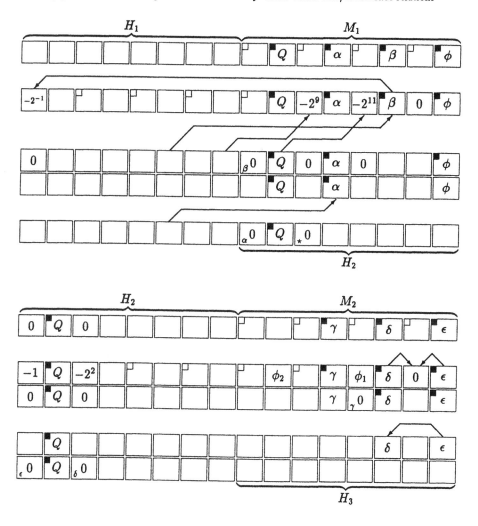

Figure 1: Schematic overview of the collisions of FFT-Hash. The state $(e_0, \ldots, e_{15})$ is depicted before and after every step.

On NIST's Proposed Digital Signature Standard

Ronald L. Rivest[*]

Laboratory for Computer Science

Massachusetts Institute of Technology

Cambridge, MA 02139

1 Introduction

The U.S. government agency NIST (National Institute for Standards and Technology) has recently proposed a public-key digital signature standard (or "DSS") [2, 3]. Although the proposal is nominally only "for government use," such a proposal, if adopted, would likely have an effect on commercial cryptography as well. There is an official "comment period" until February 28, 1992, after which NIST will respond to comments received and formulate a decision as to how (and whether) to proceed with its proposal.

In this note I review and comment upon NIST's proposal. (More correctly, it should be called the NIST/NSA proposal, since the cryptographic algorithm was designed by NSA (the U.S. National Security Agency)).

2 DSS Specifications

The following parameters are *global*, and can be shared by many users:

- a 512-bit prime p,

- a 160-bit prime q dividing evenly into $p - 1$,

- an element g of Z_p^*, whose multiplicative order is q.

- a *hash function* H mapping messages into 160-bit values.

The following parameters are *per user:*

- a *secret key* x, where $0 < x < q$.

- a *public key* y, where $y = g^x \pmod{p}$.

---

[*]Supported by RSA Data Security. email address: rivest@theory.lcs.mit.edu

So the secret x is the *discrete logarithm* of y, modulo p, with base g.

To sign a message m, a user produces his signature as (r, s), by selecting a random value k from Z_q^*, and then computing

$$r = (g^k \pmod{p}) \pmod{q}$$

$$s = k^{-1}(H(m) + xr) \pmod{q}$$

To verify a signature (r, s), one checks to see if the following equation holds:

$$r = (g^{H(m)/s} y^{r/s} \pmod{p}) \pmod{q}$$

Here division by s in the exponent is done modulo q.

3 Positive aspects of the proposal

The following positive aspects of the proposal are worth noting:

- The U.S. government has finally recognized the utility of public-key cryptography.

- The proposal is based on reasonably familiar number-theoretic concepts, and is a variant of the El-Gamal [1] and Schnorr [6] schemes.

- Signatures are relatively short (320 bits).

- When signing, computation of r can be done before the message m is available.

4 Problems with the proposed DSS

DSS is different than the de facto public-key standard (RSA). Two-thirds of the U.S. computer industry is already using RSA (current RSA users include IBM, Microsoft, Digital Apple, GE, Unisys, Novell, Motorola, Lotus, Sun, Northern Telecom, ...), and using industry-developed interoperable standards (PKCS – the "public key cryptography standard [4]). Moreover international standards organizations such as ISO, CCITT, and SWIFT, as well as other organizations (such as Internet) have accepted RSA as a standard. Adopting DSS would create a "double standard," causing difficulties for U.S. industry.

DSS has *uncertainties regarding its patent status.* While NIST claims that no patent licenses are necessary to practice the proposed DSS, Schnorr claims that DSS infringes his U.S. patent #4,995,082, and Public Key Partners asserts that DSS infringes U.S. patents #4,200,770 and #4,218,582. (To add to the confusion, NIST says it has filed for a patent on DSS, a move that has no reasonable justification.)

DSS is *buggy:* the verification process can blow up due to division by zero (when $s = 0$).

DSS is *incomplete.* A *hash function* H was not included in the original proposal, although one is clearly necessary. (This is supposed to be announced "soon.") In addition, NIST has not specified formats for certificates, which are clearly necessary is almost all applications of public-key cryptography. And of course, the proposal is only for signatures: it has no specification for privacy or key-exchange.

DSS is *slow:* the signing speed is comparable to (but slightly slower than) RSA, while the verification speed is *over 100 times slower* than RSA (where RSA uses a small public exponent such as 3, as is common in practice). This is likely to introduce unacceptable performance delays in many applications, since most applications use quite a few verification steps (to verify certificates).

DSS has *security problems*, as follows.

- The fact that many users are likely to use a shared modulus p is a security weakness, since "breaking" that one modulus can compromise the security of all users simultaneously. (Here "breaking" the modulus p means doing a precomputation that permits easy computation of discrete logarithms of any y value modulo p very easily.)

- The key size (512 bits for p) is *too short*. The standard does not permit larger key-sizes to be used, as appropriate for applications. Two experts on the discrete logarithm problem, LaMacchia and Odlyzko, have stated that 512-bit numbers offer only "marginal security." [5] Moreover, a standard should envision a 25-year lifetime, at least. With the cost of computation halving every 18 months or two years, key sizes need to increase to provide adequate security against ever more capable adversaries. At the user's discretion, key sizes exceeding 1000 bits or more should be usable within such a standard.

  To see that 512 bits is too short, we observe that the standard formula for $L(p)$ gives an estimate of approximately 2 million MIPS-years required to "break" a modulus p. Today one can estimate that MIPS-years (bought in quantity) cost about \$4 each, so the cost of breaking DSS can be estimated as about \$8M. In 25 years, this cost would decrease to only \$1400. This is clearly an inadequate level of security.

- Lenstra and Haber (in a letter to NIST) have observed that for some primes the discrete logarithm problem is "easy" for the person who created the prime (thus, a "trap-door prime"). Moreover, it seems very difficult for a user to recognize trapdoor primes. Thus, conservative users of the proposed DSS should assume that whoever creates the prime p can determine their secret keys x from their public keys y.

- The preceding analysis assumes that the cryptanalytic problem is the "discrete logarithm problem." In fact, the problem here is a variation of that problem, since g only has order q instead of order $p-1$. It is thus quite possible that breaking DSS is much easier than the general discrete logarithm problem. Without further research into this variation, it is difficult to assess the true level of (in)security provided by DSS.

- Signing with DSS requires the generation and use of random numbers (the k values). Should any k ever be disclosed, the signer's secret key k is totally compromised.

5 Summary

The proposal by NIST/NSA of a digital signature standard has some positive aspects, but is marred by serious flaws. Most notably, it is not sufficiently secure for a standard and seems vulnerable to "trapdoors."

References

[1] T. El-Gamal. A public key cryptosystem and a signature scheme based on discrete logarithms. *IEEE Trans. Inform. Theory*, 31:469–472, 1985.

[2] National Institute for Standards and Technology. Digital signature standard (DSS). *Federal Register*, 56(169), August 30 1991.

[3] National Institute for Standards and Technology. A proposed federal information processing standard for digital signature standard (DSS). Technical Report FIPS PUB XX, National Institute for Standards and Technology, August 1991. DRAFT.

[4] Burton S. Kaliski, Jr. An overview of the pkcs standards. Technical report, RSA Data Security, Inc., June 1991.

[5] B. A. LaMacchia and A. M. Odlyzko. Computation of discrete logarithms in prime fields. *Designs, Codes, and Cryptography*, 1:47–62, 1991.

[6] C. P. Schnorr. Efficient identification and signatures for smart cards. In G. Brassard, editor, *Proceedings CRYPTO 89*, pages 239–252. Springer, 1990. Lecture Notes in Computer Science No. 435.

the i-th block, we express every internal variable as a sum of the unknown constant and its exORed difference. Differences are time variant.

3. Equations of the unknown constants and the differences

We show typical equations in the bitwise analysis of FEAL-4. The number of plain-text and ciphertext pairs is N+1. We use every i-th differences (i=1,..,N) to cryptana-lyze. In the following, $a_1(i),...,a_K(i), b(i) \in GF(2)$ are known differences for the i-th block, $x_1,..,x_K \in GF(2)$ are unknown constants, and $a_j(i)x_j$ are independent for j=1,..K.

<Type (A): system of equations of the unknown constants>

The following equation on GF(2) is type (A).

$$a_1(i)x_1 \oplus a_2(i)x_2 \oplus ... \oplus a_K(i)x_K = b(i) \tag{2}$$

Where $\oplus$ shows the addition on GF(2) (i.e. exOR addition). Taking equation(2) as the i-th equation of a system of equations (i=1,..,N), we can fix K unknown constants $x_1,...,x_K$, if the rank of coefficient matrix equals K. If prob{$a_j(i)=1$}=1/2, such probability P_A is as follows.

$$P_A \doteqdot 1-(2^K-1)2^{-N} \tag{3}$$

<Type (B): equation of an unknown difference>

Let $a(i), c(i), \beta(i) \in GF(2)$ be unknown differences. If $a(i)$ and $c(i)$ be cal-culable from M unknown constants $y_1,..y_M$ and known differences, the following set of equations is type (B).

$$a(i)\beta(i)=c(i) \tag{4}$$

$$\beta(i)=a_1(i)x_1 \oplus a_2(i)x_2 \oplus ... \oplus a_K(i)x_K \tag{5}$$

We calculate $a(i)$ and $c(i)$ using the assumed values of the unknown constants $y_1,..y_M,$. If the following case A or B happens, the assumption is proved to be wrong.

case A: The case for some i, $a(i)=0$, $c(i)=1$.

case B: If $a(i)=1$, we get $\beta(i)=c(i)$. Equation(5) is type (A) equation with that $\beta(i)$. Case B is the case when the equations have no solutions.

To estimate the probability, we suppose that $a(i)$ and $c(i)$ are random and independ-ent and that prob{$a(i)=1$}=prob{$c(i)=1$}=1/2.

The true assumption will survive for any differences (i=1,..N). It never fall into case A. We get N/2 type (A) equations for some set of i. We will have true $x_1,...,x_K$ from them. For the wrong assumption, the probability to fall into case A equals 1/4 for every i. We also get N/2 type (A) equations. We can estimate the probability for their having no solutions, as a rank problem of the randomly chosen coefficients matrix.

From the previous consideration, we can fix K+M unknown constants $x_1,...x_K$, $y_1,..y_M$ with the following probability P_B.

$$P_B \doteqdot 1-(2^K-1)2^{-N/2}-2^{M+K}(3/4)^{N}2^{-N/2} \tag{6}$$

P_A (eq.(3)) and P_B (eq.(6)) converges to 1 exponentially as N increases.

A known-plaintext attack of FEAL-4 based on the system of linear equations on difference

(extended abstract)

Toshinobu Kaneko

Science University of Tokyo

2641 Yamazaki Noda Chiba Japan 278

ABSTRACT

I present a new attack of FEAL-4 blockcipher. The attack requires only 24 blocks of 8-byte randomly given known-plaintext. Using these blocks, we can break FEAL-4 with the probability of greater than 90%, in 14 seconds on a personal computer (cpu 80386 16MHz). It is based on the system of linear equations on the differences.

1. Introduction

FEAL is a blockcipher developed at NTT in 1987[1]. FEAL-4 is a 4-round one, and the standard version is now FEAL-8[2]. So far, known-plaintext attack of FEAL-4 [3][4] [5] and FEAL-6[3] have been published. The attack I present here is the refined version of reference[5]. It does not use the approximation of the S-boxes as the reference [3] do. It is a straightforward analysis on equations. Among the 3 attacks, the present one requires rather small number of blocks and the least computing time.

2. Equivalent FEAL-4

The attack does not concern the detail of the key schedule. It only requires that the extended subkeys are time invariant. Main data randomization part of FEAL is an f-function. It has two kind of S-boxes, which are S_0 and S_1.

<S-box> Let x and y be 1-byte (8 bits) inputs of S-box. The output of S_d is

$$S_d(x,y) = rot2((x+y+d) \bmod 256), \qquad d \in \{0,1\}. \qquad (1)$$

The operator rot2() denotes a 2-bit left rotation on 8-bit.

As the exOR operation is commutative, it is easy to make an equivalent FEAL-4 (fig.1) from the original one[1] by moving out the extended subkeys from f-functions. F denote zero key input f-functions. Each f-function has 4 S-boxes. The equivalent subkeys are K_l, K_r, L_l, L_r, M_l and M_r. They are linear sum of the original subkeys.

We consider all the internal variables in the system as the unknown constants, when the 0th plaintext block is processed. The unknown constant is a time invariant and unknown variable. In fig.1 X_l, X_r, Y_l, Y_r, Z_l and Z_r are the unknown constants. For

4. Difference equation of S-box

Let $z=x+y \bmod 256$. The n-th bits of z, x and y are z_n, x_n and y_n respectively. The LSB is 0-th bit. A modulo 256 sum is expressed by equations on GF(2) as follows.

$$z_n = x_n \oplus y_n \oplus d_n, \qquad n=0,1,..7 \qquad (7)$$

Where d_n indicates a carry coming into the n-th bit.

To represent the addition in S-box, we chose d_0 as d in equation(1).

$$d_n = x_{n-1}y_{n-1} \oplus d_{n-1}(x_{n-1} \oplus y_{n-1}), \qquad n=1,...,7. \qquad (8)$$

$$d_0 = d \qquad (9)$$

Let x_n, y_n, z_n, d_n be the unknown constants, and let a_n, b_n, c_n, ω_n be their differences. Differenciating equations (7),(8) and (9), we get the following.

$$c_n = a_n \oplus b_n \oplus \omega_n, \qquad n=0,1,..7. \qquad (10)$$

$$\omega_n = a_{n-1}y_{n-1} \oplus b_{n-1}x_{n-1} \oplus a_{n-1}b_{n-1} \oplus \omega_{n-1}(x_{n-1} \oplus y_{n-1} \oplus a_{n-1} \oplus b_{n-1})$$
$$\oplus d_{n-1}(a_{n-1} \oplus b_{n-1}), \qquad n=1,..7. \qquad (11)$$

$$\omega_0 = 0 \qquad (12)$$

If a_{n-1}, b_{n-1} and c_{n-1} are known differences, we can solve equation (11) for n=1 through n=7 sequentially as a type (A) equations. As a difference equation of S-box, the suffixes of c and z should be changed from n to (n+2) mod 8.

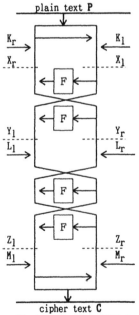

plain text **P**

cipher text **C**
Fig.1. equivalent FEAL-4
K,L,M:equivalent sub keys

Table 1. Number of unknown constants in
the difference equations

Type (B):(1st and 2nd bit position)		Type (A):(each n-th bit position. n=3,..7)	
Eq.No.	M K	Eq.No.	K
1	8 2	1	4
2	2 8	2	2
3	4 0	3	2
4	8 4	4	4
5	4 0	5	4
6	2 0	6	2
7	2 0	7	2
8	4 0	8	4

M: number of the unknown constants to be
   assumed in type (B) equations
K: number of the unknown constants to be
   solved by the system of equations

5. The attack of FEAL-4

We apply the difference equation (10) to the whole FEAL-4. Non-linearity exists only in S-box and the equation (10) is linear, so the work is not so hard. In FEAL-4 input-output is 8 bytes. Concerning to the n-th bit of each 8 byte, we wrote equations for 16 S-boxes in FEAL-4. As the n-th bit difference equations, we get 8 equations[5].

We can easily obtain the differences of the unknown constants X_l, X_r, Z_l and Z_r from plaintext and ciphertext. They are known differences. As $\omega_0=0$, we use 0-th bit difference equations to determine the 0th and 2nd bit unknown differences. Substituting (11) into the 1st and 2nd bit difference equation, we can arrange them to 8 equations of type (B). Table 1 shows the numbers K and M in those equations, when we solve them in that order. For the 3rd,..,7th bit equation, we can arrange them to type (A). We also show the number of unknown constants K in table 1.

The probability P to succeed in the cryptanalysis is the product of the probabilities of P_A or P_B with which each equation is solved.

$$P \fallingdotseq 1-6200*(3/4)^N 2^{-N/2}-273*2^{-N/2}-360*2^{-N} \tag{13}$$

If N=23 (i.e. 24 blocks of plaintext), P is greater than 90%. If N=30, then $P \geqq 99\%$.

For the unknown constants which do not appear in the difference equations, we fix them from the input-output relations of S-boxes. After all the unknown constants are fixed, we can determine the extended subkeys of the equivalent FEAL-4.

According to the algorithm, we made an attack program on a NEC PC (cpu 80386 16MHz). The experiments proved that we can determine equivalent FEAL-4 in 14 seconds by using 24 blocks of known-plaintext.(Note 1) The known-plaintext which we used, was ordinary Japanese text appeared in 'call for attack of FEAL-8'[2].

6. Conclusion

I present a new known-plaintext attack of FEAL-4. By the attack, FEAL-4 is breakable in 14 seconds on a personal computer. FEAL-4 is not secure under the known-plaintext attackable environment. But, we don't know the attacks of FEAL-8 yet.

REFERENCES
[1] A.Shimizu, S.Miyaguchi. "Fast Data Encipherment Algorithm FEAL", Trans. of IEICE Vol.J70-D No.7, pp.1413-1423, (1987)
[2] S.Miyaguchi."The FEAL-8 Cryptosystem and a Call for Attack", Lecture Note in Computer Science 435, CRYPTO'89, pp.624-627
[3] A.Tardy-Corfdir, H.Gilbert, "A Known Plaintext Attack of FEAL-4 and FEAL-6", Crypto'91, (1991)
[4] M.Matsui, A.Yamagishi. "A Study on Known Text Attack of Involution-type Cryptosystem", Tech. Rep. of IEICE ISEC91-26 (1991)
[5] T.Kaneko. "A Known plaintext cryptanalytic attack on FEAL-4", Tech. Rep. of IEICE ISEC91-25 (1991)

Note 1) Other attacks are as follows. An attack of FEAL-4 by A.Tardy-Corfdir[3] require 1000-300 blocks of known-plaintext and 0.5 hour computing time on a SUN4. An attack by M.Matui[4] needs 50-20 blocks and 130-14 seconds on a HP9425 (cpu 68040,25MHz).

Simultaneous Attacks in Differential Cryptanalysis
(Getting More Pairs Per Encryption)

Matthew Kwan - mkwan@cs.adfa.oz.au

Centre for Computer Security Research
Department of Computer Science
University of New South Wales
Australian Defence Force Academy
Canberra ACT 2600
AUSTRALIA

ABSTRACT

One aspect of differential cryptanalysis that appears to have been largely overlooked is the use of several differences to attack a cipher simultaneously. While the use of quartets and octets have been briefly described by Biham and Shamir [1], this was not carried to its logical conclusion - namely, how many different attacks can you use and still get an improvement. The issues involved are briefly covered here.

Introduction

In what I will refer to as "standard" differential cryptanalysis, a cryptosystem is attacked using some XOR difference A as input. Pairs of plaintexts which differ by A are repeatedly encrypted until a pair of outputs is obtained which has some expected XOR difference. The value A is chosen so as to have the highest probability of producing this output, thus minimizing the expected number of encryptions required.

To apply this difference, we choose some random plaintext x, then encrypt the values x and $x \oplus A$. These two values provide an input difference of A, which will be referred to as an A pair.

To generate another A pair, we choose another random plaintext y, and encrypt the values y and $y \oplus A$. In general, you get one pair for each two encryptions.

Doing Better

Now consider the XOR difference B, which has the second best probability of producing an expected output. This difference can be applied simultaneously with an A attack by encrypting the following values, where x is some random plaintext.

(1) x
(2) $x \oplus A$
(3) $x \oplus B$
(4) $x \oplus A \oplus B$

The values (1)&(2) and (3)&(4) form A pairs.

The values (1)&(3) and (2)&(4) form B pairs.

Thus we get two A pairs and two B pairs, a total of 4 pairs for 4 encryptions.

... and Better

Then consider the third best attack C. Encrypting these extra values

(5) $x \oplus C$
(6) $x \oplus A \oplus C$
(7) $x \oplus B \oplus C$
(8) $x \oplus A \oplus B \oplus C$

will give us four pairs each of A, B, and C, as follows

pair	ciphertext pairs			
A	(1),(2)	(3),(4)	(5),(6)	(7),(8)
B	(1),(3)	(2),(4)	(5),(7)	(6),(8)
C	(1),(5)	(2),(6)	(3),(7)	(4),(8)

This is a total of 12 pairs for 8 encryptions.

In general, this technique can be extended, using the best n differences, to obtain $n\,2^{n-1}$ pairs for 2^n encryptions.

This is *apparently* a factor of n improvement over standard differential cryptanalysis.

The Downside

There is an improvement, but there are drawbacks as well. For a start, there is a memory cost of 2^{n-1} ciphertexts. As an example, the attack described above using the differences A, B, and C, requires the storing of the values (1),(2),(3), and (4). These are then overwritten by the values (5), (6), (7), and (8). At any given time it is necessary to store four ciphertexts in memory. Perhaps a different algorithm exists which requires less memory, but it seems unlikely.

Also, since the differences A, B, C, *etc.* have decreasing probabilities of success, we get more and more pairs which are less and less useful as we increase n. This leads to a worsening in the signal-to-noise ratio, which eventually degrades the effectiveness of the attack.

Finally, there is some cost in actually analyzing the resulting output pairs. While the cost is minimal, requiring only the exclusive-OR of two ciphertexts and the testing of some bits, a point may be reached where the effort required outweights the expected gain, especially if the probabilities are getting very low.

The Lower Bound on Improvement

Using simultaneous attacks can be no worse than standard differential cryptanalysis. No matter what value of n you choose, you still get 2^{n-1} A pairs for each 2^n encryptions. In other words, you still get one pair for each two encryptions, which is what you get with the standard version.

The Upper Bound

Basically, you are getting n attacks performed simultaneously, at no extra cost in encryptions. Intuitively, then, you would expect the factor of improvement to be the sum of the probabilities of the best n attacks divided by the probability of the best attack.

If we denote the success probability of an attack a by $Pr(a)$, then the intuitive factor of improvement F_n is given by

$$\frac{Pr(A)+Pr(B)+Pr(C)+ \cdots}{Pr(A)}$$

Sample Values - DES

Here are some sample values of F_n for the DES cipher. Note that the input difference is the value *after* the initial permutation IP, and that the probabilities are measured in parts per 2^{64}. All the differences listed here make use of two round iterative characteristics, and while they are not *all* guaranteed to be the best characteristics, the first 10 or so certainly are.

n	input difference	prob	F_n
1	19600000 00000000	479.37	1.00
2	1b600000 00000000	479.37	2.00
3	00196000 00000000	256.00	2.53
4	000003d4 00000000	162.94	2.87
5	00001d40 00000000	100.53	3.08
6	00192000 00000000	34.17	3.15
7	4000001d 00000000	34.17	3.23
8	05f40000 00000000	34.17	3.30
9	2000001d 00000000	34.17	3.37
10	1b400000 00000000	13.42	3.40
12	000005d4 00000000	9.54	3.44
15	20000019 00000000	4.56	3.47
20	01f40000 00000000	2.00	3.51

As you can see, the improvement factor F_n levels out around 3.5 for large n.

Conclusions and Questions

Before we can find out where the actual improvement lies, between the upper and lower bounds, we have to know "Is Differential Cryptanalysis Parallelizable?". In other words, will the number of encryptions required be reduced by a factor of F_n? Is it possible to combine the information from several different attacks, and thus reduce the number of encryptions? Or is it necessary to carry out an individual attack through to completion?

Probably the answer lies somewhere between these two extremes, and varies from cipher to cipher. If the differences attack the same bits, maybe the analysis will be sped up. If they attack different bits, maybe the attack will take the same time, but more key bits will be found. There are many factors to consider, some of which are cipher-specific, so there is probably no simple answer.

If it turns out that simultaneous attacks do result in a significant improvement, the next thing to find is the optimum value of n. This is a very complex problem, dependent not only on the cipher in question (and the probabilities of its various attacks), but also on the hardware being used to mount the attack. In particular, the relative costs of encryption, XOR, and memory I/O need to be taken into consideration. Trial and error will probably be required for an accurate answer.

In conclusion, it appears that the analysis of a cryptosystem should not only measure its strength under the best differential attack, it should take into account the best several attacks. It may turn out that simultaneous attacks will reduce the search space by a significant factor, so it's something that should be taken into account.

Acknowledgements

This work has been supported by ARC grant A48830241, ATERB, and Telecom Australia research contract 7027.

References

1. E. Biham and A. Shamir, "Differential Cryptanalysis of DES-like Cryptosystems," *Journal of Cryptology*, vol. 4, no. 1, 1991.

Privacy, Cryptographic Pseudonyms, and The State of Health

Stig Fr. Mjølsnes,
Sintef DELAB,
The Norwegian Inst. of Technology *

Extended Abstract

The public health administration wants to collect records of patients systematically and extensively, thereby constructing databases that will enable researchers easy access to information vital for conquering diseases, and enable managers to run health services optimally. On the other hand, the realization of nationwide databases of this type, and keyed to person identity, may be prohibited by legal and privacy considerations, at least so in Norway [1]. A system design that alleviates this ambivalence is presented. This work extends the cryptographic pseudonym technique set forth in Ref [2].

The architecture is based on multiple autonomous database management systems, or local registers (LR), each set up in the vicinity, but organizational separate from a regional hospital. The hospital records the patient's health attributes by his or her identity number, normally a unique person number issued at birth. A selection of the attributes are forwarded to LR, but now keyed to a cryptographic pseudonym derived from the identity number. The pseudonymization of the registration must performed in cooperation with the patient. The patient provides the pseudonym and a proof (a witness) of his pseudonym (previously issued to the patient by a trusted center) to a tamper-resistant pseudonym *testing* unit at the hospital. The pseudonym tester is able to verify the correspondence between the identity (provided by the hospital) and the pseudonym (provided by the patient) without revealing anything more from the computation. The pseudonym, encrypted under LR's public key, is revealed to the hospital and sent to LR. As a result, the hospital will not know the patient's pseudonym, and LR will not know the patient's identity. The hospital cannot use the pseudonym testing unit as an efficient "pseudonym oracle" because no computation will be performed unless both the identity number and the corresponding pseudonym are input.

Each person is able to inspect his or her records anonymously, by presenting the pseudonym and prove the knowledge of the witness. Furthermore, a re-identification (from pseudonym to name) is made possible, but conditioned on an action taken by the person concerned. Hence, the identity of the person is never revealed to LR, nor is it necessary to reveal the pseudonym to the hospital.

---

*Email address: mjolsnes@delab.sintef.no

The system of distributed databases can provide extensive but controlled online access to a variety of health statistics without releasing individual attributes. Database queries are answered with locally aggregated data prepared by query servers, and subsequently collected and combined by the requesting workstation. Statistical inference control methods may be employed in addition to standard access control methods to execute the security policy at the query servers.

Means are also provided to join medical research records (identity-keyed) with registered health records (pseudonym-keyed), conditioned on active acceptance by the patient, without giving away the pseudonym to the researcher nor revealing the identity number to the database administration.

References

[1] Communications with prof. E. Boe, Dept. of public and international law, Univ. of Oslo.

[2] J. Brandt, I. Damgård and P. Landrock. Anonymous and verifiable registration in databases. Proc. of Eurocrypt'88, pp. 166–176.

Limitations of the Even-Mansour Construction

Joan Daemen

Katholieke Universiteit Leuven, Laboratorium ESAT,
Kardinaal Mercierlaan 94, B–3001 Heverlee, Belgium.

Abstract

In [1] a construction of a block cipher from a single pseudorandom permutation is proposed. In a complexity theoretical setting they prove that this scheme is secure against a polynomially bounded adversary. In this paper it is shown that this construction suffers from severe limitations that are immediately apparent if differential cryptanalysis [3] is performed. The fact that these limitations do not contradict the theoretical results obtained in [1] leads the authors to question the relevance of computational complexity theory in *practical* conventional cryptography.

1 Introduction

The Even-Mansour construction is a block cipher. Let n be the blocklength. The cipher makes use of a publicly known permutation F where it is easy to compute $F(X)$ and $F^{-1}(X)$ for any given input $X \in \{0,1\}^n$. The key consists of two n-bit subkeys K_1 and K_2. The relation between a message M and its ciphertext C is given by $C \oplus K_2 = F(M \oplus K_1)$. This is illustrated in figure 1.

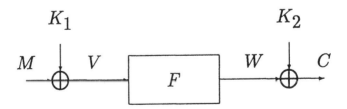

Figure 1: The Even-Mansour block cipher construction.

Traditionally the effectiveness of a cryptanalytic attack is measured by comparing it to exhaustive search over the keyspace. For instance, recently the (full 16-round version of) DES has been called broken because a sophisticated version of differential cryptanalysis would perform faster than exhaustive search. In this paper two attacks are presented that severely compromise the Even-Mansour scheme in this (real world) context.

The failure of Even and Mansour to see the obvious weaknesses of their construction lies in their one-sided complexity theoretical approach. Performing some elementary cryptanalysis provides upper limits for the security attainable with this scheme. Combination of the results from both complementary viewpoints allows for a more realistic evaluation of the possibilities of the construction.

Moreover, we give a number of arguments against the application of complexity theory in the case of *practical* conventional cryptographic schemes. In our opinion only a pragmatic approach based on existing cryptanalytic methods and the analysis of diffusion and confusion will lead to both secure and efficient conventional cryptosystems.

2 Cryptanalysis of the Scheme

The key consists of $2n$ bits, hence exhaustive search of the keyspace would take about 2^{2n-1} applications of F. The two attacks presented are significantly more efficient.

Known plaintext attack : The cryptanalyst knows two plaintext-ciphertext pairs (M_a, C_a) and (M_b, C_b). Let $V = M \oplus K_1$ and $W = C \oplus K_2$. The bitwise XOR of two blocks X_a and X_b is denoted by ΔX. The attack consists of calculating $\Delta W = F(V) \oplus F(V \oplus \Delta M)$ for all possible values of V. If $V = M_a \oplus K_1$ or $V = M_b \oplus K_1$ we have $\Delta W = \Delta C$. If F is random only a few V values will be found for which this holds. The erroneous values V' can easily be discarded by checking that $F(V') \neq M_a \oplus K_2$ and $F(V') \neq M_b \oplus K_2$. When the correct V is known the two subkeys can be easily calculated. This attack takes on the average 2^{n-1} applications of F, hence reduces the effective keylength to n bits.

Chosen plaintext attack : The cryptanalyst chooses k plaintext pairs M_i, M_i' with $M_i \oplus M_i' = \Delta M$, a constant for all i. He obtains the corresponding ciphertexts C_i, C_i'. For $k \ll 2^n$ there are about k different ΔC_i values, speeding up the search for $\Delta W = \Delta C_i$ for some i by a factor of k as compared to the known plaintext attack. The magnitude of k is limited by the memory requirement of $2nk$ bits. In the absence of memory restrictions there is an optimum for k near $2^{n/2}$. In that case the expected number of F-evaluations plus encryptions is about $2^{n/2}$, reducing the effective keylength to $n/2$ bits.

In the calculation of the complexities the permutation F is considered to be random. However, non-random properties of F can probably be used to speed up the attacks. Moreover both attacks can easily be parallelized.

Example : Let Φ be an Even-Mansour type block cipher with blocklength 64 and keylength 128. Suppose an F-processor is a dedicated chip that computes F in 100 microseconds. A cryptanalyst performing a (2 MByte) chosen plaintext attack ($k \approx 10^5$) on Φ using a parallel machine with 10^6 F-processors will recover the key in a matter of hours.

3   Discussion

In [1] computational complexity theory is employed to recommend a construction that can equally be considered as a *restriction* on the key schedule of a block cipher.

Even and Mansour prove that a polynomially bounded adversary has a negligible probability of success when attacking their scheme if permutation F is pseudorandomly 'chosen'. What this really means is that the complexity of an attack with nonnegligible probability is a superpolynomial function of the blocklength n for large enough n. Complexity theory makes abstraction of constant factors. In fact the only distinction made in [1] is that between polynomial and superpolynomial. Given a mathematical model of an attacker, complexity theory does not provide any information at all on how large the blocklength should be chosen for the scheme to be secure. Moreover, complexity theory is only applicable if the permutation F is 'chosen' in a pseudorandom way. This means that F must in some way be based on a computationally hard problem. One method to obtain this is by forming F with the Luby-Rackoff construction[4]. The random functions used in this scheme can be constructed as outlined by Goldreich-Goldwasser-Micali [5] using a random sequence generator based on some general complexity theoretic assumption. This can be used to define a concrete block cipher by the unavoidable fixing of the blocklength. Despite the involved construction, *no guarantee* on the security of this block cipher can be given. Moreover, it will definitely be very slow.

Complexity theory is a powerful tool in an important part of cryptography. However, in the case of practical conventional cryptography, we believe that its relevance is marginal. By practical conventional cryptography we mean symmetrical block ciphers, stream ciphers and cryptographic hash functions that *actually have to be implemented and used*. In our opinion the complexity theoretical way of thinking encourages poor design. The Even-Mansour construction is an example.

498

In [2] a similar argument is made concerning hash functions.

A number of statements in the Even-Mansour paper illustrate the absurdity of their point of view. In the abstract they state that the construction "removes the need to *store*, or generate a multitude of permutations." Furthermore it reads "The scheme may lead to a system *more efficient* than systems such as the DES and its siblings, since the designer has to worry about one thing only: How to implement one pseudorandomly chosen permutation. This may be easier than *getting one for each key*." (the italics are ours)

In our opinion design and analysis of conventional algorithms has to be based on diffusion and confusion properties. From this point of view permutations are not 'chosen', permutations are *implemented using realizable primitives*. The security of the algorithms is best evaluated in the light of existing cryptanalytic methods and principles. We believe that a central position herein will be taken by differential cryptanalysis. From an engineering point of view rational design implies maximum yield (= security) with minimum cost (= time, memory).

References

[1] S. Even, Y. Mansour, A Construction of a Cipher From a Single Pseudorandom Permutation, *Lecture Notes in Computer Science, Proceedings Asiacrypt '91*, Springer-Verlag 1992.

[2] J. Daemen, A Framework for the Design if One-Way Hash Functions Including Cryptanalysis of Damgård's One-Way Function Based on a Cellular Automaton, *Lecture Notes in Computer Science, Proceedings Asiacrypt '91*, Springer-Verlag 1992.

[3] E. Biham, A. Shamir, Differential Cryptanalysis of DES-like Cryptosystems, *Abstracts of Crypto '90*, 1-32.

[4] M. Luby and C. Rackoff. How to construct pseudorandom permutations from pseudorandom functions. *SIAM Journal on Computing*, 17(2):373–386, April 1988.

[5] O. Goldreich, S. Goldwasser and S. Micali, How to Construct Random Functions, *Proceedings of the 25th Annual Symposium on Foundations of Computer Science*, October 24-26, 1984.

Author Index

BOSSELAERS, A., 477

BRANDT, J., 440

BROWN, L., 36

BURMESTER, M., 360

DAEMEN, J., 82, 477, 495

DAI, Z-D., 73

DAMGÅRD, I., 440

DAVIES, D. W., 1

DAWSON, E., 299

DESMEDT, Y., 360

EVEN, S., 210

FEIGENBAUM, J., 352

FORTNOW, L., 346

GOLDBURG, B., 299

GORESKY, M., 277

GOVAERTS, R., 82, 477

HARDJONO, T., 124

HARN, L., 149, 159, 312, 450

IMAI, H., 397

ITOH, T., 331, 382

JOUX, A., 470

KANEKO, T., 485

KAWAMURA, S., 265

KIM, K., 59

KLAPPER, A., 277

KNUDSEN, L. R., 22

KUROSAWA, K., 111, 321

KWAN, M., 36, 237, 489

LAIH, C-S., 159, 450

LANDROCK, P., 440

LIN, H-Y., 149, 312

MANSOUR, Y., 210

MATSUMOTO, T., 397

MEIJERS, J., 288

MIYAGUCHI, S., 247

MIYAJI, A., 460

MIYANO, H., 51

MIZUSAWA, J., 253

MJØLSNES, S. F., 493

MORITA, H., 247

OGATA, W., 111

OHTA, K., 139, 247

OKAMOTO, T., 139, 368

OSTROVSKY, R., 352

PIEPRZYK, J., 36, 97, 124, 194, 225, 237

RHEE, M. Y., 179

RIVEST, R. L., 427, 481

SADEGHIYAN, B., 97, 194, 225

SAFAVI-NAINI, R., 170

SAITO, T., 321

SAKURAI, K., 321, 331, 382

SEBERRY, J., 36

SHIMBO, A., 265

SHIZUYA, H., 382

SHU, C., 397

SRIDHARAN, S., 299

STERN, J., 470

SZEGEDY, M., 346

TOUSSAINT, M-J., 412

VANDEWALLE, J., 82, 477

van TILBURG, J., 288

YANG, J-H., 73

YEN, S-M., 450

ZENG, K-C., 73

ZHENG, Y., 97, 124

Lecture Notes in Computer Science

For information about Vols. 1–670
please contact your bookseller or Springer-Verlag

Vol. 671: H. J. Ohlbach (Ed.), GWAI-92: Advances in Artificial Intelligence. Proceedings, 1992. XI, 397 pages. 1993. (Subseries LNAI).

Vol. 672: A. Barak, S. Guday, R. G. Wheeler, The MOSIX Distributed Operating System. X, 221 pages. 1993.

Vol. 673: G. Cohen, T. Mora, O. Moreno (Eds.), Applied Algebra, Algebraic Algorithms and Error-Correcting Codes. Proceedings, 1993. X, 355 pages 1993.

Vol. 674: G. Rozenberg (Ed.), Advances in Petri Nets 1993. VII, 457 pages. 1993.

Vol. 675: A. Mulkers, Live Data Structures in Logic Programs. VIII, 220 pages. 1993.

Vol. 676: Th. H. Reiss, Recognizing Planar Objects Using Invariant Image Features. X, 180 pages. 1993.

Vol. 677: H. Abdulrab, J.-P. Pécuchet (Eds.), Word Equations and Related Topics. Proceedings, 1991. VII, 214 pages. 1993.

Vol. 678: F. Meyer auf der Heide, B. Monien, A. L. Rosenberg (Eds.), Parallel Architectures and Their Efficient Use. Proceedings, 1992. XII, 227 pages. 1993.

Vol. 679: C. Fermüller, A. Leitsch, T. Tammet, N. Zamov, Resolution Methods for the Decision Problem. VIII, 205 pages. 1993. (Subseries LNAI).

Vol. 680: B. Hoffmann, B. Krieg-Brückner (Eds.), Program Development by Specification and Transformation. XV, 623 pages. 1993.

Vol. 681: H. Wansing, The Logic of Information Structures. IX, 163 pages. 1993. (Subseries LNAI).

Vol. 682: B. Bouchon-Meunier, L. Valverde, R. R. Yager (Eds.), IPMU '92 – Advanced Methods in Artificial Intelligence. Proceedings, 1992. IX, 367 pages. 1993.

Vol. 683: G.J. Milne, L. Pierre (Eds.), Correct Hardware Design and Verification Methods. Proceedings, 1993. VIII, 270 Pages. 1993.

Vol. 684: A. Apostolico, M. Crochemore, Z. Galil, U. Manber (Eds.), Combinatorial Pattern Matching. Proceedings, 1993. VIII, 265 pages. 1993.

Vol. 685: C. Rolland, F. Bodart, C. Cauvet (Eds.), Advanced Information Systems Engineering. Proceedings, 1993. XI, 650 pages. 1993.

Vol. 686: J. Mira, J. Cabestany, A. Prieto (Eds.), New Trends in Neural Computation. Proceedings, 1993. XVII, 746 pages. 1993.

Vol. 687: H. H. Barrett, A. F. Gmitro (Eds.), Information Processing in Medical Imaging. Proceedings, 1993. XVI, 567 pages. 1993.

Vol. 688: M. Gauthier (Ed.), Ada-Europe '93. Proceedings, 1993. VIII, 353 pages. 1993.

Vol. 689: J. Komorowski, Z. W. Ras (Eds.), Methodologies for Intelligent Systems. Proceedings, 1993. XI, 653 pages. 1993. (Subseries LNAI).

Vol. 690: C. Kirchner (Ed.), Rewriting Techniques and Applications. Proceedings, 1993. XI, 488 pages. 1993.

Vol. 691: M. Ajmone Marsan (Ed.), Application and Theory of Petri Nets 1993. Proceedings, 1993. IX, 591 pages. 1993.

Vol. 692: D. Abel, B.C. Ooi (Eds.), Advances in Spatial Databases. Proceedings, 1993. XIII, 529 pages. 1993.

Vol. 693: P. E. Lauer (Ed.), Functional Programming, Concurrency, Simulation and Automated Reasoning. Proceedings, 1991/1992. XI, 398 pages! 1993.

Vol. 694: A. Bode, M. Reeve, G. Wolf (Eds.), PARLE '93. Parallel Architectures and Languages Europe. Proceedings, 1993. XVII, 770 pages. 1993.

Vol. 695: E. P. Klement, W. Slany (Eds.), Fuzzy Logic in Artificial Intelligence. Proceedings, 1993. VIII, 192 pages. 1993. (Subseries LNAI).

Vol. 696: M. Worboys, A. F. Grundy (Eds.), Advances in Databases. Proceedings, 1993. X, 276 pages. 1993.

Vol. 697: C. Courcoubetis (Ed.), Computer Aided Verification. Proceedings, 1993. IX, 504 pages. 1993.

Vol. 698: A. Voronkov (Ed.), Logic Programming and Automated Reasoning. Proceedings, 1993. XIII, 386 pages. 1993. (Subseries LNAI).

Vol. 699: G. W. Mineau, B. Moulin, J. F. Sowa (Eds.), Conceptual Graphs for Knowledge Representation. Proceedings, 1993. IX, 451 pages. 1993. (Subseries LNAI).

Vol. 700: A. Lingas, R. Karlsson, S. Carlsson (Eds.), Automata, Languages and Programming. Proceedings, 1993. XII, 697 pages. 1993.

Vol. 701: P. Atzeni (Ed.), LOGIDATA+: Deductive Databases with Complex Objects. VIII, 273 pages. 1993.

Vol. 702: E. Börger, G. Jäger, H. Kleine Büning, S. Martini, M. M. Richter (Eds.), Computer Science Logic. Proceedings, 1992. VIII, 439 pages. 1993.

Vol. 703: M. de Berg, Ray Shooting, Depth Orders and Hidden Surface Removal. X, 201 pages. 1993.

Vol. 704: F. N. Paulisch, The Design of an Extendible Graph Editor. XV, 184 pages. 1993.

Vol. 705: H. Grünbacher, R. W. Hartenstein (Eds.), Field-Programmable Gate Arrays. Proceedings, 1992. VIII, 218 pages. 1993.

Vol. 706: H. D. Rombach, V. R. Basili, R. W. Selby (Eds.), Experimental Software Engineering Issues. Proceedings, 1992. XVIII, 261 pages. 1993.

Vol. 707: O. M. Nierstrasz (Ed.), ECOOP '93 – Object-Oriented Programming. Proceedings, 1993. XI, 531 pages. 1993.

Vol. 708: C. Laugier (Ed.), Geometric Reasoning for Perception and Action. Proceedings, 1991. VIII, 281 pages. 1993.

Vol. 709: F. Dehne, J.-R. Sack, N. Santoro, S. Whitesides (Eds.), Algorithms and Data Structures. Proceedings, 1993. XII, 634 pages. 1993.

Vol. 710: Z. Ésik (Ed.), Fundamentals of Computation Theory. Proceedings, 1993. IX, 471 pages. 1993.

Vol. 711: A. M. Borzyszkowski, S. Sokołowski (Eds.), Mathematical Foundations of Computer Science 1993. Proceedings, 1993. XIII, 782 pages. 1993.

Vol. 712: P. V. Rangan (Ed.), Network and Operating System Support for Digital Audio and Video. Proceedings, 1992. X, 416 pages. 1993.

Vol. 713: G. Gottlob, A. Leitsch, D. Mundici (Eds.), Computational Logic and Proof Theory. Proceedings, 1993. XI, 348 pages. 1993.

Vol. 714: M. Bruynooghe, J. Penjam (Eds.), Programming Language Implementation and Logic Programming. Proceedings, 1993. XI, 421 pages. 1993.

Vol. 715: E. Best (Ed.), CONCUR'93. Proceedings, 1993. IX, 541 pages. 1993.

Vol. 716: A. U. Frank, I. Campari (Eds.), Spatial Information Theory. Proceedings, 1993. XI, 478 pages. 1993.

Vol. 717: I. Sommerville, M. Paul (Eds.), Software Engineering – ESEC '93. Proceedings, 1993. XII, 516 pages. 1993.

Vol. 718: J. Seberry, Y. Zheng (Eds.), Advances in Cryptology – AUSCRYPT '92. Proceedings, 1992. XIII, 543 pages. 1993.

Vol. 719: D. Chetverikov, W.G. Kropatsch (Eds.), Computer Analysis of Images and Patterns. Proceedings, 1993. XVI, 857 pages. 1993.

Vol. 720: V.Mařík, J. Lažanský, R.R. Wagner (Eds.), Database and Expert Systems Applications. Proceedings, 1993. XV, 768 pages. 1993.

Vol. 721: J. Fitch (Ed.), Design and Implementation of Symbolic Computation Systems. Proceedings, 1992. VIII, 215 pages. 1993.

Vol. 722: A. Miola (Ed.), Design and Implementation of Symbolic Computation Systems. Proceedings, 1993. XII, 384 pages. 1993.

Vol. 723: N. Aussenac, G. Boy, B. Gaines, M. Linster, J.-G. Ganascia, Y. Kodratoff (Eds.), Knowledge Acquisition for Knowledge-Based Systems. Proceedings, 1993. XIII, 446 pages. 1993. (Subseries LNAI).

Vol. 724: P. Cousot, M. Falaschi, G. Filè, A. Rauzy (Eds.), Static Analysis. Proceedings, 1993. IX, 283 pages. 1993.

Vol. 725: A. Schiper (Ed.), Distributed Algorithms. Proceedings, 1993. VIII, 325 pages. 1993.

Vol. 726: T. Lengauer (Ed.), Algorithms – ESA '93. Proceedings, 1993. IX, 419 pages. 1993

Vol. 727: M. Filgueiras, L. Damas (Eds.), Progress in Artificial Intelligence. Proceedings, 1993. X, 362 pages. 1993. (Subseries LNAI).

Vol. 728: P. Torasso (Ed.), Advances in Artificial Intelligence. Proceedings, 1993. XI, 336 pages. 1993. (Subseries LNAI).

Vol. 729: L. Donatiello, R. Nelson (Eds.), Performance Evaluation of Computer and Communication Systems. Proceedings, 1993. VIII, 675 pages. 1993.

Vol. 730: D. B. Lomet (Ed.), Foundations of Data Organization and Algorithms. Proceedings, 1993. XII, 412 pages. 1993.

Vol. 731: A. Schill (Ed.), DCE – The OSF Distributed Computing Environment. Proceedings, 1993. VIII, 285 pages. 1993.

Vol. 732: A. Bode, M. Dal Cin (Eds.), Parallel Computer Architectures. IX, 311 pages. 1993.

Vol. 733: Th. Grechenig, M. Tscheligi (Eds.), Human Computer Interaction. Proceedings, 1993. XIV, 430 pages. 1993.

Vol. 734: J. Volkert (Ed.), Parallel Computation. Proceedings, 1993. VIII, 248 pages. 1993.

Vol. 735: D. Bjørner, M. Broy, I. V. Pottosin (Eds.), Formal Methods in Programming and Their Applications. Proceedings, 1993. IX, 434 pages. 1993.

Vol. 736: R. L. Grossman, A. Nerode, A. P. Ravn, H. Rischel (Eds.), Hybrid Systems. VIII, 474 pages. 1993.

Vol. 737: J. Calmet, J. A. Campbell (Eds.), Artificial Intelligence and Symbolic Mathematical Computing. Proceedings, 1992. VIII, 305 pages. 1993.

Vol. 738: M. Weber, M. Simons, Ch. Lafontaine, The Generic Development Language Deva. XI, 246 pages. 1993.

Vol. 739: H. Imai, R. L. Rivest, T. Matsumoto (Eds.), Advances in Cryptology – ASIACRYPT '91. X, 499 pages. 1993.

Vol. 740: E. F. Brickell (Ed.), Advances in Cryptology – CRYPTO '92. Proceedings, 1992. X, 565 pages. 1993.

Vol. 741: B. Preneel, R. Govaerts, J. Vandewalle (Eds.), Computer Security and Industrial Cryptography. Proceedings, 1991. VIII, 275 pages. 1993.

Vol. 742: S. Nishio, A. Yonezawa (Eds.), Object Technologies for Advanced Software. Proceedings, 1993. X, 543 pages. 1993.

Vol. 743: S. Doshita, K. Furukawa, K. P. Jantke, T. Nishida (Eds.), Algorithmic Learning Theory. Proceedings, 1992. X, 260 pages. 1993. (Subseries LNAI)

Vol. 744: K. P. Jantke, T. Yokomori, S. Kobayashi, E. Tomita (Eds.), Algorithmic Learning Theory. Proceedings, 1993. XI, 423 pages. 1993. (Subseries LNAI)

Vol. 745: V. Roberto (Ed.), Intelligent Perceptual Systems. VIII, 378 pages. 1993. (Subseries LNAI)

Vol. 746: A. S. Tanguiane, Artificial Perception and Music Recognition. XV, 210 pages. 1993. (Subseries LNAI)

Vol. 747: M. Clarke, R. Kruse, S. Moral (Eds.), Symbolic and Quantitative Approaches to Reasoning and Uncertainty. Proceedings, 1993. X, 390 pages. 1993.

Vol. 748: R. H. Halstead Jr., T. Ito (Eds.), Parallel Symbolic Computing: Languages, Systems, and Applications. Proceedings, 1992. X, 419 pages. 1993.

Vol. 751: B. Jähne, Spatio-Temporal Image Processing. XII, 208 pages. 1993.

Vol. 753: L. J. Bass, J. Gornostaev, C. Unger (Eds.), Human-Computer Interaction. Proceedings, 1993. X, 388 pages. 1993.